The Ribbon of Green

The Ribbon of Green

Change in Riparian Vegetation in the Southwestern United States

Robert H. Webb, Stanley A. Leake, Raymond M. Turner

Principal photography by Dominic Oldershaw

The University of Arizona Press

Tucson

The University of Arizona Press
© 2007 The Arizona Board of Regents

Library of Congress Cataloging-in-Publication Data
Webb, Robert H.
    The ribbon of green : change in riparian vegetation in the
southwestern United States / Robert H. Webb, Stanley A. Leake,
Raymond M. Turner ; principal photography by Dominic Oldershaw.
        p. cm.
    Includes bibliographical references and index.
    ISBN-13: 978-0-8165-2588-1 (hardcover : alk. paper)
    ISBN-10: 0-8165-2588-9 (hardcover : alk. paper)
    1. Riparian plants—Southwestern States. 2. Riparian ecology—
Southwestern States. 3. Repeat photography—Southwestern States.
4. Vegetation dynamics—Southwestern States. I. Leake, S. A.
II. Turner, R. M. (Raymond M.) III. Title.
  QK142.W43 2007
          2006026090

Manufactured in Canada on acid-free, archival-quality paper containing a minimum
of 10% post-consumer waste.

12 11 10 09 08 07   6 5 4 3 2 1

To Toni, Lydia, and Jeanne

Contents

Authors' Preface

"We've lost 90 percent of the riparian vegetation in the southwestern United States."[1] This statement, with its relevant statistic ranging as low as 80 percent and as high as 95 percent, has been pervasive in the news media and popular and scientific publications since its origin in the late 1970s. This perceived catastrophic loss prompted an Arizona governor to issue an executive order in 1991 "to actively encourage . . . restoration of degraded riparian areas."[2] The perception of diminishing or degraded riparian areas persists in the public's imagination, in laws, and in countless management plans for the region. Is this perception accurate?

The scientific justification for this statement has been traced to a single paper on cottonwood gallery forests adjacent to the lower Colorado River,[3] an unusual reach that is fully regulated and part of the largest river basin in the region (see chapter 27). In this book, we concentrate on the question of long-term changes, addressing whether we have endured a net loss of woody riparian vegetation. Our study area encompasses the major river valleys in parts of Utah, southern Nevada, and southeastern California as well as all of Arizona below about 5,000 feet in elevation.

Clearly, assessments of gains or losses are dependent on the time period considered, and we use a period determined by the history of photography. Long-term changes can be assessed with repeat photography, a technique that historically has been used to study landscape change in the region.[4] Repeat photography is particularly well suited to identification of changes in woody species—especially trees—in riparian zones from the beginning of photography in the latter half of the nineteenth century to the present.

Other methods, such as analysis of aerial photography or satellite imagery, are better suited for evaluating spatial changes beginning in the last quarter of the twentieth century.[5] For some areas, we compare our results to findings from analyses of both aerial photographs and remote sensing.

We interpret the changes observed in thousands of repeat images in conjunction with hydrologic data (both surface water and groundwater), previously identified periods of climatic variation, known land uses and flow regulation, and water usage. The photographs that we replicated fall into four discrete groups: photographs taken by the earliest explorers of the region, photographs taken systematically by scientists or surveyors, photographs of streamflow gaging stations, and opportunistic photographs by tourists or other travelers. In most cases, we have only one match to interpret; in a few others, a robust time series of change can be developed from the record of repeat photography. Although we acknowledge that repeat photography poses significant problems in terms of quantitative analyses, especially in comparison to aerial photography and satellite imagery, it is the only method that provides a glimpse into predevelopment conditions in our region, a glimpse that is required to address the question of gains or losses of riparian vegetation.

Although literally hundreds of plant species are present in riparian areas in the Southwest, we restricted our interpretations to about a dozen woody species and selected herbaceous perennials that are visible in photographs. Because riparian ecosystems depend on water, we devote considerable discussion to surface-water and groundwater measurements in the twentieth century and discuss the

conjunctive use of both. We further discuss the supply of water to riparian forests, from both regional groundwater flow and local streamflow. We note that although science has long recognized the interconnections between surface water and groundwater, the legal system generally does not. Whereas surface-water flow in the region is relatively well known for most of the twentieth century, groundwater levels are less known and water-use statistics may be very inaccurate. Climatic fluctuations affect water because less pumped irrigation water is needed during wet years and more is required during dry ones. Climate also affects the recurrence of floods, which have large effects on riparian ecosystems.

Accompanying our discussion of regional hydrology is a look at several value judgments and assumptions that creep into the issue of long-term change in the region's rivers. Some authors use pejorative words such as *destroyed* to describe what has happened to these watercourses historically and *lost* to describe changes in riparian vegetation.[6] That some patches of vegetation, growing as they do in a dynamic geomorphic milieu, are gone is to be expected. We consider whether they are permanently lost or merely changed. These judgments derive from some scientists' perception that humans are the ultimate cause of regional change[7] and therefore that any changes must be "bad." A different view holds that climate, in particular climatic fluctuations, has greatly influenced the rivers of the region. The truth is likely a combination of these two mechanisms, with local effects thrown in for added complexity.

Previous observers commonly conclude that humans either completely control the fate of riparian systems or are in some way completely responsible for any perceived negative changes. An example is the common evaluation of nineteenth-century photographs of southern Arizona floodplains, many of which show little or no woody vegetation.[8] Some observers invoke woodcutting to explain the absence; others believe overgrazing to be the reason. The simplest explanation—that woody plants were not present at the start of the photographic history—is the most parsimonious one. Even this simple explanation has its nuances, which we explore in our analyses.

Accompanying the pejorative perceptions are exaggerations concerning change in the largest rivers in the region. Although it is well known that steamboats once plied the reaches of the Colorado River,[9] some authors insist that the Gila and San Pedro Rivers also sustained steamboat traffic.[10] As preposterous as these claims are, they illustrate the misperception that some have concerning the amount of change in the region's riverine habitat and natural-flow regimes. These exaggerations conceal more subtle but extremely important changes that have occurred. A theme of this book is to differentiate what some people think these systems should be from what they really were and are.

Finally, we acknowledge that our look at change within a region—even one as large as the desert regions of Arizona, eastern California, southern Nevada, and southern and central Utah—has limitations when removed from the larger context of continental environmental change. Global change, whether driven by human-induced changes or natural processes, looms large over the fate of riparian ecosystems. We discuss some of the larger continental implications where appropriate, but we also stress that our look at long-term change is primarily regional in scale.

Notes on Surface-Water and Groundwater Data and Records

Throughout this book, we cite U.S. Geological Survey data on water resources of the Southwest. Unless a specific source is cited, all data are either from on-line sources or are available from either the Arizona, Utah, or California Water Science Centers of the U.S. Geological Survey.[11] The water year of October 1 through September 30 splits the fall runoff season, and a hydroclimatic water year for all annual flood series presented in this book is defined as November 1 through October 31.[12] Although most groundwater data are from the U.S. Geological Survey, some records are from the Arizona Department of Water Resources.[13] Basin characteristics, such as drainage area, are from publications of the Geological Survey.[14]

Notes on Names of Plants and Places

We selected what we believed are the most typically used common names of riparian plants, and we acknowledge that we chose, in some cases, one name from several equally valid common names. We reserve use of Latin names for instances where those names are required to avoid confusion. Otherwise, we refer readers to the Glossary of Plant Names at the end of this book for the Latin names. We generally follow the conventions of the current floras of the region.[15] Because certain names of plants and places honor people, and because some of those names are consistently misspelled in the literature, we silently correct those names. Two names commonly used and corrected here are Lee's Ferry, Arizona, named after John D. Lee but usually cited without the apostrophe, and Frémont cottonwood, named for famed explorer John C. Frémont.

Notes on Archives, Photograph Sources, and Photography

As noted in the figure captions, the historical photographs used in this book were obtained from many archives and private individuals. In our photograph captions, we refer to some of these sources in an abbreviated form (as shown parenthetically here). We thank the Bancroft Library, University of California at Berkeley (Bancroft Library); the Arizona Historical Society, Tucson; Special Collections, the University of Arizona Library, Tucson; Cline Library, Special Collections and Archives, Northern Arizona University, Flagstaff (Cline Library); Special Collections of the J. Willard Marriott Library, University of Utah, Salt Lake City (Marriott Library); the Colorado Historical Society, Denver; the Museum of New Mexico, Santa Fe (Palace of the Governors [MNM/DCA]); the National Geographic Society, Washington, D.C.; the Autry National Center/Southwest Museum, Los Angeles (Southwest Museum); the Huntington Library, San Marino, California; the Dan O'Laurie Museum, Moab, Utah; the Salt River Project Archive, Phoenix, Arizona; and the Homer Shantz Collection, University of Arizona Herbarium, Tucson. In the figure attributions that include only a name, we refer to U.S. Geological Survey contract photographers or employees who took the photographs.

Several federal government archives provided extensive photography for our effort. These archives include the U.S. Geological Survey Photographic Library in Denver, Colorado; the National Archives and Records Administration in College Park, Maryland (National Archives); and the Utah and Arizona Water Science Centers of the U.S. Geological Survey (U.S. Geological Survey). We thank the Bureau of Reclamation, Boulder City, Nevada, for several photographs of the lower Colorado River; the Historic Graphic Collection, Harpers Ferry Center, National Park Service of Harper's Ferry, West Virginia (National Park Service); and the Natural Resource Conservation Service, Washington, D.C., for several photographs of southeastern and south-central Arizona.

We particularly thank our predecessors at the U.S. Geological Survey in Tucson for their incredible foresight in preserving historical photography of gaging stations in Arizona. In addition, the U.S. Geological Survey Photographic Library has faithfully preserved the imagery of field geologists who mapped the Southwest, and these images provide invaluable information on largely inaccessible areas. The National Archives has preserved many of the earliest photographs of the region, in particular those taken during the first river expeditions on the Colorado River. The photographs preserved by these three archives form the basis for most of this book and hopefully will be a continuing resource for the future interpretation of change in the region's rivers.

Many people supplied photographs or information for river reaches. For example, Craig "Sage" Sorenson of Grand Staircase–Escalante National Monument (GSENM) provided a key photograph that his father took that fits into a time series of the Escalante River (chapter 9). Greg Christianson, also of GSENM, provided historical photographs of the Calf Creek Crossing on the same river, and Dale Wilberg of the U.S. Geological Survey supplied a historical photograph of the Escalante River gaging station from the files of the Utah Water Science Center. James Klein provided photographs from the Samuel Holsinger and Mabel Cox collections for southern Arizona (chapter 22). Other individuals who supplied useful photographs for our efforts include David E. Brown, Gretchen Luepke Bynum, George Simmons, Josef Muench, and Tom Brownold.

The quality of our repeat photography stems from the work of two individuals. Dominic Oldershaw replicated most of the photographs we matched and interpreted as part of this study, and he has well earned his status as the premier photographer in this field. Diane Boyer did the digital darkroom work that was essential for portraying and interpreting this photography. We owe them a large debt of gratitude. Other photographers who worked with us outside of Grand Canyon include Gary Bolton, Tom Brownold, Tillie Klearman, Steve Tharnstrom, Sam Walton, Tom Wise, and Steve Young; photographers who worked with us in Grand Canyon are acknowledged elsewhere.[16] All original and repeat photography is stored archivally in the Desert Laboratory Collection of Repeat Photography at the U.S. Geological Survey in Tucson.

Acknowledgments

The work reported in this book was jointly funded by Water Resources Discipline, Ground-water Resources Program, and National Research Program of the U.S. Geological Survey. We especially thank Norm Grannemann for his ongoing support of research on issues related to the hydrology of the Southwest and Bill Alley for his unwavering support of the Southwest Ground Water Project.

Many people helped us to locate camera stations for old photographs, replicate them in the field over many months of time, prepare digital photography, and interpret the results. We offer special thanks to the many field assistants who endured freezing rain, blistering heat, flash floods, and all of the other joys of working along the rivers of the southwestern United States. These individuals include scientists, students, professional river guides, and volunteers who have accompanied us on our numerous trips around the region; there literally are too many of them to acknowledge individually here. Those providing the most help include Jayne Belnap, Cassie Fenton, Keith Howard, Mimi Murov, and John Weisheit. In addition to field assistants in Grand Canyon previously acknowledged, we thank Kenny Baker, Mike Borcik, Kirk Burnett, Diane Greene, Johnny Janssen, Thom O'Dell, Andy Persio, Lynn Roeder, and Rachel Schmidt.

Many land-use agencies allowed us access to their lands or provided permits for our photographic excursions or both. Our repeat photography in Grand Canyon, which began in 1972, was aided by research permits from Grand Canyon National Park. The River Unit of Canyonlands National Park provided permits for our work, and many of our repeat photographs were taken in their company; we especially thank Charles Schelz, Steve Swanke, and Steve Young for their help. Young, in particular, matched many photographs within Canyonlands as part of an outdoors education program with Prescott College. James Aton of Southern Utah State College reviewed chapters 6 and 8.

For allowing access to the San Juan River, we offer special thanks to the River Office staff of the Monticello Field Office, Bureau of Land Management (BLM). Mark Meloy helped considerably with river-trip logistics. Gene Stevenson and Gene Foushee helped identify the locations of some obscure camera stations along the San Juan River. Similarly, the BLM office in Price, Utah, provided river-permit access to Desolation and Gray Canyons on the Green River and helped with identification of some camera stations. Tillie Klearman and Sam Walton, in particular, helped with interpreting changes along both of these river reaches. Lynn Orchard and his family accompanied us on trips on the San Juan River and in Desolation and Gray Canyons and helped with the photography as well.

We particularly thank Thomas O'Dell, formerly of GSENM, for his support of our repeat photography work along the Escalante River and Marietta Eaton of GSENM for her continued support. Richard Hereford provided encouraging comments on many chapters, especially chapters 9, 10, 11, 12, and 13. Dave Sharrow of Zion National Park enthusiastically helped us match photographs of Zion National Park, helped arrange access to the restricted area of Parunuweap Canyon in southern Utah, gave us unpublished documents on long-term change of the Virgin River, and reviewed chapter 13.

Several people helped with access to Aravaipa Canyon and provided logistical support for our work there. Tom Schnell of BLM facilitated access to the canyon. Mark Haberstich and Matt Killeen of The Nature Conservancy at Aravaipa's east end were knowledgeable, helpful, and enthusiastic, as well as a pleasure to work with, and they also provided accommodations. We especially thank Diana Hadley of the University of Arizona for help with obtaining photographs that she personally collected from current and past Aravaipa residents. Daniel Baker and David Omick helped with access and assisted with fieldwork in Hot Springs Canyon, and Lamar Smith, formerly of the University of Arizona, reviewed chapters 19 and 20.

Julia Fonseca of Pima County Floodplain Management provided considerable useful information on historical change of southern Arizona rivers and streams (chapters 19–22). Tom DuRant of the National Park Service helped us locate and properly attribute an obscure photograph of Tumacácori National Monument on the Santa Cruz River. Roy Johnson provided some unpublished information on change in riparian vegetation along the Santa Cruz River upstream from Martinez Hill, and the San Xavier District of the Tohono O'odham Nation permitted access to Martinez Hill and our camera stations (chapter 21).

Kathleen Blair, biologist with the U.S. Fish and Wildlife Service (FWS), provided significant information on long-term changes and resources at the Bill Williams National Wildlife Refuge. Andrew Hautzinger, a hydrologist with the FWS Albuquerque office, provided hydrologic data and reports on the Bill Williams River. Pat Shafroth of the U.S. Geological Survey supplied information on historical photographs of Planet Ranch. They all kindly and critically reviewed chapter 15. Marshall Brown, a hydrologist with the city of Scottsdale, provided information on his employer's ownership of Planet Ranch and its historical groundwater use. William Radke, the manager of the Leslie Canyon National Wildlife Refuge, helped with access and questions concerning our photographs there.

Many landowners, ranging from large corporations to retirees, graciously allowed us access to or through their lands to old camera stations. They include Pacific Gas and Electric Company near Needles, California (Glen Riddle); The Nature Conservancy, access to Aravaipa Canyon; the Fort McDowell Indian Reservation, access to the lower Verde River; and Dan Baker, Andy Smallhouse, Johnny LaVin, and Sheryl Roberts, access to private lands along the San Pedro River. Bert McFarland kindly allowed us to access his land for an important photograph of the San Pedro River at St. David. We particularly thank Lamar Smith, emeritus professor of Range Science at the University of Arizona, for helping us obtain permission from land owners along the middle San Pe-

dro River and its tributaries and for his comments on chapter 20.

Numerous individuals contributed to our treatment of the lower Colorado River, including Ed Glenn, Keith Howard, Gordon Mueller, Pamela Nagler, and Charles van Riper. Glenn also contributed to the data contained in table 3.1. We thank Gregory Lines and Michael Scott of the U.S. Geological Survey for their helpful discussions about riparian vegetation along the Mojave River. Both provided helpful comments on chapter 28. Tom Egan, formerly of the BLM, provided information on the tamarisk-eradication project within Afton Canyon on the Mojave River.

Others who helped us access either public or private lands include Clifford and Georgiana Wood, who contributed to our work in Aravaipa Canyon; Mike Sanders, who aided us in navigating the maze of roads to the gaging station on the Santa Maria River near Bagdad; John Utz, manager of Paloma Ranch, who allowed us to access private land near Gillespie Dam on the Gila River; Giovanni Panza and Cullen Cramer of Rancho Solano on the Cañada del Oro Wash; Donnie May of the U.S. Army

Corps of Engineers, who helped with work at Painted Rock Dam on the Gila River; and Lorrie Robinson, the office manager for Trees Ranch, who assisted us in gaining access to Parunuweap Canyon of Zion National Park.

Numerous individuals contributed their opinions and ideas to this work. In particular, we note highly productive and challenging discussions with Richard Felger, Julie Stromberg, David Goodrich, Jayne Belnap, and Ed Glenn. However, responsibility for the ideas and opinions presented here lie with the authors alone. Finally, we acknowledge those people who helped us to analyze the photography and the hydrologic data and to prepare the manuscript. Elizabeth Deliso helped interpret changes to riparian vegetation in most of the repeat photographs, and we owe her a large debt of thanks. Brian Yanites and Jesse Dickinson contributed to the analysis of hydrologic data, and Peter Griffiths assisted with the illustrations. Jayne Belnap, Diane Boyer, David E. Brown, and three anonymous readers reviewed the entire manuscript.

Robert H. Webb
Stanley A. Leake
Raymond M. Turner
January 30, 2006

The Ribbon of Green

The Ribbon of Green

1 Value of Woody Riparian Vegetation

Summary. Riparian ecosystems are highly valued as wildlife habitat, for their intrinsic value as a mesic component of the desert environment, and for their effect as filters for water quality in desert rivers. Riparian vegetation occupies floodplains desirable for agriculture, increases flow depth during floods and potential area of inundation, and uses considerable surface water and groundwater. Numerous factors can affect the stability of wetlands, including total diversion of surface water and excessive use of groundwater. From a national perspective, riparian vegetation is considered to be severely threatened, and large changes in riparian vegetation attributed to flow regulation have been described in regions outside of the Southwest.

Riparian ecosystems are a resource that is valued disproportionately to their spatial extent in the southwestern United States. Although only a small amount of this semiarid and arid region is riparian, it is heavily used by wildlife in general and neotropical migratory animals in particular, and it has a high productivity and biodiversity. Riparian ecosystems also function to stabilize riverine environments, and people like to see green trees and shade in a sea of brown desert. Riparian ecosystems consist of a large variety of plant species with different requirements for light and water and tolerances of shade and disturbance. Within the category of vascular plants found in riparian settings, we differentiate annual, perennial herbaceous, and perennial woody vegetation, and, for reasons explained later, we primarily discuss the latter.

Riparian ecosystems serve as the primary link between upland terrestrial and aquatic ecosystems[1] and, in many respects, reflect the cumulative effects of larger-scale landscape change because of their responsiveness to fluvial disturbances. Although many researchers have studied riparian vegetation mainly because of the value of its components (e.g., mammal and bird populations),[2] riparian ecosystems can be viewed as scale phenomena that range from minute properties to large-scale assemblages of biological components.[3] In this book, we concentrate on changes at scales ranging from individual reaches to entire river systems, addressing the overall question of long-term change in woody riparian vegetation.

Definitions

In order to address whether or not most of the wetlands have been lost in the Southwest, we need to define the word *wetland*. Wetlands may be anything from isolated springs to swampy areas with poor drainage (locally known as *ciénegas*), to drainages with shallow groundwater but no year-round flow, and finally to perennial rivers.[4] This book focuses on the latter two types of wetlands, also known as *riverine* environments because of their association with perennial, intermittent, or ephemeral watercourses. To round out these preliminary definitions, we differentiate *riparian* (terrestrial) from *aquatic* (subaqueous) ecosystems, both of which compose the riverine environment. Because the original proposition addresses the magnitude of wetland loss along the largest river in the Southwest, this definition is appropriate. Readers interested in ciénegas, in particular, are directed to an excellent treatise by Hendrickson and Minckley on the subject for southern Arizona.[5]

Wetlands accommodate many different types of organisms, ranging from annual plants to algae and insects to native and nonnative fish to trees, migratory birds, and resident animals. The question of regional changes in native fish populations, in particular, has already been addressed extensively[6] and thus is discussed only in passing in this book. Similarly, changes in aquatic food bases, particularly downstream of dams, are unquestionable.[7] Here, we concentrate on terrestrial riparian ecosystems, the most visible and perhaps least understood part of riverine ecosystems.

Within the woody plants inhabiting riparian ecosystems, *obligate riparian* species require a year-round dependable water supply, whereas *facultative riparian* species can also live in the more xeric uplands. In our region, the Frémont cottonwood tree is the iconic obligate riparian species, and three species of mesquite represent the facultative riparian species. These species are discussed extensively in chapter 3. On the other hand, *xerophytic* species inhabit upland areas and depend on meteoric water, but some of these species can move into the riverine environment. *Phreatophytes* are plants that use groundwater;[8] therefore, both obligate and facultative riparian species can be phreatophytes. For physiological reasons, the line between obligate and facultative can be blurred, as we discuss. Finally, we differentiate native versus nonnative species, recognizing that the latter include plants from Eurasia (e.g., tamarisk) and plants from nearby areas (e.g., fan palms). One of the major themes of this book is to address the common perception that tamarisk has taken over the riparian world in the Southwest, displacing native species (chapter 4).

Value of Riparian Vegetation

Riparian vegetation in the Southwest has high value to channel stability, water quality, and wildlife populations.[9] Some believe that the highest-quality riparian assemblages are native cottonwood-willow assemblages, although this type of riparian vegetation typically occurs below about 4,000 feet in elevation. Other qualities of riparian assemblages desirable for wildlife are diversity in structure—tall trees with an understory of species that achieve different heights—and density of foliage. Finally, wildlife populations find diversity in heterogeneous habitat types, termed *patchiness*, to be more desirable than monotypic vegetation.

Extent of Riparian Ecosystems

In terms of their biological attributes, riparian ecosystems are an extremely small component of the landscape. In California, riparian areas cover 341,000 acres or less than 0.5 percent of the state's area.[10] Similarly, in the arid Southwest, riparian vegetation constitutes 0.5 percent of the landscape or less.[11] Even with such a small amount of land area, approximately one-third of vascular plant species in the region occurs in riparian areas.[12] Because these riparian areas are so small and have such high biological value, they have a higher priority for protection than nearly any other habitat type in the region.

Channel Stability

Different types of riparian vegetation are thought to have varying effects on channel stability. Studies in Wisconsin suggest that floodplains mantled with grasses trap more sediments and are therefore more stable than those covered with trees.[13] In the Southwest, the opposite is true: woody riparian vegetation, in particular species that grow densely on low floodplains, trap sediments and thereby increase their own habitat. As a result, channel narrowing associated with vegetation encroachment on floodplains is commonly observed along rivers in the Southwest, particularly in reaches where flow is regulated.

Riparian vegetation tends to have dense root systems that bind alluvial soils. One of the largest perceived benefits of riparian vegetation is channel stabilization; proliferation of nonnative tamarisk began when canal builders deliberately planted this species because it was thought to stabilize banks better than native species (chapter 4). Dense vegetation also lowers streamflow velocities, minimizing flood damage and trapping water for infiltration into the aquifer beneath floodplains.

Low-gradient rivers meander within channel bands constrained by valley width. When the river shifts, it leaves behind low, flat-lying terraces with deep, silty, alluvial soil. These silty soils may be highly saline, particularly at lower elevations, and are able to hold more water than upland sandy soils. Some bottomlands are regularly flooded, so occupying plants must be able to tolerate standing water for long periods of time; in addition, the proximity of the river creates shallow water tables. Unlike riparian plants that grow directly adjacent to the river, plants growing in river bottomlands do not need to withstand fast-moving floodwaters, and many typical bottomland plants provide little protection against erosion.

Development of Marshes

Marshes are magnets for waterfowl and serve as nurseries for young native fish. Ciénegas, which once were locally common in the southern part of our region,[14] are a special case of marshes where surface-water flow is low; however, marshes can also form along the larger rivers. Cattail, carrizo grass, and reeds are characteristic species that line the banks of marshes, with woody riparian vegetation upslope on alluvial fill that is unsaturated near the surface.

Development of marshes along perennial rivers occurs when channel avulsions cause segments to be abandoned, forming "oxbow lakes" or wide areas of low-velocity flow. Along some rivers, marshes are ephemeral features that form during flood deposition and are destroyed by subsequent flood scour.[15] Sediment deposition by regulated flow can create marshes along rivers such as the Colorado where flood frequency is greatly reduced.[16] Along the lower Colorado River, marshes may become separated from downstream-flowing water and are subject to infilling if flow regulation eliminates the potential of scouring floods.[17]

Water Quality

Riverine riparian ecosystems are viewed simultaneously as sediment traps and biological filters that affect water quality.[18] Streamflow sediment concentrations decrease when dense riparian vegetation is present because of both the sediment-trapping influence and the root-binding influence that stabilizes channel margins. Dense riparian vegetation also affects the quantity of dissolved constituents. For example, nonvascular plants, notably algae, can bind metals to organic compounds, effectively removing them from downstream flow. Riparian vegetation also consumes nitrogen compounds and micronutrients while increasing organic carbon contents. Some consider these functions to be a major positive attribute of riparian ecosystems in helping to improve water quality.

Fish and Riparian Vegetation

The relation between native fish populations and riparian vegetation has been little studied in the Southwest, in part because native fisheries have decreased significantly and flow regulation has changed the riverine environment. Some general relations are apparent, however, when the pre-regulation aquatic environment is considered. Before dams decreased flow turbidity, large rivers were *heterotrophic* for most of the year,[19] and the food supply for native fishes was generated mostly from organic material entering the river, either falling directly into the flow or by entrainment of driftwood. This relation suggests that riparian productivity on a watershed scale was paramount, although streamside productivity likely contributed significantly. The same type of food chain has been proposed for native salmonids in the Pacific Northwest.[20]

Reptiles, Amphibians, and Riparian Vegetation

Although reptiles may fall variously in the xerophytic, facultative riparian, and obligate riparian categories, all amphibians are at least partially dependent on perennial water supplies. Arizona has a herpetofauna that consists of twenty-two species of amphibians and ninety-four species of reptiles;[21] fifteen obligate riparian reptiles are considered to be threatened.[22] Because of long-term declines in riparian species, notably amphibians, this group of animals tends to drive protections of riparian habitat in the region.

In addition to the obligate species, others use riparian zones for various purposes, notably for forage. Xerophytic reptiles prefer desert washes and avoid dense thickets with little structural diversity, in particular those dominated by tamarisk.[23] Along the Colorado River, the highest densities of lizard populations are in shoreline habitats, and the lowest densities are found in xerophytic substrate.[24] The edge effect of highly productive riparian habitat adjacent to desert creates an attractant to reptiles otherwise associated with xeric landscapes.

Mammals and Riparian Vegetation

Riparian vegetation serves as cover for local mammal species and as corridors for animal migration.[25] Mammal populations are drawn to riverine riparian vegetation, particularly when woody thickets composed of multiple structural layers are present. For example, in Grand Canyon, seventy-two species are known to occur in the national park, and twenty-seven of those (38 percent) are found along the Colorado River.[26] On the San Pedro River, 86 of the 137 mammals that live in Arizona frequent the riparian zone.[27] In some dense riparian assemblages established as a result of dams and reservoirs, small mammal populations may have lower numbers than in nearby, more xeric areas, possibly because the xeric areas are more structurally diverse.[28] For example, cottonwood habitat in Canyonlands National Park is considered to be poor for most mammals in comparison to other riparian and upland habitats.[29]

Observations during recent drought periods indicate that riparian ecosystems serve as an important buffer zone for many animals normally associated with xerophytic habitat. When water supplies and forage in upland habitat become scarce, the riparian zone, which may be unaffected by short-term drought, provides these resources and may allow some species to survive. Some species normally considered to be upland obligates use riparian zones during the dry season (e.g., bighorn sheep in Grand Canyon move to the Colorado River during summer months).

Birds and Riparian Vegetation

Riverine riparian vegetation creates the backbone of two important bird flyways in North America. The Pacific flyway allows birds to migrate from coastal and interior Mexico to the Pacific Northwest through important stops in the Colorado River delta, the lower Colorado River, and the Salton Sea. The Central flyway, allowing birds to migrate from the Sierra Madre of Mexico to the mountainous regions of the western United States, has the San Pedro River as an essential component.[30] For waterbirds, riparian vegetation arguably is irrelevant except as a riverine component that affects the aquatic food base. In other cases, riparian vegetation either directly provides a food source for frugivores or supports insect populations used by insectivores.[31] During the spring months, riparian zones can have as much as ten times more migratory birds than nearby xerophytic sites, although amounts vary.

In the region, the San Pedro River represents the most important avian habitat, with 100 species of breeding birds present and 390 species passing through or using its gallery forests. The Bill Williams River National Wildlife Refuge (NWR), at the downstream end of the Bill Williams River (chapter 15), has 343 species of birds in a relatively small area, including what is considered to be the only "complete" suite of riparian obligates in the region.[32] Neotropical migrants also use what has been termed *xeroriparian vegetation* (facultative riparian species or xerophytic vegetation) along ephemeral washes,[33] and continuous bands of riparian trees are not as im-

portant to migrating bird populations because migrating birds use both continuous gallery forests and isolated patches along their flight path.[34] Therefore, the migratory pathway of neotropical species may not be as narrowly focused on perennial rivers as is sometimes assumed.

Riparian ecosystems in the southwestern United States support five to ten times the population density and diversity of birds than in the adjacent desert landscapes.[35] Birds, in particular, are drawn to riparian ecosystems and the abundant insects and plant parts they contain.[36] Some riparian areas in the Southwest have the highest density of birds estimated in North America, owing to the unique combination of migrating neotropical species, waterbirds, raptors, and year-round residents. Perhaps 60 to 70 percent of western neotropical migratory birds depend on native stands of riparian vegetation in the Southwest, in particular cottonwood-willow forests.[37] Phenology is also important to neotropical migrants, who may time their migration to coincide with flowering periods of mesquite to take advantage of the insects on the trees.[38]

Avian populations in the Southwest cannot be accurately depicted as species per general unit area because most of the species are concentrated around riparian zones. For example, Grand Canyon National Park has 303 species of birds, including transients, and 190 of those species (63 percent) use riparian areas—either the Colorado River or its numerous perennial tributaries—exclusively or at least in passing.[39] The highest densities of birds tend to occur seasonally in habitats with the highest structural complexity[40] and highest canopy cover;[41] in other words, bird populations are greatest where a variety of plant species of different height and shape are present.

Three types of birds are drawn to riparian areas. Waterbirds seek the aquatic environment and the fish and insects that provide their food source.[42] Insectivores and fructivores seek insects and fruits, respectively, borne by riparian plants. Raptors are drawn to the high bird populations in their search of prey, either birds or small mammals. Ideal Peregrine Falcon habitat, for example, is steep terrain adjacent to a perennial river, where the heights allow diving on waterbirds, such as ducks or swallows. Bald Eagles use riparian ecosystems for food, typically fish, and nesting sites, in particular cottonwood.

Bird populations adapt to changing environments, which in the Southwest includes shifts in usage patterns away from what might be perceived as ideal habitat, both seasonally and spatially.[43] One study suggests that birds prefer monospecific cottonwood stands adjacent to agricultural fields more than nearby, mixed stands of native plants.[44] Birds in the monospecific stands are larger, which allows them to fly longer distances for food; the attraction to the cottonwood is abundant nesting sites rather than readily available food supplies. In contrast, the lower-density populations of mixed plants attract birds that forage there mostly during the breeding season.

Biodiversity and Riparian Ecosystems

Biodiversity, commonly defined as the number of native species per unit area, is at a peak in riparian ecosystems,[45] particularly in the Southwest. Bird populations in riparian areas have a higher density than measured elsewhere in the United States.[46] Plant populations in the greater riverine environment, including floodplains and the adjacent xeric landscapes that have gradients into the riparian ecosystem, have a higher species diversity than do other landscapes. Similarly, mammal populations prefer riparian areas out of proportion to the remainder of the landscape.[47] As with other valued characteristics, biodiversity in riparian zones is highest where structural diversity and foliage density is high, and where dense thickets or gallery forests are adjacent to open areas, such as grasslands or shrublands (high patchiness).

Laws

Several laws—notably the Endangered Species Act of 1973 and the Clean Water Act of 1977—are routinely applied in a regulatory fashion to questions of land use concerning wetlands in our region. Many riparian areas occur in national parks, where the wording of the Organic Act—to preserve and

protect—is not necessarily a strong enough mandate when forces of change are external to the park. In the case of the Colorado River, the Grand Canyon Protection Act was enacted to provide additional incentive to protect the riverine ecosystems downstream from Glen Canyon Dam, which are in national park units.

Other areas have more specific protections that may or may not be effective. The riparian area in part of the upper San Pedro River in southern Arizona (chapter 19) is designated as the San Pedro Riparian National Conservation Area and is under special administrative protection of the Bureau of Land Management (BLM). This ecosystem is considered to be threatened at present by increasing groundwater pumpage from nearby Ft. Huachuca and the city of Sierra Vista. Depending on a number of factors, not the least of which are the legal definition of connectivity between surface water and groundwater (chapter 2) and the amount of land-use protection, the perceived threat may eventually force major changes in water use, thereby affecting growth and economic development in the upper San Pedro River valley.

Many riparian ecosystems, because of their high value, compose the heart of NWRs. Four of these units, managed by the U.S. Fish and Wildlife Service, occur along the lower Colorado River with the mandate to encourage and protect waterbirds, obligate riparian mammals, and their habitat. Private organizations such as The Nature Conservancy own and manage riparian areas not otherwise protected by government agencies, although many of these protected areas are small. To help with external threats, The Nature Conservancy works with adjacent landowners to develop management plans that affect much larger areas.

Finally, because of the perceived negative impacts of domestic livestock,[48] grazing has greatly decreased in riparian habitat on public lands, particularly during seasons of high wildlife use or potential seedling establishment. Whether managed by the BLM, the Forest Service, other federal agencies, state and local governments, or private entities, fencing and patrols are used to exclude cattle and sheep, to repair

fences, and to minimize trespass. Grazing exclusion currently is the strongest protection used regionally to protect riparian vegetation in the Southwest, and whether this protection is warranted with respect to woody species is a subject we treat in this book.

Undesirable Qualities of Riparian Vegetation

Although riparian vegetation currently is highly valued, it once was considered not only expendable but potentially hazardous. Riparian vegetation occupies highly fertile lands adjacent to readily available water supplies; large areas were converted to produce food and forage. Because dense riparian vegetation slows streamflow, the area of inundation during floods is greater for the same discharge, increasing flood hazards upstream while decreasing flood hazards downstream. Finally, riparian vegetation uses large amounts of water that otherwise might be used for irrigation or domestic supplies. As a result, considerable effort was expended, particularly in the early and mid–twentieth century, to reduce riparian vegetation locally.

Agricultural Areas

Clearing of vegetation for agriculture has occurred along alluvial reaches throughout the Southwest. Agricultural practices in the late nineteenth century, especially at the time of settlement, had a trial-and-error approach for locating agricultural fields. For example, at Kanab, Utah, fields were planted directly adjacent to the channel of Kanab Creek, only to be abandoned within a decade owing to devastating flood damage.[49] Channel avulsions and meander swings, particularly during the period of arroyo downcutting (chapter 3), eliminated large areas of agricultural land. This process occurred repeatedly, with the expected result that most agricultural fields were removed from lands directly adjacent to rivers. Exceptions occur along the major rivers, such as the Colorado and the Green, where pumpage directly from the river sustains agricultural fields or where levees and channel stabilization lower the potential for flood damage.

In most of the Southwest, most of the agricultural fields were created from xeric habitats and watered from wells or irrigation canals. Many agricultural fields in southern Arizona remained in the distal riverine environment, commonly on the edges of floodplains. *Bosques* (riparian thickets) were cleared to create these fields, eliminating large areas of this habitat type.[50] Because of a variety of economic factors, the acreage of agricultural land under production in the Southwest is decreasing, and future conversion of riparian habitat for agricultural use is unlikely. The largest future impact of agriculture is the continued demand for water.

Reduction in Flow Conveyance

Flood damage in the Southwest was extreme during the decades before and after the end of the nineteenth century and during the period from the mid-1970s through the mid-1990s. Extensive flood-control projects, such as construction of Hoover Dam, were justified primarily to minimize flood damage and to stabilize agricultural lands and irrigation works.[51] Hydroelectric power production, once a secondary benefit from dams, became a priority as the nation's energy demands increased. Riparian vegetation thrived along irrigation canals and channels converted from natural flow to irrigation conveyance. This vegetation and notably the nonnative tamarisk (chapter 4) formed dense thickets, decreasing flow conveyance and even ponding backwaters.[52] By the late twentieth century, the objective of minimizing flood damages shifted from flood control to channel maintenance that increased flow conveyance.

Because any additional channel roughness increases depth of flow for the same discharge,[53] minimizing channel roughness became a primary goal, particularly in urban areas such as Phoenix and Tucson. Alluvial channels were straightened, and banks were protected with soil cement. For a typical alluvial channel, a 25 percent blockage of flow area by riparian vegetation may account for more than half of the channel roughness.[54] To minimize the potential for flood damage and the area of inundation, riparian vegetation

within channels was removed. Along some river reaches (chapter 21), the removal was not permanent.

Excessive Usage of Water

Riverine riparian vegetation uses considerable quantities of water that otherwise would be available for in-stream habitat, agriculture, or domestic use. When consumptive use by riparian vegetation exceeds the combined supply of streamflow and groundwater, stream reaches may temporarily dry, particularly during summer months.[55] This change not only affects domestic and irrigation supply, but also aquatic species' ability to survive. In the mid–twentieth century, "water conservation" was the justification to remove riparian vegetation and to increase water yields.[56] As a result, some have claimed that phreatophytes "are the greatest menace in areas of limited water supplies in the Southwest" and have advocated eradication of both native cottonwood and nonnative tamarisk.[57]

Some hydrologists have developed schemes for reducing evapotranspiration rates from floodplains by managed pumping from alluvial aquifers. Evapotranspiration decreases as the depth to the saturated zone increases, and, therefore, pumping of shallow groundwater to increase the depth to water has been proposed for water conservation without decimating riparian ecosystems.[58] Unfortunately, reproduction and productivity of riparian ecosystems also are a function of depth to groundwater (chapter 3), so the cost of this proposed management practice is potentially high over the long term.

Maps of Riparian Vegetation in the Southwest

Systems for classifying and delineating riparian vegetation in the United States have evolved over the past thirty years.[59] Riparian vegetation in the Southwest (fig. 1.1) has been mapped generally at a small scale, but only the state of Arizona has a detailed, although generally inaccurate, map. The first maps, published in the 1970s, did not delineate riparian assemblages,[60] but accompanying text described four wetland "formation types," including

three associated with elevations below 6,000 feet.[61] For the riverine environments we consider in this book, the woody plant assemblages would be classified as riparian deciduous forest or riparian deciduous woodland or riparian scrub, and eighteen plant associations are associated with these three formation types.[62] Through most of the region, the most common native plant communities are cottonwood-willow and mesquite bosque.

Detailed mapping of riparian vegetation is difficult owing to continuous change in this resource, and, as noted earlier, Arizona is alone in the Southwest in creating a map of riparian assemblages.[63] This mapping scheme defines nineteen associations of woody riparian vegetation occupying 673 square miles. Although these areas probably have changed since they were mapped, the most common riparian assemblage in Arizona is mesquite bosque (map unit 333.1, 263 square miles), followed by riparian deciduous woodland (map unit 333.1, 128 square miles) and riparian desert scrub communities (map unit 362.2, 90.2 square miles). The map units that explicitly mention cottonwood or black willow as a dominant total only 58 square miles.

Changes in Riparian Vegetation Outside of the Southwest

Change is the one shared characteristic of riparian ecosystems in the western United States. Regardless of the part of the country one examines or its climate or the hydrologic characteristics of sites, riparian ecosystems are changing in response to climatic variability, floods, fire, flow regulation, and other human effects. The examples that follow are culled from the large number of case studies of riparian vegetation that are available, which track the great variability in change that has been documented.

The Platte River in Colorado and Nebraska has changed from a wide, braided stream dominated by shifting sandbars to a narrow channel flanked by riparian gallery forests.[64] By the late 1960s, channel width was only 10 to 20 percent of what was observed in 1865. Between 1900 and the late 1930s, woody riparian vegetation—mostly plains cottonwood and peachleaf willow—be-

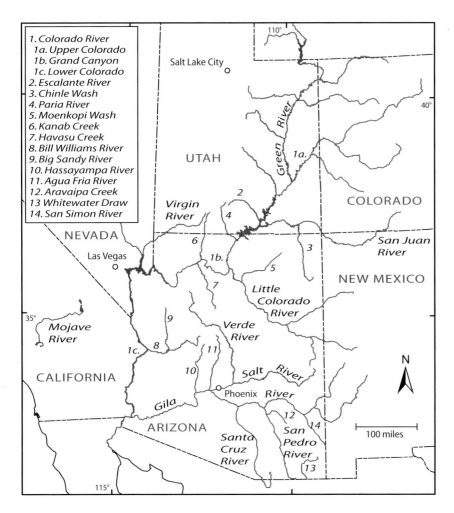

Figure 1.1 Map of the southwestern United States showing river basins discussed in this book.

came established on the formerly barren channel margins and bars.[65] The changes are related to decreases in discharge caused by upstream flow regulation, water use, and diversions,[66] and the species composition of riparian vegetation is expected to change in the future, assuming a continuation of current flow and climatic conditions. Similar changes are predicted for the upper Missouri River in North Dakota, where cottonwood forests established following flow regulation are expected to be replaced with green ash forests over time.[67]

The Cimarron River in Kansas[68] provides an example of channel change and interaction with riparian vegetation that is similar to those elements in the southwestern United States. A flood in 1914, which is the largest in the gaging record (as of 1963), widened the once-narrow channel by more than twenty times. Channel widening ceased in 1942 and was followed by floodplain deposition through the early 1950s. Riparian

vegetation became established on the newly deposited floodplain, creating a positive feedback for further vertical accretion of sediments without additional area for establishment of riparian vegetation.

The Cimarron River exemplifies one type of channel response that seems particularly appropriate for river systems with a large sand load. Large floods in eastern Colorado are known to strip riparian vegetation and to create broad, braided channels. After the occurrence of large floods, riparian vegetation becomes established on sandbars, which may coalesce into islands or merge into channel banks.[69] This coalescence is a major contributor to channel narrowing in the Cimarron[70] and is the likely mechanism for channel narrowing in the Southwest as well. One observation of change in Colorado rivers and other watercourses in the Great Plains is that channel narrowing—whether induced by or following increase in riparian vegeta-

tion—comes after the occurrence of extremely large floods.

In recent decades, the tall-willow communities in Yellowstone National Park have markedly declined, as documented using repeat photography.[71] These willow communities occur throughout the region along minor streams and moist valley bottomlands, and some researchers attribute the causes of decline to "natural regulation" by which a combination of fire suppression, primary succession, and climate change has reduced the number of willows.[72] Others believe that the loss of willows is the indirect result of population shifts in predatory bears and wolves, especially the decline in wolf populations, which has allowed elk populations to increase dramatically. Overgrazing by browsing elk may be the dominant force causing the decline in willow populations.[73]

Water in the upper Snake River in eastern Idaho is used extensively for irrigation without flood control. As a result, annual flow volumes decreased while seasonal variation in flow increased, leading to predictions that channel narrowing will occur with a concomitant increase in area for riparian vegetation.[74] Extensive flow regulation of the Snake River has resulted in stripping of fine-grained sediments from floodplains, particularly in Hells Canyon,[75] which forms the border between Idaho and Oregon.

The Owens River in eastern California has a notorious history of water development that affected riparian vegetation. Diversions currently affect 88 percent of the channel distance in this watershed, making it one of the most regulated river systems in the United States.[76] Channel width of some regulated streams has decreased as riparian vegetation has increased. Riparian vegetation along the river downstream from the main diversion point has changed from native species to a non-native assemblage dominated by tamarisk and Russian olive.[77] Owens Lake, the terminus of the Owens River, dried in the early 1900s following a nearly full diversion of streamflow. Some recovery of saline, lacustrine wetlands flanking the playa occurred between 1977 and 1992, following above-average runoff that was not diverted.[78]

2 Climate, Groundwater–Surface Water Interactions, and Woody Riparian Vegetation

Summary. In semiarid and arid environments of the Southwest, the location and stability of riparian vegetation assemblages depend on surface-water and groundwater interactions. Vegetation along stream segments with shallow groundwater or perennial flow are dependent on the configuration of alluvium and subsurface bedrock, and the complexities of alluvial stratigraphy strongly affect riparian species' ability to obtain water. Flood-frequency changes driven by climatic variation help explain long-term variation in riparian vegetation, and the combination of total surface-water diversion and groundwater overdraft is expected to be the largest reason for elimination of riparian vegetation.

The presence and stability of riparian ecosystems are determined by a complex interaction among climate, streamflow, and groundwater, with the geomorphic configuration of the channel and alluvial fill playing a secondary role (fig. 2.1). Climate affects all aspects of riverine ecosystems, from germination and establishment to productivity and biodiversity, and historical fluctuations in temperature and precipitation may help explain some changes in riparian vegetation. Surface water in most watercourses is storm related, and, in our region, only the Colorado, Green, San Juan, and Gila Rivers are known to have been fully perennial from headwaters to terminus. Groundwater outflow contributes to local perennial reaches of great importance to riverine ecosystems. Ultimately, surface-water and groundwater interactions, both overt and minute, drive the stability of riverine ecosystems.

Climate of the Region

Areas below the 6,000-foot contour in the Southwest generally receive less than 15 inches of precipitation per year.[1] In terms of land area, most of the region receives less than 10 inches of precipitation, most of which occurs as rainfall. The headwaters of many rivers in the region are above 8,000 feet and receive more than 20 inches of precipitation, most of which occurs in the winter as snowfall. Summer storms are common in the region, with the exception of the western Mojave Desert headwaters of the Mojave River. Reliable runoff generated during the winter and spring months has high sediment concentrations (e.g., 20,000 parts per million of sediment or greater); summer-storm runoff is less reliable and contains even higher sediment concentrations. As a general rule, the climatic and physiographic setting dictates that most of the runoff is generated at higher elevations, whereas most of the sediment comes from lower elevations.

In southern Arizona, precipitation is evenly split between winter and summer, with little rainfall occurring in the period April through June. Infrequent rainfall, usually from tropical cyclones or cutoff low-pressure systems, occurs from mid-September through October. Winter storms tend to yield low-intensity, long-duration precipitation, although relatively warm storms from the tropical eastern North Pacific Ocean can generate intense rainfall at all elevations. This type of storm generates some of the largest floods in the region if warm rain falls on an existing snowpack. Summer rainfall generally results from intense thunderstorms of local to regional extent, and these storms are more common in southern and central Arizona than in other parts of the region.

The seasonality of precipitation tends to be less reliable with increasing latitude, and this reliability strongly affects regional vegetation patterns. Except for the west coast of Baja California, the combination of winter and summer precipitation essentially defines the spatial extent of the Sonoran Desert.[2] Owing to the development of several persistent atmospheric steering mechanisms, most notably the Bermuda High, the Four Corners High, and the omnipresent heat-induced low pressure across the southern deserts, moisture is advected into the region from a combination of the Gulf of Mexico, the Gulf of California, and the eastern North Pacific Ocean. Much of the moisture travels northward along the western flanks of the Sierra Madre Occidental in Mexico, and during some summers moisture may penetrate as far northward as the Pacific Northwest and Canada. Therefore, summer precipitation in the Southwest is linked to continental-scale moisture-transport processes.

Tropical cyclones—a generic term that includes tropical depressions, tropical storms, and hurricanes—contribute a significant, although unreliable, amount of precipitation at any time from June through October. They form in either the eastern North Pacific Ocean, the Gulf of Mexico, or the Atlantic Ocean. Storms that originate in the Atlantic Ocean or Gulf of Mexico generally cross Central America or southern Mexico and regain energy over the eastern North Pacific Ocean; only a few storms have crossed the continental divide near the U.S.–Mexico border to inject moisture into the Southwest. Few tropical cyclones or hurricanes make landfall in the Southwest; most dissipate over the ocean, and their leftover moisture is advected into the region, steered or embedded within weak to strong circulation patterns. In the Southwest, the most significant rainfall amounts result when

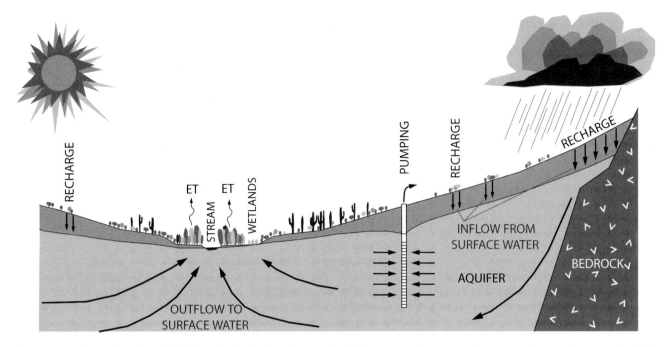

Figure 2.1 Generalized hydrologic cycle in the Southwest depicting groundwater–surface water interactions typical of many alluvial aquifer systems in the Basin and Range.

dissipating tropical cyclones coincide with cutoff lows or frontal systems from the North Pacific Ocean.[3]

Beginning at the Mogollon Rim, the climate of the Southwest changes significantly northward. The rim elevations of 7,000 feet and higher generally strip low-level moisture (below the 700-millibar pressure height) from storms, causing the highest regional storm intensities on the southern flanks of the mountains south of the rim. As a result, storms moving from the southwest and passing into the Four Corners Region may be diminished by the rainshadow effect of the rim, reducing storm amounts for some winter storms and most tropical cyclones. Similarly, storms approaching from the west encounter numerous mountains and plateaus in Nevada and southwestern Utah, creating another rainshadow effect for the plateau region of southeastern Utah.

Summer precipitation generally decreases and becomes unreliable on the Colorado Plateau in southern Utah.[4] The amount of moisture during summer is strongly dependent on the location and magnitude of those persistent steering features in the atmosphere. If, in particular, the Four Corners High does not develop or is not in its usual position for a significant period or is

stronger than normal, moisture entry into the region may be blocked. During the summer, moisture recycled into the atmosphere following storms is added to the moisture advected from the oceans.

The headwaters of the Mojave River generally respond to winter precipitation. Owing to the strong orographic precipitation associated with the interaction of storms with the Transverse Ranges, rainfall may be relatively light in the desert but strong in the headwaters, leading to large amounts of runoff. Little runoff in this river system is generated during summer storms, making the Mojave River unique in the Southwest.

The El Niño–Southern Oscillation (ENSO) phenomenon of the Pacific Ocean strongly affects interannual variation in precipitation in the Southwest.[5] ENSO effects can be separated into three general categories: warm ENSO events (commonly known as El Niño conditions), cool ENSO events (commonly known as La Niña conditions), and other conditions. Precipitation during El Niño conditions generally is high, although notable dry periods have occurred under these conditions. Winter precipitation is the most affected; the effect of El Niño on summer precipitation is weak at best.

Precipitation during La Niña conditions is reliably low, and most significant regional droughts are associated with this state. Anything can occur during other conditions, although extreme wet or dry periods are unlikely.

Despite geographic variation in processes, decadal-scale climatic fluctuations appear to affect the region in a relatively uniform fashion. Although all of the region may simultaneously experience dry or wet conditions, the magnitude and persistence of unusual climatic conditions varies. Whereas wet conditions generally are uniform from the Mojave River through southern Arizona, droughts seldom are uniform in severity or length. Orographic effects and seasonality of precipitation add to the complexity, making general statements about climatic variability difficult. Finally, average temperatures are related to precipitation; temperatures tend to be annually or seasonally high during droughts and may be relatively low or high during wet periods. This complex interaction affects riverine riparian vegetation.

The riparian vegetation encountered by early explorers undoubtedly reflected some aspect of prehistoric climatic conditions. The climate of the nineteenth century was strongly affected by cessation of the Little Ice

Age, which extended from around A.D. 1500 to around 1850.[6] The impact of this period on the Southwest is known only from tree-ring records at higher elevations, although, in general, average temperatures were lower than at present with uncertain impacts on regional patterns of precipitation. Landscape observations by Spanish explorers reflect, in part, ecosystems affected by the Little Ice Age.

Recorded observations of climate and weather began with settlement of the region by nonindigenous peoples, starting with the establishment of Spanish missions south of present-day Tucson in the late seventeenth and early eighteenth centuries.[7] The colonization of Utah by Mormons, beginning in 1847, led to settlement of the southwestern part of that state in 1860 and extending southeastward until 1880. An influx of prospectors, spurred by the California gold rush, led to boom-bust mining throughout the region; hostilities with American Indians led to the establishment of numerous army outposts throughout the region. Those outposts provided the first instrumental records of climate, although many climatologists discount most of these records as inaccurate. Reliable measurements that provide a regional perspective on climate began in the late 1880s.[8]

Analyses of climatic trends provide a general framework of decadal climatic fluctuations in the Southwest, and these analyses[9] are discussed further in chapter 30. The period 1880–1891 was generally wet, with numerous regional-scale storms that caused channel downcutting and generally led to the observation that "rainfall follows the plow." The most severe drought, and the one that affected the largest part of the region, began in the summer of 1891 and ended in 1904. In combination with overstocking of the range, the early-twentieth-century drought caused the death of half of the cattle in the region between 1891 and 1896. El Niño conditions in 1904 and 1905 ended this drought.

The wettest period in the region's history began in 1909 and extended through about 1920. The early-twentieth-century wet period had a number of lasting effects, including its influence on the overallocation of Colorado

River water and the widening of arroyo channels throughout the region. This combination of extreme events and human settlement had major effects on the stability and extent of woody riparian vegetation in the region, in particular the elimination of much of the established vegetation along alluvial channels.

Climate was regionally variable between 1920 and the early 1940s, ending with the strong El Niño conditions of 1941 through 1942. In southern Arizona, conditions were relatively dry with few significant winter storms. From the Mojave Desert through southern Utah, conditions were generally wet, punctuated with a mild drought during the Dust Bowl years of the early 1930s. Between the mid-1940s and the early 1960s, drought conditions prevailed with strong regional variation in intensity. The mid-twentieth-century drought, centered on the La Niña conditions of 1954 through 1956, was most severe in the Mojave Desert, in southern Utah, and to the east in New Mexico. Near-normal summer precipitation mitigated this drought in central and southern Arizona.

Beginning in the early 1960s and fueled by several significant El Niño periods, the climate of the region became significantly wetter and warmer. Numerous strong storms in fall and winter occurred between 1970 and 1995, leading to significant floods in central and southern Arizona and to above-average precipitation in the Mojave Desert and the Colorado Plateau. Notable periods of El Niño conditions occurred from 1978 through 1980, 1982 and 1983, and 1993 through 1995. Brief droughts interrupted this intermittent wet period in 1986 and 1989 through 1991, with the latter event having severe effects in the Mojave Desert. The current extent of woody riparian vegetation was strongly affected by this climate period, which we refer to as the *late-twentieth-century wet period.*

Despite El Niño conditions in 1997 through 1998 and 2002 through 2003, drought generally prevailed at the end of the twentieth century and the beginning of the twenty-first century. The early-twenty-first-century drought, centered on 2002, has created several record extremes, including

the lowest flow ever recorded in the Colorado River in southern Utah[10] and record low annual rainfall at many stations. The reasons for the switching of interdecadal periods between generally wet and generally dry conditions remain speculative, but research on the subject centers on hemispheric-scale, low-frequency oceanic processes in the North Pacific and North Atlantic Oceans.

Temperatures fluctuated through the twentieth century, leading to several distinct precipitation-temperature regimes. Annual temperatures generally were low from 1900 through 1930, creating a high precipitation–low temperature combination.[11] Severe freezes were common during this period,[12] with potential negative impacts on some species of riparian vegetation that otherwise would have germinated and established following winter floods. Temperatures peaked during the midcentury drought, creating a warm, dry period. Declining temperatures accompanied the earliest part of the late-century wet period, although temperatures steadily rose during the 1980s and 1990s, meaning that the late-twentieth-century wet period was also warm.

The most important characteristics of the late-twentieth-century wet period were an increase in winter floods, which create ideal conditions for germination and establishment of native riparian vegetation on larger rivers (chapter 3), and an overall increase in the length of the growing season,[13] which encourages rapid plant growth. This latter period therefore created ideal climatic conditions for germination and establishment of riparian vegetation. The early-twenty-first-century drought, which is ongoing at the time of this writing, has resulted in relatively dry conditions that are adverse to germination and establishment of native riparian vegetation, killing seedlings and even established trees in some marginal reaches.

Geomorphic Configuration and Riparian Vegetation

Riparian ecosystems are draped over a variety of geomorphic configurations in the Southwest. These settings determine how groundwater and surface water interact and, to a large extent, what species of woody vegetation can grow in a given reach. Combined with the effects of elevation gradients and climate, geomorphic setting is paramount to riparian vegetation.

Bedrock canyons refer to a complex suite of geomorphic configurations. A small number of canyons, notably those carved by Kanab Creek and the Virgin River (chapters 12 and 13), consist of a thin veneer of alluvium over bedrock. In this setting, cracks in bedrock or unstable talus slopes provide the only substrate for a sparse cover of riparian vegetation. Alluvial terraces that form locally can provide significant habitat of limited extent, as was the case along much of the Colorado River in Glen Canyon before completion of Glen Canyon Dam (chapter 14). In these settings, streamflow alone provides water for riparian vegetation; groundwater characteristics are negligible except in cases such as Tapeats and Bright Angel Creeks (chapter 11), where base flow originates from reliable springs.

Most "bedrock canyons" have a relatively thick alluvial fill that has both coarse-grained and fine-grained terraces. Grand Canyon (chapter 11), although usually considered to be a bedrock canyon, has considerable fill on the channel bed[14] and a large variety of substrates to sustain riparian ecosystems. Considerable variation in canyon width and bedrock types create mappable geomorphic settings[15] that provide a large range in possible growth sites for riparian vegetation. The North Fork of the Virgin River in Zion National Park (chapter 13) and the Verde River (chapter 23) provide the extreme cases of channels confined within bedrock walls, yet flowing over a thick alluvial fill with significant groundwater influences.

Alluvial channels form in deep sedimentary fills with little or no bedrock constraints on lateral channel migration or vertical downcutting. This general class of channels has more variation in potential riparian habitat than bedrock canyons owing to the complex interactions among surface water, groundwater, and subsurface geology. Most alluvial channels in the Southwest are *arroyos*, or channels that deeply incised into alluvium at the end

of the nineteenth century (chapter 3). The depth of these channels is dependent on regional *base-level control*, which regionally is sea level in the Gulf of California but locally can be a variety of geologic structures or units and usually those that affect the major rivers.

Alluvial channels develop terraces at varying heights and distances away from the channel thalweg because of their geomorphic history. These terraces support distinct stands of riparian vegetation.[16] The height of a terrace reflects both its depositional age and stability, and because depth to groundwater increases with the height of a terrace, the highest terraces either support more xerophytic vegetation or have old obligate species that have deep roots into groundwater. Terraces, in the broadest sense, represent a space-for-time substitution view of riparian development; the oldest stands of riparian vegetation are on the highest terraces, whereas the youngest stands are closest to the channel banks. Terraces also occur in bedrock canyons, but their relation to riparian vegetation is complicated by the configuration of alluvium and bedrock and its relation to water level.

Dependable streamflow in alluvial channels is dependent on bedrock structure, the age of alluvial terraces, variation in the particle size of the fill sediment, and regional groundwater flow. Faults create discontinuities in bedrock and alluvial aquifers and can force groundwater to the surface. Cementation of alluvial terraces, a common occurrence with increasing age or where groundwater has high concentrations of calcium carbonate, can restrict both downward water movement and root growth. Fine-grained alluvial fills can restrict downward movement of groundwater, locally raising the water level above the regional water table.

Over most of the Southwest and particularly in the Basin and Range, bedrock canyons become alluvial channels at the mountain front. On the Colorado Plateau, the complex structural geology creates a variety of transitions between alluvial and bedrock reaches. On the Escalante River (chapter 9), an alluvial channel transitions into a bedrock canyon downstream from Escalante, Utah, cre-

ating an abrupt change in the type and stability of riparian ecosystems. This type of transition is less common in the Basin and Range, but smaller-scale examples occur at The Narrows of the middle San Pedro River (chapter 19), along the Gila River both upstream from Safford, Arizona, and downstream from Coolidge Dam (chapter 17), and in the vicinity of Gillespie Dam (chapter 26). The lower Colorado River (chapter 27) represents a large-scale example of a river with alternating bedrock canyons and alluvial fills.

Surface Water

Watercourses in the Southwest have three general categories of surface water. Only the largest rivers have *perennial flow*: the Green, Colorado, San Juan, Escalante, upper Virgin, Verde, upper Salt, and upper Gila Rivers. During extreme droughts, some perennial rivers may have interrupted flow in reaches that cross deep alluvial basins. *Intermittent streams* may have periods of sustained flow, particularly during wet periods; this type of watercourse generally flows over an alluvial basin with high groundwater levels. *Ephemeral streams* flow only during storm runoff. Most mid- to low-elevation drainage basins have ephemeral streams, although intersection with local or regional groundwater systems may create short reaches of perennial flow. Owing to the combination of human influences and climatic fluctuations, intermittent and ephemeral streams have become drier historically, and certain reaches that once may have been perennial, such as the lower Gila River, are now mostly dry except during storm runoff, urban wastewater discharge, or irrigation returns.

In the intermediate-size drainages, in particular those with relatively low headwaters, channels have an alternating pattern of perennial and ephemeral reaches.[17] An excellent example is the San Pedro River, with perennial flow extending northward from the U.S.–Mexico border to about Fairbank, Arizona, and with ephemeral flow occurring northward except in the reaches near Cascabel, north of Redington, and near the confluence with the Gila River (chapter 19). Similarly, the Mojave River has perennial reaches in and north of Victorville, California, historically near Camp Cady, and within Afton Canyon (chapter 28). Moving downstream, *influent reaches* increase in surface flow owing to groundwater additions, transitioning to *effluent reaches* where flow infiltrates into the aquifer. If we use the alternative terms, *gaining streams* are influent, whereas *losing streams* are effluent.[18] Influent and effluent sections are associated with geologic structures, including faults and shallowly buried bedrock. Rivers that had this configuration at one time, such as the Santa Cruz and the lower Gila Rivers, once again have alternating perennial-ephemeral reaches owing to irrigation returns and wastewater effluent discharge.

All of the region's rivers respond strongly to storm runoff. Rivers with high-elevation headwaters, such as those crossing the Colorado Plateau and the Salt and Gila Rivers, respond annually to snowmelt or rarely to rain-on-snow events with secondary floods caused by regional-scale storms. In some cases, such as the San Juan River (chapter 8), the largest floods are storm induced, whereas most annual peaks are snowmelt related. Lower-elevation rivers, such as the Santa Cruz (chapter 21), respond strongly to seasonal rainfall, and as a result flow and flood records may be related to decadal-scale climatic processes.[19] Large rivers, such as the Colorado, span a number of climatic regimes and respond in a complex way to interannual or decadal-scale climatic processes. For example, many of the largest floods on the Colorado River occurred during El Niño conditions—for example, 1862, 1884, 1891, 1905, 1916, 1941, 1957, 1984, and 1993—but some notable floods, including the largest in the gaging record (in 1922), occurred during years not associated with ENSO (chapters 7, 11, and 27).

Several metrics describe flow in rivers. Flow is segregated by *water years*, normally defined from October 1 through September 30. Because the traditional definition splits the critical early fall season when tropical cyclones occur, we redefine the water year unconventionally as November 1 through October 31.[20] Discharges are calculated from *stage-discharge relations* at gaging stations; these relations are developed

from a combination of streamflow measurements and indirect-discharge calculations following floods.[21] Stage, the most relevant river metric that affects riparian vegetation, is the only continuously measured variable. Stage is averaged over short periods, most commonly fifteen-minute intervals; this metric is converted to *instantaneous* or *unit discharge*. Over a twenty-four-hour period, unit discharges are averaged to create *daily discharge*, the basic metric of most water-resource evaluations.

Floods and flood frequency are a slightly different subset of gaging records. The *annual flood series* represents the largest instantaneous discharge per water year of the gaging record. The *partial-duration series* represents all floods higher than a fixed base discharge (usually much higher than base flow); some years have many floods, whereas some years have only one flood or no floods that exceed base discharge. With indirect-discharge estimates, the annual flood series can be extended before the beginning of the gaging record and thus be divided into historical and gaging-station components. *Paleoflood hydrology* is a technique used for estimating the age and discharge of past floods or extreme floods within the period of streamflow gaging.[22]

The upper Gila River well illustrates the response of middle-elevation watercourses to climate (fig. 2.2). The annual flood series describes an overall *nonstationary* flood frequency,[23] where the mean and variance of the time series are a function of time, in contrast with a *stationary* annual flood series that has time-invariant mean and variance. These changes in the Gila River record occur because the size of floods tracks interdecadal variations in climate. The Gila River at the head of Safford Valley had large floods before 1920 and between 1970 and 1995. In the middle part of the twentieth century, floods were uniformly small compared with those that occurred before or after this period. In contrast, the Salt River upstream from Phoenix appears to have a stationary annual flood series (chapter 24).

If we use the partial-duration series,[24] we can see that floods caused by summer thunderstorms in the

region are demonstrably stationary over the period of record, whereas floods caused by regional storms—either in winter or dissipating tropical cyclones—are strongly related to global-scale climate.[25] Examination of the seasonal classification of peaks in figure 2.2A shows this to be the case for the Gila River as well. Separation of the annual flood series into El Niño, La Niña, and other years shows that interannual climatic variation can have a strong effect on flood frequency.[26] Because the influence of ENSO decreases northward, rivers on the Colorado Plateau show less El Niño response than do those in central and southern Arizona or the Mojave River.

Expressing the time series of flow in the Gila River using daily discharge (fig. 2.2B) yields a similar pattern to the annual flood series (fig. 2.2A). Because of the daily averaging, the peaks are lower for daily discharge than for the annual flood series, and events before the start of the gaging record are not represented. However, close examination shows that the multiple events captured in the partial-duration series but excluded from the annual flood series are apparent, as are extended drought periods. Also, duration analysis of daily discharge (fig. 2.2C) can be used to compare flow characteristics important to riparian vegetation. For the Gila River, daily discharge was significantly reduced during the mid-twentieth-century drought (around 1930–1960), whereas discharge for the period 1960 through 2003 was notably higher (fig. 2.2C).

Stage is the primary measurement made at gaging stations, and a duration analysis of daily stage (fig. 2.2D) can yield valuable information on the influence of streamflow on riparian vegetation.[27] As depicted in figure 2.2D, the commonly attained stage is 7 feet in the Gila River, which is a wide alluvial channel at the gaging station upstream from Safford, Arizona (chapter 17). To illustrate the effect of climate, the stage exceeded 20 percent of the time from 1960 through 2003 is approximately 0.5 feet higher than the stage with the same frequency before 1960. In a constant channel geometry, this difference suggests a higher water level in the alluvial aquifer and more water available for riparian vegetation.

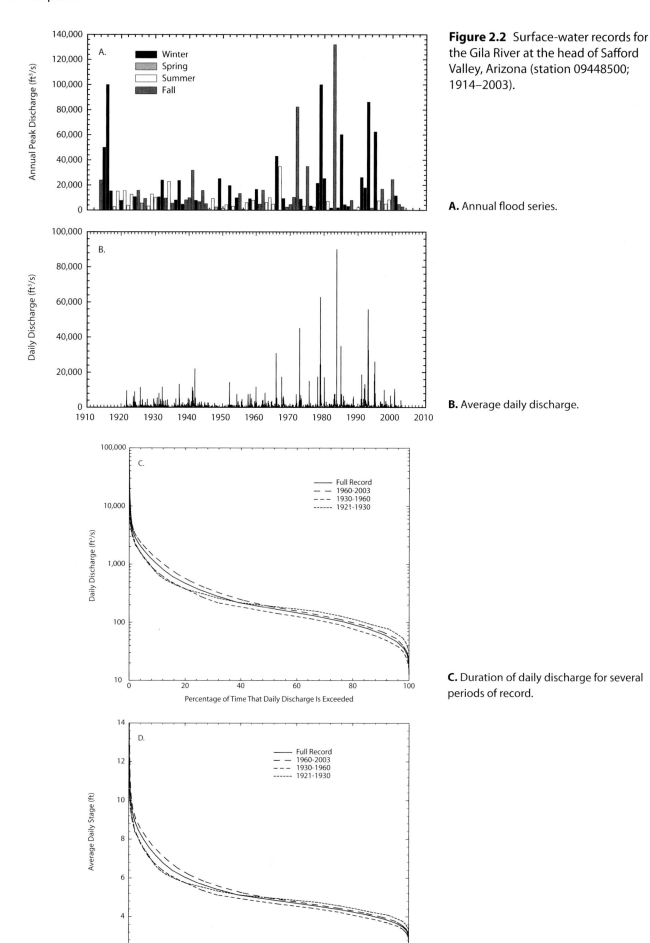

Figure 2.2 Surface-water records for the Gila River at the head of Safford Valley, Arizona (station 09448500; 1914–2003).

A. Annual flood series.

B. Average daily discharge.

C. Duration of daily discharge for several periods of record.

D. Duration of daily stage for several periods of record.

The duration of stage, combined with geomorphic setting, determines what kinds of riparian species can grow in a given reach of river. Slow-moving rivers in wide valleys have stage-discharge ratings with low slopes; velocities are low, and the wetted perimeter is large, so small increases in stage represent large changes in discharge. For rivers in bedrock canyons, the reverse is true. Variation in stage-discharge relations along a river may help explain the longitudinal variations in the extent and types of riparian vegetation.

In general, the changes in surface-water flow illustrated by the gaging record of the Gila River are common in southern Arizona. Similarities decrease northward in the region, although some general patterns remain. Large, regional winter floods occurred in 1862, 1884, 1891, 1916, and 1993; similarly, streamflow was lower during the midcentury drought than between 1960 and 1995. Differences are also significant; although floods that scoured channels and damaged riparian vegetation were common in southern Arizona in the late 1970s through the mid-1990s, these floods did not occur in southern Utah or the Mojave Desert. These regional differences in surface-water hydrology might be important factors for explaining differences in the patterns of regional change in riparian vegetation.

Groundwater

Groundwater consists of the two fundamental states of *unsaturated* and *saturated zones* delineated by the *water table.* The unsaturated zone, as the name implies, is a three-phase system of gases, water, and sediment, with both gases and water in motion. In the unsaturated zone, water generally moves downward, but it may move upward either by capillary action, vapor transport, or *hydraulic lift* by plant roots.[28] A contiguous saturated zone in permeable rocks, regardless of extent, is called an *aquifer. Recharge* of groundwater occurs when water infiltrates and enters the saturated zone from either streamflow or rainfall. *Evapotranspiration* is the combination of evaporation from bare ground surfaces and transpiration from the

saturated and unsaturated zones by vegetation.

Like surface water, groundwater moves according to elevation and pressure gradients, but its rates of movement typically are one or two orders of magnitude lower than rates of movement of surface water. The component of total energy related to the velocity of water, which is important in understanding movement of surface water, usually does not need to be considered in analyses of groundwater movement. Groundwater levels measured in wells indicate the potential energy in the aquifer, and an analysis of groundwater levels from many wells can be used to determine directions of groundwater movement within an aquifer. For this application, all water levels are referenced to a common datum, such as mean sea level.

Analyses of fluctuations of groundwater levels in wells can reveal much about aquifer response to natural conditions and to changes brought about by human activities. For this purpose, water levels in a well can be referenced to a datum such as sea level or to a measuring point at or near land surface, resulting in a level that indicates depth to groundwater below the land surface. Use of depth to water is convenient for studies in and around riparian systems because those depths can be related to possible rooting depths of phreatophytes. In recharge areas, natural fluctuations of groundwater levels can reflect changes in recharge from climate variations ranging in period from a few years to many decades. Higher-frequency variations in flow and pressure are commonly damped as water moves through the unsaturated zone to the water table. Adjacent to streams and rivers connected to aquifers, variations in groundwater levels are commonly cyclical, driven by surface-water levels and uptake of water by plants.

Surface-water levels commonly vary over annual cycles in response to conditions such as spring runoff and other yearly climate patterns. Surface-water levels in controlled streams also can vary cyclically in response to demands for release of water for power generation and irrigation. Phreatophytes also can produce a cyclical signal with greater uptake of water during the

warmer months. The example groundwater hydrograph in figure 2.3 shows that fluctuation in water level may exceed 5 to 10 feet over a period of years, and low-frequency positive trends (as shown in fig. 2.3) may be present owing to recharge beneath the river, or negative trends may signal groundwater declines, persistent drought, or increasing usage by riparian vegetation.

Groundwater development generally depends on drilling wells and extracting water by pumps. *Artesian wells*, naturally pressurized owing to subsurface geologic conditions, occurred locally in the Southwest, for example, along the San Pedro River (chapter 19). Pumping wells develop a *cone of depression*, where water levels are drawn down in the vicinity. Cones of depression may extend for a hundred feet or many miles depending on the aquifer properties and the time since pumping began. Multiple wells, pumping in proximity, can produce accelerated water-level declines with intersecting cones of depression. Over the period of our repeat-photography record, pumping and drilling technology has changed substantially,[29] leading to increases in the efficiency of water extraction from the ground. Initially, in the late nineteenth century, wells were mostly hand dug and not powered by electricity. Subsequent development of efficient turbine pumps and drilling equipment has allowed extensive groundwater development in the Southwest, notably in lower-elevation alluvial valleys.

Groundwater overdraft is considered to occur where regional pumping rates exceed the long-term average rate of natural recharge; however, substantial reduction of groundwater discharge to riparian systems can occur even in aquifers that are not in a state of overdraft according to this definition. Most recharge in the Southwest results from streamflow issuing from bedrock canyons across the mountain front, where faults or the contact between bedrock and alluvial fill or both provide a place where surface water can rapidly infiltrate. During floods, recharge along alluvial channels, termed *transmission losses*,[30] creates *groundwater mounds* beneath the watercourse; these mounds eventually dissipate into the regional

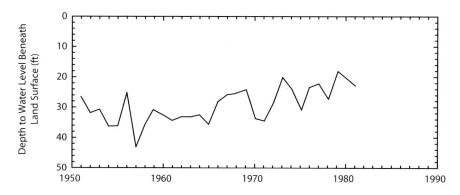

Figure 2.3 Groundwater levels for well D-06-28 31AAB1 near Safford, Arizona.

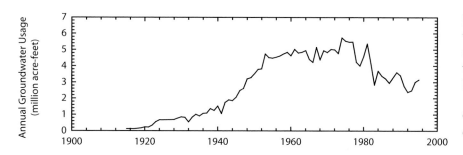

Figure 2.4 Total groundwater withdrawals in Arizona, 1915–1995 (groundwater use data for 1915 through 1990 is from Anning and Duet 1994; data for 1991 through 1995 from M. T. Anderson, U.S. Geological Survey, unpublished data, 2003).

aquifer during dry periods. Influent stream reaches can also be considered as "recharge," although a mass balance would likely indicate that the net change in the alluvial aquifer is negative as a result of direct evaporation of base flow and transpiration by riparian vegetation (in the absence of water development).

Groundwater recharge beneath alluvial channels can be locally substantial during floods, particularly if the water table is well below the stream channel and there is room in the subsurface to store water that infiltrates. For example, following the floods of December 1978, water levels rose up to 82 feet in wells in alluvial aquifers in southeastern Arizona.[31] Similar rises occurred following runoff in 1979, 1980, and 1983. Streamflow measurements during a February 1978 flood on the Gila River indicated that 112,000 acre-feet (17 percent) of the inflow recharged the alluvial aquifer; a second such measurement during January 1966 indicated that 175,000 acre-feet (29 percent) of inflow was recharged.[32] Groundwater rises attributed to flood discharges can also be a problem. In 1979, rises in groundwater levels owing to flood-related dam releases to the lower Gila River caused waterlogging in agricultural lands, removing them from production.[33]

Base flow is defined as river discharge not contributed by storm runoff. In most perennial rivers in the region, base flow ultimately is generated from groundwater discharge, either locally or, in isolated cases such as Havasu Creek (chapter 12), from a long distance away. For rivers solely benefiting from effluent discharge, base flow has little interannual variation, but it may have long-term trends owing to a variety of conditions, including decadal-scale climatic variation, groundwater pumping, and changing water use for agriculture and riparian vegetation.

In the arid and semiarid southwestern United States, groundwater has been an important source of supply for agriculture, industry, and public consumption.[34] Groundwater withdrawals in Arizona escalated rapidly in the middle part of the twentieth century when large-capacity turbine pumps became available and electricity was brought to rural areas suitable for agriculture (fig. 2.4). Trends in use of groundwater have been dynamic since then, reflecting continued growth of population, agriculture, and industry as well as the availability of surface water. For example, in 2000, approximately 80 percent of total water use in Arizona was for agriculture,[35] but that fraction declined slightly in the last

decades of the twentieth century, and, in particular, the amount of groundwater pumped for agriculture declined during that time.[36]

Many aquifer systems in the area are characterized by a large volume of water in storage but a relatively small rate of natural recharge and discharge. The recharge to the aquifers commonly comes from precipitation at higher elevations in mountains adjacent to the aquifers. Water moves slowly through the aquifers and may discharge to springs, streams, playas, and associated riparian vegetation. In spite of a large volume of water in underground storage, even a modest amount of pumping, with time, has the potential of intercepting groundwater outflow and degrading or eliminating riparian ecosystems.

Withdrawal of groundwater to the extent that natural outflow is significantly reduced is the single most important factor that can lead to the elimination of riparian ecosystems. There is little doubt that groundwater pumping can damage or kill riparian vegetation. The effects can be on a local or a reach scale; a deep cone of depression that rapidly develops around a pumping well on a floodplain can kill nearby trees in a matter of days, but the more typical situation is the pumping of a well field heavily for an

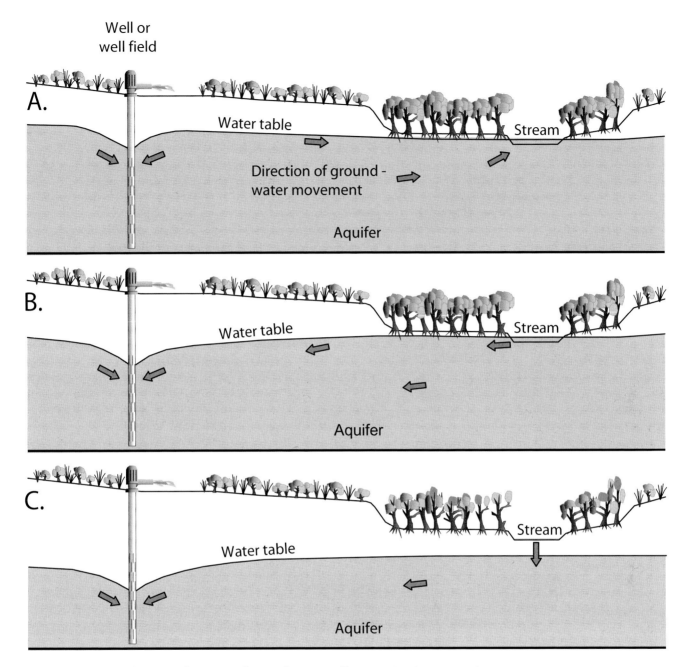

Figure 2.5 Expected stages of capture of groundwater outflow to a riparian area and stream.

A. Stage I. Near the onset of pumping, water pumped comes from storage around the well, and the stream/aquifer system functions as it did prior to pumping. Groundwater from upgradient areas supplies riparian vegetation in the floodplain as well as base flow to the gaining stream (Winter and others 2002). Periodic stream runoff events also may supply water to the near-stream aquifer.

B. Stage II. Drawdown from well has caused movement of water away from the stream and floodplain after a substantial period of pumping. Availability of water for riparian phreatophytes may not be diminished because of increased inflow from the stream, but the stream has become losing or possibly intermittent.

C. Stage III. After a substantial period of pumping in excess of rate of groundwater flow from upgradient areas, the stream and aquifer may be disconnected if streamflow cannot provide enough recharge to maintain the water table. The stream has become ephemeral.

extended period. Cottonwood provides one example of what happens in this latter situation: if lowering occurs slowly enough, in response to drought or to steady pumping, the tree's roots may be able to elongate quickly enough to keep pace and allow the tree to survive.[37] Although mesquite can withstand relatively large fluctuations in groundwater level, trees in bosques are known to become stressed when groundwater levels decline to 45–50 feet below land surface.[38] Mesquite along the middle Gila River could not survive a water-table decline of about 100 feet in the 1940s and 1950s.[39] Established plants may be able to survive rapid drawdown for short periods, but mortality in saplings may be as high as 100 percent.[40] Rapid drawdown causing mortality may be as little as 3 feet

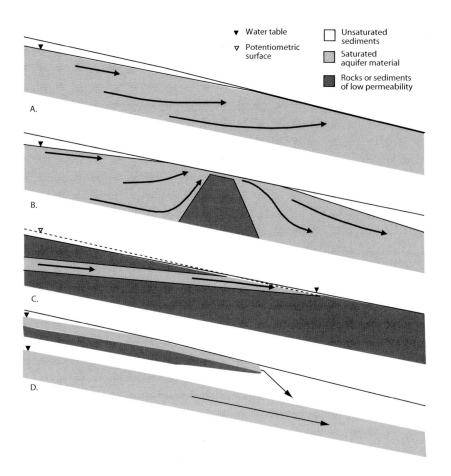

Figure 2.6 Longitudinal profiles of groundwater–surface water interactions commonly observed on rivers in the southwestern United States.

A. The channel bed intersects the water table, and effluent flow (base flow) ensues.

B. Low permeability sediments or bedrock in the subsurface forces upward movement of groundwater.

C. The channel bed intersects a confined aquifer.

D. Low-permeability sediments create a perched water table above a lower regional level.

in a matter of weeks[41] or may occur during channel downcutting during storm runoff,[42] and the rate of change, not the absolute depth to water, is the most important factor.[43]

Groundwater lowering can result from arroyo downcutting and consequent dewatering of the alluvial aquifer, excessive well pumpage, or persistent drought. Channel downcutting would create the fastest rate of groundwater decline, and most of the declines in channel bed are as much as an order of magnitude higher than the 3- to 5-foot declines that can damage floodplain trees. Most cases of groundwater overdraft likely are slow enough to allow roots to follow to a certain point (roughly 30 feet for cottonwood) before death is certain. Persistent drought likely will result in a gradual lowering of the water level that roots can adapt to, with the likely response of stem and branch dieback and low mortality.[44] The effects would be highest at the beginning of effluent reaches and at the end of influent reaches, contracting the length of these riparian ecosystems.

As illustrated in figure 2.5, the ef-

fects of groundwater pumping of an alluvial aquifer occurs in three stages. Stage I begins with the cone of depression developing at the onset of pumping, locally affecting water stored around the well and not significantly impacting the reach-scale stream/aquifer system. Stage II begins after a substantial withdrawal draws down the aquifer sufficiently to create a water-level gradient away from the stream and floodplain; water availability to riparian plants may be unaffected, but streamflow may be reduced. Finally, Stage III occurs after a long period when pumping exceeds the rate of groundwater inflow from upgradient areas; surface-water and groundwater systems may become disconnected if streamflow cannot provide enough recharge to maintain water levels in the alluvial aquifer. In Stage III, riparian vegetation can be strongly affected.

The historical perspective on groundwater resources has been that aquifers are "containers" or "reservoirs" and that water can be removed at rates up to the amounts flowing in. Removal of water by sustained pumping less than the basin recharge will eventually

be balanced by an increased inflow or decreased outflow or both.[45] In the alluvial aquifers of the Southwest, almost all of the water removed is accommodated by decreased outflow, usually to effluent reaches but also toward the end of influent reaches owing to additional recharge. In addition, the stopping of groundwater pumpage after a period of time also has the effect of reducing the cumulative volume of outflow (assuming no increased inflow) by the total quantity pumped over the period of active pumping. The time period and locations over which ultimate capture occurs are dependent on aquifer shape, hydraulic and storage properties, quantity of water pumped, and proximity of pumping to recharge and discharge features. The effect of an added withdrawal is superimposed on effects from existing withdrawals so that many small withdrawals can combine to cause a large effect on discharge of groundwater to riparian systems.

As discussed previously, surface water interacts with groundwater in a complex way in the Southwest. Discharge from bedrock or alluvial aquifers creates effluent streams that sustain woody riparian vegetation (fig. 2.6). Geologic structures, bedrock, or sediments of low permeability can force groundwater levels to the surface; abrupt termination of such features allows surface water to infiltrate back into the aquifer. Floods can recharge aquifers immediately beneath the channel, and surface water may move short distances through alluvial fill and return to the channel downstream.[46] Groundwater and surface water are inextricably linked, and the effects of either surface-water developments or groundwater developments cannot be isolated.

The geologic setting of many streams in the Southwest includes permeable younger alluvium adjacent to the stream with less permeable aquifer material beyond this alluvium, both laterally and vertically. The amount of younger alluvium surrounding streams can range in scale from tens of feet laterally and vertically to miles laterally and hundreds of feet vertically. Aquifer material beyond the younger alluvium is commonly older basin-fill material in the Basin and Range and consolidated sedimentary rocks on the Colorado Plateau. Some key points relating to groundwater movement in near-stream alluvial sediments are:

1. Most phreatophytic woody plants associated with desert streams occur within the areal extent of the saturated younger alluvium. The land surface of adjacent older sediments is likely to be at higher elevations and inca-pable of supporting phreatophytes.

2. Around perennial reaches, groundwater-flow paths are most likely complex, including flow components to or from the regional aquifer as well as local flow components in and out of the stream. The local flow components are influenced by stream geometry, streambed permeability, groundwater use by phreatophytes, and human activities such as groundwater pumping and irrigation.

3. Development of groundwater resources in near-stream younger alluvium can lead to a rapid depletion of streamflow and availability of water for riparian plants.

4. Development of groundwater resources in adjacent older aquifer materials also can lead to depletion of streamflow and availability of water for riparian plants (fig. 2.5), but timescales for these effects to occur will be longer, ranging from years to centuries.

Clearly, reduction in the amount of riparian vegetation in a riverine setting can be caused by a number of human-induced, climatic, or water-development factors. Complete elimination of riverine riparian vegetation, especially obligate riparian species, would be expected to occur if base flow were totally diverted from the channel and groundwater levels were reduced below approximately 30 feet beneath land surface. Such conditions would likely kill most if not all established individuals and prevent any seedlings from becoming established.

Several factors can mitigate riparian losses, however. In some basins with high groundwater use, wastewater effluent discharged into ephemeral channels can be a source of water for near-stream woody vegetation over downstream reaches. Reaches with riparian vegetation that are sustained in this way depend on amounts of effluent discharge and presence of subsurface perching layers to keep infiltrated water at shallow depths. For aquifers in which groundwater use is a significant fraction of the groundwater supply to riparian systems, the strategies for mitigation of loss of riparian vegetation include development of new sources of water.

The outcomes of other scenarios are not so obvious and are explored in the volume within the context of regional water utilization and flow regulation. A common scenario observed along large rivers is flow regulation characterized by flood control and reduction of hydrologic variability; in other words, the river behaves as if base flow has been increased and flood damages are reduced or eliminated. Such a scenario would be expected to cause increases in biomass and life span of established individuals, but it may be the precursor to ecosystem collapse if germination and establishment do not occur. Purposeful flood releases are one management remedy suggested for these types of rivers.

3 Woody Native Riparian Vegetation

Summary. Common woody riparian species in the southwestern United States include cottonwood, black willow, netleaf hackberry, coyote willow, arrowweed, seepwillow, and several other species of shrubs and trees. The stability of riparian ecosystems depends on these species' individualistic response to a variety of biotic and abiotic factors, including competition for water, shading, depth to reliable water, and response to flooding. A number of factors have strongly affected native riparian species, including the occurrence of large floods, arroyo downcutting, trapping of beaver, and increases in the length of the growing season.

Riparian vegetation in the southwestern United States either depends on permanent supplies of shallow water or adapts, through fluctuations in density or individual plant sizes, to the boom-bust cycles characteristic of ephemeral-stream environments. Whether trees or shrubs, these species—in particular obligate riparian plants—are expected to react most strongly to changes in hydrologic conditions, whether caused by humans, climate, or other factors. Facultative riparian species are more adaptable and may be expected to show more flexibility in the face of environmental changes. Some species normally associated with upland desert ecosystems establish in riparian zones and function as riparian species. Plant functional type is a critical characteristic of riparian species and has some deterministic capacity to predict how the species might react to climatic and hydrologic changes.[1]

Woody riparian vegetation is adapted to extremely harsh conditions. In June, midday air temperatures exceed 100°F, and humidity typically is less than 15 percent, which severely stresses plants that do not have root access to reliable water. In part of this region, the summer monsoon reliably brings relief through lower temperatures and higher humidities, but many riparian species grow in the Mojave Desert and the northern Colorado Plateau, where the monsoon is not as reliable as in Arizona. The large environmental variation over the region suggests that plants that span its breadth will show physiological variation.[2] Many species, notably mesquite, are limited by low winter temperatures; in the lower Colorado River reaches, winter temperatures are high enough to allow growth all days of the year. Although many of the native riparian species are adapted to stress brought about by periodic droughts, even these species cannot withstand the effects of dewatering of shallow groundwater tables, which can result from several natural or human-caused processes.

Common Native Woody Species

Cottonwood

Frémont cottonwood is the largest native tree in the Sonoran Desert,[3] growing to heights of 50 to 90 feet in prime habitats.[4] Named for explorer John C. Frémont, but also known by its Mexican name, alamo, cottonwood is the most widespread native riparian tree in the Southwest, extending from the Green and the upper Colorado Rivers downstream to the Colorado River delta.[5] Within the complex of cottonwood species in the West, Frémont cottonwood has the most southerly distribution with the driest ambient climate.[6] Although many species and subspecies may collectively compose what is known as Frémont cottonwood,[7] we treat all of these trees that we have seen in the region as one species.

Individual trees are fast growing in some hot desert settings, and this species can live to several hundred years.[8] The largest known individual in Arizona, on Sonoita Creek near Patagonia, is 92 feet high and 42 feet in circumference, and it has a crown diameter of 108 feet.[9] Because the wood of this tree is light and subject to rot, it has limited value as firewood or building material.

Willows

Willows are one of the most common woody plants found in riparian settings in the southwestern United States. With twenty-five species in Utah and sixteen in Arizona, the genus *Salix* is difficult to differentiate, particularly when flowers are absent.[10] Two species—black willow (Goodding willow) and coyote willow—are the most important members of this group and are widely distributed in the Southwest. Bonpland willow is locally abundant in southern Arizona and generally has a different geographical distribution from black willow.[11] Yewleaf (silverleaf) willow[12] is a largely tropical species with a distribution from southeastern Arizona to Guatemala. It is closely related to coyote willow and is locally common along the driest reaches in tributaries of the San Pedro River (chapter 20) and Rillito Creek (chapter 22). Peachleaf willow, uncommon along the Green and upper Colorado Rivers and absent elsewhere, is briefly discussed in chapter 7.

The combination of Frémont cottonwood and black willow is considered to be the most valuable riparian habitat in the Southwest and the most threatened.[13] Black willow is one of the most widespread trees along low-desert rivers in the Southwest. With an elevation range of 200 to 4,000 feet, this species is most abundant along the lower Colorado River, where it forms dense stands. Its range extends into southwestern Utah along the Virgin River.[14] It reaches a maximum height of about 45 feet and reportedly is the most widespread riparian tree in Sonora, Mexico.[15] This tree tends to occur in slightly more mesic sites than Frémont cottonwood.[16] It is a conspicuous tree in a desert region and has numerous uses in construction and basketry.[17]

Coyote (or sandbar) willow is one of the most common species found along perennial rivers in the Southwest. This species typically occurs as a shrub, but older individuals may attain the height of small trees. Widespread in Utah, this species is found at altitudes as high as 9,500 feet.[18] An obligate riparian species, coyote willow typically occurs on the riverward side of floodplains in the wettest sites; it seldom is found in ephemeral reaches. Along with cottonwood, willows appear to be a favorite food of beaver.[19]

Bonpland willow is distributed in the middle-elevation reaches of the Salt, Gila, Santa Cruz, and San Pedro Rivers.[20] With a height of 30 to 35 feet and leaves that are silvery on the undersides, Bonpland willow is easily distinguished from either black or coyote willow. Unlike coyote willow, it is generally solitary or in low-density groups, and its range only slightly overlaps that of black willow.[21] It is particularly abundant in Sabino Canyon (chapter 22).

Sycamore

Sycamore is one of the most stately trees in the riparian zone. Arizona sycamore occurs south of the Mogollon Rim in Arizona.[22] This species is distinctive in historical photographs, with its maximum height of about 80 feet, mottled white and reddish brown bark, and large, palmately lobed leaves. Although its elevation range is 1,000 to 6,000 feet, sycamore is not as common as cottonwood or black willow and is most commonly found in higher-elevation canyon settings along either ephemeral or perennial streams.[23] The variety found in California (*Platanus racemosa* v. *racemosa*) occurs mostly in the Transverse and Peninsular Ranges or the Sierra Nevada, but individuals occasionally grow in the desert along the Mojave River.

Arizona Ash

Arizona ash, or fresno, is a common obligate riparian tree along the low- and intermediate-elevation rivers in the region. In the best-watered sites, such as Havasu Canyon (chapter 12), Arizona ash attains a height similar to that of Frémont cottonwood. It extends into southern Utah along the Virgin River and is particularly common in Zion National Park (chapter 13).[24] Subspecies *Fraxinus velutina coriacea* extends into the desert along the Mojave River in California (chapter 28).[25] Arizona ash grows along perennial and intermittent streams with shallow groundwater tables, and it is most common in canyon reaches. It is common, for example, along the San Pedro River and its tributaries (chapters 19 and 20).

Mesquite

Three species of mesquite are present in the region discussed in this book. Two, western honey and velvet mesquite, are facultative riparian species. Western honey mesquite is perhaps the most common, occurring in the Mojave Desert, Grand Canyon, southeastern Arizona, and along the lower Colorado River.[26] Subspecies *Prosopis glandulosa* v. *glandulosa* occurs to the east in New Mexico and Texas and is seen in our photographs of Animas Creek (chapter 16). Velvet mesquite is commonly found in southeastern Arizona, particularly in the Santa Cruz and San Pedro River basins. Screwbean mesquite is the least common of the three species and is widely distributed along the lower Gila and Colorado Rivers, as well as along the middle reaches of the Rio Grande.[27] Western honey mesquite, velvet mesquite, and screwbean mesquite can occur together along some river reaches, in particular the Gila River and its tributaries in central and western Arizona.

Honey mesquite and velvet mesquite are common constituents of bottomland vegetation, forming large thickets known as bosques. These two species are also among the most common woody species that have increased in the desert grassland. The northern distribution of mesquite is limited by a combination of low temperatures and availability of suitable riparian habitat, although geographic limits are affected by ecotypic variation.[28] In southern Arizona, both honey mesquite and velvet mesquite were killed to the ground by temperatures as low as 0°F in December 1978.[29] Bosques can develop quickly because honey mesquite grows quickly in riparian settings, but the trees can be quite old in stable settings. Individual honey mesquite trees in Grand Canyon are known to be in excess of one hundred years old.[30]

Velvet mesquite is widespread in central and southern Arizona, although it does not occur along the lower Colorado River.[31] Like honey mesquite, velvet mesquite can live more than a century.[32] Reportedly, it can attain a maximum height of 50 feet, although the height of most trees is less than 20 feet.[33] Bosques of velvet mesquite were once extensive along the Santa Cruz River in southern Arizona (chapter 21) and remain common in the San Pedro River basin (chapter 19) and along other river systems in southern Arizona. Screwbean mesquite is the most shrublike of the three species, although it can grow to a height of 30 feet.[34] Today it occurs primarily along the lower Colorado River and in central Arizona along ephemeral and perennial rivers, particularly in slightly saline soils.[35] In the past, it was more widespread and was collected along the Santa Cruz River of Tucson in 1904 (specimen, University of Arizona Herbarium).

The three species of mesquite are impossible to distinguish in historical photographs. We lump them simply as "mesquite" in our interpretation of historical changes, except in reaches where there is definitely only one species present. Mesquite wood is hard and valued as a building material, as firewood, and as a source of charcoal. Currently, large trees are harvested to make fine furniture. Historical woodcutting associated with mining activities or the need for fuel for steamboats on the lower Colorado River may have depleted some mesquite bosques. Bosques are one of the riparian ecosystems most frequently described as destroyed in historical accounts.[36]

Catclaw

Catclaw is a common species along low-desert rivers.[37] A facultative riparian species, catclaw typically occurs within or upslope of mesquite bosques or other dense sections of riparian vegetation along perennial streams, such as the Colorado, Gila, and Salt Rivers. Along the Colorado River in Grand Canyon, catclaw commonly is the only riparian tree for long reaches, and it can be distinguished from mesquite in photographs because of its smaller leaves and grayish bark.[38] In riparian settings, it can grow to a height of 21 feet or more. Catclaw wood has properties similar to those of mesquite, although catclaw is less desirable for utilitarian purposes because of its small stature and stem diameter.

Netleaf Hackberry

Netleaf hackberry is widely distributed in Utah and Arizona below 6,000 feet.[39] This tree is more common and widespread in the more arid parts of the northwestern United States and the intermountain West.[40] A typically gnarly tree up to 20 feet tall, netleaf hackberry characteristically has insect galls in the central part of nearly every leaf. It is long-lived; in Cataract Canyon, one individual is more than two hundred years old.[41] This species is absent from the more arid parts of our region but is relatively common in canyon reaches. In Cataract Canyon and Grand Canyon (chapters 7 and 11), netleaf hackberry grows mainly on rocky slopes at or above the old high-water line.[42] It is particularly common in the bedrock canyon of the San Juan River (chapter 8) but also occurs infrequently throughout the region, especially in southern Arizona.

Shrubs

Several shrubs, mostly from the Composite family, are important to riverine riparian ecosystems. Arrowweed, a straight, grayish-leaved shrub also known as cachanilla, is common in low-desert river reaches in Arizona and in the low-elevation reaches of the Virgin and Colorado Rivers in southern Utah.[43] As the name implies, this plant was used by Native Americans

for arrow shafts. Along the Gila River, the Pima used it as a roof-thatching and basket-making material.[44] Both Native Americans and settlers used the fast-growing plant as an integral part of brush dams to divert river flow into canals or, in the case of the lower Colorado River, as a material in the failed attempt to repair the 1905 breach near Yuma (chapter 27). Arrowweed accumulates salt in its leaves almost to the extent of tamarisk[45] (chapter 4), and because of its salinity tolerance and size, it has much in common with tamarisk. Dense, seemingly impenetrable thickets of arrowweed occur along the lower Colorado River and its delta.[46]

Although many species from the genus *Baccharis* are present in the region, we consider two to be functionally synonymous and indistinguishable in photographs.[47] Collectively known as seepwillow,[48] these two species are similar to arrowweed in their range, their tendency to form dense thickets, and their use by Native Americans.[49] Seepwillow generally occurs where depth to groundwater is less than 7 feet,[50] and it has a lower salinity tolerance than arrowweed. Squaw baccharis is less common and more closely associated with permanent water. Finally, desert broom is a facultative riparian species that also is a pioneering species of disturbed ground in uplands. In Grand Canyon, desert broom composed a locally significant part of the old high-water zone (chapter 11); in the ephemeral washes of southern Arizona, it is one of the first species to recolonize flood-disturbed channel margins of ephemeral streams.

Burrobrush is widely distributed in the southwestern quarter of Arizona.[51] Common in sandy washes in the low deserts, it is unpalatable to livestock,[52] and therefore it may be able to increase in grazed areas at the expense of more-palatable species. Like desert broom, it quickly colonizes bare, ephemeral channels following floods. Burrobrush tends to occur in more mesic settings than desert broom, although the two appear to be functionally similar. In well-watered settings, such as in the Bill Williams River system, burrobrush can grow to the size of a small tree.

Carrizo Grass

Carrizo grass—also known as carrizo, cane grass, reedgrass, or sawgrass—grows locally and in large, monospecific stands along perennial streams below 6,000 feet.[53] Although it is an herbaceous species, we include it in our discussion of change in woody riparian vegetation because it is readily identified from photographs. One of the most widespread plant species in the world,[54] carrizo grass is common in Utah and typically occurs around springs or adjacent to slow-moving reaches of rivers.[55] It is widespread in hotter areas of the Mojave and Sonoran Deserts, although it also occurs along the Colorado River and extends into southern Utah.[56] Along the lower Colorado River, it forms impenetrable thickets next to the water, most notably in the reach near Picacho Peak (chapter 27), once known as Canebrake Canyon.[57] In the Colorado River delta, this species can grow from a water depth of 3 feet,[58] making it one of the few perennial plants in the region that can tolerate submergence or saturated ground. It is difficult to distinguish carrizo grass from giant reed, an invasive species (chapter 4), without carefully examining the inflorescence. Because that is impossible in photographs, we recognize that we may have confused the two species in some cases.

Cattail

Cattail is another exception, albeit limited, to our consideration of woody riparian vegetation in this chapter. Two very similar species occur in our area and cannot be distinguished without close examination;[59] we combine them in our observations of change. Cattail typically grow in standing water, although they may also grow in saturated soils.[60] Because of habitat preferences, they grow along sloughs and other areas of slow-moving water but seldom occur along flowing rivers. Because these species tolerate brackish conditions, they are the most common plant along parts of the lower Colorado River, particularly in the Ciénega de Santa Clara, where agricultural wastewater with high salinity is discharged into the eastern part of the Colorado

River delta. Cattail frequently occur with carrizo grass, although the latter can better tolerate moving water.

Other Species

Several species more common outside of the region considered here are of local importance. Box elder, found at higher elevations in Utah, Colorado, and other parts of the West,[61] appears along the Green and Colorado Rivers in central Utah. Western soapberry is locally common along both perennial and ephemeral streams at elevations of 2,500 to 6,000 feet in southern Arizona.[62] Unpalatable to livestock, this typically understory tree is seldom abundant. Arizona walnut is widespread in the low deserts, extending north to Havasu Canyon;[63] this distinctive species seldom forms pure stands and instead tends to be relatively uncommon locally. Wild grape grows along perennial and ephemeral streams in the region, forming dense networks of vines, usually as understory plants in the intermediate elevation range. Desert olive occurs along the Colorado River in southern Utah and is locally important in the predam high-water zone.[64]

Life History of Native Riparian Species

Phenology and Seed Production

Frémont cottonwood blooms in late winter to early spring, most commonly in February and March.[65] Seed production in most native riparian species is prodigious. One plains cottonwood tree can produce 28 million seeds,[66] and a single mesquite can produce 140,000 seeds.[67] In the early spring, Frémont cottonwood seeds are produced in a brief pulse that is shorter than it is for other species, in particular seepwillow.[68] Frémont cottonwood, seepwillow, and willows in general are *dioecious*,[69] meaning that trees are either male or female, but other common riparian species are *monoecious*, with male and female flower parts on the same plant.

Seeds are commonly wind dispersed over several miles,[70] but water dispersal during late-winter floods is perhaps the most effective way for cottonwood to become established. Similarly, floods greater than a five-year event disperse mesquite seed to low floodplains.[71] Seed production therefore appears to be timed to coincide with late-winter floods.[72] In most species of cottonwood, seed viability decreases dramatically several months after shedding.[73] One summary suggests that under natural field conditions, seeds remain viable for only one to three weeks.[74]

Germination and Establishment

The seeds of most native riparian species germinate at rates of 50 percent or less and at temperatures that peak between 60 and 80°F.[75] With the exception of sycamore, which requires saturated soil for germination, most riparian species require unsaturated soil with high water content. Except for screwbean mesquite, most native riparian vegetation cannot germinate where salinity is high or even moderate. Germination occurs primarily in the late winter and spring,[76] although late summer and fall germination also is significant for mesquite.[77]

Germination and establishment of Frémont cottonwood are closely related to hydrologic characteristics, in particular the occurrence of floods, and the postflood flow history.[78] As a result, alluvial terraces can be dated from the establishment age of cottonwood growing on them because even-aged stands suggest a cohort that coincided with formation or major modification of the terrace.[79] Other species, such as seepwillow, require sustained saturated conditions for early-spring germination,[80] indicating a strong dependence on winter flooding. Seasonality of flooding can determine the species composition of riparian vegetation.[81]

Whether a seed that germinates becomes an established plant is a complex function of soil moisture, depth to groundwater, weather following germination, and the occurrence of floodplain disturbances. For example, seedling roots of Frémont cottonwood elongate at a rate of about one-third of a foot per week,[82] and survival may depend on those roots reaching the capillary fringe of the alluvial aquifer before the overlying sediments dry in the late spring and early summer. If

beaver, other wildlife, or domestic livestock use the floodplain, the seedlings may be consumed; if floods occur after germination, seedlings may be scoured away, or, alternatively, their survival may be assured with the addition of soil moisture.

In general, native riparian vegetation in the Southwest requires disturbance openings for germination and establishment. Most native species are capable of both sexual and vegetative reproduction, the latter usually from flood-transported branches. Numerous researchers have shown that establishment success of cottonwood, in particular, requires open patches of wet alluvium.[83] Subsequent mortality of seedlings depends on subsequent precipitation and flood history as well as the stage at which establishment occurs—the farther from permanent water supplies, the higher the mortality.[84] One study suggests that as long as four years may be required before seedlings become phreatophytes.[85]

Germination and establishment requirements for mesquite are well known because this species is increasing in areas formerly occupied by grassland in the region.[86] Mesquites produce abundant seeds in pods during the late spring and early summer,[87] and rodents store and eat them. The thick walls of the seed pods contain germination inhibitors, and the heavy, waterproof seed coatings inhibit water absorption. As a result, the seeds must be released from the pods and scarified before germination can occur. Passage through the stomachs of cattle has been suggested as a reason why mesquite has shifted from drainages to uplands. However, rodent dissemination of seeds occurred before the introduction of livestock, and expansion of mesquite to uplands more likely resulted from fire suppression.[88] Germination of mesquite occurs at higher temperatures than for other native species[89] and is more likely to occur during late summer or early fall instead of during winter or spring.

Rooting Depth

Obligate riparian species in the Southwest have variable root architecture, depending on the trade-offs of depth to water, competition with other species, and ability to withstand floods. Some species produce diffuse, interconnected root systems designed to withstand scour; these species tend to require a constant, shallow source of water and tend to occur at either springs or the edges of perennial rivers. Other species have deep rooting systems, and some have multilevel root systems that allow movement of water from deep to shallow depths. In general, riparian species tend either to produce lateral root systems or to have a deep taproot with lateral roots distributed at a variety of depths.[90] In some areas, trees use water from deeper sources that are unaffected by surface-water interactions with the alluvial aquifer.[91]

Riparian species use their roots to rearrange soil water by means of hydraulic lift.[92] Deep roots pump water up to shallow roots, where water is temporarily exuded and stored in the surrounding soil mass. Soil moisture beneath perennial grasses in riparian settings is known to increase even during periods of maximum growth,[93] and the probable mechanism for this increase is hydraulic lift. As a result, certain species, such as black willow, may not deplete soil moisture beneath them, whereas other riparian species, such as cottonwood and mesquite, deplete soil moisture in the unsaturated zone despite the close proximity of the groundwater table. Like nonnative species (chapter 4), native riparian plants cause diurnal fluctuations in shallow water levels as water use increases in the daytime and decreases at night.[94]

Seepwillow, arrowweed, and coyote willow tend to have the shallowest root systems. Seepwillow, in particular, is noted for its dependence on shallow water tables less than 3 feet deep, with most of its roots in the upper foot of alluvium.[95] This species can withstand saturated soils, reproduce from cuttings, and produce *adventitious roots* on buried stems. Coyote willow produces annual rings in above-ground stems, but ring anatomy changes with burial during sediment accumulation,[96] suggesting one adaptation to the fluvial system that favors production of adventitious roots and shoots.

Arrowweed is noteworthy for its interconnected root systems that appear to provide a structural advantage in resisting stresses induced during flood flows. Other species have flexible branches that lay over during floods, minimizing exposure to flows and reducing damage. For example, coyote willow has long flexible branches that allow it to withstand floods because its branches lay over in response to the force of floodwaters and may later take root. The stems of black willow are more flexible than mesquite, as determined through measurement of the force required to flex stems to a forty-five-degree angle.[97]

Cottonwood root structure is dependent on growth history related to fluctuations in groundwater levels.[98] Mature Frémont cottonwood trees generally are associated with groundwater levels of 3 to 12 feet below land surface,[99] and cottonwood-willow assemblages dominate over nonnative species where surface water is present,[100] suggesting that these species require high water levels in alluvial aquifers. Their roots do not necessarily reach permanent water, instead stopping 1 to 2 feet above the water table to utilize the capillary fringe of the unsaturated zone.[101] Frémont cottonwood is known to switch water sources between ground, surface, and meteoric water, depending on hydrologic and climatic conditions.[102] Depth to groundwater and whether nearby watercourses are ephemeral or perennial may influence seasonal water usage.[103]

Suitable habitat requires depth to groundwater of greater than zero and less than a maximum that varies among the native species. Without this buffer zone, root aeration would not occur, and death would follow; cottonwood and black willow cannot withstand significant periods of inundation.[104] Flooding that involves long periods of root inundation may cause growth suppression.[105] Mesquite is well known for long roots that probe for water to depths of more than 150 feet.[106] Mesquite bosques, however, become established where depth to groundwater typically is in the 15 to 30 foot range.[107] The ratio of biomass allocated above and below ground probably is dependent on depth to reliable water supplies.

Water Use

Water use by riparian vegetation involves a complicated, multilevel extraction process from alluvial aquifers. Structural diversity of typical stands of riparian vegetation makes estimation of water use difficult, and no single method is considered better than others for estimation of evapotranspiration and water use.[108] Among the many methods available, the most commonly used in recent studies include sap-flow measurements of stems, eddy-covariance towers, energy-balance equations, and remote sensing.[109] A summary of measured water use by native and nonnative species is given in table 3.1.

Cottonwood, mesquite, and black willows along a perennial stream in southern Arizona do not use summer precipitation but instead rely on groundwater; cottonwood uses some summer precipitation along an ephemeral reach, and mesquite shifts between groundwater and precipitation-recharged soil moisture depending on season and hydrologic changes.[110] Daily and hourly evapotranspiration in the summer is strongly dependent on a number of factors, including temperature and humidity; and the arrival of the Arizona monsoon strongly affects evapotranspiration rates in southern Arizona.[111] Along the Mojave River, cottonwood primarily use groundwater, and plant health is related to depth to the water table.[112] Measurements from the middle Rio Grande in New Mexico indicate that a gallery forest of cottonwood uses 39 inches per year, or 0.11 inch per day on average.[113]

Water use by cottonwood is a function of tree size as well as canopy exposure to the atmosphere.[114] Position on floodplains—whether on overflow channels or adjacent to the main channel—may be important, at least seasonally.[115] According to one estimate made during July in southern Arizona, the average transpiration rate was 0.19 inch per day for cottonwood and varied considerably with tree structure.[116] Transpiration by black willow is less than cottonwood primarily because black willow has a higher ground cover.[117] Leaf area, which controls the amount of potential transpiration, is a function of relative humidity at the time of growth,[118] the size of new leaves, and changes as the cool, moist spring turns to hot, dry summer to summer monsoon. Of the native

Table 3.1 Estimates of Water Use by Native and Nonnative Species of Woody Riparian Vegetation.

Assemblage Type or Species	River Basin or Region	Water Use (in./yr)	Source
Cottonwood/willow	Santa Cruz	67	Unland and others 1998
Mature cottonwood assemblage[1]	Middle Rio Grande	48	Dahm and others 2002
Closed canopy mature cottonwood[2]	Middle Rio Grande	39	Dahm and others 2002
Cottonwood	Middle Rio Grande	39–47	Nagler, Cleverly, and others 2005
Cottonwood	San Pedro	50	Scott, Shuttleworth, and others 2000
Dense mesquite	San Pedro	30	Scott, Shuttleworth, and others 2000
Dense mesquite woodland	San Pedro	36–48	Nagler, Scott, and others 2005
Less-dense mesquite shrubland	San Pedro	4–28	Nagler, Scott, and others 2005
Mesquite	Santa Cruz	33	Unland and others 1998
Medium-density mesquite	San Pedro	19	Scott, Shuttleworth, and others 2000
Willow[3]	San Pedro	47	Scott, Shuttleworth, and others 2000
Black willow	Lower Colorado	36–54	Nagler, Scott, and others 2005
Giant sacaton grass	San Pedro	20–30	Nagler, Scott, and others 2005
Dense grasses	San Pedro	38	Scott, Shuttleworth, and others 2000
Arrowweed	Lower Colorado	16–26	Nagler, Scott, and others 2005
Tamarisk	Lower Colorado	20–42	Nagler, Scott, and others 2005
Dense tamarisk	New Mexico	44–48	Dahm and others 2002
Dense tamarisk	Gila	56	Culler and others 1982
Less-dense tamarisk	New Mexico	29–30	Dahm and others 2002
Less-dense tamarisk	Gila	43	Culler and others 1982

[1] *Populus deltoides* with tamarisk and Russian olive understory.
[2] *Populus deltoides*.
[3] Species of willow is not specified.

species, coyote willow may have the highest transpiration rate per unit leaf area,[119] which partly explains its usual presence adjacent to perennial water.

The various species of mesquite are facultative riparian species, and the availability of shallow groundwater affects mostly individual plant size and density. Water use by mesquite is highly variable, in part because stand structure and water-use requirements shift according to the position of the trees on the landscape.[120] Also, the plants can shift between ground and meteoric water and likely have root systems that can extract from a variety of different levels depending on need. Mesquite can use meteoric water or moisture from the capillary fringe of groundwater tables.[121] Water use by mesquite is dependent on depth to groundwater, air temperature, season, and the occurrence of freezing temperatures. In June, water usage can average 0.42 inch per day.[122] Annual water use ranges from 16 to 27 inches per year.[123] One study reported that leaf-

corrected water use of mesquite was higher than that of nonnative tamarisk (chapter 4).[124]

Factors Affecting the Stability of Native Riparian Species

Riparian plant communities are highly dynamic, and change is an integral part of their long-term life-history strategies. As a result, definition of what is natural in riparian areas of the Southwest is fuzzy and requires a temporal context. Many human-induced hydrologic changes and land uses affect the health, productivity, or survival of riparian vegetation in the Southwest.[125] Although livestock grazing is frequently cited as a force for alteration or damage to riparian zones, another animal user of riparian zones, the beaver, was extirpated from some river reaches, and its numbers were greatly reduced in others, a change that may also have affected riparian vegetation. Dams have greatly altered riparian ecosystems downstream, but

modification of dam operations are now proposed as a means of restoring some riverine ecosystems.

Livestock Grazing

Grazing by domestic animals has direct and indirect effects on riparian ecosystems[126] that some have labeled "disastrous."[127] Livestock heavily use river channels because of the high productivity of herbaceous species and the readily available water and shade. Effects that diminish riparian habitat include consumption and trampling of native-plant seedlings, soil compaction, destabilization of channel banks, increased streamflow sediment, and displacement of wildlife.[128] Although climate, hydrology, and elevation have stronger effects,[129] livestock grazing has secondary effects that shape species composition of riparian vegetation.

Domestic grazing animals have also been implicated for watershed changes that may have led to, or at least may have affected, arroyo cutting at the end of the nineteenth century, woody plant encroachment on ranges formerly occupied by grasslands, and the spread of many nonnative species in upland environments; cumulatively, these changes might have indirectly affected riparian ecosystems (discussed later). Unless riparian areas are fenced, livestock will use them disproportionately to other parts of the landscape.[130] The impacts are even stronger when native ungulates heavily use riparian zones along with domestic animals.[131]

Cattle grazing in particular has wide-reaching effects on riparian ecosystems and uplands. For plains cottonwood, various ages of seedling cohorts have higher densities in ungrazed sites.[132] It is claimed that Frémont cottonwood reproduction nearly ceases in grazed riparian habitats,[133] and some heavily grazed sites reportedly developed senescent stands of cottonwood owing to removal of seedlings.[134] Some species that vegetatively reproduce, such as sycamore, may be unaffected by grazing, whereas seedling decimation by livestock grazing can eliminate cottonwood in some heavily grazed reaches.[135]

In particular, herbaceous species are generally the most severely impacted, followed by seedlings of woody trees and shrubs, in particular willow species.[136] Some shrubs, including seepwillow, may actually increase under grazing pressure, however.[137] Some researchers have found that bird populations decrease in heavily grazed areas because of the impacts on woody plants' seedlings and the implications for long-term regeneration, but that they may rebound following cessation of grazing.[138]

Livestock were introduced to the Southwest over a broad period, beginning around 1700 in southern Arizona[139] and as late as the 1880s in parts of southern Utah and northern Arizona. Before water development became widespread in the middle of the twentieth century, grazing was restricted to short distances from perennial water, so riparian ecosystems are thought to have been disproportionately impacted.[140] Construction of artificial watering areas associated with windmills that pump groundwater or stock ponds that store surface runoff allowed dispersion of grazing animals over a larger area of the landscape.[141]

Floods

Floods, whether natural or the result of dam operations, have the single largest short-term impact on riparian ecosystems in the Southwest.[142] In alluvial reaches, major floods cause bank scour and remove at least some of the riparian vegetation.[143] Many plants are uprooted in response to detrital loading on their upstream sides, or whole terraces may be removed during lateral avulsions, channel widening, or meander propagation.[144] In particular, mesquite bosques that are on terraces may be removed during flood-induced channel meandering.[145]

High-shear stresses on banks frequently induce lateral channel changes at the expense of floodplain area (and floodplain vegetation). Dense riparian vegetation slows flow velocity, protecting floodplains from erosion during overbank flooding, and floodplains with sparse vegetation may be subject to what has been termed *unraveling*.[146] Some floods cause channel downcutting, which may drain alluvial aquifers (chapter 2) and change a floodplain from mesic to xeric habitat.

Large amounts of woody riparian vegetation become established within or immediately following the years of large floods.[147] Overbank flooding adds water to the local alluvial aquifer and deposits nutrient-rich sediments.[148] Flood scour and deposition create openings in typically dense riparian assemblages that enable germination and recruitment of shade-intolerant species. Cottonwood recruitment, in particular, is known to be widespread following large floods.[149] Channels that have not experienced large floods tend to have low reproduction.[150]

Floods have a major role in shaping the species composition and productivity of riverine riparian vegetation.[151] Reduction in the size of flash-flood peaks, as has occurred in the Southwest, has an effect similar to that occurring on a river with little variation in flood peaks: the riparian vegetation on floodplains faces long periods of low disturbance and tends toward senescence.[152] Extreme floods can reset riparian communities to the stage of initial establishment.

Particularly with streams subject to large stage fluctuations ("flash floods"),

aquatic species and channel-margin vegetation are constantly adjusting to the occurrence and aftermath of such floods on timescales of days to months.[153] Long-term directional changes may occur as well. On the Gila River in central Arizona, a 1905 flood removed much of the original cottonwood and seepwillow stands,[154] but they recolonized by the mid-1930s. However, between the 1930s and 1960s dense tamarisk thickets replaced these native species (chapter 17). The question is: Will this tamarisk be replaced by an assemblage consisting of native species, or will tamarisk simply change the riverine conditions to favor its own presence (chapter 4)?

In bedrock canyons, gradients tend to be steeper and channel widths are narrower, resulting in higher flow velocities and stream power during floods. As a result, riparian vegetation is likely to be sparse, although it is locally abundant in low-velocity zones. Cottonwood tend to occur on river floodplains and low terraces in wide reaches,[155] and germination, establishment, and mortality are all flood related. Alluvial substrate is typically coarse in narrow bedrock canyons, in particular those subject to occasional debris flows. Some species, such as netleaf hackberry, preferentially grow in coarse substrates and have life-history traits—such as adventitious sprouting of new individuals from widespread lateral root systems—that simultaneously anchor plants and promote root resprouting after floods.

Riparian plants are arranged both in response to and because of floods. Germination and establishment of native riparian species is highly dependent on winter and spring floods; plants form distinct bands that reflect both depth to groundwater and time since the most recent flood. In bedrock canyons, plants are arranged according to their tolerance of flood damage. Along the Escalante River in southern Utah, coyote willows are found in the narrowest reaches[156] where velocities are highest during floods. They tend to occur on the outside of river bends along the upper Green and Colorado Rivers (chapters 6 and 7) and are inundated during the annual flood.[157] Floods in narrow canyons create ongoing disturbance that does not allow senescence in riverine riparian vegetation.[158]

Arroyo Downcutting

Beginning around the mid-1880s and extending through the early 1940s, the frequency and severity of floods increased in the Southwest.[159] These floods caused downcutting in essentially all of the alluvial channels in the region. Determining the cause for historical arroyo cutting has been geomorphologists' goal for more than a century.[160] Making a determination is difficult because the major variables that might have triggered alluvial downcutting cannot be separated. Although the precise reason why arroyos formed in the Southwest may never be known, some of the details give us insight into the processes of long-term change in riparian vegetation.

In the early and middle parts of the twentieth century, many observers asserted that overgrazing caused arroyo downcutting. This conclusion was based partly on circumstantial evidence, partly on observation, and partly on assumption. The primary evidence was the close temporal association of overgrazing and downcutting. Although cattle had been introduced in southeastern Arizona in the 1700s, large-scale grazing began in the early 1880s.[161] A series of large floods and channel downcutting began in the mid-1880s, and extensive channel change had occurred on all major streams in the region by 1909. A drought occurred between 1891 and 1904 that generally was severe,[162] but varied in duration and intensity across the Southwest. As noted earlier, this drought, combined with overgrazing, depleted the range so heavily that reportedly half of the livestock died in much of the region. The poor range conditions became associated with arroyo downcutting.

Some scientists hypothesized that climatic conditions, combined with livestock grazing, induced arroyo downcutting. Because arroyos had repeatedly downcut and backfilled in the Holocene without significant numbers of grazing animals on the landscape, climatic fluctuations were assumed to be the only unifying cause of the downcutting.[163] Early proponents of the climatic-fluctuations hypothesis believed drought, combined with grazing, deteriorated watersheds to increase runoff. Other proponents of the drought theory hypothesized that riparian vegetation on floodplains died, reducing roughness and thus allowing downcutting.[164] According to this hypothesis, arroyos would have formed even if grazing animals were not on the landscape. As discussed in chapter 2, groundwater response to climate is sufficiently slow that riparian ecosystems are largely buffered from short-term drought in many (but not all) settings, reducing the likelihood of this possible cause for arroyo downcutting.

The period of historical arroyo downcutting is known to have been a period of unusually large storms of regional extent, which caused some extraordinary floods.[165] Periods of prehistoric arroyo downcutting are associated with periods of increased El Niño activity.[166] A complex set of factors contributed to arroyo cutting,[167] but recent studies suggest that seasonality of precipitation has changed through the twentieth century with strong regional patterns.[168] Decadal climatic variation includes periods of intense and heavy precipitation and seasonal shifts in the relative amounts of summer and winter precipitation.[169] Although the dates of incision for many arroyos remain unknown, most arroyos in the Southwest—from southern Arizona to central Utah—began downcutting in the forty-year period from around 1862 through 1909, with most of the change beginning in the 1880s and early 1890s. Although the floods did not occur simultaneously in the region, they occurred during a period distinct in climatic history. Climate therefore created an apparent synchroneity in geomorphic processes—in a geologic, not an event, sense.

Several human-introduced factors exacerbated the climatic forcing of arroyo downcutting. Livestock grazing undoubtedly had local impacts, possibly on a watershed scale and certainly on floodplains. Irrigation canals and roads channeled flow on floodplains, plotting the locations of future gullies and arroyos. In-stream gravel mining may renew channel downcutting and knickpoint migration.[170] Irrigation dams frequently failed, releasing flood pulses that might have been larger than the flood alone would have been. One of the most intriguing hypotheses concerning the cause of arroyo downcutting focuses on the decimation of beaver populations by trappers in the nineteenth century (see the next section).[171] Beaver dams served as local base levels, mitigating the tendency for downcutting during floods; decimation or extirpation of beaver along rivers such as the San Pedro might have allowed floods to cut more readily through the channel bottoms.

The result was downcutting of narrow channels through floodplain alluvium, followed by channel widening. Downcutting occurred during discrete events, although several events were required to coalesce arroyos over long alluvial valleys. The alluvial aquifer, which was high prehistorically, drained through the newly eroded banks, further inducing destabilization of the floodplain and likely leading to bank collapse and additional channel widening. By the mid-1940s, channel widening ceased regionally because of decreases in the sizes of floods, and low floodplains were deposited, representing the start of arroyo filling.[172] As discussed in chapter 2, flooding resumed in central and southern Arizona from 1977 through 1995, causing the removal of low terraces along many rivers in the southern part of the region.

The processes of arroyo downcutting and widening greatly affected riparian vegetation.[173] Rapid downcutting quickly dropped the alluvial water level, with dieback or death occurring in woody riparian vegetation. Channel widening removed plants that survived the water-level drop. Development of low terraces above the saturated zone encouraged establishment of woody riparian species, which then trapped sediments during overbank flows. This positive feedback mechanism, in the absence of additional decadal-scale climatic fluctuations favoring large floods, may enhance the tendency toward arroyo filling.

Beaver

Beaver (*Castor canadensis*) once were widespread throughout the Southwest[174] and doubtless had profound effects on prehistoric distributions of riparian vegetation.[175] Although it is improbable that a cow can kill a mature cottonwood, beaver can easily do so. Beaver use woody riparian vegetation for both food and building materials, and ponds impounded by their dams can kill inundated trees upstream within a year.[176] The combination of unlimited livestock grazing and beaver use can "expedite the collapse of riparian forests"[177] because of beaver depredation of adult trees and livestock consumption of seedlings. However, beaver-dam control of stream gradient slows water flow and increases saturation of alluvium. Some researchers have suggested creating habitat for beaver not for the sake of the beaver, but to improve riparian habitat for plants and animals.[178]

Beaver are the largest rodents in the United States and are vegetarian, eating cottonwood, willow, mesquite, and other species, notably nonnative tamarisk (chapter 4).[179] With respect to cottonwood, they unwittingly may be selective about what they eat on the basis of genetics, preferring Frémont cottonwood over narrowleaf cottonwood and hybrids between the two because of the chemistry of the bark.[180] They build at least two types of dens, depending on habitat and the availability of woody vegetation. Along small rivers, beaver build dams and lodges; along large rivers, such as the Colorado, they build dens into banks.[181] Materials used for constructing dams and lodges vary widely, ranging from cottonwood and willow branches along many rivers to cattail on the lower Bill Williams River (see chapter 15).

Beaver prefer relatively quiet water, and although their dams can withstand modest floods, they avoid building cross-channel dams along rivers that have large floods. In one case, a robust beaver population in western Washington abandoned a reach after a series of floods in 1997.[182] A similar response was claimed for the Gila River in western Arizona following the 1891 flood.[183] Flood control without significant flow reduction, therefore, creates ideal beaver habitat. Whereas beaver were not observed to be common along the Colorado River in Grand Canyon before Glen Canyon Dam fully regulated flow in 1963 (chapter 11), this species is now common.

Although beaver were once ubiquitous along rivers in the Southwest, nineteenth-century trappers significantly reduced their numbers, extirpating them from some reaches. The Spaniards who entered the Southwest with Francisco Coronado in the early 1500s undoubtedly noticed the beaver populations, but they were more interested in gold. Trappers entered the region in the 1820s from the north and east. In 1825, James Ohio Pattie made the first of several incursions into the Gila River drainage, with one trip involving a dangerous exploration of the Colorado River delta.[184] Jedediah Smith explored the Virgin and lower Colorado Rivers as far as present-day Laughlin, Nevada, and then across the Mojave Desert and along the Mojave River in 1826 and 1827. The Bill Williams River is named for William Sherley (Old Bill) Williams, who spent much of his life trapping throughout Arizona.[185] Many other trappers were known to work the region until the mid-1830s, when the demand for beaver pelts for hats diminished.[186] Trapping continued with diminished intensity until well after the beginning of the twentieth century.

It is difficult to assess the impact of beaver extirpation on riparian ecosystems, although it is expected to be large. Beaver use both herbaceous and woody plants for food and building materials. They require open water and occupy only perennial rivers. The presence of beaver does not mean that they completely suppressed woody riparian vegetation, but it does suggest that they culled it regularly. Beaver were not extirpated from any rivers; instead, their numbers were greatly reduced, and since trapping has ceased, they have gradually repopulated the parts of their range where perennial flow remains. Their low numbers in the middle of the twentieth century may have been a factor in the growth of gallery forests, and their increases and reintroductions may affect the future stability of riparian ecosystems.

Introduction of Nonnative Species

The introduction of plants from other continents as ornamentals has caused large changes in the riparian landscape of the Southwest (chapter 4). Once nonnatives become established, they use considerable amounts of water, produce deep shade that minimizes the germination and establishment of some native plants, and change the number and species composition of animals that use riparian ecosystems. In addition, some species exude salt that may decrease the potential for native species to grow nearby. Some nonnative species, particularly tamarisk, may create the potential for fire-induced cycles that perpetuate their dominance on floodplains (chapter 4).

Fire

Before fire suppression began in the early twentieth century, wildfires—set either deliberately by aboriginals or accidentally by lightning strikes—were common in the upper elevations of the Southwest.[187] Fires could be generated in riparian ecosystems or could sweep out of adjacent grasslands or shrublands into the riparian zone. Although little evidence exists to suggest that prehistoric stands of woody riparian vegetation were suppressed by frequent fires, the possibility remains. With widespread human presence in riparian ecosystems of the Southwest, fires are more frequent.

Along the lower Colorado River, fires in one reach burned more than one-third of the area occupied by riparian vegetation.[188] Although nonnative tamarisk responds positively to fire, native species also resprout. Along the Green and Colorado Rivers in Canyonlands National Park (chapters 6 and 7), desert olive can resprout with tamarisk and potentially maintain its position in the riparian landscape.[189] At lower elevations, arrowweed can dominate in the early stages of natural recovery to revegetate burned areas.[190] Crown fires that kill mature cottonwood and black willow may create an irreversible change in community composition along rivers regulated for flood control owing to the lack of postfire germination.

Growing Season and Favorable Climatic Conditions

As discussed in chapter 2, the twentieth-century climate fluctuated between wet periods and drought. Two other factors affecting plant growth—increased temperature and carbon dioxide content of the atmosphere—have also favored high growth rates in riparian plants.[191] Primarily because of increases in nighttime temperatures and winter temperatures overall, the growing season has increased by as much as sixty days in the Southwest. This increase in the growing season is felt primarily at higher elevation sites, where killing or growth-retarding frosts are now less numerous. Growth in conifers has increased,[192] for example, but the effects of these two factors on riparian plants have not been quantified.

The direct effects of increased atmospheric carbon dioxide are less certain. Most riparian plants, and all of the woody plants considered in this book, use the C3 photosynthetic pathway. As for most upland plants in southern Arizona, C3 plants benefit from higher concentrations of carbon dioxide.[193] Because the increases have been relatively modest as of 2000, it is unlikely that they have affected the status of the woody riparian vegetation discussed in this book. The amount of atmospheric carbon dioxide is likely to increase indefinitely, so growth of riparian vegetation may be enhanced in the future. This positive effect will add to the increased growing-season effects that encourage woody plant growth.

Watershed Modifications

Watershed manipulations designed to increase runoff were popular in the 1950s and 1960s and continue to be suggested as a means of increasing water yield. Mechanical removal of trees or applications of herbicides were recommended to induce a variety of watershed changes that presumably increased grazing lands, upland wildlife habitat, and water yield. Removal of woody vegetation, particularly by fires, increases flood peaks for at least several years, but also increases base flow owing to decreased transpiration demands. The result is a benefit to riparian vegetation downstream from burned or otherwise manipulated areas. This benefit may be short-lived; twenty years after a fire, the density of riparian trees along one watershed in

central Arizona was three times larger than in nearby similar but unburned watersheds.[194]

Groundwater Withdrawals

Use of groundwater that results in persistent water-level drawdown is the single most important factor that can lead to the rapid elimination of woody riparian vegetation. Although flood control and flow diversions have large effects spread over time, the effects of groundwater withdrawals can be immediate. Effects can be local; a deep cone of depression rapidly developed around a pumping well on a floodplain can kill nearby trees in a matter of days. The more typical situation occurs where well fields heavily utilize groundwater in an overdraft status for large-scale agricultural or municipal use.

Cottonwood provides a prime example of how native trees respond to the lowering of water levels.[195] If lowering occurs slowly enough, in response to drought or to steady pumping, roots may be able to elongate quickly enough to keep pace; the tree may suffer leaf or stem dieback but will survive.[196] Rapid water-level drawdown may induce cavitation in stem xylem, leading to general losses in hydraulic conductivity between roots and branches. Established plants may be able to survive for short periods, but mortality in saplings may be as high as 100 percent.[197] Rapid drawdown resulting in such mortality may occur with as little as 3 feet of water-level decline in a few weeks, such as might occur after channel downcutting from a storm runoff event.[198] The rate of water-table decline, not the absolute depth to water, is the most important factor.[199]

Mesquite can withstand larger fluctuations in groundwater level. However, bosques, where the density of mesquite is high, require shallow groundwater for their long-term stability.[200] Trees in bosques become stressed when groundwater levels decline to 45 to 50 feet below land surface, although rainfall may alleviate this stress, at least temporarily. Mesquite along the middle Gila River did not survive a water-table decline of about 100 feet in the 1940s and 1950s.[201]

As discussed earlier, water-table lowering can result from arroyo downcutting and consequent dewatering of the alluvial aquifer, from excessive well pumpage, or from persistent drought. Arroyo downcutting would create the fastest rate of groundwater decline, and most of the declines in channel bed are as much as an order of magnitude higher than the 3- to 5-foot declines that can damage floodplain trees. Most cases of water-table lowering are likely slow enough to allow roots to follow to a certain point (roughly 30 feet for cottonwood) before death is certain. Persistent drought would likely result in gradual water-level lowering that roots can adapt to, with the likely response of stem and branch dieback and low mortality.

Flow Regulation

Hydrologists have long known that dams have large effects on downstream riverine resources.[202] The response of riparian vegetation to flow regulation is a function of water use and control, which ranges from complete dewatering of rivers used extensively for irrigation to sediment-deprived flood flows released from run-of-the-river hydroelectric facilities. Flood-control operations greatly reduce peak discharges, creating conditions dissimilar to any climatic effects known during the Holocene.[203] Examples of this situation include the Colorado River in Grand Canyon (chapter 11) and the lower Colorado River (chapter 27).

After construction of a dam, a period of geomorphic adjustment to the changed flow and sediment-transport regimes is expected, followed by an adjustment in riparian vegetation.[204] Along large rivers in the Great Plains, geomorphic adjustments occur at a much more rapid pace than do the responses of riparian vegetation. However, smaller rivers do not necessarily respond in the same way as larger ones.[205] In the Southwest, the relative rates of these changes would be difficult to predict owing to the potential mitigating influences of other factors, such as floods, sediment additions from tributaries, and the effects that riparian vegetation has on trapping sediments and narrowing channels.

A complex interrelation occurs between climate, flooding, and flow regulation. Decadal-scale changes in hydrology induced by climatic fluctuations may have effects that confound or add to those caused by flow regulation.[206] Clearly, extended drought conditions, as occurred in the mid–twentieth century (chapter 2), can reduce surface-water supplies by the same magnitude as irrigation diversions. Extended drought coupled with flow regulation can completely eliminate surface flows. Similarly, extended wet periods, as occurred from the mid-1970s through the mid-1990s, can offset the negative effects of flow diversion. Climate can be thought of as an amplification of flow-regulation effects on riparian vegetation.

Flow regulation and large-scale water use induce complex responses in riparian vegetation. Flood control, which is a common justification for dams, reduces peak discharges and the amount of water available to riparian systems on floodplains. As a result, riparian species on abandoned floodplains are considered to be disconnected from the river.[207] Species dependent on flooding for germination and establishment can only rarely reproduce.[208] Because flood peaks are lower, groundwater tables in the alluvial floodplains are also lowered, killing or retarding growth of established cottonwood and other native trees. Flood control decreases nutrient cycling because detritus accumulates on floodplains that are not regularly inundated.[209] Long-term flood control may cause conversion of formerly mesic floodplains to xeric shrublands.[210]

However, elimination of floods, coupled with lowered variability in daily discharge, greatly benefits riparian species along the river's edge. Some regulated rivers, such as the Colorado River through Grand Canyon, have little interannual flow variability compared with predam conditions, and plants established in the predam era with deep taproots can take advantage of the more stable water supply.[211]

Low interannual variability, with no definite seasonal peak, may retard establishment of some species immediately adjacent to the stream owing to periodic inundation of fragile seedlings.[212] Conversely, overbank flooding becomes infrequent or nonexistent, eliminating the possibility of germination and establishment on the higher-elevation floodplains.

Flow diversions, whether partial or total, strongly affect riverine riparian ecosystems by reducing recharge to shallow aquifer systems. Water use for agriculture can also lower river flows, particularly during summer months, and thereby stress riparian plants in the season when they need water most. Following such diversions, plants may respond by decreasing their total leaf area and by reducing their water use and growth rates.[213] In addition, scouring floods are reduced, thus further encouraging tamarisk growth, reducing the possibility of germination and establishment of native species, and leading to the senescence of riparian systems that require periodic disturbance for renewal.

Many researchers view the potential for change in dam-operating regimes as a tool for river restoration.[214] Human-created floods can be designed to have mostly positive effects on native riverine systems, while minimizing some negative effects. For example, many nonnative species germinate during warm-weather conditions (chapter 4), and release of floods during fall or winter months may minimize germination of these nonnatives while encouraging germination of native species.[215] Discharges can be specified that allow inundation of low floodplains, thereby recharging the alluvial aquifer, fostering germination of native species, while minimizing terrace-removing channel change.

Change in Riparian Ecosystems

Major philosophical disputes arise when the factors causing change in riparian vegetation are discussed.

Some researchers view the forcing factors of change to be abiotic, caused by hydrologic and geomorphic factors, whereas others view changes as primarily biologically driven. Some view the evolution of woody riparian ecosystems to be a successional process[216] that progresses from a colonization stage (which may now be dominated by nonnative species) to a climax vegetation. Progression from colonization of bars to attainment of gallery forests requires increasing time of stability and, to a certain extent, distance from active channels.[217] Because of occasional flood damages, some view this succession as continual to the point that climax vegetation cannot be attained,[218] except in the case of flow regulation that involves flood control.

Everything from the life-history strategies of the species involved to the strong abiotic forces acting in riverine environments suggests that woody riparian vegetation should form dynamic plant assemblages. Disturbances are periodic but frequent, and the plants capitalize on any available opening for germination and establishment. Large floods may decimate riparian vegetation, particularly along alluvial watercourses, but they also provide impetus for growth and renewal. Flow regulation may benefit riparian vegetation initially by providing a more stable hydrologic environment, but a cost is paid in reduced opportunity for germination and recruitment. Rapid change in groundwater levels, particularly by groundwater overdraft, is expected to have the largest negative impact, either retarding growth or killing plants. Complete diversion of flow, coupled with flood control, would be the most severe form of flow regulation. Drawdown of the alluvial aquifer, in the absence of pumpage, would require considerable time because of the partial decoupling of surface-water and groundwater systems.

4 Woody Nonnative Riparian Vegetation

Summary. Woody (and selected herbaceous) nonnative species are thought to have displaced native riparian vegetation through much of the Southwest, raising concern about irreversible changes in riparian ecosystems of the Southwest. For example, tamarisk spread rapidly after its introduction to the region in the late nineteenth century, although its spread from initial horticultural sites was complicated. Russian olive is a significant colonizer, particularly at higher elevations, but does not appear to be rapidly expanding at lower elevations. Both species successfully colonize riparian areas and may or may not be part of a process of orderly replacement by native species with increasing time following disturbance. Although water use by tamarisk is roughly equivalent to that of native species, reduction or elimination of tamarisk is advocated to increase water supplies and for ecological reasons. Several control mechanisms are available that may limit future expansion or even cause contraction of the amount of nonnative vegetation.

Nonnative species, introduced species, exotics, invasives, and *aliens* are terms or words used interchangeably to describe plants or animals that humans have purposefully or inadvertently moved from one habitat to another. In the Southwest, the most widespread nonnative woody species in riparian settings are tamarisk and Russian olive, although other species, including Athel tamarisk, giant reed, tree-of-heaven, and tree tobacco, are locally abundant. These species arrived in the Southwest at different times, and some species are well-enough established to be considered naturalized into the riparian ecosystem. The threat posed by nonnative species is highly controversial, with some observers predicting partial or total elimination of native species

and others downplaying the potential long-term impact.[1]

Nonnative species are considered to be the largest threat to riparian ecosystems, and eradication programs of various scales have been proposed and enacted to suppress or eliminate these species. One widely held reason for eradication is the belief that nonnative species aesthetically do not belong in natural areas, particularly in national parks, regardless of their benefits or detriments. At least one economic reason propels control measures: tamarisk removal can increase water availability,[2] although only if the tamarisk is replaced by native species with a lower leaf area.[3] It is estimated that complete removal of tamarisk from the West, an unlikely outcome, would save as much as sixteen billion dollars in water over half a century.[4] Finally, nonnative species also are thought to create biologically sterile habitat, and their removal is thought to enhance the native wildlife's ability to use riparian ecosystems.

Discussion of nonnative species follows the dual paths of scientific objectivity and value judgment. Some of the terms or names used to describe species such as tamarisk are pejorative ("the Devil's Own"),[5] reflecting these species' rapid expansion into their new range ("infested"),[6] possibly to the detriment of native plants. Other terms are more ambivalent, suggesting either objectivity or that introduced species are deemed harmless. In all cases, the terms imply value judgments—some negative, some neutral. Here, we use the neutral term *nonnative species* to discuss woody plants that have been introduced to riparian ecosystems in the Southwest. As is noted, at various times these species are valued for some purpose, such as erosion control, and then are reviled for expanding rapidly

at the perceived expense of native species.

At least two classes of nonnative species are found in riverine habitats. Some nonnative species, such as date palms, appear to pose little threat to wetlands at present. Other nonnative species quickly colonize riparian zones and therefore are problematic for land managers. This second type generally expands into habitats disturbed by floods or fire, or into habitat newly stabilized by flood control. Here, we discuss the nonnative species with an emphasis on their life-history strategies and potential for control. Tamarisk has become naturalized along most watercourses below about 6,000 feet in the western United States and particularly in the Southwest. Because so much research and invective has been invested in tamarisk, we review what is known of this species as well as what is known about Russian olive, giant reed, and other nonnative riparian plants.

Spread of Common Woody Nonnative Species

Tamarisk (Saltcedar)

Any discussion of tamarisk must begin with a review of the confusing taxonomic literature on this group. Initially, the species of tamarisk introduced into the United States from Africa and Asia were believed to include *Tamarix pentandra, T. tetrandra,* and *T. aphylla.*[7] *T. aphylla,* known as Athel tamarisk, is discussed in the next section. *T. tetrandra* is not thought to be an invasive species. The confusion concerns the taxonomic name of the highly invasive species, variously known as tamarisk or saltcedar, that commonly grows along rivers throughout the arid and semiarid regions of the United States.

At one time, *T. gallica* was thought to be this invasive species, but *T. gallica* appears to be confined to the Gulf Coast of Texas.[8] Some botanists consider *T. pentandra* to be the invasive species and synonymous with *T. gallica*.[9] Others list four species of tamarisk in the southwestern United States, but lump *T. pentandra* and *T. gallica* under *T. ramosissima* and consider *T. parviflora* as another invasive riparian species.[10] *T. chinensis* is thought to be synonymous with *T. ramosissima* (or not), and both names are frequently used.[11] One study suggests that *T. chinensis* can most easily be distinguished from *T. ramosissima* using color aerial photography.[12] Adding to the confusion, hybrids between *T. ramosissima* and *T. chinensis* and a third species, *T. canariensis*, are thought to be common throughout the western United States.[13] Here, we choose to use the name *tamarisk* to refer to *T. ramosissima, T. chinensis*, and their hybrids.

Some researchers believe tamarisk may have become naturalized prehistorically or was introduced by the Spaniards as early as the seventeenth century.[14] Part of this belief stems from translation of the Spanish word *taray*, which can mean either "tamarisk" or other riparian shrubs, in particular willows.[15] Similarly, the Spanish word *teraque* is ambiguous.[16] The evidence is overwhelming that Americans introduced tamarisk to the Northern Hemisphere well after the period of Spanish conquest. Tamarisk first was imported into the United States in the early 1800s, and by the 1850s California nurseries sold it as an ornamental.[17]

In southern Utah, tamarisk was first planted about 1870 at Adairville on the Paria River, was in cultivation in St. George in 1880, and was present in Green River, Utah, in about 1900.[18] Although it first escaped cultivation sometime in the 1870s,[19] the first botanical collection of a naturalized specimen from the region was made at Kanab, Utah, in 1909. Also at that time, tamarisk had become established along the Virgin River to the west.[20] It may have been present at the Confluence of the Green and Colorado Rivers in 1914.[21] After 1925, it became well established along the Colorado River and its tributaries in Utah.[22] Biological surveys from 1933 and 1938 found it along the Green, Colorado, and San Juan Rivers on the Colorado Plateau,[23] and photos record its presence along the Colorado and Green Rivers in 1941.[24]

Because tamarisk roots stabilize banks, it was planted in the early 1900s to stem erosion along western waterways and canals.[25] Its spread in New Mexico dates from 1912 and was rapid in the 1930s.[26] In central Arizona, the first definitive report of tamarisk introduction was in around 1898 along an irrigation canal near Gila Bend.[27] In 1892, tamarisk was collected near Tucson. The 1916 floods in Arizona deposited tamarisk propagules downstream of their original plantings, leading to widespread establishment on newly eroded channel bottoms.[28] Despite some claims to the contrary,[29] there are no records of tamarisk along the lower Colorado River before 1916.[30]

Deltas of small or large reservoirs provide large areas for colonization of tamarisk.[31] As a result, the spread to free-flowing reaches frequently begins from colonizers at the heads of reservoirs. Gillespie Dam on the Gila River was completed in 1921, and a small delta supporting dense tamarisk thickets quickly formed.[32] Tamarisk spread upstream on the Gila River between 1900 and 1937, although floods and heavy groundwater use in the 1960s through 1970s reduced its distribution.[33] Similarly, tamarisk spread upstream of San Carlos Reservoir on the Gila River.[34] Flow regulation, in particular flood control, encouraged tamarisk expansion by creating newly stable, barren substrate available for colonization. Because most flow regulation in the Southwest began during a period of infrequent winter floods and frequent summer events, tamarisk was able to germinate first on the open riverine habitat.

The spread of tamarisk was rapid following floods.[35] In 1920, it covered 10,000 acres in the western United States; by 1961, it covered 900,000 acres,[36] with a spread rate of 2,500 acres per year.[37] By 1987, the area covered reportedly had increased to 1.5 million acres.[38] The sources of propagules were in towns and irrigation ditches associated with agriculture; on the Colorado River, the colonization proceeded down tributaries leading from settlements and then upstream or downstream from the tributary junctures.[39] The spread likely started slowly, with seeds blown or floated onto wet, bare soil. Channel-scouring floods, common in the Southwest between 1880 and the 1930s,[40] were the most likely reasons. The combination of newly scoured channels and banks with flood-deposited branches made widespread dissemination and establishment possible. It is extremely unlikely that tamarisk propagated upstream on the Colorado River from its delta toward its headwaters.[41]

Athel Tamarisk

Athel tamarisk is much different from the species that has colonized extensive areas of riparian habitat. Unlike tamarisk, Athel tamarisk is an evergreen, ornamental tree used in extensive plantings as windbreaks and shade trees. The natural distribution of this species is a broad band from Morocco across North Africa and the Middle East to western India.[42] Individuals in southern Arizona were imported from Algeria by J. J. Thornber, a botanist with the University of Arizona College of Agriculture.[43]

Athel tamarisk is thought to be unable to reproduce sexually outside of cultivation because few seedlings have been observed in either the Sonoran Desert or its natural range.[44] However, its seeds germinate and seedlings grow under laboratory conditions,[45] and seedlings have been observed in the Southwest.[46] Moreover, the discovery of recent hybrid crosses between *Tamarix aphylla* and *T. ramosissima*[47] suggest that sexual reproduction may be more common in the future. The maximum-known rooting depth of Athel tamarisk is 30 feet.[48] It reproduces vegetatively, and individuals found along low-desert watercourses are thought to have been established from driftwood transported during floods. Low temperatures that cause frost damage and mortality reportedly limit Athel tamarisk in its northward distribution.[49]

Russian Olive

Russian olive is locally dense along the Green, San Juan, Escalante, Owens, and Colorado Rivers. This thorny tree has gray leaves that contrast with its reddish bark; it occurs on reaches with very slow-moving water and is common near towns. This species, native to Eurasia,[50] was introduced to North America before 1900 and spread rapidly in the eastern United States.[51] Promoted for erosion control, windbreaks, and wildlife habitat, it is now considered to be a major management problem in the western United States, especially in Colorado, Utah, and northern New Mexico.[52] It became established outside of horticultural settings between 1900 and 1930 in Utah[53] and is particularly common in dam-regulated reaches.[54]

In northern Utah, Russian olive is widespread, particularly in the wetlands of the Uinta basin[55] along the Green River upstream from Desolation Canyon. Planted extensively as an ornamental in Moab, Russian olive has spread down Pack and Mill Creeks, and Courthouse Wash in Arches National Park is clogged with Russian olive near its juncture with the Colorado River (chapter 7). Although Russian olive is spreading quickly along the San Juan River, it is spreading very slowly, if at all, along either the Green River or Colorado River corridors. Its spread in the Moab area must have occurred quickly; it is not listed as a major invasive species of concern in one account of Arches National Park, where currently it is widespread along perennial streams such as Courthouse Wash (chapter 7).[56]

In Arizona, Russian olive is along channels on the southern Colorado Plateau, such as Chinle Wash; in Canyon de Chelley;[57] in Oak Creek Canyon;[58] and at various points along the Colorado, Salt, and Gila Rivers.[59] In Grand Canyon, Russian olive was a target for eradication in the 1970s,[60] and in 2003 only one tree was observed downstream from the Paria River confluence. Similarly, a young individual was observed near Rapid 18 in Cataract Canyon (chapter 7). Russian olive occurs along the Mojave River near Victorville, California, although it is not common.[61] It is not widespread in river reaches of the low deserts and appears to be restricted from high-velocity zones in bedrock canyons.

This species may compete directly with Frémont cottonwood with respect to germination and establishment. Compared with cottonwood, Russian olive has a wider range of conditions under which germination occurs and a lower mortality of seedlings.[62] Unlike tamarisk, it prefers soils with relatively low salinity.[63] This species grows better in open conditions, but, unlike tamarisk, it can germinate and establish in shade.[64] Its low tolerance of salinity may be another reason it is sparse in the lower-elevation desert areas, where alluvial soils of the riparian zone tend to be salty. Despite its heavy colonization of riparian areas, in 1984 many state agencies—except in Utah—continued to offer this nonnative species for erosion control and to create windbreaks and wildlife habitat.[65]

Other Species

Tree-of-heaven, also known as garbage tree,[66] is a large tree, with a maximum height of 80 feet or more, that is native to China.[67] This fast-growing plant appears to be restricted to roadsides and riparian areas at elevations between 2,600 and 6,000 feet. It commonly moves from landscape settings to wildlands in Utah and Arizona. In Utah, tree-of-heaven is present in the Virgin River basin;[68] in Arizona, it is present along the Verde River but does not occupy large areas of riverine habitat. Despite its restricted distribution, it is considered a "problem species."[69] For example, it was recently eradicated from Grand Canyon.[70]

Tree tobacco, a native of Argentina and Chile,[71] became naturalized in the Southwest, perhaps during the period of Spanish occupation. Widespread in disturbed areas or along ephemeral rivers, it reportedly was established in California in 1848.[72] It is locally dense along the rivers of central and southern Arizona. In the Namib Desert, tree tobacco is considered to be a major invasive species.[73]

Giant reed, a species of grass native to Europe, was planted as windbreaks, then escaped to become common along irrigation ditches in the Southwest.[74] This species is uncommon in riverine riparian settings; it occurs mostly adjacent to standing or slowly moving water, and one is more likely to find this species in a horticultural setting or along the sloughs or "lakes" of the lower Colorado River. As discussed in chapter 3, it is difficult to differentiate giant reed from carrizo grass, except when the two are in flower.[75] Tracy and DeLoach have written an excellent treatise on the impact of giant reed, a minor invasive species in the Southwest.[76]

Several other species have at times raised concerns about changes in riparian ecosystems. Siberian elms, planted as ornamentals, are now occasionally found along rivers in the southern Colorado Plateau.[77] To date, these species have not created a significant problem. Various nonnative herbs and grasses originally from cultivated sites in northern Arizona have become established in riparian settings along the Colorado River in Grand Canyon.[78] Some, including camelthorn, are noxious and arise from perennial rootstocks. These species are not readily identifiable in photography, and their distribution in the region cannot be reconstructed using repeat photographs.

Life-History Characteristics of Woody Nonnative Species

As is evident in the species descriptions, the nonnative species considered to be problematic in Southwest riverine ecosystems have many different characteristics. Because tamarisk is considered the largest threat, it has been extensively studied[79] and represents an example of nonnative species' general behavior. Tamarisk quickly colonizes bare or disturbed soil, either through vegetative propagules or seeds; Athel tamarisk is apparently reproductively sterile, and vegetative propagation is its only means of dispersal. Branches of both tamarisk species readily root, making flood transport of propagules a significant mechanism for the spread of these species.[80] The ability to propagate vegetatively makes both species of tamarisk ideal colonizers on barren surfaces along rivers, particularly after floods.

The major question some researchers have posed concerns whether tamarisk can disrupt natural processes,[81] particularly owing to its wider range in environmental tolerances than most native species. Tamarisk can grow in a wide range of soil-salinity conditions and is able to withstand higher water stress than most native species, thereby increasing its potential for taking over riparian ecosystems.[82] It occupies so much riparian habitat in the Southwest now that some have declared it to be a permanent type of riparian unit in the region.[83] Tamarisk's life-history strategy explains a great deal of its invasiveness and ability to dominate riparian zones. Its deep roots, its ability to survive in seemingly waterless environments, and its physiological capability to withstand more xeric conditions than native species have convinced some to propose that it is a facultative riparian species.[84]

Seed Production and Germination

Tamarisk produces prodigious amounts of seed.[85] The numbers are staggering: one study indicates that a density of one hundred viable seeds is produced per square inch of ground from April through October.[86] Put another way, a mature tamarisk can produce in excess of a billion seeds per year.[87] Experimental work indicates that seed viability may be lower than 33 percent,[88] which is offset by the massive seed production. The period of viability is short, typically two to four months,[89] but this reproductive restriction is offset by the extremely long period of flowering and seed production (April through October, depending on location).

Tamarisk's small seeds bear a feathery plume that allows them to float in the water or air for long distances. Tamarisk has even become established in an isolated volcanic crater in northwestern Sonora, Mexico, which has an extremely arid climate and is 9 to 10 miles from the nearest seed source.[90] The presence of tamarisk around isolated springs in the Mojave Desert indicates wind-borne seed transport as well.[91]

Seeds can germinate in the warm season within twenty-four hours on wet sands,[92] and carpets of tamarisk seedlings are common following the annual flood of the Colorado River in southern Utah. Tamarisk has a low percentage of germination during the winter months,[93] a factor of great importance to the timing of flow re-

leases specifically designed to thwart its expansion. Germination and establishment in the wild is complicated by the short period of seed viability, which precedes the winter rainfall in the Middle East, where some species of tamarisk are native.

Tamarisk seeds can germinate in saltier substrates than native species. The native arrowweed and non-native tamarisk have an advantage over other native species on floodplains where salt buildup occurs,[94] either as a result of evaporation of saline water, irrigation practices that accumulate salts, or addition of salt exudates from tamarisk leaves. Mature tamarisk can tolerate as much as 35,000 parts per million (ppm) of soil salinity.[95] If periodic flooding is eliminated by flow regulation, and rainfall is low, floodplain soils can accumulate sufficient salts to retard or prevent seed germination of native species, which typically can tolerate only up to 1,500 ppm of soil salinity. Tamarisk has a broad range of salinity tolerance, making its functional role in riverine ecosystems context specific and variable.[96] One experimental study did not verify the presence of seed-toxic substances beneath Athel tamarisk.[97]

Vegetative Reproduction

Both tamarisk and Athel tamarisk can reproduce vegetatively, which is an extremely important characteristic for expansion during floods. Branches torn from mature trees can root in saturated alluvium. Growth of adventitious roots, which allow vegetative (asexual) propagation, results from the ability of surficial stem cells to propagate into roots under high-humidity conditions.[98] Survival of these propagules is much lower than for individuals established from seeds,[99] but this process provides a potent mechanism for expansion of these species. In contrast, Russian olive is not dependent on flood disturbances[100] and tends to grow on floodplains away from high-velocity channel margins.

Growth and Salinity Tolerance

Tamarisk is known for its fast-growing seedlings, which elongate at a rate of up to 2 inches a day[101] or 12 feet per

year.[102] Roots can reach a depth of 30 inches in ten weeks under experimental conditions.[103] Although cattle will eat young tamarisk, its rapid growth eventually eliminates grazing animals from thickets.[104] In addition, tamarisk creates dense thickets that exclude light from the understory, reducing the possibility for germination and growth of some other species, in particular cottonwood, but not necessarily coyote willow.[105]

Tamarisk appears to produce annual rings in trunks and roots reliably, and burial of stems results in changes in ring anatomy that blur annual growth.[106] This burial response shows a shifting of cellular growth from stem to root and reflects this species' ability to reproduce vegetatively. Because of its annual-ring production, tamarisk provides sedimentologists with a method for dating mid- to late-twentieth-century deposition along rivers in the region.[107]

Tamarisk excretes salts onto its leaf surfaces,[108] and litterfall contributes to the salinity of soil surfaces. Most of the excreted salts are sulfates, and tamarisk leaves are high in sodium, magnesium, and calcium.[109] Tamarisk can tolerate as much as 8,000 ppm of total dissolved solids,[110] indicating a high tolerance of groundwater salinity. The amount of sodium chloride excreted is a function of groundwater salinity.[111] Therefore, in certain cases, particularly where groundwater salinity is high and overbank flooding is minimal, tamarisk can modify its environment to inhibit germination of native species.[112] However, if soils get too salty, even tamarisk will not germinate, leaving barren soils.

Individual trees have long, downward-branching tap roots (an average of 7.5 feet, with a maximum of 35 feet) that extend into the saturated zone.[113] Although tamarisk becomes established in any type of substrate (it is found growing from bedrock in some canyons), it is more likely than native species to grow within areas of finer-textured floodplain alluvium.[114] Lateral root spread has been described as minimal in clay-rich soils[115] and as complex in sand dunes.[116] Tamarisk roots do not necessarily compete with the roots of other, more shallow-rooted species, although competition for water between tamarisk and black willow has been documented.[117] Downward root growth of tamarisk is rapid, and seedlings are known to survive drops in groundwater levels that native species cannot survive;[118] however, even tamarisk can be killed by large drops in groundwater level.[119] Despite extended periods of dryness, tamarisk quickly responds to the renewed presence of water,[120] although some studies suggest that it does not use summer rainfall, at least in the Mojave Desert.[121]

Because of its deep, isolated roots and preference for sandy soils,[122] tamarisk may be more vulnerable to destruction during floods than are seepwillow, arrowweed, or coyote willow, which have shallow, interconnected root systems.[123] This vulnerability may be one reason why tamarisk tends to be sparse or absent from high-velocity zones in bedrock canyons[124] or occurs in protected sites behind rocks or bedrock projections.[125] In alluvial reaches, banks exposed to higher velocities are mantled with willows, whereas tamarisk generally grows in backwaters and the upstream edges of eddies.[126] Native species can tolerate more inundation than tamarisk, suggesting that controlled flooding along regulated rivers may reduce tamarisk's competitive advantage.[127] Tamarisk reportedly can withstand inundation for as long as seventy days,[128] although many individuals were killed by extended inundation along the Colorado River between 1983 and 1986.[129]

Carbon Balance

Because of its fast growth rate and different life-history strategy, tamarisk may alter the carbon balance in riverine ecosystems. Cottonwood produces more litter than tamarisk,[130] which suggests that conversion of dense native vegetation to tamarisk may slow carbon cycling. For riparian settings with sparse native vegetation, the reverse may be true. Because of its high leaf-litter production, tamarisk may overwhelm carbon production and decomposition rates in desert streams.[131] These changes propagate through the aquatic food web, indicating that tamarisk may suppress native aquatic insects and fish and potentially promote other nonnative aquatic species.[132]

Water Use

As described in chapter 1, water use by nonnatives, in particular tamarisk, has caused considerable concern owing to the value of water for domestic supplies, agricultural use, industrial use, and in-stream flows.[133] Tamarisk water use is reportedly high enough to dry up certain springs;[134] it has long been known that shallow water tables along streams fluctuate as much as a foot diurnally in response to tamarisk water use.[135]

Early measurements of transpiration by tamarisk were extraordinarily high.[136] When measurement is corrected for leaf-surface area, it shows that a tamarisk tree generally uses about the same amount as or less water than native species (table 3.1),[137] despite some claims to the contrary.[138] It depends on how one measures water use, but tamarisk water use is within the range of variation of common native species, although its water-use efficiency is higher.[139] Transpiration is a function of season, daily temperature and solar radiation, soil salinity, wind speed, elevation, and elapsed time since rainfall.[140] Water use is also highly variable, even within stands that appear to be relatively homogeneous.[141]

Water use by tamarisk is affected by depth to water[142] and floodplain salinity;[143] variability among individuals is high and dependent on stand density and canopy overlap.[144] However, the larger surface area of tamarisk leaves increases water usage by about 30 percent.[145] For the Gila River upstream from San Carlos Reservoir, evapotranspiration from tamarisk-dominated vegetation averaged 43 inches per year and up to 56 inches per year in dense stands; removal of tamarisk lowered the evapotranspiration to 14–26 inches per year, depending on the replacement species.[146] In New Mexico, water use in tamarisk stands ranged from 29 to 48 inches per year, which is comparable to mature cottonwood stands and mixed stands of native and nonnative vegetation.[147]

Flood Tolerance

Tamarisk has a relatively low resistance to damage during floods, particularly if high-velocity flow impinges on the

plants. In mechanical tests, it has been found to have about the same flexibility as black willow,[148] and although young individuals can be laid over in the flow, older plants are destroyed, either as a result of substrate mobilization, which undermines the roots, or buildup of debris on the upstream side, which increases flow resistance. Along large alluvial rivers, such as the Green River (chapter 6) and the Colorado River (chapter 7), undercutting of low deposits induces bank failure and undermines both tamarisk and coyote willow. As noted earlier, floods can induce both vegetative propagation of tamarisk through deposition of stems in wet sediments and, if floods occur in summer, germination and establishment from seeds.

Frost Tolerance

Tamarisk is sensitive to extreme minimum temperatures.[149] This limitation is most significant where minimum temperature is less than about -4°F,[150] which, in the region discussed in this book, occurs at the northernmost margins and at elevations above about 5,000 feet. Individuals at colder sites have a greater biomass in roots to support winter survival.[151] The occurrence of Russian olive, in contrast, decreases with increasing minimum temperatures,[152] which helps to explain why this species is uncommon in the low deserts of Arizona and California.

The Effects of Nonnative Woody Vegetation

Tamarisk and Salt Accumulations on Floodplains

Tamarisk is known to increase the salinity of floodplain sediments through dissolution of salts accumulated on leaves and leaf fall. As previously discussed, the increase in floodplain salinity may be one way that tamarisk modifies its habitat to ensure its long-term dominance. Although this increase may occur where flow regulation prevents overbank inundation, as in the lower Colorado River, it may not be effective on rivers where the flood regime is intact. The effects of tamarisk removal on the alluvial aquifer's water quality may be masked by the overwhelming influence of water-quality changes induced by streamflow floods and evapotranspiration.[153] This result suggests that periodic leaching by floods may reduce floodplain salinity to "natural" levels, eliminating tamarisk's competitive advantage. It also indicates the importance of controlled flooding on regulated rivers if native species are to be encouraged.

Nonnative Vegetation and Channel Narrowing

As is shown later in this book, areas of woody riparian vegetation are generally wider than they were at the time of the first photographs taken of rivers in the region. This increase in area is not the result of increase in alluvial deposits with proximity to groundwater; instead, several environmental and human-caused changes—in particular arroyo downcutting and decreases in flood discharges—created new streamside habitat, leading to channel narrowing. Channel downcutting resulted in water-table declines, followed by deposition of low terraces within the channel walls (chapter 2), creating valley bottoms without vegetation and with unsaturated sediments. Tamarisk is a major reason for this change because it colonized the newly open banks, trapping sediment and building stable floodplains.[154] Alternatively, it may simply have colonized newly available habitat created by the narrowing process and caused by reduction in flood control and sedimentation.[155]

Whatever the reason, the widening of the riparian zone at the expense of the river has narrowed channels throughout the region, stranding remaining native vegetation far from the river's edge.[156] The process can be as simple as stabilization of point bars, followed by tamarisk encroachment and sediment trapping,[157] or as complex as the conversion of a formerly wide and deep channel to a braided system as a result of the complex interaction between flood control and riparian vegetation encroachment.[158] As discussed in chapter 3, ripple effects minimize the original native species' ability to reproduce or even to survive. Clearly, nonnative riparian species—especially tamarisk—played a major role in this change, but abiotic factors, in particular reduction in flood discharges, cannot be discounted.

Nonnative Species and Insect Populations

In some river reaches, native insects avoid tamarisk, resulting in a lower diversity on tamarisk than on willow.[159] However, the density of insects, especially nonnatives such as honeybees, can be significantly higher on tamarisk than on native vegetation.[160] Early studies on the Canadian River in Oklahoma and Texas showed that insect numbers were higher in tamarisk stands than on equivalent native vegetation.[161] Fifty-eight insect species were captured in one study in Arizona aimed at finding natives that might serve as biological controls; one species of leafhopper (*Opsius stactagallus*) was omnipresent on tamarisk.[162] The increase in tamarisk in the region is thought to have promoted Apache cicada (*Diceroprocta apache*) populations, thereby attracting birds.[163]

Tamarisk represents a paradox for populations of native insects. Its flowering period is longer than that of comparable native species, thus providing a longer period of resource availability to insect populations. However, only adult insects use tamarisk; no earlier life stages have been found on these trees, indicating either that insects have not adapted to the presence of these trees or that tamarisk trees are simply not preferred for egg laying or larval growth.[164]

Nonnative Species and Bird Populations

Avian use of either tamarisk or native species is species specific and dependent on a number of factors. Along the lower Colorado River, ground feeders, granivores, and species with food sources outside the riparian zone prefer, use, or at least do not avoid tamarisk, whereas most other birds—in particular insectivores and frugivores (fruit and berry consumers)—avoid tamarisk.[165] Other studies suggest that birds use tamarisk more than native arrowweed.[166] In the tamarisk-dominated riparian zones of some rivers, such as the middle Rio Grande and the Pecos Rivers in New Mexico, summer use by avian insectivores is about the same as for equivalent native riparian vegetation.[167] The number of bird species that use either cottonwood or tamarisk stands is similar; the differences lie in species composition. The thin branches of tamarisk minimize or eliminate use by specialists, such as woodpeckers and raptors; the latter group generally cannot penetrate the dense thickets in search of food. Some bird species that nest in willows and cottonwood may use tamarisk less than they use comparable stands of native species.[168]

The situation changes with increasing structural diversity brought about by mixtures of native and nonnative vegetation, which occurs more frequently in less saline soils or at higher elevations.[169] In these settings, many birds use and nest in tamarisk habitat, including the Southwestern Willow Flycatcher, Mourning Doves, and Black-chinned Hummingbirds.[170] For example, the Black-chinned Hummingbird, which is common in the southwestern United States, nests almost exclusively in tamarisk trees along the Colorado River in Grand Canyon.[171] Both Mourning Doves and White-winged Doves nest in tamarisk along the lower Gila and Colorado Rivers.[172] Lucy's Warbler and Yellow Warbler preferentially use insects specific to tamarisk in Grand Canyon.[173]

The Southwestern Willow Flycatcher is an unobtrusive bird that has been substantially reduced in its historic range in riparian areas of southern California because of extensive urban development.[174] This flycatcher has moved eastward into dense riparian vegetation, including tamarisk.[175] This change has resulted in the unusual situation of an endangered subspecies utilizing a nonnative species. For example, Southwestern Willow Flycatchers preferentially nest in tamarisk in Grand Canyon and along other river systems in northern Arizona.[176] The attraction, deemed neutral to beneficial for this species, appears to be the increased and more reliable insect populations present because of the longer flowering season.[177] Use of the nonnative species may have a significant cost; flycatchers that live in tamarisk have smaller body biomass, lower fat stores, and reduced density when compared to birds in adjacent willow and cottonwood stands.[178]

The Southwestern Willow Flycatcher and its dependence on riparian forests provide another interesting example involving an endangered species. During the post-1880s period of arroyo downcutting, large areas of primary sparrow grassland were converted to mesquite bosque or cottonwood-willow forests.[179] Populations of sparrows, such as Bairds' and Botteri's, experienced sharp declines.[180] Two types of sparrow, therefore, have become species of concern following grassland losses, and restoration of their habitat would require conversion of woody to herbaceous riparian vegetation, likely at the expense of obligate birds such as the Southwestern Willow Flycatcher.

When Russian olive was first introduced, wildlife managers specifically recommended planting it to enhance animal populations. Unlike tamarisk, this nonnative species creates habitat that is comparable to stands of native species. Monospecific stands of Russian olive have intermediate populations of bird species when compared with native stands of riparian vegetation and upslope communities.[181] At least fifty species of birds and animals use Russian olive for food or cover.[182] In some areas, bird populations may benefit from Russian olive because it adds an extra element of vertical structure to communities with few species in the height range between coyote willow and cottonwood.

Nonnative Species and Wildlife Populations

It is frequently stated that nonnative species, in particular tamarisk, create habitat that excludes wildlife.[183] For example, early site-specific studies indicated that bird species avoid tamarisk in favor of native vegetation, specifically in low-desert river reaches, except where tamarisk is the only species available.[184] The presence of tamarisk has altered many aspects of riparian zones in the Southwest, especially through the formation of seemingly impenetrable thickets that increase difficulty in accessing water by bighorn sheep, deer, and other large mammals.

Compared with many native species, tamarisk lacks palatable fruits and seeds, and, as a result, few animals forage in tamarisk thickets. Minimal structural diversity, at least compared with cottonwood-willow stands, was thought to reduce bird use of tamarisk for cover and nesting. More recent work suggests that wildlife usage of this nonnative species is a much more complicated function of elevation, the number and type of other species present, and available habitat.

Larger animals may be excluded from the densest thickets. Vertebrates, in general, avoid dense vegetation with little structural diversity, whether it is native or nonnative. Reptile density and diversity is higher in native vegetation than in tamarisk stands.[185] As discussed in chapter 3, beaver populations may have had a major impact on native riparian vegetation before widespread trapping reduced their numbers. In some areas, beaver prefer native vegetation over tamarisk as a food source; however, along parts of the Colorado and San Juan Rivers, they consume or use tamarisk along with native species.[186] According to one study, they may actually encourage growth of tamarisk and Russian olive by culling cottonwood at a disproportionately higher rate, reducing the competition effects on nearby nonnative species.[187]

Control of Nonnative Species

Control of nonnative species is predicated on numerous assumptions, among them that restoration is possible and that a conversion from nonnative to native species is both possible and desirable.[188] Some have argued against control of tamarisk on the grounds that the extent of human modification of riparian systems is so great that restoration projects are doomed to failure.[189] A naturalized nonnative species such as tamarisk, with its potential for recolonization, may be impossible to eliminate unless all individuals are removed from the region. A long-term monitoring plan to find and eradicate seedlings is of paramount importance to tamarisk control because direct elimination of every individual is impossible and the remaining individuals produce huge

quantities of seed. Complete removal is extremely unlikely.

Several methods have been proposed to check the spread of nonnative species or even to control or eliminate the populations. Control methods can be grouped into several categories: restoration of more-natural hydrologic processes; spraying or injection of herbicides; burning, cutting, and uprooting plants; biological control; and natural replacement by native species.[190] The latter method occurs throughout the Southwest, and the other active control methods have been applied in various places at various times.

Artificial Floods

Restoration of hydrologic processes is frequently cited as the most appropriate control mechanism.[191] Artificial floods with discharges significantly higher than ordinary dam releases would theoretically scour nonnative vegetation, potentially favoring native species. Restoration of floods would also provide open sediments for germination of native species—in particular cottonwood—and leach salts from floodplains. Seasonal fluctuations in flow that mimic natural conditions would reduce the hydrologic certainty under which tamarisk has assumed dominance in many areas. The problem with hydrologic restoration is that tamarisk continues to remain even along rivers that have had large natural floods in the region. Also, flood damage creates open habitat that tamarisk can readily colonize, particularly following summer floods, which raises the specter that tamarisk naturalization cannot be reversed or even impeded by natural processes.

Herbicide Applications

At various times, chemical control mechanisms have been used to kill mature tamarisk. In the 1950s, the now banned herbicide 2–4T was broadly applied to tamarisk thickets in Arizona and New Mexico. In 2003, tamarisk control in New Mexico included aerial spraying of an herbicide.[192] Application of herbicides, followed years later by burning, has been shown to kill most of the tamarisk in the treated area.[193] However, significant negative side effects are possible with herbicide application, in particular aerial application, ranging from death of native riparian species to spread of carcinogens into water supplies.

Controlled Burning

Because of the high amount of fine fuel present in the form of dry leaves and twigs, fire commonly sweeps through tamarisk thickets,[194] suggesting the possibility that this species may create a disturbance that perpetuates its dominance. Tamarisk resprouts quickly following burning, thus eliminating burning as a primary control option.[195] The presence of tamarisk has increased the frequency and intensity of fires along the river corridor because its fine foliage ignites easily, and the thick litter layer beneath the trees burns readily. Moreover, deep ash, salty soils, vertical and horizontal stranding from the water table, and thick, newly sprouted tamarisk may prevent or inhibit the fast return of native plants.[196] In fact, fire—whether natural or human caused—promotes tamarisk, at least initially. Used with either herbicide treatments or mechanical removal, fire aides eradication of tamarisk by removing its considerable biomass from the landscape.

Mechanical Removal

The most common means of tamarisk eradication, especially within national parks, is cutting or uprooting by crews with rudimentary hand tools. Cutting of above-ground branches typically occurs, but some devices allow whole plants to be pulled from the ground. Disposal of slash is the largest problem with this technique, and plants will resprout if herbicides are not applied to the freshly cut stumps. Bulldozers have been used over wide areas outside of parks and natural areas. Bulldozing tamarisk stands, an expensive alternative, has been attempted and failed on several occasions.[197] Because tamarisk can vegetatively resprout, mechanical removal must get all plant parts out of the soil or be repeated for maximum success. Herbicide application, whether painting of stumps or direct injection, is considered a necessary part of this control

technique.[198] Broad-scale application of herbicides has occurred in the past,[199] but it is now severely restricted owing to negative environmental effects, particularly to wildlife.

Control of Russian olive consists of direct removal of plants by bulldozers, tractors, or hand.[200] Because this species resprouts after cutting, eradication of it requires a concerted effort. Stumps are either burned or sprayed with an herbicide in a manner similar to tamarisk control, and control might require several seasons or years of persistent effort.

Biological Control

Biological control currently is the most promising active method for suppressing nonnative species, in particular tamarisk. Natural biological control might occur if beaver populations were widely reestablished; small rodents, such as pocket gophers, also are known to kill tamarisk.[201] With the exception of a few leaf-sucking aphids, no native insects eat or harm tamarisk. Biological control efforts focus on a future program of nonnative insects to control giant reed[202] and an extensive current program for tamarisk.[203]

Biological control of tamarisk using nonnative insects, especially the leaf beetle (*Diorhabda elongata deserticola*), has been demonstrated as feasible and is claimed to be safe to nontarget species.[204] Control would be accomplished by releasing insects that specifically target tamarisk without harming native plant or animal species. The leaf beetle is able to defoliate adult tamarisk within a season, kill them within a few years, and refrain from eating native species.[205] It is most effective against the invasive tamarisk, but much less so against Athel tamarisk.[206] If biological control is enacted, scientists expect gradual increases in cottonwood-willow habitat as the colonization of tamarisk is slowed.[207]

Several factors possibly will mitigate the expected negative effects of biological control of tamarisk. First, not all tamarisk would be killed, and, second, those trees heavily attacked would take several years to die, thus allowing time for native species to begin the replacement process. In reaches with natural flood regimes, the most likely outcome

would be recolonization by native shrubs, in particular coyote willow.

Biological control has been proposed for decades, and the nonnative insects are only now being released in selected isolated areas. None of these release sites is in the region considered in this book. The recent studies concerning bird use of tamarisk, in particular use by the endangered Southwestern Willow Flycatcher, have postponed this control mechanism in some areas. In addition, the potential side effects that the release of any new, nonnative organism might have on native plants have raised concerns about biological control. No biological control has been proposed for Russian olive or for most of the other common nonnative species in riparian areas of the Southwest.[208]

Competition with Native Species

In a river with some semblance of a natural flooding regime, the mere passage of time may be an effective control mechanism as native species replace nonnative ones.[209] Under conditions when floods are timed with periods of seed availability, seedlings of plains cottonwood and coyote willow will outgrow seedlings of tamarisk when all three species germinate during the same event.[210] For some reaches of perennial rivers in the Southwest, coyote willows are increasing on the streamward side of riparian stands despite the adjacent tamarisk thickets.[211]

Willows are more likely than tamarisk to establish along banks adjacent to higher-velocity flow, especially if disturbances, such as floods, occur when more willow than tamarisk seed is available. Coyote willow competes directly with tamarisk, but the interrelation between the two species may result in increased numbers of willow and decreased numbers of tamarisk.[212] Coyote willow reproduces vegetatively, is shade tolerant, and can become established under mature tamarisk trees. As old tamarisk trees die, they may be replaced by native willow in some areas. If increasing biodiversity does serve as a barrier to the entry of nonnative species,[213] then tamarisk may be excluded (and reduced) as native species increase.

At the northern edge of tamarisk

in Wyoming, shading by plains cottonwood is thought to limit tamarisk expansion.[214] Frémont cottonwood appear to be replacing established tamarisk along the San Pedro River in southern Arizona (chapter 19) in a process described as "self-repair" of Sonoran Desert riparian systems.[215] Similarly, black willows are increasing within tamarisk thickets in Grand Canyon (chapter 11), and cottonwood are doing the same thing along the Virgin[216] and Bill Williams Rivers (chapters 13 and 15). Established tamarisk trees do poorly in the full shade of larger trees, notably cottonwood. They appear to be slowly dying under closed cottonwood galleries. One possible control mechanism would be to create openings in tamarisk thickets artificially, then plant and cultivate cottonwood or other trees of stature larger than tamarisk.

The ultimate control of tamarisk comes from the same mechanism that has eliminated native stands of riparian vegetation. Complete diversion of surface water, combined with high usage of groundwater, will kill tamarisk or Russian olive. At various times, death of tamarisk owing to water development has occurred, most notably near Gila Bend, Arizona, in the 1950s.[217] An extreme method for conversion of monospecific stands of tamarisk to native vegetation would involve dewatering of an aquifer to kill the tamarisk, followed by restoration of water levels and streamflow that can sustain planted native species.

5 Repeat Photography and Riparian Vegetation Change

Summary. Repeat photography has been used extensively to document landscape changes in the southwestern United States. Photography in this region began in 1863 and became extensive in the late nineteenth and early twentieth centuries. The large number of historical photographs of gaging stations, in particular, provides a basis for interpretation of change in riparian vegetation. Unlike analysis of aerial photography or satellite-based remote sensing, this method provides data on long-term, species-specific changes in riparian vegetation. Within the largest collection of repeat photography in the world, 3,067 historical photographic comparisons have been matched that show changes in a variety of riparian settings in the region. These photographic comparisons, while biased toward reaches with gaging stations and limited in terms of spatial representation, provide evidence of change that can be interpreted for ephemeral and perennial rivers in the region.

The history of landscape photography in the western United States is, in one sense, the history of government-sponsored expeditions to assess the potential for landscape development. As the science and art of photography developed in the mid–nineteenth century, application for its use in the West arose as proposed transportation routes required verification of feasibility.[1] Railroad and steamship operation provided the first need for photographic documentation, accompanied by the ever-demanding needs for scientific illustration of the landscape. In the process, thousands of photographs were taken of riparian ecosystems in the Southwest, providing ample fodder for repeat photography.

The First Photographs of the Southwest

The art of photography began in 1839, with the announcement by Louis Jacques Mandé Daguerre that he had created the first image recorded on silver iodide media.[2] The images became known as daguerrotypes, and the technology was eventually used in our study area. The famous explorer John C. Frémont pioneered photography on the Colorado Plateau during his 1853–1854 expedition, although the daguerreotypes exposed by his photographer, Frederick von Egglofstein, were destroyed in a fire at the Frémont residence.[3] Lieutenant Joseph Ives took von Egglofstein with him on his 1857 steamboat trip up the lower Colorado River, but no images were obtained in the United States.[4] Ives also brought an artist, Heinrich Balduin Möllhausen, whose watercolors provide the first vivid (although somewhat inaccurate) images of the lower Colorado River (chapter 27).[5]

In 1851, a major advance in photographic media fueled the future of nineteenth-century photography. The wet-plate, or collodion, process uses a glass plate coated with a transparent emulsion and dipped in a solution of silver nitrate. Before it dries, the plate is quickly placed in a camera and exposed.[6] The sensitivity to light decreases to nearly nothing if the plate is allowed to dry before exposure, meaning that these nineteenth-century photographers were bound to a darkroom tent as well as to a water source. Therefore, most of the earliest photographs were taken of water bodies, including riverine riparian systems.

When Arizona became a territory in 1863, it had no registered photographers.[7] The first known surviving photographs of the Southwest were taken during an expedition of the California Geological Survey from San Bernardino to Needles, California, in the winter of 1863. Traveling downstream along the Mojave River before traveling the old Mojave Road to the Colorado River, photographer Rudolf D'Heureuse captured images of the Mojave River (chapter 28) and the Colorado River at Fort Mohave (chapter 27).[8] The latter photograph is the earliest known surviving landscape photograph taken of Arizona and its riparian systems, and the whole group is the oldest known set of photographs of the Southwest still in existence.[9]

Photographers slowly entered the desert Southwest with various agendas. Some, like D'Heureuse, wanted to document landscapes and mining claims. Others required scientific documentation to accompany their descriptions of the region and its resources. A third group worked to demonstrate the feasibility of transportation routes. William A. Bell[10] was in this latter group and served as the photographer of an expedition seeking a southern railroad route across New Mexico and Arizona. In 1867, he photographed Aravaipa Canyon (chapter 20), a tributary of the San Pedro River.[11] Both D'Heureuse and Bell used the collodion process on glass plates, and, as with early photographers in general, most of their camera stations are located close to perennial water supplies.

Two expeditions and some freelance photographers vied for the next landscape photographs taken of the region. At about the same time in 1871, Lieutenant George Wheeler and Major John Wesley Powell, leaders of competing topographic surveys, pioneered the use of the camera as a tool for scientific documentation. Wheeler had Timothy O'Sullivan take photographs in Death Valley and western Grand Canyon, while Powell had E. O.

Beaman photograph the Green and Colorado River canyons upstream from Lee's Ferry.[12] O'Sullivan lost most of his glass plates in a stagecoach robbery near Wickenburg, Arizona, late in 1871,[13] but most of Beaman's glass plates have survived. William Bell replaced O'Sullivan as the Wheeler Expedition photographer in 1872, then O'Sullivan returned in 1873.[14] Similar changes occurred in the Powell Expedition, as Beaman left in January 1872, to be replaced first by Clem Powell, then by James Fennemore, and finally by John K. "Jack" Hillers.

Hillers, who became the official photographer of the U.S. Geological Survey, was one of the most prolific nineteenth-century photographers of the Colorado Plateau.[15] In the publications of the Powell Survey, many of his photographs were used to illustrate canyon topography, plateau landscapes, geologic features, or Native Americans, but usually they were rendered into lithographs for ease of publication.[16] In some cases, the original photograph no longer is available in archives, and the lithograph was interpreted instead (see chapter 10, Moenkopi Wash).

Although Hillers was trained in the field, the other photographers of this period owed their initial training in some way to either Matthew Brady, noted Civil War photographer, or William Henry Jackson, who ran a photographic business in Denver.[17] Jackson was a pioneer landscape photographer in the West, working for the Hayden Survey in 1870 and 1871 north of our study area. He ventured into the San Juan River country in 1875.[18] One of Jackson's protégées, George Benjamin Wittick, photographed in Havasu Canyon and the canyon of Diamond Creek in 1873 and 1875,[19] and some of his images ended up in Jackson's collection, giving the false impression that Jackson went to Grand Canyon as well. Trading of landscape images was pervasive in the late nineteenth century, adding confusion to who took what photographs when.

The period 1880–1900 spawned a veritable explosion of photography in the Southwest. Dry plates that did not require field preparation were introduced, making field photography easier and eliminating the need to set up a darkroom for every photograph.[20] Carleton E. Watkins, noted photographer of Yosemite in the 1860s,[21] visited Arizona in the 1880s and took a few photographs. Surveyors then assumed photographic dominance, led by Robert Brewster Stanton's railroad surveys along the Green and Colorado Rivers in 1889 and 1890.[22] Stanton's photographer, Franklin A. Nims, pioneered the use of flexible emulsion film in the region, replacing the heavy glass plates and eliminating the need for a field darkroom. George Roskruge photographed southern Arizona while surveying for the General Land Office in the 1890s.[23] As a result, every subsequent survey, exploration, or scientific party carried a camera as standard equipment.

Photographic images of Arizona riparian areas from the early twentieth century are widely available. As part of the regular operation of surface-water gaging stations in Arizona, U.S. Geological Survey hydrographers took thousands of photographs to document channel conditions. The photographs show many features of the channel that are important to accurate streamflow gaging, including shifts in channel thalwegs (the deepest point in the channel), changes in low-water controls, and changes in channel roughness. Photographs have been taken since gaging stations were first established in Arizona in the 1910s and continued to be taken as gages were established in the 1920s and 1930s.

Incidental replication of views occurred as early as the fall of 1872, when both the Powell Expedition and the Wheeler Expedition photographed extensively in Kanab Canyon.[24] Repeat photography as a scientific tool began in 1888 as a technique to document changes in European glaciers.[25] In 1911, brothers Ellsworth and Emery Kolb began the long-standing practice of repeat photography in our region by replicating views of the Green and Colorado Rivers originally taken by Beaman and Hillers.[26] In Africa in the late 1950s, Homer Shantz made the first large-scale application of repeat photography to the question of vegetation change.[27] Since 1960, repeat photography has been used extensively to document vegetation change in various parts of the United States[28] as well as Africa.[29]

Equipment for Original and Repeat Photography

During the period encompassed by the original photography used in this study, both cameras and film evolved radically. Pioneer photographers invariably used *large-format cameras* (4-by-5-inch images or larger), some capable of taking 11-by-14-inch images on glass plates and others designed for stereographic images on 8-by-10-inch media. Many twentieth-century photographers used *medium-format cameras* (2¼-by-3¼-inch or 6-by-7-centimeter images or slightly larger) equipped with flexible roll film. Cameras using the 35-millimeter film format gradually gained favor with photographers in the middle of the twentieth century. The result is a large variation in the resolution of the original images as well as differences in the captured visible spectrum.

Hillers, who photographed over a wider area than most pioneer photographers, inherited his equipment from Beaman. He primarily used a stereoscopic camera, which produced stereoscopic 4-by-5-inch images on an 8-by-10-inch sheet of glass, but he also had an 11-by-14-inch large-format camera that produced a very high-resolution single image. Either camera could use a *normal lens*, with a focal length equivalent to a 50-millimeter lens on a 35-millimeter camera, or a *wide-angle lens*, matched by modern lenses with focal lengths of about 24 to 28 millimeters on a 35-millimeter camera. Before around the turn of the twentieth century, lenses did not have shutters; to expose the negative, the lens cap was removed and then replaced after a timed period. O'Sullivan's equipment was similar to that used by Hillers. Owing to the weight of glass plates, these expeditions carried thousands of pounds of photographic equipment, and typically either a wagon or a boat was devoted to the photographer.[30]

Because Nims worked for Jackson, he was fully aware of the state of the art in cameras and film available in 1889. He chose an Albion camera, with a format of 6½ by 8½ inches, as his primary tool, and he equipped it with shutterless, normal, and wide-angle lenses.[31] To shed weight and save space, Nims chose to use newly available roll film that could be developed after the trip. Whether glass plates or roll film was used, however, the photographic media was *orthochromatic*, and these blue-sensitive emulsions, when properly exposed for typical landscapes, severely overexposed the sky. As a result, the images look significantly different from modern *panchromatic* black-and-white films.

The central challenge in nineteenth-century photography was to decrease weight while enhancing the ability to create reproducible positives. Glass plates, although heavy, were stable and created minimal distortion. Negatives on flexible film could be printed several ways, depending on the film base. For paper-stripping film, used in the 1880s, the easiest method was to oil the paper, making it translucent but decreasing the film's longevity. The highest-quality negatives required stripping the emulsion from the paper, a delicate and laborious process, and mounting the emulsion on glass or translucent plastic. Development of flexible, transparent film in 1889 created the ideal combination of low weight and permanent translucence, although it took until the mid-1890s to work flaws out in the original design.[32]

Because of the remoteness and rough conditions encountered on the pioneering expeditions, photographers had to resort to some extreme measures to make sure their images came out of the field. Glass plates were fragile, and all the users of collodion media lost negatives to breakage, either in the field or subsequently. Especially during river travel, extreme measures were taken to keep negatives dry and secure. Powell had waterproof compartments built into his boats to store food, scientific instruments, and photographic equipment. Stanton soldered his exposed film into tin boxes and stored duplicate rolls on different boats.[33] Some of the largest problems were with documentation after the fieldwork was completed. Whereas Stanton accurately numbered his two thousand images in downstream order after his expedition ended, Hillers, who both inherited images from his predecessors and produced his own, provided minimal documentation, so, as a result, John Wesley Powell and his associates mislabeled many of Beaman's and Hillers's views.[34]

In the early twentieth century, cameras became smaller as films gained resolution and reliability. With the exception of the Kolb brothers, who stuck with glass plates while at the same time pioneering motion pictures, most photographers used medium- and large-format cameras, usually with a film size of no larger than 4-by-5-inch images. The exposed film rolls, although smaller and lighter, were no less vulnerable to the elements; Raymond Cogswell, photographer for Julius Stone, lost whole rolls of film in a boating accident in Cataract Canyon.[35] The earliest films were cellulose-nitrate based, a flammable substance that created a fire hazard for tightly rolled films. Photographic manufacturers started to replace nitrate film with cellulose-acetate film in the 1920s. This type of film was subsequently improved. Polyester-based films, introduced in the 1960s, are the predominant type in use today.

Proper interpretation and analysis of repeat photography require both knowledge of the equipment and film that the original photographers used and employment of appropriate modern equipment to take the matches. The camera used for repeat photography does not have to be outfitted with a lens of the same focal length as the original camera, however, as some have proposed.[36] Instead, the field of view, which varies with film format (e.g., 35 millimeter versus 4-by-5 inches) and the focal length, is the most important characteristic. We typically aimed for the center of the original view using a lens that meets or exceeds the original field of view.

Although inexpensive and widely available, 35-millimeter cameras and their digital equivalents do not provide the resolution of medium- or large-format cameras in repeat photography. Ability to use film that can be processed in the field is an additional benefit available to cameras with formats larger than 35 millimeter. The development of small, battery-powered printers that print images directly from digital cameras reduces the differences in format. At present, medium- and large-format cameras provide higher film resolution than typical 35-millimeter cameras or the digital equivalents and are more comparable to historical cameras.

To replicate historical views, we used manual medium- and large-format cameras equipped with a variety of lenses. Our medium-format cameras exposed either normal or wide-angle views onto 2¼-by-3¼-inch film in eight-exposure rollbacks. The large format cameras expose 4-by-5-inch sheet film. We used instant film to evaluate the camera station, its orientation, and the exposure setting. Instant film is particularly useful for setting camera heights and angles. When persistent foreground features such as rocks, plants, or fence posts were present in both the original view and in the match, we used a ratio between the distances separating persistent foreground and background features to adjust camera placement.

We used black-and-white negative, color positive, and color negative films to document each view. The black-and-white film we used was primarily TMax 100, which is panchromatic and extremely fine grained, and we occasionally used filters, typically Wratten yellow 8, as a haze filter to increase contrast and bring out background scenery.[37] Color films included Vericolor III and VC negative films and Fujichrome Provia 100 and Velvia 50 transparency films. Color film, not used by most of the original photographers, primarily adds current information for future interpretation of changes. Small-format digital images were obtained for many of the views that were blocked by encroaching riparian vegetation or could now be matched only from boats or were taken from positions now under water.

Repeat photography offers some technical difficulties in closely occupying the original camera position, making an acceptable match, and then assessing changes.[38] In the field, we attempted to relocate the original camera position as accurately as possible. Sometimes this was difficult owing to channel change, obstruction of the view by vegetation growth, lack of background features, or loss of the original camera station owing to road or other construction. When possible, we attempted to duplicate lighting, shadows, and seasonal condition of perennial vegetation as depicted in the original photograph; however, logistics of repeat photography for many sites did not allow us even to duplicate original seasons for many photographs. This limitation was not a major problem because the primary subjects were woody riparian species, which are persistent throughout the year.

All camera stations were assigned a unique number, which we refer to in this book (in the figure captions) as the *stake number.* Locations of camera stations were documented using hand-held global-positioning-system devices (latitude, longitude, elevation, and estimated position error) and usually were marked permanently using rebar stakes, aluminum angle iron, X marks on rocks, or bolts in rocks or cairns. The geometry of the view (camera height, azimuth, and tilt) and the camera settings (shutter speed, f-stop, and film type) were also recorded.[39]

The Legacy of Repeat Photography

Repeat photography has been used to document landscape change, including change in riparian vegetation, in eastern Colorado, the Great Basin, and other areas in the western United States.[40] A 1984 annotated bibliography of repeat-photographic publications describes more than 450 studies undertaken worldwide between 1888 and 1984.[41] Following these early works, several investigators in the American West made extensive use of repeat photography to evaluate vegetation change.

Spatial Distribution of Repeat Photography

The Desert Laboratory Collection of Repeat Photography in Tucson, Arizona, houses the largest collection of historical and replicate images in the world: 6,466 camera stations for historical photographs have been reoccupied, and 9,144 matches at these stations are contained in the collection.[42] Most of the camera stations are in the southwestern United States and northern Mexico (fig. 5.1), although 109 of the camera stations are in Kenya.[43] Of the 5,523 camera stations available for the Southwest, 3,103 sets of repeat photographs show riparian vegetation and are interpreted in this book (table 5.1).

The geographic region with the most repeat photography of it and the most published materials[44] on it is Grand

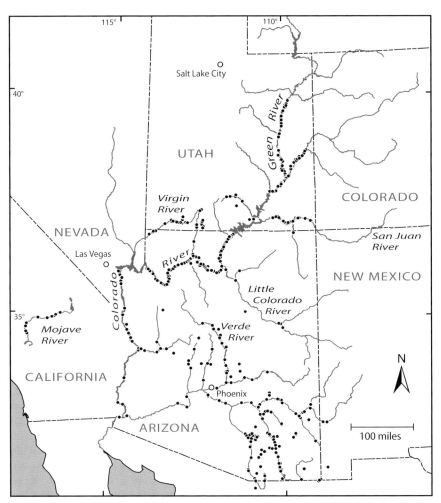

Figure 5.1 Map of the southwestern United States and northern Mexico showing the locations where photographs of riparian vegetation, contained in the Desert Laboratory Collection of Repeat Photography, were taken.

Canyon, with 1,367 camera stations that document riparian vegetation (table 5.1). Less repeat photography is available for other river reaches, but most reaches reported in this book have more than 10 camera stations. The distribution of camera stations varies widely; in Grand Canyon, owing to Stanton and Nims's systematic photography, camera stations are uniformly distributed along the river corridor, whereas for other river reaches (e.g., the Hassayampa River) the only available repeat photography is clustered at gaging stations.

As previously discussed, the earliest photograph that has been matched is the 1863 D'Heureuse view of the Colorado River at Fort Mohave, Arizona (fig. 27.2). As shown in figure 5.2, the number of views that show riparian vegetation is high for the 1890s, when 18 percent of the original photographs were taken. A second peak period of historical photography, in the 1920s, resulted from the estab-

lishment and photography of gaging stations within Arizona as well as from large-scale expeditions conducted by the U.S. Geological Survey. The photographers with the highest number of photographs that we have matched are Stanton and Nims, with a total of 503 photographs that show riparian vegetation (table 5.2). As a group, U.S. Geological Survey field geologists took 608 of the original photographs that we matched, followed by the hydrologic technicians and other members of the Arizona District of the U.S. Geological Survey, who took 581.

Benefits and Limitations of Repeat Photography

The aggregate research shows that repeat photography is an excellent means of documenting many aspects of long-term ecological change, complementing other techniques. Remeasurement of permanent vegetation plots surpasses

Table 5.1 Historical Photographs Matched or Analyzed for Change in Woody Riparian Vegetation.

River	Region	Discussed in Chapter	Number of Matched Photographs	Earliest Photograph Date
Agua Fria River	Central Arizona	25	12	1940
Animas River	Southwestern New Mexico	16	1	1913
Aravaipa Creek	Southern Arizona	20	25	1867
Bill Williams River	Western Arizona	15	42	1923
Cañada del Oro Wash	Southern Arizona	22	12	1912
Chinle Wash	Northern Arizona	8	7	1935
Colorado River	Professor Valley	7	37	1905
Colorado River	Meander Canyon	7	10	1914
Colorado River	Cataract Canyon	7	207	1871
Colorado River	Glen Canyon	11	31	1871
Colorado River	Grand Canyon and tributaries	11	1,251	1871
Colorado River	Lower River	27	92	1863
Escalante River	Southern Utah	9	24	1872
Gila River	Central Arizona	17, 26	191	1908
Green River	Desolation Canyon	6	44	1871
Green River	Gray Canyon	6	18	1871
Green River	Labyrinth Canyon	6	23	1871
Green River	Stillwater Canyon	6	27	1871
Guadalupe Canyon	Sonora, Mexico	16	1	ca. 1885
Hassayampa River	Central Arizona	25	27	1933
Havasu Creek	Grand Canyon	12	79	1885
Kanab Creek system	Utah, Arizona	12	91	1872
Little Colorado River	Northern Arizona	10	43	1923
Mojave River	California	28	37	1863
Paria River	Northern Arizona	9	21	1872
Rillito Creek	Southern Arizona	22	81	1890
Salt River	Central Arizona	24	40	1917
San Carlos River	Central Arizona	18	9	1935
San Francisco River	Southern Arizona	18	18	1891
San Juan River	Southern Utah	8	127	1875
San Pedro River system	Southern Arizona	19, 20	109	1883
San Simon River	Southern Arizona	16	16	1931
Santa Cruz River	Southern Arizona	21	133	1880
Sonoita Creek	Southern Arizona	22	15	1895
Verde River system	Central Arizona	23	63	1917
Virgin River	Southern Utah	13	121	1873
Whitewater Draw	Southern Arizona	16	12	1930
Miscellaneous reaches	Throughout	various	70	1891

repeat photography in accuracy but necessarily is limited spatially; repeat photographs of permanent plots reveal changes beyond the plant transects.[45] Repeat photography is unsurpassed in its ability to provide species-specific changes in woody vegetation for time spans exceeding a century. It also is use-ful for documenting geomorphic and habitat changes that are concurrent with ecological changes.

Despite the many contributions repeat photography has made to studies of landscape change, the technique has some noteworthy deficiencies. Perhaps the most often noted limitation is its myopic view of only a narrow slice of landscape. The need for a clear, if narrow, unobstructed view when matching photographs obviously limits the technique's use in forested habitats or riparian areas that have undergone significant expansion. In these cases, the historical view is obstructed or cameras

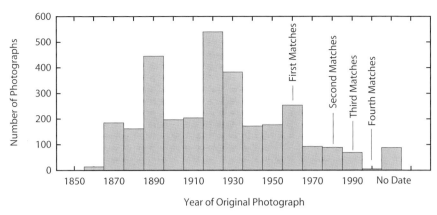

Figure 5.2 Histogram showing the temporal distribution of the original dates of historical photography used to document long-term change in riparian vegetation in the Southwest. *First matches* refers to the decade of the first systematic set of repeat photography in the region (1960s). *Second, third,* and *fourth matches* refer to the approximate decades of time-series documentation of change in riparian vegetation; see subsequent chapters for examples.

Table 5.2 Photographers with Credit for the Most Photographs That Show Woody Riparian Vegetation.

Photographer	Number of Matched Photographs
R. B. Stanton	416
E. C. La Rue	347
J. K. Hillers	118
R. A. Cogswell	88
F. A. Nims	87
E. and E. Kolb	67
J. A. Baumgartner	65
W. E. Dickinson	44
G. Roskruge	37
E. O. Beaman	35
Unknown	294
Others	1,505

must be placed on elevated prominences overlooking the tree-covered landscapes. Even though a forest's interior may be clearly shown in a photograph taken on some early date beneath a forest canopy, that photograph's follow-up after several decades may well show tree branches only inches from the lens of the properly aimed camera. Most applications of repeat photography are understandably in relatively arid regions where the view usually is unobstructed.

Another limitation is the variation in areal representation of change among photographs. Although some photographs from canyon rims can be rectified to calculate changes on a per-unit-area basis,[46] most cannot. In riparian settings, the ability to interpret changes decreases with distance into the views, precluding a per-unit-area interpretation of changes. Although changes cannot be assessed on an areal basis in most cases, they can be evaluated based on what is visible in the view.

In order to broaden the field of view, some researchers have suggested that the technique should be augmented or even replaced by repeat aerial photography.[47] In the southwestern United States, aerial photography first became widely available in the mid-1930s, well after the times when many landscape changes began, and its resolution usually does not allow assessment of changes in species composition or evaluation of presence-absence.[48] The inability to identify accurately plants that appear in the scenes creates problems; for example, when one tree species becomes established beneath another, the new tree ultimately replaces the original at the same location.[49] This situation cannot be identified using aerial photography. Aerial photography complements repeat photography for evaluations of twentieth-century changes, allows extrapolation over relatively long channel reaches, and can be used to assess assemblage changes on a per-unit-area basis.[50] Combination of repeat ground and aerial photography provides species-specific information on changes over short channel sections for long periods of time; many photographs provide a systematic dataset on changes in riparian vegetation within the region.[51]

Multispectral satellite imagery provides an excellent technique for monitoring and evaluating ecosystem changes; however, although satellite imagery provides large spatial coverage and collects visible and nonvisible spectral data, it provides only limited information on changes in species composition and became available only after 1974.[52] Satellite-based imagery provides quantitative data on general changes in riparian vegetation averaged over 30-meter pixels for the period of 1974 to the present without the benefit of species-specific data.[53] Changes documented using satellite imagery thus supplement the results from repeat photography and provide the additional benefit of determining the areal extent of changes (see the discussion of satellite images in chapter 19 in reference to the San Pedro River).

Interpretation of repeat photography to document change in riparian areas has focused on changes in woody species, in particular trees and shrubs. Species recognizable in old photographs include Frémont cottonwood and other obligate phreatophytes, such as black willow, coyote willow, velvet ash, box elder, brickellbush, tamarisk, and Russian olive. Black or Goodding willow is distinct from cottonwood trees because of its smaller size, upright branching pattern, and smaller leaves; coyote willow is generally a shrub. Mesquite forms dense stands, known as bosques, along some desert rivers and has a distinct branch architecture. Tamarisk, ubiquitous in modern riparian views, can be identified in historical photographs by its generally larger height than coyote willow, the feathery appearance of its leaves, and the fine stem architecture, which contrasts with black willow and cottonwood.

Even though plant identification to species is often possible when viewing large perennial plants, we have avoided overreaching the technique's capabilities. In particular, we did not identify most herbaceous plants in the views or interpret changes in this type of riparian vegetation. Grasses and forbes in general are often clearly shown in grassland views or views of disturbed patches, but identification of the plants to species level usually is not possible.

Summary. The Green River across the Colorado Plateau, the major drainage of northeastern Utah, is valued for its water, its aquatic resources, its whitewater and flatwater recreation, and its riparian corridor through a desert region. Main-stem flow is regulated by Flaming Gorge Dam upstream from the Colorado Plateau, but unregulated flow in the Yampa River, which joins the Green upstream from the plateau, provides some natural-flow variability. The channel of the Green River has narrowed owing to the increase of native and nonnative species. Although Frémont cottonwood has generally increased in Desolation Canyon, it appears to be declining along the lower reaches of the Green River. Other native species, in particular coyote willow, have greatly increased, and desert olive, once the dominant species on the banks in Labyrinth and Stillwater Canyons, is now obscured behind a wall of tamarisk.

The Green River drains 44,850 square miles upstream from Green River, Utah, and begins with headwaters in the Wind River Mountains of Wyoming, the northern Rocky Mountains of Colorado, and the Uinta Mountains of northern Utah (fig. 6.1). Here, we consider reaches of the Green River on the Colorado Plateau, specifically between Jensen, Utah, and the Confluence of the Green and Colorado Rivers in Canyonlands National Park. The Green River flows past Jensen and through the Uinta basin, then enters Desolation Canyon downstream of Ouray, transitions immediately from Desolation Canyon into Gray Canyon at the mouth of Range Canyon, flows out into a wide alluvial section in the vicinity of Green River, Utah, and then reenters the bedrock canyons of Labyrinth and Stillwater Canyons before reaching the Confluence. The Green River therefore traverses some of the most complex sedimentary terranes of any river in the Southwest, with geologic formations ranging from Tertiary sandstones, mudstones, and shales to Paleozoic limestones.[1]

Like most other rivers in the Southwest, the Green River has a diversity of uses and valued resources, some of which conflict with others. Flaming Gorge Dam, completed in 1965, produces hydroelectric power and stores water as part of the Colorado River Compact of 1922 among Wyoming, Utah, Colorado, New Mexico, Nevada, Arizona, and California. Utah's water allocation under the compact is taken primarily from the Green River and its tributaries. The river and its major tributary, the Yampa River, have the largest remaining populations of humpback chub (*Gila cypha*), Colorado River pikeminnow (*Ptychocheilus lucius*), and eight other species of native fish.[2] Native fish are threatened mostly by nonnative fish and by the presence and operations of Flaming Gorge Dam (fig. 6.1), which alters water temperature, sediment concentrations, peak discharges, and seasonality of flow. Nonnatives outnumber native fish species by nineteen to ten.[3]

The Green River represents the startling contrast of an arid environment with a perennial river. In 1869, John Wesley Powell had this impression of the first canyon downstream from present-day Jensen: "The walls are almost without vegetation; a few dwarf bushes are seen here and there clinging to the rocks. . . . [W]e are minded to call this the Canyon of Desolation."[4] Most visitors now have difficulty seeing some of that "desolate" scenery through the dense riparian vegetation. Flaming Gorge Dam, since its construction in 1962, has reduced the magnitude of floods and altered the riverine environment of the Green River, inadvertently promoting some elements of the riparian ecosystem and damaging other parts, especially native fish.

Early Observations of Riparian Vegetation

Few historical observations of riparian vegetation were made on the Green River, but the large amount of historical photography associated with river expeditions (see the next section) provides ample evidence of the distribution and abundance of Frémont cottonwood and other common species. In 1938, Norman Nevills made his first commercial river trip, launching from Green River, Utah, and disembarking at Lake Mead. His passengers were botanists Elzada Clover from the University of Michigan and her graduate assistant Lois Jotter. They made observations at two sites in Labyrinth and Stillwater Canyons.[5] Their plant list shows a sparse riparian ecosystem consisting of Frémont cottonwood, seepwillow, coyote willow, peachleaf willow, and tamarisk.

Beaver and river otter attracted trappers and were noted by members of river expeditions. Enigmatic trapper Denis Julien, whose inscriptions carved in bedrock walls are the only documentation of his Green River travels in 1836, sought the pelts of these animals.[6] At least three crew members of both Powell Expeditions, in 1869 and 1871, reported seeing or killing beaver and river otter for food upstream from and within Desolation Canyon.[7] Legendary trapper and boatman Nathaniel Galloway plied the waters of the Green seeking beaver pelts between 1895 and 1909.

History of Photography of the Green River

Because of its importance as a transportation corridor, the canyons of

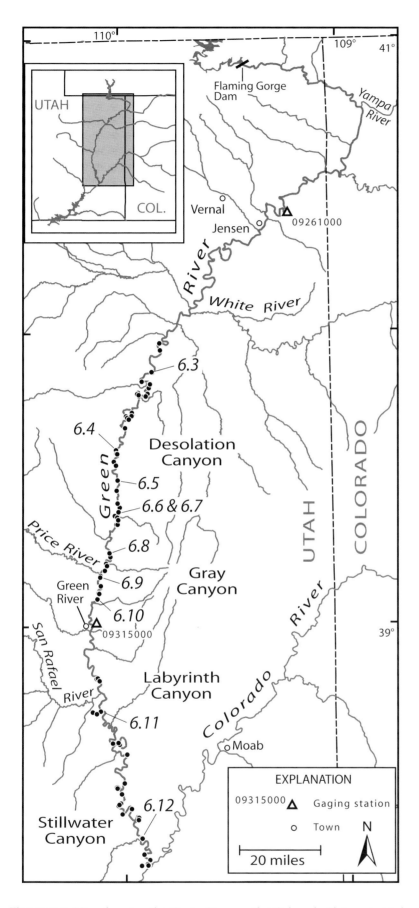

Figure 6.1 Map showing the Green River on the Colorado Plateau in Utah.

the Green River have been frequently photographed. E. O. Beaman, the photographer of the second Powell Expedition in 1871, took numerous views of Desolation, Labyrinth, and Stillwater Canyons; for some reason, none of his views shows Gray Canyon. Franklin Nims, photographer of the Brown-Stanton Expedition in May 1889, captured views of Labyrinth and Stillwater Canyons. In 1909, Raymond Cogswell, photographer for the Stone Expedition, took the largest number of photographs of the four canyons. U.S. Geological Survey expeditions, notably the ones in 1922, also provided useful imagery that documents changes in riparian vegetation.

Repeat photography of these canyons began with Emery and Ellsworth Kolb in 1911. They used Powell's book *Canyons of the Colorado* as a river guide and replicated Beaman's views at several sites, including Bowknot Bend.[8] Scientific use of historical photography began with the Stephens-Shoemaker trip of 1968; Hal Stephens replicated many Beaman photographs.[9] Repeat photography has since been used to document channel narrowing on the Green River,[10] change in riparian vegetation,[11] and changes in habitat that may affect native fish.[12] A substantial increase in riparian vegetation occurred between 1922 and 1936, with smaller increases from 1936 to 1963, and substantial increases from 1963 through 1993.[13]

Floods, Flow Regulation, and Channel Change

The annual flow volume for the gaging station Green River at Green River (fig. 6.1) is 5.3 million acre-feet, which corresponds to an average daily discharge of 6,200 cubic feet per second (ft^3/s). Tree-ring reconstruction of flow at this gaging station, from A.D.1570 through 1895, indicates that the long-term annual flow volume is 4.48 million acre-feet;[14] the higher flow volumes of the gaging record reflect the unusually high runoff from 1906 through the early 1930s. The highest flow volume, in around A.D. 1621, was 9 million acre-feet; other years with high annual flow volumes were 1843, 1853, and 1841, with 7.48, 7.23, and 6.71 million acre-feet, respectively.

Floods on the Green River are not as large as those on the Colorado River (chapter 7), but they are still a major force in shaping the channel and riparian ecosystems on the Colorado Plateau. At Jensen and at the town of Green River, the largest floods in the annual flood series were 40,000 and 68,100 ft³/s in 1984 and 1917, respectively (fig. 6.2). At Green River, the mean annual discharge and two-year flood values have decreased by 28 and 30 percent since the middle of the twentieth century, in part because of Flaming Gorge Dam operations, but the influence of climatic fluctuations cannot be discounted. The history of peak discharges shows a decrease in flood magnitude beginning around 1928.[15]

The annual sediment load of the Green River at the Green River gaging station is 15.6 million tons, with a range from 1.8 to 43.4 million tons in 1934 and 1937, respectively.[16] Construction of Flaming Gorge Dam is thought to have reduced the sediment load by about 35 percent. However, climatic fluctuations, notably in summer precipitation, may also have affected sediment delivery to the river.[17]

Flaming Gorge Dam, completed in November 1962, was the first of the dams authorized under the Colorado River Storage Act of 1956. Located near the Utah-Wyoming border, this 502-foot-high dam stores water and generates hydroelectric power. Although Flaming Gorge Dam completely regulates the Green River downstream from its base, the Yampa River, an unregulated tributary with headwaters in the northern Rocky Mountains of Colorado (fig. 6.1), provides some natural-flow variability.

Three factors—flow regulation, climatic fluctuations, and the introduction of tamarisk—have contributed to channel narrowing on the Green River. The combination of repeat photography and aerial photography suggests that the channel in Labyrinth and Stillwater Canyons has narrowed by about 25 percent.[18] In the vicinity of the Green River gaging station, the channel narrowed by 5 percent between 1930 and 1940, at the end of the period of high flood frequency, and another 14 percent after 1959, which may reflect mostly the effects of flow regulation by Flaming Gorge Dam.[19]

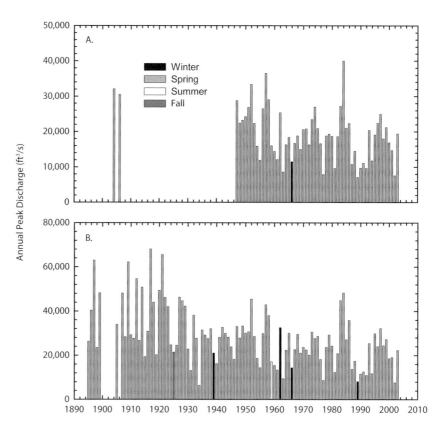

Figure 6.2 Annual flood series for the Green River.

A. The Green River near Jensen, Utah (station 09261000; 1904, 1906, 1947–2003).

B. The Green River at Green River, Utah (station 09315000; 1895–1899, 1905, 1907–2003).

Changes in Riparian Vegetation

We have matched a total of 112 photographs of the Green River, including 44 of Desolation Canyon, 18 of Gray Canyon, 23 of Labyrinth Canyon, and 27 of Stillwater Canyon. They include matches added to those made by the Kolb brothers in 1911 and by Stephens in 1968. In general, riparian vegetation has increased in all of these canyons, especially tamarisk, but also coyote willow and other native species. Frémont cottonwood has a variable history of change in these reaches that might reflect a response to flood control by Flaming Gorge Dam.

Desolation Canyon

In Desolation Canyon, riparian vegetation has increased at nearly all of the camera stations (86 percent), with large increases apparent in twenty-two views and decreases apparent in only one. This increase likely follows the introduction of tamarisk and the reduction in flood size. Frémont cottonwood is visible in three-quarters of the photograph matches; it has increased in about one-third of the views, has decreased in about one-tenth of the views, and either is about the same or had variable changes in one-third of the views. Crown damage, which may herald future dieback, is apparent in about nine photograph matches.

Downstream from Jensen, the Green River meanders through the Uinta basin and past Ouray before entering a canyon with low walls (fig. 6.3). The channel gradient is low and bottomlands are common; a diversity of woody riparian species lines the river in this reach. Upstream from Sand Wash, which is considered to be the river-running start to Desolation Canyon, groves of Frémont cottonwood are common, and cottonwood has increased in many photographs taken of the whitewater reaches (fig. 6.4). Cottonwood has increased on debris fans at the head of rapids (fig. 6.5), and both coyote willow and tamarisk may also be present, giving these sites a structural diversity that may be attractive to bird populations.

Figure 6.3 Photographs showing the Green River at Long Bottom (river mile 99.6).

A. (October 1, 1909.) This downstream view on the Green River upstream from Sand Wash shows a grove of Frémont cottonwood and box elder trees. The three boats of the Stone Expedition of 1909 are moored against the barren bank in the foreground. The far bank appears to be nearly vertical and sparsely covered with what probably is coyote willow. (R. A. Cogswell 97.24. 212, Stone Collection, courtesy of the Cline Library.)

B. (November 5, 1999.) The original camera station is about 15 feet into the view in the river because the bank used for the original camera station in 1909 has eroded. Despite this shift, it is clear that the channel has narrowed significantly here and that riparian vegetation has changed. The view is now partially blocked by tamarisk. The fate of the groves of trees shown in photograph A cannot be determined because the increased vegetation on the left bank obscures the former riverbank. (S. Young, Stake 3955.)

A

B

C

Figure 6.4 Photographs showing the Green River at Fretwater Falls, Desolation Canyon.

A. (August 14, 1871.) This view across Fretwater Falls, a relatively minor rapid in Desolation Canyon, shows a debris fan in the foreground and Frémont cottonwood trees upstream from the rapid. Utah juniper and pinyon pine dot the slopes in the midground. (E. O. Beaman 538-2, courtesy of the U.S. Geological Survey Photographic Library.)

B. (August 6, 1968.) Cottonwood have become established on the debris fan, partially blocking the view. The trees are damaged, suggesting that this site may be a marginal one for this species. The pinyon and juniper have increased on the slope in the background. (H. G. Stephens T-4.)

C. (October 11, 1998.) The number of cottonwood in the immediate foreground has decreased, possibly because of a large flood on the Green River in 1984, which may have removed them. However, another cottonwood's branches enter the view from the right, and still more appear beyond the persistent tree on the left. The pinyon and juniper continue to increase on the far slope. (D. Oldershaw, Stake 1346.)

Figure 6.5 Photographs showing the Green River at Three Canyon in Desolation Canyon.

A. (October 7, 1909.) This upstream view across the top of Three Canyon Rapid, a relatively minor riffle, shows a grove of Frémont cottonwood trees. These trees appear to have been damaged in the past and to have resprouted from near the base. The low shrubs beyond the bare sand around the cottonwood are likely coyote willow. (R. A. Cogswell 97.24. 243, Stone Collection, courtesy of the Cline Library.)

B. (October 14, 1998.) The cottonwood have died, and the fallen trunks of the tree in the foreground of photograph A appear on the left side. Tamarisk blocks most of this upstream view. Riparian species now present here include rubber rabbitbrush, skunkbush, and tamarisk. (D. Oldershaw, Stake 3751.)

Islands are common in Desolation Canyon, and they appear in ten of the forty-four historical photograph matches (fig. 6.6). Seven of these islands have an increased vegetation cover, changing from barren to a high density over most of the surface. The colonizing species generally are a combination of tamarisk and coyote willow, although other species, such as seepwillow, may also be present. At least one of these islands has aggraded vertically, a common process on the Green River.[20] Possible channel narrowing was observed in nineteen pairs of repeat photographs, but channel widening could be seen in ten pairs. Although channel narrowing has occurred in most reaches downstream from the town of Green River,[21] narrowing is not as common in Desolation Canyon. One study indicates that channels have narrowed by 19 percent in Desolation Canyon.[22]

Historical photography suggests that bare sand was once a common feature in Desolation Canyon. Now, riparian vegetation is established on most of the formerly exposed sand and coarser-grained deposits (fig. 6.7). The combination of channel narrowing and island attachment to the banks inevitably leads to a reduction in the area backwaters, thereby reducing potential habitat for young native fish.[23] Riparian vegetation encourages this reduction by trapping sediments and reducing flow velocities along the banks, causing what appears to be a visible increase in the amount of fine-grained sediment along the narrowed river (fig. 6.7B).

Gray Canyon

Desolation and Gray Canyons meet in an area of low hills. Coal Creek Rapid is the most significant drop in either canyon, and several historical photographs captured this whitewater reach (fig. 6.8). The magnitude of increase in riparian vegetation is similar to that observed upstream, although changes in Frémont cottonwood may be more variable. At Coal Creek Rapid, persistent cottonwood appear in a setting of increased riparian shrubs.

Change in cottonwood is highly variable in Gray Canyon. A grove of cottonwood that once graced the mouth of the Price River has disappeared, although at least three trees persist over an eighty-nine-year period (fig. 6.9). At Gunnison Butte, the end of Gray Canyon, a once-barren terrace now supports a dense stand of tamarisk and coyote willow with a new small grove of cottonwood behind (fig. 6.10). As for Desolation Canyon, riparian vegetation has increased in most of the repeat photography of Gray Canyon, although change in cottonwood is more variable.

Labyrinth Canyon

The Green River meanders through the wide valley at the town of Green River, where woody riparian vegetation has increased historically.[24] Tamarisk initially was planted in towns before escaping cultivation, either by wind-borne or water-borne seeds or branches that can take root. Although dense stands of tamarisk grow along the river in the vicinity of the town of Green River, native species are common here as well. Downstream, the river abruptly enters a bedrock canyon with low walls that gradually increase in height. Although Labyrinth Canyon begins just downstream from the bridges south of town, the meandering that inspired its name begins at Trin-Alcove Bend (fig. 6.11), where Navajo Sandstone constrains the valley width. The channel has narrowed as low bars and islands have attached to the banks, and riparian vegetation has increased.[25]

Labyrinth Canyon is 82 river miles long, and over most of that distance the channel is lined with either narrow terraces or wide bottomlands. Photographs taken between 1871 and 1909 suggest that the most common species on the banks were coyote willow and skunkbush in the upper half of the canyon and coyote willow and desert olive along the lower half. Many individuals of skunkbush and desert olive persist along the banks, hidden behind the wall of tamarisk. Gambel oak remains common, and many have persisted in dense thickets over the past century, notably upstream from Trin-Alcove Bend (fig. 6.11).

Although cottonwood has increased in side canyons, such as Three Canyon in Trin-Alcove Bend, fewer of these trees are now found along the river corridor.[26] Lower-statured species, such as tamarisk and coyote willow, have increased. Previous research suggests that riparian vegetation either induces[27] or follows[28] channel narrowing. The time series of photographs suggests that colonization follows channel narrowing, and vertical aggradation occurs after plants become established and trap sediments. The presence of riparian vegetation inhibits lateral bank erosion.

Stillwater Canyon

Downstream from Mineral Bottom, the Green River flows through relatively open terrain before entering Stillwater Canyon. Like Desolation and Gray Canyons, the separation point between Labyrinth and Stillwater Canyons is obscure. Perhaps the best point to divide the two canyons is the contact between Mesozoic and Paleozoic strata, which occurs at about mile 36, where the White Rim Sandstone arises several miles upstream from Millard Canyon. Stillwater Canyon is the last canyon of the Green River, extending to its confluence with the Colorado River (fig. 6.1).

The name "Stillwater Canyon" comes from the slow velocity of flow there, which is related to debris-flow aggradation of rapids downstream in Cataract Canyon. The rapids serve as weirs that raise the bed elevation, backing water upstream and encouraging deposition of fine-grained sediment in the channel. Tributary channels, which are graded to the Green River but at a lower bed elevation, are flooded with fine-grained sediments deposited overbank during floods. This environment is conducive to the formation of bottomlands, which typically have high groundwater levels and thriving assemblages of riparian vegetation. Because of the high water tables, evaporation is also high, leading to high salinity in soils and a tendency for more salt-tolerant species—such as greasewood, fourwing saltbush, and tamarisk—to grow.

Overlooked in the concern over nonnative species along the Green River are the changes in native species. Desert olive, which is prominent in old photographs (fig. 6.12A), still mark the

Figure 6.6 Photographs showing the Green River at Joe Hutch Canyon in Desolation Canyon.

A. (1922.) This upstream view from a ridge towering over river left of the Green River shows the debris fan at the mouth of Joe Hutch Canyon *(left side)*. Two emergent patches of sand appear as islands in the view. A broad, nearly barren island appears on the left side of the river, with a narrow channel separating it from the left bank. A second, smaller patch of sand is completely barren upstream in the left-hand bend. Frémont cottonwood are apparent throughout the view, and thick patches of what appears from this distance to be coyote willow are on the downstream side of the island and on the left bank before it disappears behind the foreground cliff. Other than several isolated cottonwood, the right bank is largely free of woody riparian vegetation. (R. R. Woolley 32245, courtesy of the Marriott Library.)

B. (October 15, 1998.) The right bank now has a continuous band of riparian vegetation consisting of tamarisk, coyote willow, and the occasional cottonwood or box elder tree. The smaller island of 1922 is now fully vegetated with coyote willow, but remains detached from the bank. The lower, larger island of 1922 is now connected to the left bank except at high river flows and is fully vegetated with mostly coyote willow and a line of tamarisk at the downstream end. (S. Walton, Stake 3663.)

Figure 6.7 Photographs showing the Green River downstream from Joe Hutch Canyon in Desolation Canyon.

A. (October 7, 1909.) This view, from the banks of an eddy below Joe Hutch Rapid, shows an open grove of riparian trees along an otherwise barren bank with a low gravel bar adjacent to the river. The boats of the Stone Expedition of 1909 are parked in this eddy against a bare sandbank. These trees appear to be young Frémont cottonwood, although some might be box elder, and the shrubs at the downstream end of the reach might be skunkbush. (R. A. Cogswell 97.24.253, Stone Collection, courtesy of the Cline Library.)

B. (October 15, 1998.) The gravel bar is now mostly covered with sand, and riparian vegetation has increased in the view. The foreground sandbar now supports coyote willow on its upslope edge. The species that have increased the most across the river are cottonwood, coyote willow, and tamarisk, and several of the cottonwood trees present in 1909 appear to persist. (S. Walton, Stake 3664.)

Figure 6.8 Photographs of Coal Creek Rapid, Gray Canyon.

A. (October 9, 1909.) Coal Creek Rapid is the most significant whitewater on the Green River in Desolation and Gray Canyons. Large rocks in the river, as shown throughout this downstream view across the river from the right bank, create holes at higher water levels that are notorious for flipping boats. Riparian vegetation here is sparse, although several small Frémont cottonwood are present on the near bank in the right foreground. Several cottonwood appear in the hazy distance downstream from the rapid on river left. (R. A. Cogswell 97.24.275, Stone Collection, courtesy of the Cline Library.)

B. (October 17, 1998.) Riparian vegetation, in particular coyote willow with one tamarisk, blocks most of the original view. In the view shown here, the camera station was raised to show at least part of the rapid over the foreground coyote willow. The cottonwood on the right side persist and are not much larger than they were in 1909. Riparian vegetation—mostly coyote willow with a line of tamarisk behind—has colonized the once-barren debris fan across the river, and one of the cottonwood in the distance persists. (D. Oldershaw, Stake 3757x.)

A

B

Figure 6.9 Photographs showing the confluence of the Price River and the Green River in Gray Canyon.

A. (1909.) This upstream view across the Green River shows the mouth of the Price River on the left. At this time, an open grove of Frémont cottonwood was established on the narrow bottomlands on river left and across the river on both sides of the Price River; the shrubs in the foreground are greasewood and skunkbush. The river is not visible, but its barren banks appear on the left side. (M. O. Leighton 160, courtesy of the U.S. Geological Survey Photographic Library.)

B. (October 18, 1998.) The river has migrated to the left and now is clearly visible in the view, and a dirt road that dead-ends 2 miles upstream crosses the view. Three large skunkbushes grow on the river side of this road at left. The three largest cottonwood on river left *(center)* persist, but those that once decorated the mouth of the Price River are gone, replaced with tamarisk and coyote willow thickets. Bare sand is generally not found in the vicinity now; most of the substrate has been colonized by riparian vegetation. (T. Klearman, Stake 3709.)

A

B

Figure 6.10 Photographs showing the Green River at Gunnison Butte at the mouth of Gray Canyon.

A. (October 16, 1911.) This view across the Green River at the mouth of Gray Canyon shows Gunnison Butte, a prominent landmark upstream from Green River, Utah. The photograph was taken by either Emery Kolb or Ellsworth Kolb from a moving boat, and one side of the second boat of the trip appears at lower right. The far bank *(right)* of the Green River is mostly bare, with a few scattered shrubs that might be either saltbush or greasewood. A low island begins at right center and extends to the left out of the view. (Kolb 568–1064, courtesy of the Cline Library.)

B. (October 17, 1998.) Channel narrowing, which is common on the Green River, is obvious here, and the former island is now connected to the right bank. Tamarisk forms a barrier on the bank, with coyote willow on the riverward side and several cottonwood behind *(right)*. Another gravel bar appears on the extreme right that may reflect water lower at the time of this photograph than it was in 1911. (D. Oldershaw, Stake 3764.)

A

B

C

Figure 6.11 Photographs showing the Green River at Trin-Alcove Bend in Labyrinth Canyon.

A. (September 8, 1871.) During his second expedition in the summer of 1871, Major John Wesley Powell and his crew climbed up through a side canyon and crested the inside wall of the bend at Trin-Alcove. E. O. Beaman's upstream view shows the river corridor, confined within Navajo Sandstone, approaching this bend. The river corridor is wide with emergent sandbars. The riparian vegetation includes coyote willow *(nearest the water surface)* and box elders *(taller trees);* the dark clumps back from the terrace edge likely are Gambel oak. Solitary Frémont cottonwood are barely discernable in the distance. An island covered with coyote willow indicates that channel narrowing has begun in this reach. (E. O. Beaman 728, courtesy of the National Archives.)

B. (August 17, 1968.) The island has merged with the bank. The right bank *(left side)* is now fully vegetated with a mixture of tamarisk and coyote willow; the tamarisk with its feathery branches contrasts with the other species along the water edge. Groves of box elder are now more prominent, and the round patches of Gambel oak are particularly obvious. The surface of the floodplain, which appears to be covered with low shrubs in 1871, now has greasewood and saltbush. At least one cottonwood has died, but more can be seen in the distance of this clearer photograph. (H. G. Stephens V-7.)

C. (October 9, 1999.) The tamarisk along the right bank are much taller in this view, and the presence of emergent sandbars near the right bank suggests that channel narrowing, apparent in the 1968 view, continues. (W. Graf [1978] documented channel narrowing along the Green River using a combination of repeat and aerial photography.) The box elders have grown, as has the dense coverage of shrubs and low trees on the floodplain, making foot traffic on this bottomland extremely difficult. (R. H. Webb, Stake 3892.)

A   **B**

Figure 6.12 Photographs showing the Green River in Stillwater Canyon.

A. (September 13, 1871.) This downstream view shows a characteristic scene in Stillwater Canyon. The camera station is 14.5 miles upstream of the confluence of the Colorado and the Green Rivers, and the walls are limestones of the Honaker Trail Formation. The continuous band of vegetation on river right *(right side)* is a combination of coyote willow and desert olive. (E. O. Beaman 743, courtesy of the National Archives.)

B. (October 14, 1999.) The camera station for the original view is blocked by tamarisk and coyote willow, and floodplain deposition has rendered the camera height at less than one foot above the ground, so the camera has been moved toward the river and into the open. The channel has narrowed, and the wall of vegetation—much taller than in 1871—is tamarisk. Desert olive continues to mark the old bank along the lower Green River, and coyote willow is also abundant, particularly on the formerly barren bank on river left in the distance. (C. Schelz, Stake 3911x.)

line of the old channel banks. Fires are now common along the Green River, and most are started by careless visitors burning toilet papers or allowing unattended fires to rage out of control. Dense tamarisk (fig. 6.12B) provides ready fuel for hot, fast-moving fires, and vegetation over entire bottomlands has been consumed in the flames. A recent fire at the mouth of Jasper Canyon burned tamarisk and desert olive. Although many plants died, many individuals of both species resprouted afterward.

Summary. The Colorado River upstream from its confluence with the Green River is only partially regulated by dams on the Gunnison and Dolores Rivers. The annual spring flood still results from snowmelt runoff from the Rocky Mountains. The channel has narrowed in some reaches, but the narrowing is less than along the Green River, possibly because annual flood sizes on the Colorado River have not decreased as much. Riparian vegetation has increased in this reach, with tamarisk and coyote willow accounting for most of the increase. Frémont cottonwood have declined, possibly because of fire and because overbank flooding is reduced. Downstream from the Confluence with the Green River, regulation minimally affects flow through a whitewater reach, and netleaf hackberry, tamarisk, and coyote willow have increased. Lake Powell inundates the lower end of Cataract Canyon, creating a delta in a narrow bedrock canyon dominated by tamarisk.

The Colorado River enters the state of Utah near the geometric center of the Utah-Colorado border and abruptly flows into the bedrock-controlled reach of Westwater Canyon. After exiting that narrow defile, the river flows through a broad alluvial valley, then gradually becomes more confined within bedrock walls as it passes through Professor Valley. At Moab, Utah, the river flows a short distance across the broad Spanish Valley, where significant and valued stands of riparian vegetation are present, before entering another bedrock canyon at the Portal. The river is confined in the colorful Meander Canyon in Canyonlands National Park downstream to the Confluence of the Green and Colorado Rivers (fig. 7.1). The river upstream from the Confluence is the least regulated of any reach of the Colorado River in the semiarid or arid Southwest and sustains a thriving riparian ecosystem along its flatwater sections.[1] It also once had a unique riparian assemblage containing desert olive, a species with restricted distribution that also occurs along the Green River (chapter 6).

Downstream from the Confluence, the river enters Cataract Canyon, a reach of significant whitewater that is the last free-flowing section upstream from Lake Powell. Cataract Canyon contains another of the Southwest's unique riparian assemblages, which consists primarily of coyote willow and netleaf hackberry. Large stage changes during the annual flood, combined with high velocities adjacent to the banks, create harsh conditions for most riparian species, including tamarisk. Several "parks" have formed in wider reaches, allowing limited establishment of Frémont cottonwood, box elder, and peachleaf willow.

From Professor Valley through Cataract Canyon, the Colorado River represents a patchwork of management that affects riparian vegetation. Some reaches have large amounts of development along the banks, ranging from guest ranches and subdivisions to recreational boat ramps to the mining operations at Potash. The reaches through Canyonlands National Park are managed for their near-wilderness quality, and three endangered fish— humpback chub, Colorado River pikeminnow, and bonytail (*Gila elegans*)— are present in Cataract Canyon.[2] At Moab, the Matheson Wetlands, owned by The Nature Conservancy, is a magnet for birds and aquatic wildlife. Finally, the river and its riparian resources contrast with the red rocks of the Mesozoic and Paleozoic sections, providing the classic scenery of the Colorado Plateau.

Observations of the Riparian Ecosystem

Surprisingly, there are few historical observations of riparian vegetation and associated wildlife along the upper Colorado River, either upstream or downstream from its confluence with the Green River. The first botanical survey of Cataract Canyon was conducted by Elzada Clover and Lois Jotter in 1938, who found only sparse riparian vegetation.[3] Historical photographs provide the only information about the distribution of riparian vegetation before 1938.

Clearly, beaver attracted trapper Denis Julien to Cataract Canyon in 1836.[4] In 1869, John Sumner of the first Powell Expedition saw river otter tracks near the Big Drops in Cataract Canyon;[5] noted trapper Nathaniel Galloway sought beaver in Cataract Canyon in the 1890s and 1900s and found few. A river otter was recently sited in this canyon, and beavers remain common along all the reaches of the upper Colorado River.[6] Muskrat is also known from these reaches, which makes this river and the Green unique because the three primary aquatic mammals in the region remain in their waters.

History of Photography of the Colorado River

The Colorado River has long reaches that have little photographic documentation and shorter reaches that have a rich photographic history.[7] No nineteenth-century photographs are available for Meander Canyon because

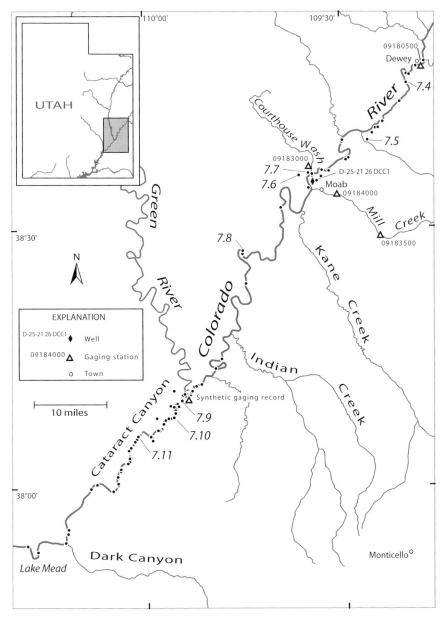

Figure 7.1 Map of the Colorado River from the Utah-Colorado border through Cataract Canyon, Utah.

Nims of the Brown-Stanton Expedition followed in May and June 1889, and Raymond Cogswell took numerous photographs in 1909, although many were lost in a boating accident at Rapid 18. The Kolb brothers matched Beaman photographs and made new views in 1911, and they were boatmen on the 1921 U.S. Geological Survey Expedition. In addition to Emery Kolb, the photographers for this expedition included Sidney Paige and Eugene C. La Rue, and the presence of so many photographers resulted in the largest collection of photography of the upper Colorado River from a single year.

Floods and Flow Regulation

The Colorado River provides the largest volume of water for irrigation and domestic supplies in the Southwest, and its main stem provides more water than the Green and San Juan Rivers, the two other primary sources entering the Colorado in Utah (chapters 6 and 8). Its headwaters receive about 50 inches of annual precipitation, and the precipitation near the Cisco gaging station is about 8 inches.[8] The prodigious floods in this river result from snowmelt runoff in May and June (fig. 7.2), and the annual peak discharge is highly correlated with the annual flow volume. After 1950, annual peak discharges have decreased by 29 to 38 percent near the Utah-Colorado border,[9] and snowpacks throughout the West are melting earlier in the spring, resulting in an earlier date for the annual flood.[10] Although dams partially regulate flow in this watershed, we cannot separate the effects of flow regulation from interannual climate variations in the annual flood series.

Two methods using tree rings have been employed to estimate prehistoric flow in the Colorado River. Netleaf hackberry trees growing in Cataract Canyon use river water, producing annual tree rings that respond to variation in flow volumes.[11] Conifers in the headwaters, which grow annual rings proportionally to the amount of precipitation stored in snowpacks, have been used to estimate annual flow volumes at Lee's Ferry.[12] Both methods use correlation to relate ring width to river flow measured at gaging stations,

early river expeditions did not use it as a transportation corridor. Instead, the first photographs of this canyon were taken by Eugene C. La Rue in 1914. By the 1950s, photographs were commonly taken, particularly during river trips with a science mission, and we have replicated ten of them. Upstream in Professor Valley, several U.S. Geological Survey geologists and hydrologists captured expansive views of the river corridor between around 1905 and 1930. Local residents took photographs in the vicinity of Moab,

primarily to document new bridges or other constructed features near the river. In all, we have forty-seven repeat photographs of the Colorado River upstream from the confluence. We found few useful historical photographs of Westwater Canyon and the 16-mile reach upstream from Dewey Bridge.

For Cataract Canyon, 208 matches of historical photographs are available. E. O. Beaman of the second Powell Expedition took the first photographs here in September 1871. Franklin

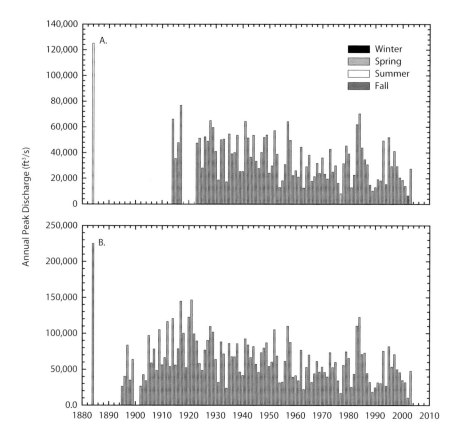

Figure 7.2 Annual flood series for the Colorado River.

A. Colorado River near Cisco, Utah (station 09180500; 1884, 1914–1917, 1923–2003).

B. Colorado River in Cataract Canyon. This record (station 09315005; 1895–1899, 1902–2003) is developed by summing the daily discharges for the Colorado River near Cisco (photograph A), the Green River at Green River (fig. 6.2B), the Dolores River near Cisco (station 09180000), and the San Rafael River near Green River, Utah (station 09328500).

and flow volumes are estimated for the length of the tree-ring records. The flow reconstruction using netleaf hackberry, which begins in 1853, shows that the highest annual flow volume was in 1869, and years with known high peak discharges (e.g., 1884 and 1921) had significantly lower runoff volumes.

The annual flow volume for the Colorado River near Cisco was 6.71 million acre-feet from 1906 through 2003.[13] Use of headwater tree-ring variation to estimate flow at this station resulted in the estimate of an annual volume of 5.85 million acre-feet.[14] As with the Green River (chapter 6), the twentieth-century flow volume is higher owing to the above-average runoff from 1906 through 1930. From A.D. 1640 through 1905, the largest annual flow volumes in 1720, 1868, 1791, and 1869 were 9.39, 9.12, 8.56, and 8.22 million acre-feet, respectively.

The 1884 flood, which peaked at 125,000 ft³/s at the gaging station near Cisco and 225,000 ft³/s in Cataract Canyon,[15] occurred in July but actually was a snowmelt runoff flood. This flood, the largest known for the upper Colorado River, occurred in a year with an annual flow volume (11.2 million acre-feet) that is low considering the magnitude of the annual peak. In the twentieth century, the largest annual peak discharges near Cisco and in Cataract Canyon were 76,800 and 147,000 ft³/s in 1917 and 1921, respectively. In contrast, the average daily flows near Cisco and in Cataract Canyon were 7,260 and 12,900 ft³/s, respectively, for 1918 through 2003.

The Colorado River was named for its "reddish" (colorado in Spanish) appearance, which refers to the large sediment load it carries. At the Cisco gaging station, the annual sediment load from 1930 through 1982 was 11.4 million tons.[16] Perhaps more telling, the range in measured annual sediment load is 2.0 to 35.7 million

tons in 1981 and 1938, respectively. The sediment load is generated mostly from the lower-elevation Mesozoic terrane in the drainage basin, whereas the water is generated from headwater snowpacks. Construction of dams, primarily on the Gunnison River, is thought to have decreased the annual sediment load by 40 to 65 percent,[17] although this decrease could also have been affected by climatic variation.

Because the river is confined mostly within a bedrock canyon from the Dolores River to Lake Powell, the alluvial aquifer is mostly shallow and responsive to seasonal recharge from the Colorado River. Although ranches adjacent to the river extract groundwater to irrigate crops, primarily alfalfa and fruit trees, the effects are likely local in scope. At Moab, where the fault-controlled Spanish Valley is relatively deep, groundwater levels have fluctuated without an apparent trend (fig. 7.3). Moab's water supply is primarily surface water from Pack Creek, although many people use groundwater for domestic supplies and irrigation.

Changes in Riparian Vegetation

Dolores River to Moab

Repeat photography of the Colorado River upstream from Moab shows that both native vegetation and nonnative vegetation have increased along its banks. All of our matches now have tamarisk in the view, compared to roughly 40 percent tamarisk in the original photographs, which span the period from 1871 through 1968. Native species, in particular coyote willow and skunkbush, have also increased. As on the Green River, Frémont cottonwood appears to be declining in most places where it was present in historic photographs (fig. 7.4). Comparison of 1951 and 1980 aerial photographs of the Colorado River between Glenwood Springs and Grand Junction, Colorado, showed a nearly 18 percent decrease in cottonwood, which probably included both Frémont and narrowleaf.[18] On the Dolores River, riparian vegetation has benefited from the flood-control operations of McPhee Dam, although the increases have been mostly in coyote willow, not in narrowleaf cottonwood.[19]

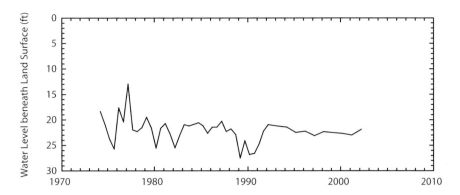

Figure 7.3 Water-level data for well D-25-21 26DCC1 near Moab, Utah.

Lack of overbank flooding is likely the cause of cottonwood decline because old individuals are dying without replacement. Lack of channel change may be another reason why cottonwood is not reproducing along the upper Colorado River. Channel narrowing has occurred in some reaches (fig. 7.5), although not to the extent as on the Green River[20] and on the Colorado River near Grand Junction, where the channel has narrowed by an average of about 60 feet.[21] Although flow regulation does not appear to have significantly dampened flood peaks, the combination of tamarisk occupying formerly barren substrate and even modest reduction in flood size may significantly reduce cottonwood establishment.

The River Corridor across Spanish Valley near Moab

The Colorado River crosses Spanish Valley north of Moab in less than 3 river miles. Riparian vegetation is particularly abundant in this wide alluvial reach (fig. 7.6). Three tributaries enter the Colorado River in Spanish Valley: Courthouse Wash drains parts of Arches National Park north of the river, and Pack and Mill Creeks enter from the south. Despite severe disturbances associated with bridge construction and floodplain development, riparian vegetation has increased along lower Courthouse Wash. This increase includes native species, such as Frémont cottonwood, as well as tamarisk and Russian olive. Russian olive remains a cultivated ornamental tree in Spanish Valley despite its tendency to dominate floodplain environments in the region.

Pack Creek and Mill Creek are two small tributaries that flow through Moab and enter the Colorado River in Spanish Valley (fig. 7.1). Repeat photography on Mill Creek has shown increases in riverine riparian vegetation upstream from Moab.[22] These creeks join the Colorado River in a hydrologically confused area known as the Moab Sloughs, where high groundwater levels adjacent to the Colorado River create a valued wetland. In high-water years on the Colorado River, the Moab Sloughs back up, flooding the lower parts of Moab. In 1990, The Nature Conservancy purchased property in the Moab Sloughs, renaming it the Scott M. Matheson Wetlands Preserve and managing it for its riparian attributes, including two hundred species of migratory and resident birds.[23] Although cottonwood and willow are common in these wetlands, repeat photography shows that the largest increase is in tamarisk (fig. 7.7). The river channel has shifted and narrowed as well.

Several sets of repeat photographs (figs. 7.6 and 7.7) show a number of changes in the riverine environment through Spanish Valley. Development, ranging from riverside businesses to mineral-extraction facilities, has claimed some of the former vegetation along the riverbanks. Fire has increased owing to the permanent human presence and large visitation here, and has occasionally swept through parts of the riparian ecosystem. Tamarisk and Russian olive have become established in large numbers, colonizing formerly barren areas as well as sites beneath the once-scattered cottonwood trees. Three net changes have occurred from these

human-induced factors and other natural ones: the total amount of woody riparian vegetation has increased, cottonwood is less abundant, and the channel through Spanish Valley is narrower than it was at the beginning of the twentieth century.

The Portal to the Confluence of the Green and Colorado Rivers

Between Moab and the Confluence of the Green and Colorado Rivers, the channel gradient through Meander Canyon is low and the river is wide, resulting in slow-moving water. The low gradient can be attributed to the river's backup because of bed-elevation rise in Cataract Canyon.[24] Slower water causes less scouring, and despite occasional lateral channel change and undercutting of riparian plants, the channel is wider, and the riparian assemblage is better developed in this reach. Bottomlands and flooded tributary mouths provide wide floodplains locally, which is similar to the reach through Spanish Valley but unlike the reach through Professor Valley.

In the past, banks along this slow-moving reach were dominated by native species that may have included up to three species of cottonwood, six species of willow, water birch, box elder, and desert olive.[25] With the exception of desert olive and coyote willow, none of these species was abundant historically. Most are now uncommon, and those that remain—in particular desert olive—still mark the old banks but are inundated in a sea of tamarisk and coyote willow. Tamarisk became established in this reach before 1952, and channel narrowing has occurred locally (fig. 7.8).

Seepwillow appears in the old views as well, but it is much less obvious now than previously because of the increase in tamarisk and coyote willow. Peachleaf willow is uncommon but appears to be increasing, and box elder may be declining in its limited distribution in the mile upstream from the Confluence. Netleaf hackberry is another native tree that is difficult to see through the wall of tamarisk; it is sparse but common, particularly within 10 miles of the Confluence.

A

B

Figure 7.4 Photographs of the Colorado River downstream from Dewey Bridge (mile 89.7).

A. (1925.) This downstream view across the Colorado River shows a small grove of Frémont cottonwood on the right bank of a slow-moving reach. The extensive bare sand and driftwood are likely indicative of the large floods that were common on the Colorado River in the first quarter of the twentieth century. The lower shrubs present are a combination of squawbush, rubber rabbitbrush, greasewood, and coyote willow. (W. T. Lee 2954, courtesy of the U.S. Geological Survey Photographic Library.)

B. (July 4, 2000.) The grove of cottonwood is now gone, with little evidence of its former presence remaining. Tamarisk now line both banks of the narrower river channel. Access to river right *(right side)* is difficult, minimizing the possibility that the trees were cut or burned purposefully. (D. Oldershaw, Stake 1605C.)

A

B

Figure 7.5 Photographs of the Colorado River in Professor Valley.

A. (1926.) This view, across and upstream on the Colorado River, shows the river approaching the end of Professor Valley. The camera station is on an unnamed butte west of Porcupine Canyon, and Castle Creek (not in view) passes beneath the camera station. Several small groves of Frémont cottonwood are present in the right side of the view, but most of the woody riparian vegetation is native shrubs, in particular squawbush and coyote willow. (A. A. Baker 163, courtesy of the U.S. Geological Survey Photographic Library.)

B. (September 8, 2001.) The Colorado River is now narrower at several points, largely because once-barren point bars have been colonized by nonnative tamarisk and several native shrubs. The once-discontinuous line of woody riparian vegetation is now nearly continuous with two breaks at right center; this line is mostly tamarisk. The persistent islands also have gained vegetation, and several new ones are present. Cottonwood remain numerous and may have increased at certain points, particularly in the bends on the right side of the view. Utah Highway 128 crosses the scene, providing access to recreational sites and ranches, like the one at right center. (D. Oldershaw, Stake 3646.)

A. (1936.) This view, across Courthouse Wash with the Colorado River in the background, shows a desolate foreground with lush riparian vegetation in the background. A grove of young Frémont cottonwood trees lines the Colorado River, showing through the steel-arch bridge leading south to Moab. The trees are leafless, indicating that this photograph was taken during the winter months. The bridge across Courthouse Wash was built just before this photograph was taken, replacing a steel structure that was destroyed in a flash flood. (I. Provonsha PR-9, courtesy of the Dan O'Laurie Museum, Moab, Utah; Stake 1339.)

B. (1950s.) This photograph shows Courthouse Wash and its bridge from a different camera position than photograph A. The channel is still mostly barren, but several young cottonwood trees have become established by this date. The trees lining the Colorado River are larger and in full leaf. What appears to be tamarisk forms low clumps on the river side of the cottonwood trees in the distance. (B. Boulden 34-19, courtesy of the Dan O'Laurie Museum, Moab, Utah.)

Figure 7.6 Photographs of Courthouse Wash.

C. (December 17, 1997.) Both bridges have been replaced (the Courthouse Wash Bridge in 1992), and the road is Utah Highway 191, the main north-south road in southeastern Utah. This view matches photograph B and shows that the road has been realigned and moved south *(to the right in this view)*. The cottonwood have declined along the river, although numerous leafless young trees appear in the right foreground. Several factors have contributed to the decline of cottonwood here, including fire—one occurred on the approach to the Colorado River bridge in 2003, for example—and construction of businesses, such as the one at right center. Tamarisk and coyote willow are now the most common species along the river, although cottonwood remains, and Russian olive appears to be increasing. Upstream from this camera station, Courthouse Wash has a lush riparian ecosystem composed of cottonwood, coyote willow, Russian olive, and tamarisk. (D. Oldershaw, Stake 1338.)

A

B

Figure 7.7 Photographs of the Colorado River at the Portal downstream from Moab.

A. (June 21, 1905.) This downstream view on the Colorado River shows its channel entering from the Portal, the beginning of the bedrock canyon that leads to Cataract Canyon, about 70 miles downstream. The river is at flood stage, and some trees appear to be partially submerged. Scattered Frémont cottonwood trees dot the floodplain; gallery forests were not present at this time. Across the river, a shallow water table is suggested by the scattered pools. Such conditions, which are present along many bottomlands of both the Green River and the Colorado River, are not conducive to extensive woody riparian vegetation. (C. W. Cross 800, courtesy of the U.S. Geological Survey Photographic Library.)

B. (July 7, 2000.) The Colorado River has shifted south, away from the view, and although cottonwood remain, they are not obvious in the sea of tamarisk that dominates the riparian ecosystem here. The Scott M. Matheson Wetlands Preserve (also known as the Moab Sloughs), a riparian area owned by The Nature Conservancy and established as a natural area of 890 acres in 1990, appears at left center. The Atlas Tailings Pile, an abandoned and controversial relic of uranium mining in the region, is the barren area in the left midground. Several Utah juniper and big sagebrush persist in the foreground, although overall woody vegetation has increased on the slopes and next to the river. (D. Oldershaw, Stake 2514.)

A

B

Figure 7.8 Photographs of the Colorado River in Meander Canyon.

A. (July 1952.) This view across the Colorado River in Meander Canyon shows Deadhorse Point in the distance and an unnamed small tributary confluence in the foreground. The vegetation on the banks is mostly tamarisk, although coyote willow may be present on the left side. (G. Simmons 6, courtesy of the photographer.)

B. (October 12, 2001.) Tamarisk now blocks the view of most of the landscape except Deadhorse Point. The channel narrowed in many of the wider reaches of Meander Canyon as islands became attached to the banks. In this case, it is uncertain whether the original camera station is on an island or photograph A was taken from a moving boat. In either case, what once was open water is now dry land covered with tamarisk and coyote willow. (R. H. Webb, Stake 3190.)

Cataract Canyon

Because of its rich photographic record, the history of riparian vegetation in Cataract Canyon is well known.[26] Frémont cottonwood, which appear in certain low-velocity reaches, have declined (fig. 7.9). In Spanish Bottom, human-caused fire killed mature trees. At Tilted Park, cottonwood numbers are decreasing without replacement. Desert olive is also common in low-velocity reaches but is mostly obscured behind walls of new tamarisk and willow. Unlike cottonwood, desert olive can resprout following fires, but, unlike tamarisk, it also experiences heavy

mortality. Apache plume, a shrub that normally occurs at higher elevation, marks the high-water line locally, and its population appears to be stable in Cataract Canyon, not declining as it is in Grand Canyon (chapter 11).

La Rue's photographs at the Confluence in 1914 are the first that show tamarisk (those trees are gone). Within Cataract Canyon, the steep, narrow canyon walls and steep gradient result in fast-moving water and large annual stage fluctuations, so that the wide, continuous band of riparian vegetation found upstream is replaced by a thin, discontinuous line of vegetation, and tamarisk is restricted mostly to

the calmest part of eddies. At Rapid 3, tamarisk became established on a formerly barren beach in 1941 (fig. 7.10). Thirty-five plants were cut in 1994, and their stumps were painted with herbicide, but about half of the plants survived. Coyote willow, which was not in the dense grove of tamarisk in 1994, now dominates the vegetation at this site.

After tamarisk, netleaf hackberry[27] is the most common tree in Cataract Canyon, marking the high-water line along canyon slopes (fig. 7.11). Hackberry produces annual rings that both reveal the trees' age and record river flow.[28] The oldest living netleaf hack-

Figure 7.9 Photographs of the Colorado River at Spanish Bottom in Cataract Canyon.

A. (May 31, 1889.) This downstream view shows a surveyor with the Brown-Stanton Expedition of 1889. Water in the Colorado River is close to the annual peak for this year. Spanish Bottom appears across the river and is marked by a line of riparian vegetation, including prominent Frémont cottonwood. Although it is difficult to determine the species from this distance, the large clumps *(extreme right)* appear to be desert olive, and the shoreline plants appear to be coyote willow. (F. A. Nims 37 and P0197 55:5:21, courtesy of the Marriott Library.)

B. (July 22, 1991.) Considerable change has occurred at this site owing to the combination of tamarisk and fire. Tamarisk forms a nearly impenetrable barrier on river right above the waterline. In the 1980s, a fire that was accidentally started on channel left (downstream from this camera station) swept upstream beyond this point; then embers blew across the river and started another fire in Spanish Bottom. A dead Frémont cottonwood, established after 1889, appears on channel left, and a combination of snags and trees that survived the fire are on river right. Cottonwood increased on Spanish Bottom before the fire and remains more numerous here than in 1889. Desert olive may not have been as fortunate; stumps are common along the top of the terrace at Spanish Bottom. Jack Schmidt, a noted geomorphologist from Utah State University, appears in the foreground. (R. H. Webb, Stake 2339.)

A

B

Figure 7.10 Photographs of the Colorado River below Rapid 3 in Cataract Canyon.

A. (September 17, 1921.) This upstream view shows Rapid 3 from the large, boulder-strewn island on the left side. This island is inundated at much higher flows. Almost no riparian vegetation is visible here, although very close inspection of the left banks in the distance *(center)* shows four to five netleaf hackberry trees. (S. Paige 1393, courtesy of the U.S. Geological Survey Photographic Library.)

B. (August 29, 2000.) Tamarisk, coyote willow, and netleaf hackberry have increased in this view. On river right in the middle of Rapid 3, a grove of tamarisk was cut and poisoned with herbicide in 1994; by means of ring counts, it was determined that the largest tamarisk became established in 1941. Following this local eradication, many of the tamarisk stumps resprouted, but coyote willow also began growing where it had not been present. The small grove of hackberry has increased in size to the point that individual trees cannot be recognized, and downstream *(right side)* a mixture of tamarisk and netleaf hackberry appears prominently on the high-water line. (S. Young, Stake 3011.)

A. (1952.) This view across Rapid 15 in Cataract Canyon shows a prominent line of netleaf hackberry on river left above the rapid. The low sun angle is producing prominent shadows, which makes identification of individual trees difficult. A patch of Apache plume appears on the extreme right side. Although tamarisk was well established in Cataract Canyon by the 1950s, no individuals of this nonnative species are apparent. (G. Simmons, courtesy of the photographer.)

B. (June 12, 2002.) On the far bank, the number and density of netleaf hackberry have increased, an increase that has occurred wherever hackberry is present in Cataract Canyon. A few tamarisk are now intermixed with the netleaf hackberry, and the patch of Apache plume appears to be about the same size despite the occurrence of a small debris flow, which deposited coarse-grained levees on the upstream side of the patch. (S. Young, Stake 4234.)

Figure 7.11 Photographs of the Colorado River below Rapid 15 in Cataract Canyon.

berry in Cataract Canyon germinated in around 1804, and rings extracted from stumps indicate that some now dead trees germinated in the early eighteenth century. Netleaf hackberry has increased in Cataract Canyon, with recruitment higher than mortality by five to one in the twentieth century.

The free-flowing reach of the Colorado River is only 16 miles long through Cataract Canyon. The rest of the canyon, as well as Narrow Canyon, the next one downstream, is inundated beneath Lake Powell (chapter 14). The upstream extent of the lake is dependent on the reservoir level, which at present (2004) is down more than 50 feet. Fluctuations in water level affect not only the species established on deltaic sediments—mostly nonnative tamarisk—but also the native species established at the highest lake elevation. Repeat photography from this interface between free-flowing river and high lake stand indicates that lake-level changes have removed the native species, notably cottonwood and hackberry, and tamarisk forms mono-specific stands on deltaic sediments.[29]

8   The San Juan River

Summary. The San Juan River in south-eastern Utah is an important source of water for agriculture and domestic use, is home to several endangered fish species, is valued for whitewater recreation, and supports significant riparian habitat. In its wide alluvial reaches, the San Juan River had a large increase in woody riparian vegetation following decreases in flood magnitudes in the middle of the twentieth century. Although much of the increase is in nonnative tamarisk and Russian olive, Frémont cottonwood and coyote willow also have increased in these reaches. Channel narrowing occurred after 1941, as it has on other rivers in the region, and establishment of new riparian vegetation has accelerated in recent decades. Operation of Navajo Dam and Reservoir has affected the San Juan River in complex ways, including attenuation of flood peaks, storage of sediment, and changes in seasonal flow, all of which might have influenced riparian vegetation. Little vegetation change has occurred in the bedrock canyon except the establishment of tamarisk and an increase in coyote willow.

The San Juan River is a major tributary of the Colorado River that flows westward from its headwaters in the San Juan Mountains of southwestern Colorado (fig. 8.1). This river drains 23,000 square miles of the Four Corners Region upstream from Mexican Hat, Utah, and cuts a course through northern New Mexico and southern Utah, passing through the Mesozoic sedimentary rocks of the Colorado Plateau. As the river flows west between Bluff and Mexican Hat, it cuts perpendicularly through a series of uplifted monoclines and anticlines in Paleozoic rocks and becomes entrenched deep in the limestones of the Honaker Trail Formation and the Paradox Formation. Along the way, the San Juan is joined by numerous tributaries, and in southern Utah the most important of them are McElmo and Montezuma Creeks and Comb, Cottonwood, and Chinle Washes.

Because the San Juan River is one of the major tributaries of the Colorado River upstream from Lee's Ferry and Grand Canyon, flow in it affects irrigation supplies in Colorado, New Mexico, and Utah and is an important source of water to the lower basin states of Arizona, California, and Nevada. Diversions are common locally on the river between Navajo Dam in northern New Mexico and Bluff, Utah. Once home to thriving populations of native fishes, this river has important populations of Colorado River pikeminnow and razorback suckers (*Xyrauchen texanus*), both of which are endangered species.[1] Finally, the extensive stands of riparian vegetation along the San Juan makes this river unusual in the region and a valued resource. Extensive historical photography documents changes in riparian vegetation, which are relatively well known for this river compared with other rivers in the Southwest.

Photography of the San Juan River

Historical photography of the San Juan River extends back to two of the pioneers of Western landscape photography. Timothy O'Sullivan photographed Chinle Wash in Canyon de Chelley in 1873 while serving as the principal photographer for the Wheeler Expedition. William Henry Jackson took the first photographs of the San Juan River in 1875 during the Hayden Expedition.[2] Before 1900, the largest number of photographs was taken by noted landscape photographer Charles Goodman, who worked in the region in the early 1890s documenting placer mining operations in San Juan Canyon and life in Bluff.[3] While searching for potential dam sites in 1921, a U.S. Geological Survey river expedition took approximately 75 photographs of the river corridor in the remaining free-flowing reach, primarily within the bedrock canyon; we have matched 58 of these views. Geologists interested in the exposed strata, notably the famous Goosenecks of the San Juan, took many photographs from the 1910s through the 1930s. Finally, river runners photographed extensively in the middle and latter parts of the twentieth century. We matched a total of 137 historical photographs between Aneth, Utah, and Clay Hills Crossing on Lake Powell.[4]

Floods and Flow Regulation

Floods and Channel Changes

Humans have occupied the San Juan River valley for most of the past eleven thousand years, and archaeological sites along the river are common.[5] Near Bluff, several archaeological sites are present along the river and its tributaries,[6] and several important archaeological sites occur downstream in the bedrock canyon. Eolian activity is high in the valley, and active dunes overlie much of the higher alluvial terraces along the river.

From 1923 through 2003, the annual flow volume for the San Juan River near Bluff (the gaging station is at Mexican Hat; fig. 8.1) is 1.65 million acre-feet, which corresponds to an average daily discharge of 2,275 ft³/s. Tree-ring reconstruction of flow at this gaging station, from A.D. 1661 through the 1920s, indicates that the long-term annual flow volume is 2.2 million acre-feet.[7] The years with highest flow

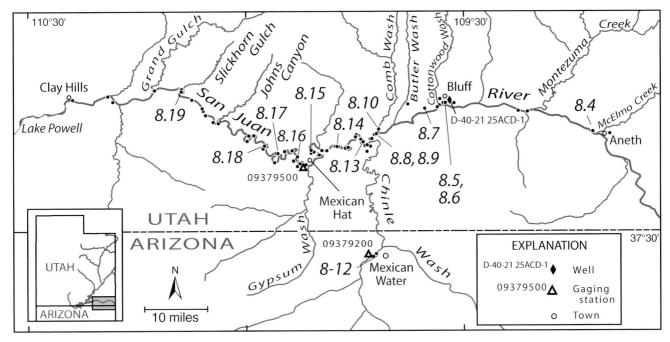

Figure 8.1 Map of the San Juan River in southern Utah and northern Arizona.

volume are 1816, 1869, and 1746, with 4.57, 4.23, and 4.01 million acre-feet, respectively. The highest flow volume in the twentieth century was also 4.01 million acre-feet in 1920.

Settlement of the lower reaches of the San Juan River basin occurred in 1880, when Mormon settlers traveled from Escalante, Utah, through the Hole-in-the-Rock crossing of the Colorado River.[8] Immediately upon reaching the present-day location of Bluff, the settlers built a town, constructed an irrigation dam, and commenced floodplain agriculture. The twentieth-century gaging record near Bluff records a long-term average flow before completion of Navajo Dam (1915–1917, 1927–1961) of 2,460 ft³/s, and that magnitude of reliable water supply made the site attractive to the agrarian Mormons. Settlers described the river as having "reed swamps" and the floodplain as being "covered with cottonwoods" on a terrace about 6 feet above the channel.[9]

Floods plagued the irrigation project and immediately threatened the existence of the new settlement. Particularly damaging floods occurred in 1884 and 1891, and one report estimates that the peak discharge of the March 16, 1884, flood was greater than 50,000 ft³/s.[10] Both the March and June 1884 floods caused severe damage to

the settlements of Montezuma Creek, which was abruptly abandoned, and of Bluff, where dams and waterwheels were washed away in the torrent and fields were buried in sand and mud.[11]

Several floods early in the twentieth century are notable for their size, their effect on channel geometry, and their damage to human structures in the watershed. During the 1905 flood, local newspapers reported that Bluff was "washing away."[12] The summer of 1907 was particularly damaging, with the combination of snowmelt and rainfall producing destructive floods; one giant cottonwood toppled into the flow.[13] The 1909 flood occurred in early September and was attributed to heavy and continuous rains in the San Juan's headwaters. Long-time residents of nearby Farmington, New Mexico, stated that the flood was the largest since the settlement of the area in 1880.[14]

That flood was quickly exceeded. A very large flood occurred on the San Juan River on October 6, 1911, causing severe damage along the river's length.[15] The discharge for this flood was first estimated to be 150,000 ft³/s;[16] a recent study estimated the peak discharge to be 148,000 ft³/s in the bedrock canyon.[17] The 1911 flood clearly is the largest peak discharge known in the twentieth century for the San Juan River (fig. 8.2). This flood destroyed

most of the bridges on the main-stem San Juan River, including the first Goodridge Bridge at Mexican Hat. The Goodridge had no piers, and its deck was 39 feet above the river, indicating that the stage had to be at least that high. Driftwood piles along the river in the early 1920s were described to be as much as 35 feet above the water surface and covering "several acres."[18] Clearly, the 1911 flood caused severe damage to riparian ecosystems throughout the entire length of the river.

The net result of the floods was that perhaps 1,000 acres of agricultural land was eroded from the vicinity of Bluff.[19] Much of that destruction occurred between 1909 and 1925, when whole fields were washed away.[20] Another flood in 1933 inundated parts of Bluff.[21] Most of the woody riparian vegetation that was along the river and noted by Mormon residents was swept downstream as well. From 1937 through 1980, there was no further erosion of the high-terrace deposits in this reach.[22] Instead, the active channel has gradually narrowed to a fraction of its nineteenth-century width.

Many researchers have hypothesized that the increase in livestock grazing associated with Mormon settlement and land-use changes on the Navajo Indian Reservation are responsible for landscape changes.[23]

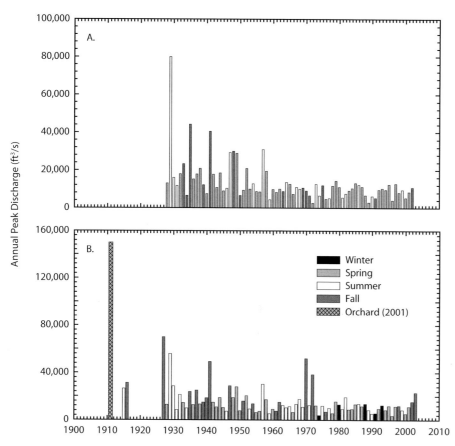

A. The San Juan River near Shiprock, New Mexico (station 09368000; 1928–2002).

B. The San Juan River near Bluff (Mexican Hat), Utah (station 09379500; 1914–1916, 1927–2003).

Figure 8.2 Annual flood series for the San Juan River.

One concept, which mirrors the regional hypothesis, states that livestock were overstocked on ranges in the 1880s and 1890s; overgrazing and the drought between the 1890s and 1904 (which peaked about 1896 in southern Utah) caused severe disruption of rangelands, watersheds, and riparian vegetation, which led to large floods and widespread erosion. Although overgrazing no doubt contributed to the size of floods, the storms during this period and the annual precipitation totals were unusually large.[24]

Gaging Records

Several long-term gaging records are available for the San Juan River because of its importance to irrigation and water supply within the Colorado River Compact. In New Mexico, the station San Juan River near Shiprock has recorded flow and floods since 1928 (fig. 8.2A), thereby missing the larger floods before that year. The 1929 flood is the peak of record for this gaging station at 80,000 ft³/s. Floods on the San Juan River near Bluff have

been recorded from 1914 to 1916 and 1927 to the present. Following the 1911 flood, which is not part of the official gaging record (fig. 8.2B), the largest flood near Bluff had a discharge of 70,000 ft³/s on September 10, 1927; the stage of this flood was about 31 feet. Flood frequency at both gaging stations has decreased through the twentieth century. After 1962 and through 2002, the largest flood on the San Juan River was 38,500 ft³/s, caused by Tropical Storm Norma on September 6, 1970.[25] Most of the runoff generated by this storm was downstream from Navajo Reservoir. Historically, the largest floods on the San Juan River occurred between late summer and early fall (August–October).

Flow Regulation

Other than small diversion structures for irrigation, the watershed was unregulated before June 28, 1962, when Navajo Dam was completed. Operated by the Bureau of Reclamation and located in northern New Mexico, this structure is 402 feet high and operates

as a flood-control and hydroelectric power structure. As suggested in figure 8.2, the dam's flood-control operations reduce peak discharges downstream on the San Juan River. Several major tributaries—such as the Animas River, Montezuma Creek, McElmo Creek, Chaco Wash, and Chinle Wash—are unregulated, and storm runoff from these tributaries contributed to the 1970 flood. The lower 75 miles of San Juan Canyon have been inundated by Lake Powell, which began filling in 1963 after completion of Glen Canyon Dam (chapter 14).

Changes in Riparian Vegetation

New Mexico–Utah Border to Bluff

From the New Mexico–Utah border, the San Juan River flows west through a broad alluvial valley that was settled in 1880. Groundwater levels are generally high along the river (fig. 8.3), despite extensive areas of floodplain agriculture. Two photographs by William Henry Jackson in 1875 near the mouth of Montezuma Creek (fig. 8.1) show

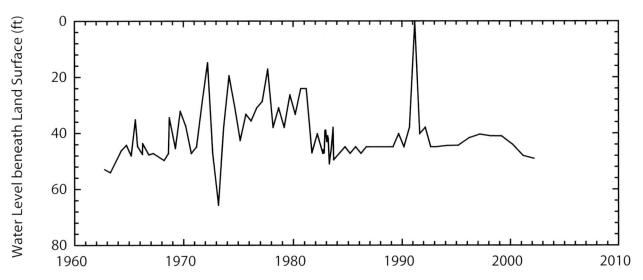

Figure 8.3 Groundwater levels for well D-40-21 25ACD-1 near Bluff, Utah.

scattered Frémont cottonwood along a wide river channel well before the arrival of livestock.[26] Because the river was wide and barren in 1875, these photographs show that overgrazing could not have been the principal cause for sparse riparian vegetation along the San Juan River in the nineteenth century. Similarly, recent reductions in grazing intensity occurred after increases in riparian vegetation had begun.

Five sets of repeat photographs document change in riparian vegetation in this reach. Near Aneth and the mouth of McElmo Creek, the San Juan and this major tributary had wide, barren channels and only scattered trees in 1928 (fig. 8.4A). Downcutting and subsequent backfilling of the channel of McElmo Creek upstream in southwestern Colorado occurred three times during the past fifteen hundred years,[27] including creation of the current arroyo in the nineteenth century. This channel instability may explain the barren floodplains that appear in early-twentieth-century photographs. Now, riparian vegetation is so dense that the channel is not visible from the original camera station. Other repeat photography also documents the conversion from a barren, wide channel to one filled with riparian vegetation.[28]

Bluff to San Juan Canyon

Considerable photography documents change in riparian vegetation in the vicinity of Bluff back to 1880. Twenty-one photographs were matched between the town and the mouth of Chinle Wash, 13 miles downstream. Several views from vantage points on top of or within the cliffs north of town show the San Juan River in the distance beyond Bluff (figs. 8.5 and 8.6). The river channel is easily visible as a wide barren swath in the original photographs (1890 and 1921), and the channel is no longer visible from this vantage point.

Early-twentieth-century photographs of the river corridor consistently show a wide floodplain with sparse riparian vegetation (fig. 8.7). The channel is so featureless that the position of the thalwegs containing low flow cannot be readily determined. Part of the reason for few trees near Bluff was that cottonwood were cut to clear land for agriculture as well as to create rip-rap for flood protection.[29] However, no evidence exists that documents the previous extent of cottonwood stands, only that trees were cut near Bluff.

Repeat photographs from the late 1990s and early 2000s show a large increase in both native and nonnative vegetation. Russian olive and tamarisk, in particular, occupy most of the formerly barren channel, but a line of native riparian species, including Frémont cottonwood and coyote willow, is typically close to flowing water. All of this vegetation is newly established, indicating that native species can germinate and become established within the sea of nonnative riparian species.

Comb Wash, one of the major tributaries of the San Juan River between Bluff and the mouth of San Juan Canyon, has its headwaters in the Abajo Mountains of southern Utah and flows south. Repeat photography along this wash upstream from the San Juan River shows that riparian vegetation here has increased considerably.[30] At the confluence of Comb Wash and the San Juan River, the San Juan has narrowed appreciably, and tamarisk and Russian olive are extremely dense (figs. 8.8 and 8.9). However, cottonwood and willow have increased as well, particularly on low terraces near the river. No cottonwood visible in the original view persists, indicating that reproduction in native species is occurring. One striking view from the early 1950s (fig. 8.10), repeated in the late 1990s, shows that channel narrowing and the increase in riparian vegetation accelerated in the latter half of the twentieth century during the period of flow regulation by Navajo Dam.

Chinle Wash

Chinle Wash is a major tributary of the San Juan River in northern Arizona and southern Utah (fig. 8.1). This 3,650-square-mile watershed originates in mountainous terrain of the Navajo Indian Reservation, notably the Carrizo and Chuska Mountains southwest of the Four Corners Region and the northeast side of Black

A

B

Figure 8.4 Photographs of the San Juan River at the mouth of McElmo Creek.

A. (1928.) This upstream view across the mouth of McElmo Creek, with the San Juan River channel in the distance and at right midground, shows what appears to be an abandoned agricultural field in the right foreground and a Pleistocene river terrace on the left. The channel is perhaps a quarter-mile wide and barren; Frémont cottonwood are scattered on both banks and on midchannel islands. (H. E. Gregory 558, courtesy of the U.S. Geological Survey Photographic Library.)

B. (September 21, 2003.) Utah Highway 163 has replaced the one-lane dirt track present in 1928, and the town of Aneth, Utah, is around the bend and across McElmo Creek. The channels of both McElmo Creek and the San Juan River are fully vegetated with tamarisk, Russian olive, and Frémont cottonwood. (D. Oldershaw, Stake 4730.)

Figure 8.5 Photographs of the San Juan River at Bluff, Utah.

A. (Early 1890s.) This downstream view, with the Navajo Twins rock formation at left, shows Bluff in the early 1890s and the wide and denuded channel of the San Juan River in the middle distance. Scattered groves of trees line the channel, which in places is about 1,300 feet wide. Bluff was founded in 1880, more than a decade before this photograph was taken. A photograph from a different camera station but approximately matching this view shows that the channel was denuded in 1909 as well. (C. Goodman P0068:064, courtesy of the Marriott Library.)

B. (July 2, 2000.) The channel of the San Juan River is no longer visible from this vantage point. The floodplain is clogged with riparian vegetation, both native and nonnative species. Cottonwood are locally abundant in this reach, but the most widespread species are coyote willow and nonnative tamarisk; Russian olive is locally abundant. Trees have grown in town, blending with the riparian vegetation. There is a similar view from a different camera station (Hindley and others 2000, pp. 102–3). (D. Oldershaw, Stake 2858.)

Figure 8.6 Photographs of the San Juan River at Bluff, Utah.

A. (July 15, 1921.) This view, taken in approximately the same direction as figure 8.5, but upstream from the Navajo Twins, shows that the channel of the San Juan River is free of riparian vegetation in 1921. The trees in Bluff in 1921 are taller than in 1890, and the town appears to be considerably larger. (R. N. Allen 18, courtesy of the Dan O'Laurie Museum, Moab, Utah.)

B. (July 2, 2000.) The San Juan River channel is completely obscured by riparian vegetation. Both native and nonnative species densely cover the former floodplain. (D. Oldershaw, Stake 3524.)

Figure 8.7 Photographs of the San Juan River west of Bluff.

A. (Ca. 1925.) West of Bluff, the San Juan River flows beneath bluffs of Page Sandstone. This view, from the old road between Bluff and Mexican Hat, faces due south from a terrace above the right bank. Flow is from left to right. The channel is broad and almost entirely free of vegetation, with numerous secondary channels; the main channel is probably directly beneath the distant bluffs. (W. T. Lee 3172, courtesy of the U.S. Geological Survey Photographic Library.)

B. (March 22, 1998.) The channel of the San Juan River has narrowed considerably, and all secondary channels have disappeared. Most of the vegetation is tamarisk and Russian olive, although coyote willows line the active channel. Other researchers have also reoccupied this camera station and photographed the view (Hindley and others 2000, pp. 104–5). (R. H. Webb, Stake 3522a.)

Figure 8.8 Photographs of the San Juan River at Comb Wash.

A. (Ca. 1927.) The original view was taken from the mouth of Comb Wash looking southwest toward the Mule's Ear diatreme *(center)*. The flowing water is almost indistinguishable from the sandy channel, but the river is flowing beneath the cutbank in the middle distance. The channel here is wide and vegetation free except for small Frémont cottonwood at right. (H. E. Gregory 536, courtesy of the U.S. Geological Survey Photographic Library.)

B. (March 23, 1998.) The view is now blocked by riparian vegetation, including tamarisk *(taller trees)* and rubber rabbitbrush *(branched with white feathery fruits)*. The cottonwood trees, leafless in March, appear in the right midground. The cottonwood in the original view are no longer present, but new cottonwood are growing in approximately the same location. (R. H. Webb, Stake 3525.)

A. (March 1953.) This upstream view of the San Juan River shows the mouth of Comb Wash *(left midground, just beyond the jeep)*. Short-statured riparian plants with the appearance of coyote willow line both the San Juan and the upstream side of Comb Wash. This photograph was taken nine years before completion of Navajo Reservoir. (G. C. Crampton P0197:52:1:21, courtesy of the Marriott Library.)

B. (August 26, 1997.) Considerable riparian vegetation is now present throughout this view, and Comb Wash drains into a shallow distributary channel north *(river right)* of the main river channel. Both native and nonnative riparian vegetation has increased; coyote willow generally is the shrub closest to the river channel, with Russian olive and tamarisk just behind. Cottonwood trees are present on the closest terraces to the river on both sides. (D. Oldershaw, Stake 2285.)

Figure 8.9 Photographs of the San Juan River at Comb Wash.

Figure 8.10 Photographs of the San Juan River at "San Juan Hill."

A. (March 1953.) The photographer is standing on "San Juan Hill," a promontory just upstream from the confluence of Comb Wash *(right foreground),* and the view of the San Juan River is downstream. The Mule Ear diatreme is at left center, and the entrance to San Juan Canyon is at right center; Chinle Wash joins the San Juan River at about the point where the river disappears from view. The channel of the San Juan, which was obviously much wider, has narrowed recently; note the sharp cutbank across the river in the center of the photograph as well as the low cutbanks and floodplains. Low-statured riparian vegetation is becoming established along both drainages; a lone cottonwood appears at lower center. (G. C. Crampton P0197:52:1:36, courtesy of the Marriott Library.)

B. (March 23, 1998.) The channels of both the San Juan River and Chinle Wash have narrowed considerably. In this reach, the channel of the San Juan is divided through a series of islands densely covered with riparian vegetation. Cottonwood trees are difficult to identify in this late-winter view, but they are present along most of the channel, particularly along river right. (D. Oldershaw, Stake 3558.)

Mesa. Downstream from the Chuska Mountains, headwater streams pass through Canyon de Chelley National Monument, the largest U.S. National Park Service unit on the Navajo Indian Reservation. Replication of the extensive historical photography of Canyon de Chelley (the first photographs were taken in 1873) shows large increases in Frémont cottonwood, coyote willow, tamarisk, and Russian olive.[31]

As with other smaller streams in the region, floods occur in every season except spring in response to warm storms in winter or thunderstorms in summer or a variety of storm types in the fall months (fig. 8.11). Chinle Wash downcut into its floodplain at an unknown time at the end of the nineteenth century, and the history of channel change here is poorly known. At one point in the watershed, the arroyo depth was 100 feet in 1913.[32] In the latter part of the twentieth century, filling occurred on most channels in this region.[33]

Near the gaging-station site at Mexican Water, downcutting of the arroyo of Chinle Wash has left the channel with a bedrock floor, creating a local base level and a control for a gaging station just upstream of Arizona Highway 160. Seven historical photographs show that riparian vegetation just downstream has increased within the arroyo and on its high banks since 1935 (fig. 8.12). Although tamarisk is the most common riparian species in the Chinle Wash watershed, Frémont cottonwood and coyote willow also occur in significant numbers. Russian olive occurs in low numbers on the floodplain as well, creating the possibility of a future expansion.

At its confluence with the San Juan River, Chinle Wash flows through a broad valley within an arroyo that is relatively shallow compared with what occurs upstream (fig. 8.10). Dense tamarisk with cottonwood and coyote willow occurs at the mouth, and cottonwood is scattered upstream within a nearly continuous stand of tamarisk and Russian olive. As with the San Juan River upstream from San Juan Canyon, the increase in riparian vegetation occurred in the last half of the twentieth century.

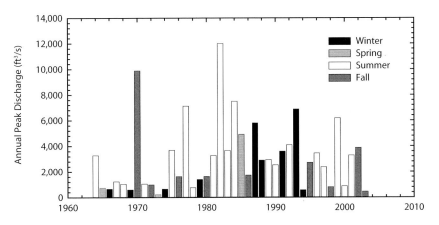

Figure 8.11 Annual flood series for Chinle Wash near Mexican Water, Utah (station 09379200; 1964–2003).

The Entrance to San Juan Canyon to Mexican Hat

Downstream from Bluff, the San Juan passes between sandstone bluffs and gradually becomes a bedrock-controlled channel. Below its confluence with Chinle Wash, it plunges through the Lime Ridge Monocline[34] and enters a bedrock canyon in Paleozoic sedimentary strata, notably the Honaker Trail Formation. As noted in chapter 7, the Honaker Trail Formation consists of alternating limestone, dolomite, and mudstone beds, and the type section is at the Honaker Trail in lower San Juan Canyon. The steep canyon walls and stable limestone substrate attracted would-be dam builders such as Eugene C. La Rue, who photographed a potential damsite just downstream from the entrance to San Juan Canyon (fig. 8.13). Unlike channel conditions immediately upstream, the San Juan River has not narrowed appreciably, and although riparian vegetation has increased, the increase is relatively small compared with that in the alluvial reaches.

Even though the increases in riparian vegetation are modest, the fact that they are occurring reflects a widespread directional change in riparian vegetation along this river. Within the narrowest parts of the bedrock canyon, both native and nonnative species are increasing (fig. 8.14). Only in relatively wide reaches, such as the reach upstream from Mexican Hat and downstream from the Raplee Anticline, are channel narrowing and large increases in riparian vegetation apparent (fig. 8.15). The increase in coyote willow in these reaches is nearly as large as the

increase in tamarisk, and Russian olive, abundant in the alluvial reaches upstream, is uncommon in the relatively wide reaches and absent from most of the narrow reaches.

San Juan Canyon from Mexican Hat to Lake Powell

At Mexican Hat, the San Juan River passes from its relatively wide valley into a narrow bedrock gorge. At the Goodridge Bridge, the channel narrows considerably; historical photography here (not shown) dates the establishment of noticeable amounts of tamarisk to the 1950s. For approximately the next 30 river miles, the river flows in a section without rapids and meanders gently through several scenic bends, notably Mendenhall Loop (fig. 8.16) and the Goosenecks (fig. 8.17). Expansive views of the channel from the rim indicate that woody riparian vegetation—probably coyote willow—was established near the low water surface in the 1890s, but that vegetation was replaced by barren sand in views taken in the 1920s owing to the occurrence of large floods, in particular the 1911 event. Both tamarisk and willow are common along the river now, and other native species, including rubber rabbitbrush and netleaf hackberry, higher on the banks. Russian olive is present in relatively wide reaches, such as the outside of the bend in the Mendenhall Loop.

Small netleaf hackberry trees are common in small groves along the river. This species is not distinct enough in any of the repeat photographs to allow interpretation of change. Because hackberry has increased in Cata-

Figure 8.12 Photographs of Chinle Wash near Mexican Water, Utah.

A. (December 1935.) This downstream view, taken from perhaps 200 yards north of Arizona Highway 160 on the Navajo Indian Reservation, shows a relatively wide channel and low floodplains with incipient riparian vegetation. Some of the vegetation appears to be nonnative Russian thistle, an annual species. Dead trees, probably Frémont cottonwood, are prominent on the right bank in the foreground. An open grove of cottonwood appears in the midground, and a second grove of trees appears at far left. (M. S. Snow, number Ariz.-920, courtesy of the Natural Resources Conservation Service.)

B. (September 14, 2002.) The channel has widened, and the low alluvial terraces within the arroyo are now more prominent. Cottonwood now grow within the arroyo, and some are persistent on the right bank (the view of most of them is blocked by the prominent cottonwood). Both rubber rabbitbrush and coyote willow are present in low numbers. Several nonnative species are prominent in this reach, including tamarisk, which lines the low-water channel, and Russian olive. (D. Oldershaw, Stake 4326.)

A

B

Figure 8.13 Photographs of the San Juan River downstream from the entrance to San Juan Canyon.

A. (July 19, 1921.) Eugene C. La Rue, a U.S. Geological Survey hydrologist, took this photograph of the San Juan River about 2 miles downstream from the entrance to the canyon to document a potential dam site. This upstream view, which is the left center cropped from La Rue's wider view, shows the hand-drawn dam on the photograph. Very little riparian vegetation is present in this reach, and the point bar across the river appears scoured. (E. C. La Rue 827, courtesy of the U.S. Geological Survey Photographic Library.)

B. (March 24, 1998.) At this low water level of less than 1,000 ft³/s, a line of pour-overs marks the first rapid of note in San Juan Canyon. The formerly denuded point bar is covered with riparian vegetation, much of which is nonnative tamarisk. (D. Oldershaw, Stake 3560a.)

A

B

Figure 8.14 Photographs of the San Juan River at First Narrows.

A. (July 19, 1921.) This downstream view shows the entrance to First Narrows and is just downstream from Eight-Foot Rapid. The scattered riparian vegetation *(left side)* appears to be mostly netleaf hackberry trees. (R. N. Allen 27, courtesy of the Dan O'Laurie Museum, Moab, Utah.)

B. (March 24, 1998.) The original camera station is buried under a new sandbar; as a result, the new camera station is about 4 feet too high. Coyote willows line the river, with tamarisk just visible behind the willows. Beaver had culled some of the willows from the sandbar shortly before our visit. (R. H. Webb, Stake 3529.)

Figure 8.15 Photographs of the San Juan River at Mexican Hat, Utah.

A. (December 1894.) This downstream view of the San Juan River is taken from a point on a cliff of Moenkopi Formation just upstream from Mexican Hat. A gold dredge is visible in the center of the river in the midground. The channel is relatively wide and braided around sandbars in this reach. Other photographs taken at this time indicate that the riparian vegetation on the terraces *(both sides of the river)* is native coyote willow, leafless in this winter photograph. (C. Goodman P0068:78, courtesy of the Marriott Library.)

B. (July 3, 2000.) Riparian vegetation has increased greatly on both sides of the river, resulting in significant channel narrowing. This reach is only partly bedrock controlled and has responded similarly to the reach from Bluff to the head of the canyon, both in terms of increases of riparian vegetation and channel narrowing. The vegetation consists of native coyote willow closest to the river channel and on the point bar in the middle distance where the dredge once was; a band of dense tamarisk occurs behind the willows. The floodplain consists of a mixture of native shrubs and smaller tamarisk. (D. Oldershaw, Stake 1790.)

A

B

Figure 8.16 Photographs of the San Juan River at Mendenhall Loop.

A. (Ca. 1894.) This upstream view of the San Juan Canyon shows a reach downstream from Mexican Hat, Utah. Alhambra Butte, a volcanic neck next to the road between Monument Valley and Mexican Hat, appears in the distance. Little riparian vegetation is present along the banks, which are sandy. A band of trees in the center midground appears to be either clumps of coyote willow or small netleaf hackberry trees. (C. Goodman P0068:048, courtesy of the Marriott Library.)

B. (July 2, 2000.) Despite higher water in 2000, several important changes are apparent in the San Juan River. The amount of sand has decreased, owing both to encroachment of riparian vegetation and to erosion, as illustrated by the increase in visible areas of gravel. Riparian vegetation has increased along this reach, mostly tamarisk and coyote willow. A few Russian olive trees are in the bend in the distance. (D. Oldershaw, Stake 2366.)

A. (Ca. 1894.) The Goosenecks of the San Juan River have attracted photographers since their discovery in the late nineteenth century. These fabulous river meanders in Honaker Trail Formation show a symmetry that is unmatched on any other river in the Southwest. The view also shows what appear to be sandwaves in the channel, which is discontinuously lined with small shrubs that likely are coyote willow. (C. Goodman P0068:045, courtesy of the Marriott Library.)

B. (1921.) This view is part of a larger panoramic photograph taken from a different camera station. More shrubs—probably coyote willow—line the riverbanks, but the riparian vegetation remains short and discontinuous. (E. C. La Rue 802, courtesy of the U.S. Geological Survey Photographic Library.)

C. (July 2, 2000.) This view matches photograph A and is from a viewpoint that was established as part of Goosenecks State Park. The river channel is now almost continuously lined with tamarisk and coyote willow, a marked change from conditions in the early 1890s and 1921. (D. Oldershaw, Stake 1363.)

Figure 8.17 Photographs of the San Juan River at the Goosenecks.

A

B

Figure 8.18 Photographs of the San Juan River near the bottom of the Honaker Trail.

A. (December 1894.) Charles Goodman, pioneer photographer, made a living documenting mining claims during the brief mining boom along the San Juan River in the mid-1890s. This upstream view, taken from the right bank about a mile downstream from the foot of the Honaker Trail, shows a placer mining operation and two miners digging into gravels along the low-water channel. Little in the way of riparian vegetation is present in this view; what is present on the right side of the view may be small hackberry trees or seepwillow. (C. Goodman P0068:073, courtesy of the Marriott Library.)

B. (April 29, 2002.) All trace of the miners' activities no doubt were removed during the 1911 flood and other large-flow events that occurred in the intervening 108 years. Tamarisk is now common here, along with coyote willow. This low density of riparian vegetation is common in San Juan Canyon; significant woody riparian vegetation occurs only in wide sections, which are uncommon, and at tributary junctures. (D. Oldershaw, Stake 4319.)

Figure 8.19 Photographs of the San Juan River at Slickhorn Rapid.

A. (August 7, 1921.) Slickhorn Rapid is a minor reach of whitewater on the San Juan River and forms around a debris fan from Slickhorn Gulch. In this view, very little riparian vegetation is present, and it appears to be mostly coyote willow. The channel of Slickhorn Gulch appears to have a continuous line of coyote willow. (R. N. Allen 66, courtesy of the Dan O'Laurie Museum, Moab, Utah.)

B. (August 30, 1997.) Slickhorn Rapid is now the last rapid upstream from Lake Powell. The 1921 view was first matched in 1973, and the vegetation along the river appeared to be remarkably similar; tamarisk had not become very common (Baars 1973). Tamarisk lines the banks of the river and disrupts the former line of coyote willow, which is still common. On some banks *(lower left)*, riparian vegetation is very dense. Cottonwood trees, which once were present a short distance upstream in this side canyon, were swept away during a flood in the late 1990s. (D. Oldershaw, Stake 1994.)

ract Canyon (chapter 7), it would be expected to have increased along the San Juan River as well. The trees along the river have mostly small-diameter trunks, suggesting they are young, and few older trees were observed. It is possible that the 1911 flood damaged this species significantly, and what remains along the river are root-crown sprouts from individuals whose above-ground biomass was damaged or destroyed.

Photographs within the narrow San Juan Canyon show the limited available space for riparian vegetation, both in the 1890s and at present (fig. 8.18). The substrate along the channel is typically sandy with low water-holding capac-

ity, and the stage variation during even small floods is high; both of these factors limit the riparian species that can grow in this setting. Only where wide debris fans are present, such as at the mouths of Johns Canyon and Slickhorn Gulch (fig. 8.19), can significant amounts of riparian vegetation grow. In these settings, slowing of river flow upstream from the constriction allows deposition of fine-grained sediments along the channel margin, and growth of riparian vegetation becomes lush. Downstream from rapids, riparian vegetation can be considerable on the upslope parts of sandbars (fig. 8.19B).

Little is known of nineteenth-

century beaver trapping along the San Juan River, although poor descriptions suggest that James Ohio Pattie may have trapped in San Juan Canyon in the 1820s.[35] Beaver remain common in San Juan Canyon, as well as in the alluvial reaches upstream. Riparian vegetation used by beaver, as suggested by stumps and damaged branches along the river, includes tamarisk, willow, and netleaf hackberry. As in other bedrock canyons, beaver cull riverine riparian vegetation—both native and nonnative species—and limit its potential for establishment and growth.

9 The Escalante and Paria Rivers

Summary. The Escalante and Paria Rivers have both alluvial and bedrock-controlled reaches that drain the scenic Tertiary and Mesozoic terranes of southern Utah. Although flow is diverted from both rivers, neither is regulated to the extent that floods are reduced in size, and, with only local exceptions, neither river has been channelized. These rivers have a number of contrasts between them, but riparian vegetation has increased significantly along both in the reaches documented by historical photography. Both watercourses once supported extensive stands of low-statured riparian shrubs—primarily coyote willow—that were removed during large floods and arroyo cutting. These stands were replaced at first by tamarisk, with a later infilling of native species and Russian olive. The increases are particularly noteworthy in the bedrock canyon reach of the Escalante River, which was mostly devoid of woody riparian vegetation in the first half of the twentieth century. The increases are in native and nonnative species, in particular tamarisk and Russian olive, both of which are locally dense on the Escalante River and on the upper reaches of the Paria River. Along both rivers, cottonwood was not common earlier in wide reaches where it now forms galleries; it probably colonized those reaches from seed produced in smaller tributaries.

The Escalante and Paria Rivers share common headwaters in the plateau region of south-central Utah (fig. 9.1). They drain 1,730 and 1,410 square miles, respectively,[1] and these watersheds drain the colorful Tertiary and Mesozoic terranes of this region, which are known as the Grand Staircase.[2] These drainage basins have common morphological characteristics: both rivers have headwaters in mesas that range from 8,000 to 10,000 feet, and just downstream both rivers flow through alluvial reaches dominated by arroyos. Finally, both rivers enter bedrock canyons downstream from the arroyos that control the amount of historical channel change while restricting growth of riparian vegetation. The Escalante River supports a more extensive riparian system than does the Paria River because of two factors: its basin generally is higher in elevation than the Paria River basin, and the latter basin has a higher proportion of Cretaceous rocks, which are clay-rich and highly saline. Seventeen historical photographs were matched along the Escalante River, in both the alluvial and bedrock reaches; thirty photographs of the Paria River were matched, but all were near the gaging station and the confluence with the Colorado River.

The Escalante River

The headwaters of the Escalante River are on the Aquarius Plateau, yet this river appears to begin on the saddle between the Table Cliffs and the northern end of the Kaiparowits Plateau (fig. 9.1). This broad flat, known as Upper Valley, gathers several tributaries to form Upper Valley Creek, which flows eastward through an alluvial valley. About 5 miles west of the town of Escalante, Utah, Upper Valley Creek combines with Main Canyon and North Creek to form the Escalante River. Downstream from the town of Escalante, the river abruptly enters a sandstone canyon and joins with Pine Creek, coming from the north; the combined streams flow east, then south, to the Colorado River, passing through some of the most spectacular scenery in North America.

Gaging-station data for the Escalante River and Pine Creek show perennial flow and relatively small floods within the period of record (fig. 9.2). Pine Creek has a mean annual discharge of 5.29 ft³/s; the Escalante River, which includes the flow of Pine Creek, has a mean annual discharge of 12.6 ft³/s, indicating that about 7.3 ft³/s come from upstream of Escalante. Although the flood of record on the Escalante River is 4,550 ft³/s on August 24, 1998 (1944 through 2002), larger floods have occurred in the past. Flood deposits preserved about 2 miles downstream from the gaging station were correlated with the floods of August 1932 and September 1909, which yielded discharges of 21,000 and 20,000 ft³/s, respectively.[3] These floods are among the largest preserved in a two-thousand-year paleoflood record.

Surface water is diverted mostly into canals downstream from the junction of Upper Valley Creek, Main Canyon, and North Creek, and this water enters Wide Hollow Reservoir, an irrigation-storage facility north of the river. Similarly, some water is diverted from Pine Creek for limited irrigation in that valley north of the town of Escalante. Groundwater is not used extensively near the Escalante River, and from 1979 to 2002 water levels varied between 9 and 15 feet below land surface at one well about 2 miles west of Escalante (fig. 9.3). This suggests that groundwater levels remain high along the Escalante River in its alluvial reaches.

There is no evidence that Upper Valley Creek or the other channels that coalesce to form the Escalante River had significant numbers of Frémont cottonwood before settlement in 1875.[4] During a Powell Expedition excursion in June 1872, Almon Harris Thompson describes Upper Valley as "swampy" and mentions that, despite nearly continuous rain, "every time a little stream gathers it sinks into

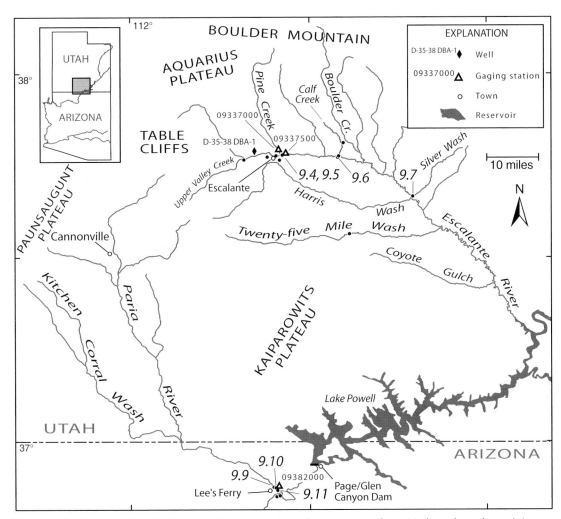

Figure 9.1 Map showing the Paria and Escalante River basins in southern Utah and northern Arizona.

the sand."[5] No riparian plants taller than coyote willow are visible in the first photograph of the river (fig. 9.4). However, another description from the same expedition, from a point 2 miles downstream from present-day Escalante in the bedrock canyon, stated, "[we] found large cotton wood trees, some pine and box elders."[6] Ring counts on four cottonwood in this reach in 1985 indicated that the trees were in excess of 150 years old, and the oldest tree was 283 years old.[7]

The channel from Upper Valley through the town of Escalante was a discontinuous arroyo with low banks at the time of settlement.[8] Beginning with a large regional storm at the end of August 1909, floods scoured the floodplain, cutting a continuous arroyo. This episode of arroyo cutting was the most recent in a series of late Holocene downcutting events, although its cross-sectional area is much larger than others preserved in the alluvial stratigraphy.[9] In photographs

taken at the end of a series of floods in 1932, the channel of the Escalante River appears scoured and devoid of riparian vegetation.[10] Three matches of these photographs—in 1984, 1999, and 2002—show that the width of the channel has decreased and that riparian vegetation has increased, despite the 1998 flood. The scoured condition was still apparent at the Escalante River gaging station in the 1940s and 1950s (fig. 9.5). As shown in figure 9.5, the vegetation increased between the 1950s and the present.

Downstream from Escalante, perennial flow increases, primarily because of contributions from tributaries draining the Aquarius Plateau. At Calf Creek Crossing, the river's floodplain was at one time wide and barren (fig. 9.6). In the later part of the twentieth century, the floodplain was colonized with cottonwood, coyote willow, tamarisk, and Russian olive. The channel has narrowed by about one-fifth of its former width. Similarly, downstream

at the mouth of Harris Wash (fig. 9.1), riparian vegetation increased after the middle of the twentieth century, causing substantial decreases in channel width (fig. 9.7). Grazing animals were removed from this reach in the 1980s, well after encroachment by both native and nonnative species.

The Paria River

The Paria River drains both the Aquarius and Paunsaugunt Plateaus of south-central Utah (fig. 9.1). Most of the drainage area of this river occurs in Cretaceous rocks that are fine grained and laden with salts. As suggested by the rugged terrain of Bryce Canyon National Park, on the eastern side of the Paunsaugunt Plateau, and the extensive badlands within the Tropic Shale, sediment production and transport are high in the Paria River basin. Several streamflow-sediment samples collected at the Lee's Ferry gaging station have the highest sediment

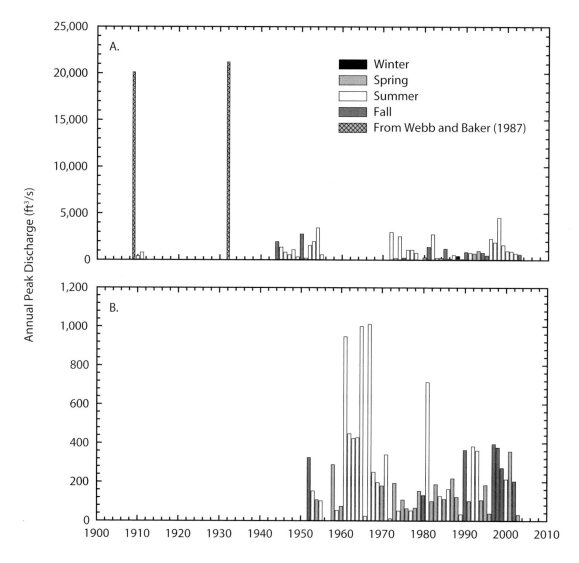

Figure 9.2 Annual flood series for the Escalante River basin.

A. Annual flood series for the Escalante River near Escalante, Utah (station 09337500; 1909–1911, 1943–1955, 1972–2003).

B. Annual flood series for Pine Creek near Escalante, Utah (station 09337000; 1952–1955, 1958–2003).

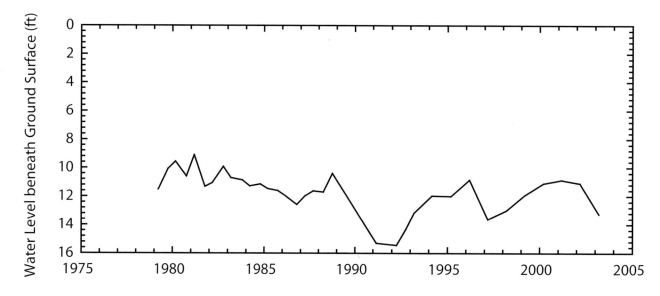

Figure 9.3 Groundwater level data for well D-35-38 DBA-1 west of Escalante, Utah.

concentrations measured in the United States.[11]

Floods and Flow Regulation

The annual flood series of this river shows a striking pattern of decreased flood peaks,[12] despite the fact that regulation and diversions have not significantly changed in the basin (fig. 9.8). Ongoing construction and maintenance of stock ponds might be one reason for the decrease in flood frequency. At Lee's Ferry, the Paria River has a mean annual discharge of 28.6 ft³/s. The peak flood in the gaging record is 16,000 ft³/s on October 25, 1925; analysis of paleoflood deposits indicates that higher discharges occurred in the past, including a flood of a little less than 17,000 ft³/s in September 1909.[13]

Observations of Riparian Vegetation

The first written accounts of the confluence of the Paria and Colorado Rivers, made during the Domínguez-Escalante Expedition of 1776,[14] do not describe riparian vegetation or conditions along the Paria River. Mormon explorers and military expeditions of the mid–nineteenth century tersely reported riverine conditions within the context of the otherwise-described desolation of this locale. Thales Haskell, on an exploratory expedition with Jacob Hamblin in October 1859, saw "considerable beaver sign" and cottonwood along the Paria River in the 5-mile reach upstream from the confluence.[15] In early March 1869, Edwin G. Woolley, looking down from the cliffs above Lee's Ferry, saw "a few cottonwoods" but "no wood . . . of any consequence" along the Paria River.[16] Most nineteenth-century visitors, notably members of the Powell Expe-

ditions, mention the dense willows along both the Paria and Colorado Rivers, among which they established camps.

Settlement of the Paria River basin began in 1870 with farming and ranching in the headwater area,[17] but downstream of the alluvial reaches near Cannonville the river enters a series of bedrock-controlled reaches or canyons. South of the Utah-Arizona border, the river enters a labyrinthine canyon in Triassic-Jurassic rocks, primarily Navajo Sandstone, as it winds its way to the Colorado River at Lee's Ferry. This historic site was first visited by the Domínguez-Escalante Expedition of 1776; then by Mormon pioneer Jacob Hamblin, first in 1858 and several other times in the 1860s; then by the Powell Expedition in 1869 and 1871; and then again by the notorious John D. Lee, who arrived here to settle on Christmas Day in 1871.[18]

Lee, who was executed in 1876 for his involvement in the Mountain Meadows Massacre,[19] was earlier sent to Lee's Ferry as an exile from the southern Utah settlements, as the Mormon presence at the main crossing of the Colorado River, and as a ferryman for traffic between Utah and Arizona. Lee's Ferry became a focal point of travel from southern Arizona to the Arizona Strip and southern Utah, affording the only reliable and easy place to cross the Colorado River between the foot of Grand Canyon and Moab, Utah. Because of its importance, Lee's Ferry was photographed extensively, except by the photographers of the Powell Expedition, who were the first on the scene but inexplicably did not record its scenery. William Bell, one of two photographers of the Wheeler Expedition, took the first photographs here in 1872.[20] Timothy O'Sullivan, Bell's predecessor and replacement, took several more in 1873.[21]

Repeat Photography

O'Sullivan took many photographs of the Lee's Ferry area, including a classic view across the Colorado River toward the mouth of the Paria, which was first matched in 1972.[22] The original photograph shows the Colorado and Paria Rivers lined with coyote willow and arrowweed, but no trees; the match shows a sea of tamarisk. However, Bell took panoramas about 2 miles up the canyon, probably in November 1872.[23] Individual frames of this series show that the canyon upstream and downstream sustained a sparse riparian system composed of scraggly shrubs (probably coyote willows) and what appear to be struggling, small cottonwood trees (fig. 9.9). This reach now has extensive stands of tamarisk, with scattered, relatively large cottonwood.

Lee struggled to develop the water of the Paria River to irrigate his lands at Lonely Dell Ranch. His dam repeatedly washed out, and channel downcutting forced him to move the dam farther and farther upstream.[24] As in other cases of arroyo cutting in the region, the channel widened and deepened, eventually shifting the confluence with the Colorado River about one-quarter of a mile downstream from the location photographed by O'Sullivan. The riparian vegetation once present was scoured (figs. 9.10 and 9.11), although historical photographs show newly established shrubs along the channel's ragged margin. By the early 1980s, the channel had narrowed, at least partly in response to the thickets of tamarisk that became established along the banks of the Paria and Colorado Rivers. Cottonwood, coyote willow, and other shrubs are also common but easily overlooked in the sea of tamarisk.

A

B

C

Figure 9.4 Photographs of the Escalante River Canyon.

A. (June 2, 1872.) While searching for the mouth of the Dirty Devil River to retrieve a boat cached the previous summer, members of the Powell Expedition traveled into the Escalante River basin. They took this downstream view to document the change from alluvial valley to what they perceived as an impassible bedrock canyon. Pine Creek enters from the left. The dense riparian vegetation nearly obscuring the channel appears to be mostly coyote willow, and sagebrush covers the floodplain. (J. Fennemore 583, courtesy of the National Archives.)

B. (October 14, 1988.) The channels of both Pine Creek and the Escalante River are lined with a mixture of cottonwood, tamarisk, and coyote willow. No Russian olive was observed in the reaches visible from this camera station. The Escalante River gaging station shown in figure 9.5 is at lower center and downstream from the confluence. (R. H. Webb.)

C. (June 7, 2002.) Although some trees have disappeared, particularly at the most upstream point that is visible, most of the trees present in 1988 are now larger. Russian olive is relatively common here in the form of young individuals, suggesting that this nonnative species is spreading downstream from dense stands near Escalante. (D. Oldershaw, Stake 1245.)

Figure 9.5 Photographs of the Escalante River at its gaging station.

A. (October 24, 1957.) This upstream view shows the gaging station on the Escalante River just downstream from the mouth of Pine Creek. The discharge is 3.2 ft³/s. With the exception of one tamarisk tree at far left, riparian vegetation is absent. (W. McConkie, courtesy of the U.S. Geological Survey, Utah District.)

B. (November 5, 2002.) Although the trees are mostly leafless in this early-winter view, several cottonwood trees are now apparent. Nonnative tamarisk and Russian olive are common, and coyote willow as well as other native, more xeric shrubs are apparent. A recently established Ponderosa pine appears at far left. (R. H. Webb, Stake 4425.)

A

B

Figure 9.6 Photographs of the Escalante River at the Calf Creek Crossing.

A. (1948.) This downstream view shows the Utah Highway 12 Bridge over the Escalante River just upstream from its confluence with Calf Creek. The floodplain has two cottonwood trees, but otherwise is mostly devoid of woody vegetation (J. Breed, courtesy of *National Geographic* magazine). This view is similar to another view (not shown) taken in 1951 (Brownlee 93.37.1088, Cline Library.)

B. (1959.) Although not an exact match, this view shows much of the same reach as photograph A. Note that the number of cottonwood trees has increased, and small plants that appear to be tamarisk form a line adjacent to the channel, which recently flooded. Craig "Sage" Sorenson, now retired from his job as a recreation planner for Grand Staircase–Escalante National Monument, appears with his mother and is the taller of the two children in the view. (C. Sorenson, private collection.)

C

D

C. (October 5, 1984.) Cottonwood have increased in size and density on the floodplain as the width of the channel has decreased; coyote willow is also common adjacent to the active channel. Half of the bridge span is blocked from view by the increased vegetation. Nonnative tamarisk is dense here, and Russian olive is relatively uncommon. Toni Yocum appears in the view. (R. H. Webb.)

D. (June 7, 2002.) The bridge is barely visible owing to the extensive growth of riparian vegetation. Russian olive is now common here and appears in a dense stand just behind the visible part of the bridge. Tamarisk is now mostly beneath a closed canopy of cottonwood, and coyote willows form a band between the tamarisk and the open water of the river. Dominic Oldershaw appears in the view. (R. H. Webb, Stake 1244.)

Figure 9.7 Photographs of the Escalante River above the mouth of Harris Wash.

A. (1953) This downstream view shows the Escalante River above the mouth of Harris Wash, which enters from the right, behind the low mesa at center. Several small cottonwood are present on the broad, mostly barren channel and low terraces. (D. Brownlee and A. Brownlee, 93.37.1170, courtesy of the Cline Library.)

B. (October 7, 2004.) The channel is now choked with riparian vegetation, primarily cottonwood, coyote willow, young black willow, Russian olive, and tamarisk. About half of the plants along the channel are Russian olive, and coyote willow increases in abundance with distance away from the channel center but below the first higher terrace. Tamarisk is locally abundant in groves but is in much lower quantity than cottonwood and coyote willow. (D. Oldershaw, Stake 4806.)

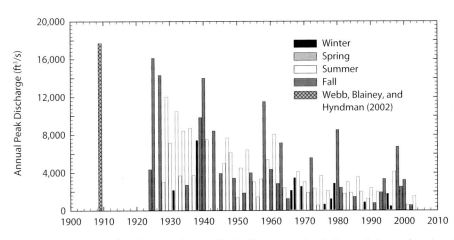

Figure 9.8 Annual flood series for the Paria River near Lee's Ferry, Arizona (station 9382000; 1909, 1924–2003).

A. (November 1872.) This view is one of four photographs in a panorama from north to south of the channel of the Paria River about 2 miles upstream from Lee's Ferry. In the midground, the Paria River is lined with small Frémont cottonwood that are mostly leafless. Smaller shrubs, probably coyote willow, line the bank of a relatively narrow channel. (W. A. Bell 267, courtesy of the National Archives.)

B. (May 25, 1992.) The riparian vegetation along the channel in the distance is now mostly tamarisk with some coyote willow. Cottonwood are no longer present in this view, but they remain in this reach. (R. H. Webb, Stake 2254a.)

Figure 9.9 Photographs of the Paria River upstream from Lonely Dell Ranch at Lee's Ferry, Arizona.

A

B

Figure 9.10 Photographs of the Paria River at Lee's Ferry, Arizona.

A. (1921.) This view, from a camera station on the cliff above the gaging station, shows the Paria River flowing onto its delta in the Colorado River. In 1912, the mouth of the Paria River shifted to below the delta (Hereford 2004), and in this view the upper channel is filled with sediment and covered with a dense thicket of willows and arrowweed. The only cottonwood that appear in this view are around Lonely Dell Ranch at lower right. (Kolb 568–676, courtesy of the Cline Library.)

B. (May 5, 1994.) The channel of the Paria River has narrowed and become lined with tamarisk and coyote willow in the intervening seventy-three years. The difference in stature between native willows and nonnative tamarisk is particularly apparent in a comparison of the left midground, where the Paria River flows away from the cliff on the left. (D. Oldershaw, Stake 2854.)

Figure 9.11 Photographs of the Paria River at Lonely Dell Ranch, Arizona.

A. (September 1, 1915.) This right half of a panoramic photograph documents the change in riparian vegetation along the Paria River. The river, which flows from right to left, has a broad, nearly denuded floodplain in the right foreground, and the right bank is vertical at lower right. (E. C. La Rue 299, courtesy of the U.S. Geological Survey Photographic Library.)

B. (November 11, 1999.) In the intervening eighty-four years, the channel of the Paria River has narrowed considerably, in part because sediment has been stored in low floodplain deposits within the higher channel banks, and possibly because the channel deepened (Hereford 2004). The center of the channel has shifted slightly to the left, and riparian vegetation—mostly dense tamarisk with willows along the channel margin—has encroached on the formerly wide channel, filling in the right floodplain. (D. Oldershaw, Stake 3990.)

Summary. The Little Colorado River drains much of the southern Colorado Plateau and sustains the largest extent of riverine riparian vegetation in northeastern Arizona. Moenkopi Wash, one of its tributaries, is one of the major, low-elevation watercourses draining the Navajo and Hopi Reservations. Alluvial reaches of the Little Colorado River once sustained groves of cottonwood trees that locally were described as continuous gallery forests or bosques, but their spatial distribution is not known. Following a series of floods and arroyo downcutting, most of the riparian vegetation along the Little Colorado River was removed or died from drainage of the alluvial aquifer. Twentieth-century changes include growth of woody riparian vegetation, especially tamarisk, with lesser amounts of native species, on terraces and within the arroyo channel. Owing to high soil and groundwater salinity, tamarisk favors riparian zones in this region.

Upstream from the gaging station that is downstream from Cameron, Arizona, the Little Colorado River drains 26,500 square miles of the southern Colorado Plateau in northern Arizona and extreme northwestern New Mexico (fig. 10.1). The main stem has its headwaters in the White Mountains of eastern Arizona, flowing northward from the Mogollon Rim toward its confluence with the Rio Puerco at Holbrook. The combined watercourses flow west-northwest into Grand Canyon, where the Little Colorado River is the principal tributary of the Colorado River. Like a miniature version of the Colorado River, the Little Colorado receives much of its base flow from snowmelt runoff from the rim country that divides the Colorado Plateau from the Arizona central highlands, but it derives its high sediment load from the sedimentary rocks of the Colorado Plateau.

The southern Colorado Plateau has an arid to semiarid climate and a landscape dominated by exposed sedimentary bedrock, mostly limestone, sandstone, and shale. These rocks are generally sedimentary strata of Mesozoic and Tertiary age that weather to barren slopes and produce high sediment yields.[1] Soil salinity is high in this region owing to the combination of ancient marine-deposited sediments and the current arid and semiarid climate. Eolian activity also occurs high on the southern Colorado Plateau, creating large sand sheets. Some of these sheets alternate between stability and instability owing to the combination of land uses and climatic fluctuations. The rates of alluviation are high in this region, and many of the watercourses flow through arroyos deeply incised into their floodplains. Many rivers and streams are perennial or intermittent and depend on groundwater discharge from bedrock aquifers for their flow.

Riparian vegetation in this environment faces the combined effects of a harsh environment, unstable channels that strongly respond to arroyo cutting and filling, high salinity of surface and groundwater, and high streamflow sediment concentrations that encourage high rates of deposition on floodplains. In addition to these factors that discourage riparian vegetation are the combined effects of water diversions, groundwater withdrawal, and clearing of floodplains for agriculture. These characteristics suggest why tamarisk in particular is so abundant along watercourses in this region. It was likely established by the 1930s and became locally abundant by 1960.[2]

Historical Observations

The Little Colorado River shares a history of channel change that generally applies to the entire Colorado Plateau. Spanish explorers in search of the fabled Cities of Gold crossed the Little Colorado River in 1583 in the vicinity of present-day Winslow, Arizona.[3] In the sixteenth century, the river was described as containing groves of willows (likely coyote willow) and Frémont cottonwood, and the Spaniards named it Río Alameda (the River of Cottonwood Groves). Unfortunately, the diaries of the Spaniards did not describe either the spatial extent or the density of cottonwood.

By the mid–nineteenth century, the river flowed through a channel 8 to 10 feet deep with scattered cottonwood trees. In 1857, Lieutenant E. F. Beale described the confluence of the Little Colorado River and the Rio Puerco: "There is abundance of large cottonwood trees in the bottom, which resembles very nearly the bottom of the Rio Grande."[4] Beale did not mention cottonwood trees near present-day Winslow, but he did note abundant grasslands, high groundwater levels, and the presence of beaver. In 1864, Pratt Allyn observed bosques of cottonwood "at least a half a mile across."[5] Clearly, considerable stands of cottonwood were present along this river, particularly in the wide, alluvial reaches between Holbrook and Winslow.

Mormon settlers had a difficult time actually finding the channel near present-day Holbrook at the time of settlement in the 1870s. Descriptions upstream from Grand Falls tell of a channel lined with cottonwood of varying ages and supporting a population of beaver. The channel remained narrow until floods increased in the early 1880s, culminating in the 1891 event. An even larger flood occurred in 1923, exacerbated in part by failure of a reservoir upstream near St. Johns (fig. 10.1). Wider channel conditions

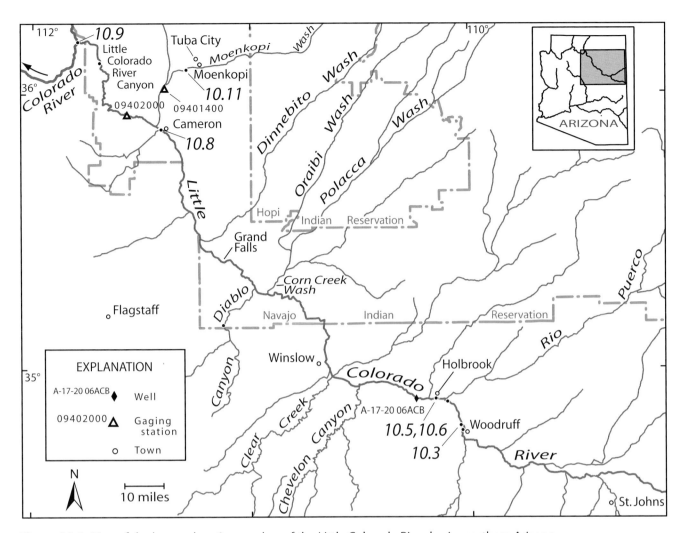

Figure 10.1 Map of the lower-elevation reaches of the Little Colorado River basin, northern Arizona.

persisted through the early 1940s, when deposition of low floodplains began to narrow the channel.[6] Riparian vegetation, which appears to have been generally sparse on the Colorado Plateau, declined in response to the channel changes; many cottonwood trees were swept downstream in the 1891 and 1923 floods, but others remain on a terrace level now referred to as the Cottonwood Terrace.[7]

Cottonwood that germinated in the late-nineteenth- and early-twentieth-century floods are present on a high terrace that predates 1880.[8] During the drought between 1891 and 1904, the Navajo fed cottonwood to their cattle.[9] In 1936, the river "flows, when it does flow, through a wide treeless valley" although "a few old gnarled cottonwood trees grow near the river bank."[10] Tamarisk was likely introduced into the basin in the late nineteenth century and is known to have been planted next to the river at Winslow in 1909. By the

mid-1930s, it was present along the river,[11] but the channel was wide and devoid of vegetation before the 1950s. By 1952 and especially by 1956, tamarisk was established in dense groves along the Little Colorado River.[12]

Floods and Flow Regulation

From the time of Mormon settlement through the twentieth century, the Little Colorado River was known both for its floods and for its periods of low flow. Floods caused irrigation-diversion structures to fail; for example, thirteen dams failed at St. Joseph, and ten failed at Woodruff before the turn of the twentieth century.[13] The flood of February 1891 on the Little Colorado River caused extremely high water in Holbrook.[14] At the Woodruff gaging station, the flood of record is 25,000 ft^3/s in December 1919 (fig. 10.2), compared with an annual runoff of 51 ft^3/s. However, the flood of Septem-

ber 19, 1923, peaked at 50,000 ft^3/s at Woodruff (not included in the official record) and 120,000 ft^3/s at Cameron (fig. 10.7), and the peak discharge of this flood may have been increased by dam failures upstream. The largest floods on the Little Colorado River occur in fall and winter, although the second-largest flood recorded at Woodruff occurred in July 1940 (fig. 10.2). Floods in recent decades have been small in comparison to those that occurred in the early twentieth century.

Only two dams exert any control over flow of the Little Colorado River, and neither has a large influence unless it fails during floods. Lyman Lake, impounded by an unnamed dam, has a long history of failure and reconstruction. The original dam, built in 1886, impounded Salado Reservoir but failed during a flood in 1903.[15] Rebuilt and renamed Lyman Dam, it failed again during the regional floods of 1915.

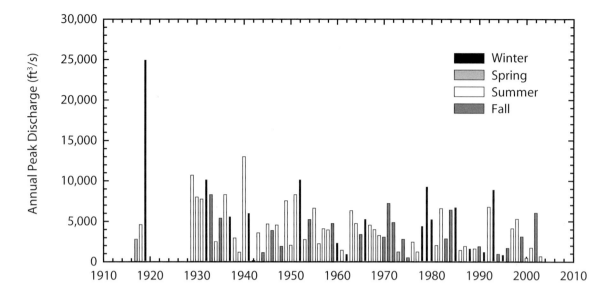

Figure 10.2 Annual flood series for the Little Colorado River at Woodruff, Arizona (station 09394500; 1917–1919, 1929–2003).

Rebuilt a third time, it also reportedly failed at least partially during the 1923 flood and may have contributed to the extremely large discharge of this event. Among other small impoundments in the upper part of the watershed, St. Johns Reservoir supplies water for agriculture near St. Johns, Arizona.[16] The amount of agricultural acreage upstream from Lyman Lake is relatively small and confined mostly to the river's floodplain.

Changes in Riparian Vegetation

No historical photographs show the condition of riparian vegetation in the nineteenth century to support the historical observations. At and downstream from Woodruff, twenty-nine historical photographs of the gaging operations document changes in riparian vegetation. The original photographs show conditions that are typical in the region at this time: the arroyo is wide and barren of riparian vegetation, and scattered cottonwood and other riparian vegetation occupy the high banks (fig. 10.3A). Riparian vegetation, mainly tamarisk, has increased in this ephemeral reach.

Just upstream from Holbrook (fig. 10.1), the Rio Puerco joins the Little Colorado River, and the combined watercourses form a wide, sandy channel. Groundwater levels are generally high in the agricultural areas downstream from the confluence,[17] allowing devel-

opment of a significant agricultural area. Springs are also common in the Holbrook area, again indicating that groundwater levels are high in this part of the watershed. Large fluctuations in groundwater levels have occurred during recent drought periods (fig. 10.4).

Because of the considerable sediment load of the Rio Puerco, the channel of the Little Colorado River is sandy and braided at and downstream from Holbrook (figs. 10.5 and 10.6). The flood of September 1970, caused by Tropical Storm Norma, was larger in the Rio Puerco watershed; bridges in and downstream from Holbrook were damaged, including the two bridges shown in the repeat photography (seven views). Although tamarisk has increased in the channel of the Little Colorado River at Holbrook, and Russian olive is present, large increases have also occurred in native species, in particular cottonwood, coyote willow, and seepwillow (figs. 10.5 and 10.6).

The Little Colorado River flows northwest from Holbrook through a wide valley past Winslow to the Navajo Indian Reservation. Grand Falls of the Little Colorado River (fig. 10.1) is the result of a natural dam formed by a Pleistocene lava flow that originated several miles south of the falls. Downstream from Grand Falls, the Little Colorado River passes through a short bedrock canyon and then flows through a wide alluvial valley constrained by low bedrock walls. Between

Woodruff and Cameron, the drainage area increases from 8,072 square miles to 26,459 square miles, including the major tributaries Chevelon and Clear Creeks from the Mogollon Rim country as well as Dinnebito, Corn Creek, and Moenkopi Washes (fig. 10.1). As a result, the gaging record at Cameron shows considerably larger floods that occurred mostly in fall and winter. The size of these floods appears to have decreased in magnitude through the twentieth century (fig. 10.7).

At Cameron, photographs taken at the old bridge show a wide, barren channel in 1914 (fig. 10.8). Four photographs at this site show that the channel narrowed considerably in the twentieth century. This reach now supports considerable riparian vegetation, particularly tamarisk, with less abundant Frémont cottonwood, coyote willow, and mesquite.

Downstream from Cameron, the gorge of the Little Colorado River deepens through the well-known Paleozoic sequence of Grand Canyon. All but the last 13 miles of this reach are ephemeral and subject to extreme stage fluctuations during floods. Beginning at a series of springs, of which Blue Springs and the Sipapu are the best known, the Little Colorado River is perennial. Spring discharges are high in dissolved calcium carbonate, like flow in Havasu Creek (chapter 12), and travertine dams are common. Near the mouth, numerous photographs,

Figure 10.3 Photographs of the Little Colorado River at Woodruff, Arizona.

A. (June 1939.) This upstream view shows the ephemeral channel of the Little Colorado River downstream of Woodruff. The channel at this point is a shallowly incised arroyo, and floods that occurred before this photograph was taken scoured the channel clear of riparian vegetation. A lone Frémont cottonwood appears on the left bank of the arroyo. The low, wide shrubs in the center midground cannot be identified. (H. S. Hunter 2562, courtesy of the U.S. Geological Survey.)

B. (September 22, 2003.) Channel erosion has continued, with perhaps 3 feet of downcutting in the years intervening between the original photograph and the match. The cottonwood has died, and the stump remains on the far bank, but coyote willow is now present. Tamarisk is now extensive in this reach, as illustrated by the prominent individual in the foreground. (D. Oldershaw, Stake 4736.)

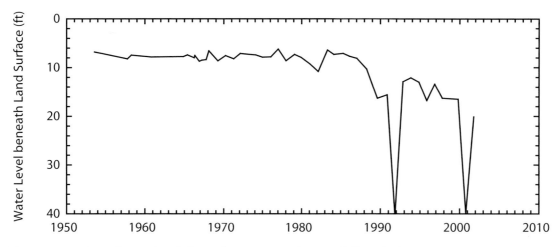

Figure 10.4 Groundwater levels for well A-17-20 06ACB near Holbrook, Arizona.

A. (July 28, 1949.) This upstream view, taken from a bridge for the Apache Railroad, shows a bridge over the Little Colorado River at Holbrook. The Little Colorado River here is ephemeral and has a sand bed. Bank protection at the extreme left protects the right bridge abutment. The riparian trees on the extreme right are not shown with resolution sufficient for positive identification, but it is likely that they are young Frémont cottonwood. (R. B. Sanderson, courtesy of the U.S. Geological Survey.)

B. (October 3, 2003.) Both bridges have been replaced, possibly following the 1970 floods, and a new pipeline crossing is present downstream from the new highway bridge. The channel is now filled with riparian and xerophytic shrubs, including coyote willow and rubber rabbitbrush. Cottonwood is prominent at right, and both tamarisk and Russian olive are present in this reach. Bank protection has been installed on both sides of the channel, narrowing its width. (D. Oldershaw, Stake 4699.)

Figure 10.5 Photographs of the Little Colorado River at Holbrook, Arizona.

Figure 10.6 Photographs of the Little Colorado River at Holbrook, Arizona.

A. (July 28, 1949.) This view, from the bridge shown in figure 10.5, shows downstream channel conditions on the Little Colorado River at Holbrook. The channel is devoid of riparian vegetation as far downstream as is visible, and only a few small cottonwood trees appear on the left side. (R. B. Sanderson 768A, courtesy of the U.S. Geological Survey.)

B. (October 3, 2003.) Owing to floods, notably the one in September 1970, new revetment that protects the city of Holbrook appears on the right side. The original bridge has been replaced, and the new railroad bridge appears in the distance behind the pipeline bridge. The new riparian vegetation in the channel is mostly coyote willow and tamarisk, but cottonwood have grown on river left *(left side),* and rubber rabbitbrush is locally dense in the floodplain. A few individuals of Russian olive are in this reach as well. (D. Oldershaw, Stake 3714.)

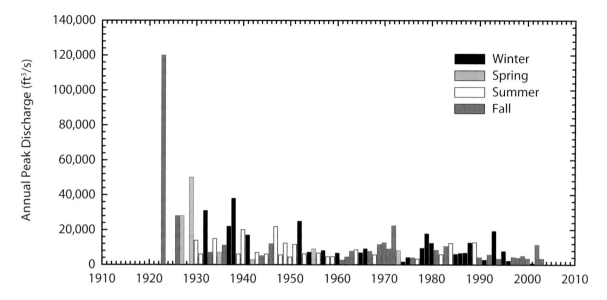

Figure 10.7 Combined annual flood series for the Little Colorado River at Cameron, Arizona (09402000), and the Little Colorado River at Grand Falls, Arizona (station 09401999; 1922, 1926–2003).

beginning with views taken by Jack Hillers of the Powell Expedition of 1872, document changes in riparian vegetation along both the Colorado and the Little Colorado Rivers. Riparian vegetation—mostly tamarisk with some mesquite—has increased along the channel since 1890 (fig. 10.9).

Moenkopi Wash

Moenkopi Wash is a major tributary of the Little Colorado River on the northern side of its drainage basin. Draining more than 1,600 square miles of lands on the Navajo and Hopi Reservations, this watercourse is unregulated, with only about 500 acres in the watershed under cultivation. The annual flood series for this watercourse is a composite of gaging records from three

locations (fig. 10.10); its flood history is similar to that of the Little Colorado River. Floods in the early twentieth century were larger than those later, with resulting adjustments to channel geometry and increases in riparian vegetation.

Although several important expeditions, including parties of Mormon emissaries seeking peaceful trade with the Hopi, crossed Moenkopi Wash, few individuals noted its presence. John Wesley Powell described Moenkopi Creek near the town of Moenkopi as having "a little fringe of green willows, box-elders, and cotton-woods."[18] Drawings made at the time of this description indicate that Hopi farmers had cultivated most of the floodplain between the bedrock cliffs but left an indistinct river corridor with barely

discernable riparian vegetation. Arroyo downcutting in this reach occurred at an unknown time after this description and probably eliminated this riparian ecosystem.

Eleven historical photographs at or near the former gaging station near Tuba City were evaluated for changes in riparian vegetation (fig. 10.11). This arroyo system, which developed at an unknown time before 1930 (and likely in the late nineteenth century), appears to have had a wide, barren channel in the early 1930s. Development of a low floodplain, likely after 1942,[19] and channel widening changed the riverine environment here considerably. Tamarisk is the dominant species along this channel, although scattered groves of cottonwood occur upstream and downstream from this site.

Figure 10.8 Photographs of the Little Colorado River at Cameron, Arizona.

A. (September 1914.) In 1914, Herbert Gregory paused at the north *(river right)* abutment of the bridge over the Little Colorado River at Cameron and took this downstream view. At this point, the river is in a shallow canyon cut through the Triassic Moenkopi Formation. The broad, denuded floodplain is typical of the condition of floodplains throughout this region at the end of the nineteenth century. (H. E. Gregory P00134:7, courtesy of the Marriott Library.)

B. (July 9, 2000.) Despite intensive land uses of the floodplain, including ongoing livestock grazing and agricultural clearing, both tamarisk and native species have colonized the open space. The native species present include cottonwood trees *(left midground and center foreground)* and mesquite *(distal margin of the floodplain)*. Flood frequency on the Little Colorado River decreased as the twentieth century progressed, and the channel has narrowed in response both to this decrease and to the increase in riparian vegetation. (D. Oldershaw, Stake 3582.)

A

B

Figure 10.9 Photographs of the Little Colorado River near its confluence with the Colorado River.

A. (January 20, 1890.) During a well-documented expedition down the Colorado River (see Webb 1996), Robert Brewster Stanton stopped at the mouth of the Little Colorado River and took this upstream view from a talus cone on channel left. The view shows low-water conditions in this perennial reach. Distinct mesquite individuals line the bank closest to the camera station. (R. B. Stanton 374, courtesy of the National Archives.)

B. (February 24, 1993.) Tamarisk is now the dominant riparian tree along the lower Little Colorado River. Photographs taken by river runners in the 1950s show that tamarisk was well established in the mouth of the Little Colorado River but relatively uncommon along the main-stem Colorado River. Mesquite remains in this reach, and the dead branches protruding from the trees at lower left show that mesquite may have recently died back. (S. Tharnstrom, Stake 1428.)

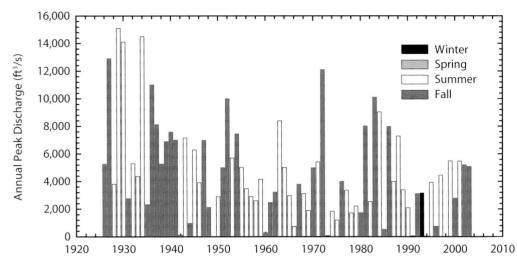

Figure 10.10 Annual flood series for the combined record of Moenkopi Wash near Tuba City and Cameron, Arizona (station 09401399; 1926–1979), and of Moenkopi Wash at Moenkopi, Arizona (station 09401260; 1980–2003).

A. (August 12, 1932.) Moenkopi Wash is an intermittent stream in this reach near Tuba City. With a drainage area fully within the Navajo Indian Reservation, this gaging station *(left midground)* provides the only flow data in the watershed. This upstream view shows an arroyo channel denuded of perennial vegetation, and low terraces have begun to form between the arroyo walls. What appears to be annual Russian thistle, a nonnative species, is on the foreground terrace. (R. E. Cook 6017, courtesy of the U.S. Geological Survey.)

B. (November 7, 2001.) Floods have altered the configuration of arroyo walls, low terraces, and location of the low-water channel, which is now much narrower. Tamarisk is ubiquitous in this wash, forming a nearly monospecific stand, and Russian olive is present nearby. A solitary individual of rubber rabbitbrush appears on the extreme left. (D. Oldershaw, Stake 4304.)

Figure 10.11 Photographs of Moenkopi Wash near Tuba City, Arizona.

11   The Colorado River in Grand Canyon

Summary. Grand Canyon is one of the crown jewels of the national park system. Before regulation by Glen Canyon Dam, the Colorado River supported a sparse riparian ecosystem above the old high-water stage of 100,000 ft³/s. This old high-water zone, which variably consisted of xerophytic species concentrated at densities not found in the surrounding desert today, has a decreasing density of riparian plants with a high mortality of mesquite. The new high-water zone, once a monospecific stand of tamarisk, now contains a cosmopolitan mixture of native and nonnative species, with many native species increasing, possibly at the expense of tamarisk. Unlike plants in the old high-water zone, new high-water zone vegetation benefits from flood-control operations of Glen Canyon Dam and from the hydrologic stability afforded by higher low flows throughout the year. Tributary canyons also have significant increases in riparian vegetation.

The Colorado River through Grand Canyon (fig. 11.1) is the most highly visited reach of whitewater in the southwestern United States. Regulated by Glen Canyon Dam since 1963, the reaches of free-flowing river include 15 miles of Glen Canyon immediately downstream from the dam, 61.5 miles of Marble Canyon, and about 178 miles of Grand Canyon upstream from Lake Mead. The river once had barren banks below the stage reached by 100,000 ft³/s, a level known as the old high-water mark or stage. Now, nonnative and native riparian species clog the banks, forming dense thickets on some banks and scattered stands on others. Cattle and sheep have never grazed in most of the canyon reaches, and feral burros were confined to specific places.[1] Change in riparian vegetation along the river corridor through Grand Canyon is representative of what would be expected to result from flood controls and increases in low flows while maintaining the mean annual discharge.

Floods and Flow Regulation

Before operations of Glen Canyon Dam began in 1963, the Colorado River through Grand Canyon was known for exceptionally large floods. Measurement of discharge began at Lee's Ferry and near Grand Canyon in 1923, with estimates of peak discharge beginning in 1921 (fig. 11.2). From 1923 through 2002, the average daily discharge at the Grand Canyon gage before and after construction of Glen Canyon Dam was 16,900 and 14,100 ft³/s, respectively.

When the largest flood occurred depends on the time frame considered, and our knowledge of past floods with evidence in the geologic record extends only over the past several thousand years. Evidence of the largest flood in the paleoflood record, which has a length of 4,500 years, is preserved 2 miles downstream from Lee's Ferry at a place informally known as Axhandle Bend. As determined from radiocarbon dating of organic material in flood deposits, this flood occurred between A.D. 350 and 750, and the height of the flood deposits indicate that its discharge was about 500,000 ft³/s.[2]

As discussed in chapter 7, three other notable floods occurred before the onset of streamflow records. Downstream at Topock, Arizona, a flood with an approximate discharge of 400,000 ft³/s was estimated long after the fact for the flood in 1862 (chapter 13). An exceptionally large flood of unknown size left a high driftwood pile in Cataract Canyon (chapter 7) that appears fresh in an E. O. Beaman photograph taken in 1871,[3] suggesting it was deposited after the first Powell Expedition of 1869. The largest flood since establishment of Lee's Ferry in 1872 occurred in 1884. Jerry Johnson, son of ferry operator Warren Johnson, rescued a rabbit from an apple tree surrounded by the floodwaters, later estimated to be 300,000 ft³/s.[4] Other historic floods not included in the gaging record may have occurred in 1891, 1912, 1916, and 1917. The 1912 flood is notable because it was probably the event that caused the mouth of the Paria River to shift from the upstream to the downstream side of its massive debris fan (chapter 9).

The largest flood in the gaging record, estimated at 220,000 ft³/s, occurred June 18, 1921. This flood was almost 100,000 ft³/s larger than the second-largest recorded flood, which peaked on July 1, 1927. The largest pre-dam floods peaked between mid-May and early July; the flood of September 1923, which peaked at 112,000 ft³/s, is the one annual peak that did not result from snowmelt.[5] Most of the largest floods on the Colorado River occurred during El Niño years, including those in 1884, 1891, 1905, 1917, 1941, and 1957, although notable floods also occurred during years without El Niño conditions, including in 1921, 1927, and 1952, the first being the year of the largest flood in the gaging record.

Construction of Glen Canyon Dam took seven years and cost nearly three hundred million dollars.[6] At a height of 710 feet, the dam impounds the world's longest lake, inundating about 285 miles of the Colorado River and uncounted miles of its tributaries. When its gates were closed on March 13, 1963, Glen Canyon Dam completely changed the nature of the Colorado River through Grand Canyon. The Bureau of Reclamation operates Glen Canyon Dam to release a long-term average of 8.23 million acre-feet of water per year.

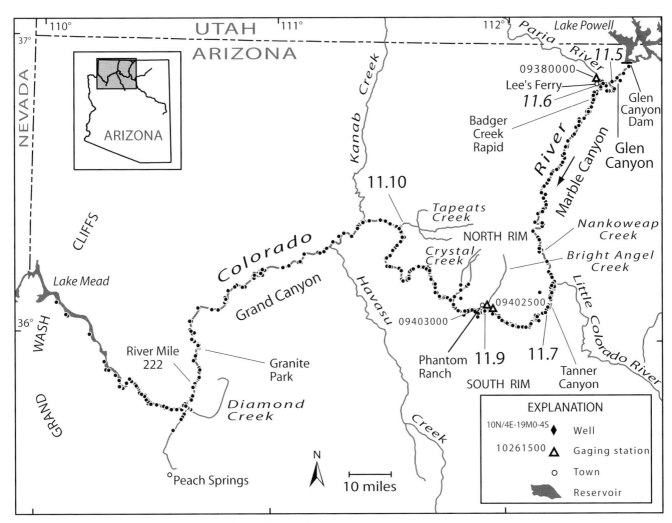

Figure 11.1 Map of the Colorado River through Grand Canyon.

Most of the water released from the dam passes through the hydroelectric power plant, which has a maximum capacity of about 33,500 ft³/s depending on water level in the reservoir. Releases since 1963 typically have fluctuated diurnally with a complex history over the forty years of operation.[7] After regulation, the typical annual peak discharge was around 31,000 ft³/s; the largest dam release—97,300 ft³/s at the Grand Canyon gage—resulted from unusually high runoff into Lake Powell during the El Niño conditions of 1983 (fig. 11.2).

Operation of Glen Canyon Dam has profoundly changed the hydrology of the Colorado River. Before completion of the dam, the two-year and ten-year floods were 76,000 and 139,000 ft³/s, respectively; flood-control operations decreased these floods to 30,000 and 50,800 ft³/s, respectively. Dam releases passing Lee's Ferry average 5 percent of the predam sediment

concentrations, and sediment in the regulated river comes from tributary inputs, primarily from the Paria and Little Colorado Rivers and from bed and bank scour. Diurnal fluctuations, which were relatively minor during the predam period, became a daily occurrence, with few periods of steady flow releases. One extremely important hydrologic change was in the duration of the daily discharge curve; the peaks were decreased, minimizing scour of riparian habitat, and the lowest discharges were raised, providing a less variable water supply to plants growing on the banks (fig. 11.3). As a result of dam operations, the nature of riparian vegetation in Grand Canyon has changed substantially.[8]

Photography of the Colorado River in Grand Canyon

Most of the evidence for change in riparian vegetation in Grand Canyon

comes from repeat photography. The first image of the canyon—a watercolor by Heinrich Balduin Möllhausen made in 1858—shows fantasy riparian vegetation at Diamond Creek and no vegetation along the Colorado River.[9] The first person to photograph Grand Canyon remains debatable,[10] but Timothy O'Sullivan of the Wheeler Expedition generally is given credit, with photographs taken during a grueling up-river trip to Diamond Creek in 1871. E. O. Beaman, Powell's original photographer, followed in early 1872, and James Fennemore, Beaman's replacement, photographed Lava Falls Rapid in April 1872. Finally, William Bell, who replaced O'Sullivan as expedition photographer for one year, photographed from the rim at Toroweap Point, within and at the mouth of Kanab Canyon, and in the vicinity of Lee's Ferry in October and November 1872.[11]

The Powell Expedition pioneered

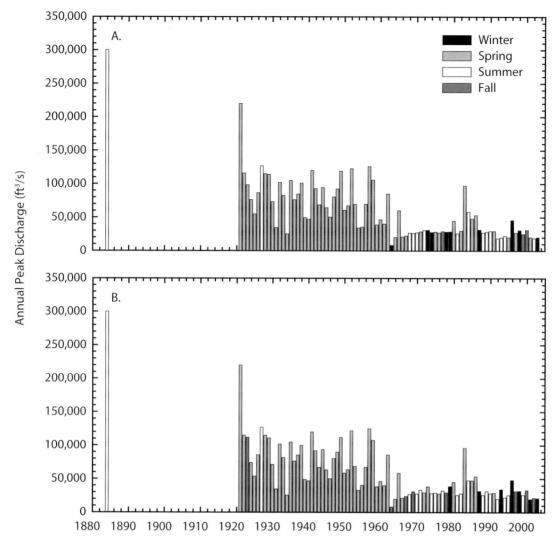

Figure 11.2 Annual flood series for the Colorado River.

A. Colorado River at Lee's Ferry, Arizona (station 09380000; 1884, 1921–2003).

B. Colorado River near Grand Canyon, Arizona (station 09402500; 1884, 1921–2003).

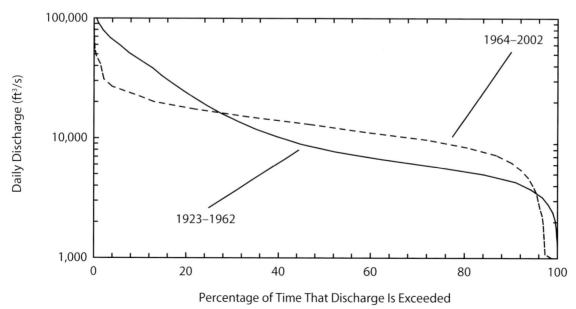

Figure 11.3 Duration of daily discharge of the Colorado River at Lee's Ferry, Arizona (09380000), for the periods of pre- and postcompletion of Glen Canyon Dam.

photography of the Colorado River from boats in 1871 and 1872. J. K. Hillers apprenticed under Beaman, Fennemore, and Clem Powell, and he became Major Powell's photographer at the start of the second expedition into Grand Canyon in 1872. Hillers is generally credited with most of the views taken by the Powell Expedition despite the fact that his first negative was not exposed until April 1872; he took a total of 46 photographs of Marble and Grand Canyons.[12]

Of the numerous photographers who followed Powell, Franklin A. Nims and Robert Brewster Stanton are noteworthy.[13] The two systematically photographed the river downstream from Glen Canyon Dam, creating an astounding collection of 445 photographs that document the predam (and predevelopment) river. Other photographers who created significant imagery of Grand Canyon include Raymond Cogswell, the photographer for the Stone Expedition of 1909; the brothers Ellsworth and Emery Kolb, who took river trips in 1911 and 1923; Eugene C. La Rue, one of the photographers for the U.S. Geological Survey Expedition of 1923; and P. T. Reilly, who photographed the river from boats and aircraft between 1949 and 1964. We matched 1,447 historical photographs of these and other photographers in the remnant of Glen Canyon and Grand Canyon, and 1,391 of these matches show changes in riparian vegetation.[14]

As discussed in chapter 5, the Kolb brothers were the first purposefully to replicate photographs in the region, including ones in Grand Canyon.[15] P. T. Reilly, whose first trip through Grand Canyon was in 1949, replicated Powell and Stanton photographs in 1953, 1964, and 1967.[16] In 1968, Eugene Shoemaker and crew commemorated the Powell Expedition by replicating Beaman and Hillers views ninety-seven years later, including twenty-eight replicates in Grand Canyon.[17] Raymond Turner began systematic repeat photography of the canyon in the early 1970s as an attempt to document changes in riparian vegetation following the start of Glen Canyon Dam operations; the original photographer of most of these matched views was E. C. La Rue.[18]

Before December 1989, Turner had replicated about one hundred views of Grand Canyon, including seven by Nims and Stanton.

Change in Riparian Vegetation

Early Observations of Riparian Vegetation

Dr. Elzada Clover of the University of Michigan led the first botanical survey of the Colorado River in June 1938. Guided by legendary commercial river runner Norman D. Nevills,[19] Clover, with the help of assistant Lois Jotter, made plant lists and collected specimens along the way.[20] Their survey is the only one that systematically described predam conditions in the riparian zone.[21] Clover and Jotter described five habitat types in the canyons, including the "margin of moist sand" adjacent to the river's edge. Most of this zone was devoid of any plants, which could not survive the scour by periodic floods.[22] Above the margin of moist sand was a dense thicket of shrubs and trees clustered above the 100,000-ft^3/s stage. Before operations began on Glen Canyon Dam, the riparian zone in Grand Canyon consisted of western honey mesquite, catclaw, Apache plume, and various obligate or facultative riparian shrubs. This zone is now known as the "old high-water zone" (fig. 11.4).

Clover and Jotter also observed tamarisk, which they reported to be common in Grand Canyon, although it was absent from large reaches of Marble Canyon. Their notes reveal few actual records of tamarisk at the sites they sampled.[23] Tamarisk was known to be in Grand Canyon in the mid-1930s, with "thickets" reported in eastern Grand Canyon, in tributary mouths and the wide canyon reach between Nankoweap Creek and Tanner Canyon, and along the river at the mouth of Bright Angel Creek.[24] In 1938, tamarisk was observed in a cave upstream from Lee's Ferry and at Badger Creek Rapid (fig. 11.1).[25] It appeared in dense thickets along the lower reaches of the Little Colorado River in the 1950s, but it was not present in significant numbers upstream or downstream from the mouth of the Little Colorado River in the predam era.[26]

In the late 1960s, Dr. Paul S. Martin of the University of Arizona collected plants along the river corridor to aid in paleoenvironmental interpretations.[27] In the course of his collecting, he found that Clover and Jotter's "margin of moist sand" had been colonized aggressively by riparian plants in the five years following completion of Glen Canyon Dam. Tamarisk was "abundantly distributed" along the river corridor.[28] It was not the only species that increased: coyote willow, black willow, cottonwood, and several species of shrubs were aggressively colonizing the newly available substrate. Repeat photography in 1968 and from 1974 through 1977 verified his observations.[29] By the late 1970s, tamarisk was distributed throughout the river corridor, in many places completely covering what formerly was barren substrate.[30]

The increase in riparian vegetation along the Colorado River is the most obvious change of the past century in Grand Canyon; this increase is linked to operation of Glen Canyon Dam.[31] Once a stark corridor, largely devoid of trees and riparian plants, the riverbanks are now occupied by thickets of tamarisk, two species of willows, arrowweed, and other native shrubs. This "new high-water zone"—where the number of breeding birds increased five to ten times[32]—is considered a critical resource along the river corridor in Grand Canyon (fig. 11.4).[33] The old high-water zone—at and just above the old high-water stage—is facing a slow death from hydrologic drought.

The Old High-Water Zone

The old high-water zone supports several important types of riparian plant assemblages. In the narrow gorges from Glen Canyon Dam to about river mile 38, this zone typically is indistinct, and native trees are uncommon (fig. 11.5). The most common species are Apache plume, netleaf hackberry, live oak (*Quercus turbinella*), and, more rarely, Utah juniper. More typically, as illustrated by predam conditions at Lee's Ferry (fig. 11.6), the banks were lined with coyote willow with no trees. Downstream of mile 38, catclaw is the most common tree of

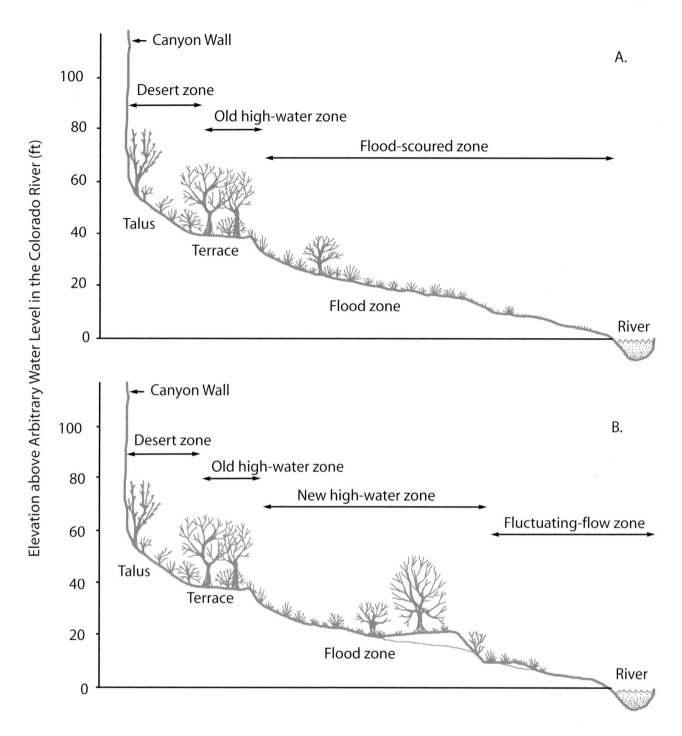

Figure 11.4 Schematic cross sections of the Colorado River in Grand Canyon before and after completion of Glen Canyon Dam (adapted from Grinnell 1914 and R. Johnson 1991).

A. Before the dam was built, the old high-water zone supported the only assemblage of riparian trees that lined the river. This zone, which consists of plants that grow in more xeric settings in the southwestern United States, benefited from annual flooding. Plants received additional water in early summer, and inundation may have promoted seed germination. Below the lower margin of the old high-water zone, floods would scour the seedlings of perennial vegetation that attempted to become established. Perennial plants became established only in wide, stable reaches, where an open gallery forest of Frémont cottonwood would grow in wide, slow-moving reaches. Photographs from before 1900 generally show no perennial vegetation established below the old high-water zone.

B. Flow regulation, in particular flood-control operations, have allowed establishment of a new high-water zone and the slow breakup of the old high-water line. Lowering of flood discharges and raising of base flow in the river have created a more stable hydrologic environment that allows creation of marshes and other wetlands. Conversely, the zone of saturated substrate is lower, and plants in the old high-water zone do not receive as much water as they did before flow regulation. Moreover, germination of mesquite in the old high-water zone is now infrequent.

A

B

Figure 11.5 Photographs of the Colorado River in Glen Canyon.

A. (August 29, 1915.) This upstream view at river mile -10.2, or a little more than 10 miles upstream from Lee's Ferry, shows a potential dam site in Glen Canyon. The banks are mostly free of riparian vegetation, and discontinuous lines mark the old high-water stage of 100,000 ft^3/s. The plants at or above this stage are xerophytic shrubs. (E. C. La Rue 251, courtesy of the U.S. Geological Survey Photographic Library.)

B. (October 29, 1992.) Tamarisk became established in this reach in the 1930s and is now common, particularly below the old high-water stage. (T. Wise, Stake 2639.)

the old high-water zone; in the narrow, steep Inner Gorge, where this zone is indistinct, catclaw typically is the only species. Between miles 40 and 77 and between miles 167 and 225, mesquite joins catclaw to form another conspicuous assemblage of the old high-water zone. Coyote willow appears locally adjacent to the river in some historical views. Netleaf hackberry is locally abundant, particularly just downstream from Lee's Ferry and near Parashant Canyon (river mile 198).

Old high-water zone plant assemblages thrived along the predam Colorado River. Flood discharges exceeded 50,000 ft³/s in 80 percent of the years; in one-third of the years, floods exceeded 100,000 ft³/s. High water lasted for weeks, allowing plenty of time for bank alluvial fill to become saturated. The seasonally plentiful water supply created an environment where trees and shrubs that ordinarily would be dispersed in desert settings crowded together in dense bands.

Glen Canyon Dam radically altered the hydrologic regimen of the old high-water zone. Since flow regulation, 100,000 ft³/s has been approached only once (in 1983), and 50,000 ft³/s has been exceeded less than 1 percent of the time. Saturated substrate now occurs in a zone 20 to 50 feet below the old high-water zone. Plants stranded on the old high-water zone cannot reach this saturated soil, despite their long taproots, and they cannot easily reproduce. This problem caused significant changes in the species composition of this zone. For example, the once-distinct line of Apache plume shrubs in Marble Canyon is disintegrating into discrete patches of individual plants.[34] From 1890 to 1990, mesquite increased in density in 51 percent of the views, but decreased in 28 percent of the views (fig. 11.6). The increase likely resulted from high runoff in the Colorado River, combined with higher winter temperatures.[35]

Flood-control operations of Glen Canyon Dam are thought to decrease the probability of reproduction in certain species, in particular mesquite and netleaf hackberry. Although hackberry may no longer be able to reproduce owing to lack of flooding, growth of existing trees is more consistent now owing to the more stable water supply.[36] Many once-large mesquite trees are dead at and just upslope from the old high-water zone, victims of the combination of flood control and the early-twenty-first century drought. Finally, catclaw, the most xerophytic and opportunistic of the old high-water zone species, increased in 17 percent of the views, with little indication of dieback. The prognosis is that the old high-water line will eventually become similar in species composition to the desert upslope, a process that may be hastened by periodic meteoric drought.[37]

The New High-Water Zone

In the postdam era, Clover and Jotter's "margin of moist sand" is rarely inundated, and most of the silt-rich predam deposits have been scoured by large dam releases. These once-barren substrates now support disparate plant assemblages comprised of nonnative and native trees and shrubs as well as grasses and other herbaceous species. The increase is so striking that one scientist claims that the new high-water zone of Grand Canyon is the only major riverine habitat with increases in riparian vegetation and associated animal populations in the desert regions of the Southwest.[38]

The overall trajectory is toward increasing vegetation despite reversals associated with changes in flow releases from Glen Canyon Dam, which included unintentional high releases in the mid-1980s, an experimental flood in 1996, and generally constrained releases following 1992.[39] In addition, two dam-related substrate changes are thought potentially to limit riparian vegetation, in particular native species.[40] Flood-size reduction lowers the frequency of floodplain inundation, which limits the potential for flushing of salts accumulated under tamarisk, and newly deposited substrate is better sorted and coarser than predam sediments, which lowers water- and nutrient-holding capacity.

Photographs taken between 1963 and 1980 verify the rapid spread of tamarisk into the new high-water zone. Tamarisk appears to have spread downstream from the major tributaries, most notably the Paria and Little Colorado Rivers, but also Bright Angel Creek. The native species that are so abundant today, including arrowweed, coyote willow, seepwillow, carrizo, and longleaf brickellbush, were formerly abundant only in the tributaries. In 1990s matches of 1889–1890 photographs, tamarisk was apparent in 71 percent of the paired views of the river corridor.[41]

The development of a continuous line of dense riparian vegetation increased the potential for fire, particularly given the high human visitation that Grand Canyon receives. There is no evidence that fire occurred along the river corridor before the recent vegetation expansion; the potential for fire was particularly low owing to the twin factors of sparse vegetation upslope and low potential for lightning strikes in the narrow river corridor. Recent fires at Saddle Canyon and Whitmore Wash (river miles 47 and 188, respectively) are probable precursors of a larger problem already occurring upstream from Lake Powell (chapters 6 and 7).

Before Glen Canyon Dam, the only marshes with perennial vegetation were well developed in spring-fed areas above the old high-water zone. Elimination of the annual flood created conditions for marsh and backwater development in wide reaches with large eddies (fig. 11.7). The thick vegetation, with its roots mostly in water, creates a protective cover for many species of nesting birds and animals, some of which were unknown in Grand Canyon before Glen Canyon Dam.[42] These new marshes would have been quickly eliminated by the unregulated Colorado River and in fact have suffered damage from dam-released floods.[43] The current marshes benefit from the increased hydrologic stability afforded by operations of Glen Canyon Dam.

Although the new high-water zone was dominated by tamarisk at least initially, the increase in riparian vegetation created additional habitat for wildlife. Bird populations are thought to have increased in the new high-water zone, particularly in marshes.[44] Both Southwestern Willow Flycatcher and Black-chinned Hummingbirds are attracted to the new habitat.[45] Animals—in particular deer—use the new riparian thickets, despite the fact that many of the stands are dense tamarisk.[46] Of seventy-four species of mammals known to be present in Grand Canyon National Park, fourteen are restricted to the river corridor, adjacent slopes, or tributary canyons, with perhaps three-quarters of the species present using riparian zones for at least a transient period.[47]

In some reaches, species characteristic of the old high-water zone are slowly becoming established in the new high-water zone. Mesquite seeds do not readily germinate in the old high-water zone, but many plants have become established in the new high-water zone. Catclaw also is expanding slowly into this zone. As the old high-water line degenerates, a new line of mesquite and catclaw may eventually become established downslope. Given these trees' slow establishment and growth rates, their maturation in the new high-water zone will require decades if not centuries.

Although tamarisk initially dominated the new high-water zone, it is now joined by many native species, especially seepwillow and coyote willow. Like many riparian species, in particular initial colonizers following floods, tamarisk is unable to germinate in shade. In contrast, coyote willow can root sprout in areas where neither species can germinate. One scenario for future change is that coyote willow will replace tamarisk trees after they die by this mechanism.[48] This scenario may be set back in the event of a catastrophic event, such as a large flood, that would again create bare ground that favors tamarisk reestablishment. Small, intentional floods released from Glen Canyon Dam, such as occurred in 1996, have limited effects on riparian vegetation.[49] Although the 1996 flood buried some woody species—notably tamarisk, seepwillow, and coyote willow—in sand, the one-week peak discharge of 45,000 ft^3/s, which was less than the annual predam flood, did not substantially reduce tamarisk or native woody species, despite scouring of some marshes.

In recent years, black willow has become established in noticeable numbers in the wide reaches of Grand Canyon (fig. 11.7C). Before completion of Glen Canyon Dam, the only black willows known along the river corridor were a single, old plant at Granite Park (river mile 209) and a small grove of young trees at river mile 222.[50] Following the drawdown of Lake Mead in the mid-1990s, black willow became established in abundance in the delta area within Grand Canyon, indicating that this species may eventually have a much larger presence there. In contrast, although Frémont cottonwood have produced seedlings at various places along the river corridor, beaver have killed most of these plants, including a two-foot-diameter tree at Seventy-five Mile Wash and a smaller one at Olo Canyon (mile 145.6).

Establishment of nonnative vegetation continues to be a concern in Grand Canyon. In 2002 and 2003, a goal of tamarisk eradication was set for selected tributaries through cutting down plants and applying herbicides. Other potentially noxious weeds have been removed.[51] Selection of which nonnative species and individuals are to be removed and which are to stay remains a subjective exercise. For example, the National Park Service cut down a date palm tree at Christmas Tree Cave (river mile 135), but date palms remain at Phantom Ranch and Diamond Creek. Tamarisk trees are valued for bird habitat along the Colorado River but are eliminated from some tributaries. Management of riparian vegetation along the Colorado River in Grand Canyon well illustrates the lack of a consistent policy toward nonnative species in the region.

Changes in Small Tributaries

Extensive riparian habitat exists in the tributaries of the Colorado River in Grand Canyon. The major tributaries are the Little Colorado and Paria Rivers and Havasu and Kanab Creeks. Riparian vegetation in these tributaries is discussed in other chapters. In addition to the large drainages, 740 tributaries ranging in drainage area from 0.04 to 365 square miles enter the Colorado River between the Paria River and the Grand Wash Cliffs (fig. 11.1). These tributaries produce both debris flows and streamflow floods,[52] and as a result their riparian zones are frequently scoured.

Bright Angel Creek, which joins the Colorado River at river mile 87, drains 101 square miles of rugged terrain mostly on the North Rim. It is traversed by the North Kaibab Trail, the major north-south route across Grand Canyon, and has a pipeline that transfers water from Roaring Springs in its headwaters to the South Rim. Despite the water transfer of about 900 acre-feet per year, the creek has a mean annual discharge of 35 ft^3/s and an annual yield of 25,400 acre-feet (1927–74), mostly in steady base flow. By means of dendrochronological reconstructions beginning in A.D. 1753, the annual flow volumes is estimated to be 31,300 acre-feet per year, owing to high runoff periods around 1870 and from 1906 through 1920.[53] Both debris flows and other floods are common in this creek, often occurring together during large runoff events (fig. 11.8). Large floods occurred in 1936, 1966, and 1995.

Near the Colorado River, Phantom Ranch at the mouth of Bright Angel Creek provides accommodations, a campground, and a restaurant for visitors accessing this site by trail or river. Hikers are attracted by these amenities as well as by the numerous cottonwood trees planted around the buildings, providing shade during the heat of summer. Although tamarisk occurs along the stream and particularly at the Colorado River, much of the vegetation is native coyote willow and cottonwood (fig. 11.9). An 1890 photograph shows cottonwood established at the mouth of the creek, making the creek one of the few places where cottonwood occur near the Colorado River.

Other tributaries with significant riparian vegetation sustained by perennial spring flow include Nankoweap, Kwagunt, Lava Canyon, Unkar, Crystal, and Tapeats Creeks, as well as the large, ephemeral-flow tributaries in western Grand Canyon that have varying amounts of riparian vegetation. Flow in Tapeats Creek comes from several large springs, notably Thunder River and Tapeats Caves, and varies seasonally with snowmelt runoff. Cottonwood are a large component of the riparian vegetation in these tributaries (fig. 11.10), which also have large populations of longleaf brickellbush, seepwillow, coyote willow, mesquite, catclaw, and tamarisk.

A. (1873.) This view, which shows the former mouth of the Paria River from the left side of the Colorado River, is the first of a time series that documents riparian vegetation at this site. In 1873, coyote willows line the Colorado River; the only cottonwood trees are scattered individuals upstream on the Paria River (chapter 9). (T. O'Sullivan 348, courtesy of the National Archives.)

B. (December 28, 1889.) This view was taken sixteen years later from a point slightly downstream from photograph A. A flood from the Paria River has deposited sand and mud, increasing the delta area at the top of the Paria Riffle and decreasing the stand of coyote willow. Cottonwood and fruit trees planted by John D. Lee and Warren Johnson have grown in the left midground, giving the impression of a natural increase in riparian trees. (R. B. Stanton 261, courtesy of the National Archives.)

C. (1915.) This view is from a different position and angle, but it shows two important aspects of the right bank of the Colorado River. Owing to floods in both the Paria River and the Colorado River, the mouth of the Paria shifted from its 1889 position about a quarter-mile downstream. Overbank sedimentation during Colorado River floods, notably the one in 1912, filled in the former channel mouth. Coyote willows still dominate the right bank, and the trees planted around Lonely Dell have grown. (H. E. Gregory 285, courtesy of the U.S. Geological Survey Photographic Library.)

Figure 11.6 Photographs of the Colorado River at Lee's Ferry.

D. (August 22, 1972.) This match of the 1873 view, as previously described (Turner and Karpiscak 1980, pp. 38–39), shows dense tamarisk where coyote willows once grew. In the midground, the National Park Service (Glen Canyon National Recreation Area) has built infrastructure associated with the boat ramp and parking lots *(out of view to the right)*. Overall, considerably more riparian vegetation, including planted trees, is present than in 1873. (R. M. Turner.)

E. (September 5, 2001.) The National Park Service used bulldozers to clear the dense tamarisk thicket in 2000. Frémont cottonwood, which never grew here historically, was planted with coyote willow and other species as a "restoration" project. Tamarisk *(background)* continues to increase along the Paria River and in the bottomlands. (D. Oldershaw, Stake 706.)

A. (January 23, 1890.) This upstream view at river mile 71.3 in Grand Canyon shows the Colorado River through a wide section of Grand Canyon at the mouth of Cardenas Creek. Except for scattered mesquite and what appear to be clumps of willow, little riparian vegetation is present below the old high-water stage. (R. B. Stanton 396, courtesy of the National Archives.)

B. (February 26, 1993.) What once was sand and gravel is now a valued marsh, a wetland feature that was not present along the unregulated Colorado River. Most of the new vegetation is tamarisk, although willow, arrowweed, and other native species also are present. The mesquite along the banks and in the sand dunes is decreasing in density, with many dead individuals in this reach. (T. Wise.)

C. (March 19, 2003.) Black willow increased significantly in the decade 1993 through 2003. This species now provides additional structure in the Cardenas Marsh, presumably with benefits to the valued bird populations found here. The increase in this species suggests that tamarisk may be gradually replaced by native species. (D. Oldershaw, Stake 1440.)

Figure 11.7 Photographs of the Colorado River at the mouth of Cardenas Creek.

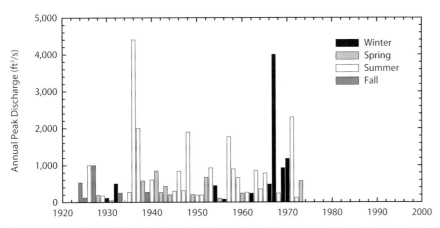

Figure 11.8 Annual flood series for Bright Angel Creek near Grand Canyon (station 09403000; 1924–1973). Flood peaks in 1936 and 1966 likely had debris-flow components, as was the case for the flood of March 5, 1995 (not shown or measured).

A

B

Figure 11.9 Photographs of Bright Angel Creek.

A. (August 1965.) This slightly out-of-focus upstream view shows Bright Angel Creek between Phantom Ranch and the Colorado River. Two cottonwood trees and a river lined with coyote willow are evident. The Bright Angel Creek campground is out of view to the left. (G. Luepke Bynum, courtesy of the photographer.)

B. (March 24, 2003.) The original view is nearly blocked by cottonwood trees, which have grown up despite two debris flows (in 1966 and 1995) that wreaked havoc on riparian vegetation and human structures lining Bright Angel Creek. This section is now channelized with gabions (rock encased in wire cages), and the young cottonwood are also caged to minimize damage from beaver. (R. H. Webb, Stake 4473.)

Figure 11.10 Photographs of Tapeats Creek.

A. (September 7, 1872.) This downstream view on Tapeats Creek, about one-half mile upstream from the Colorado River, shows abundant cottonwood trees and coyote willow. The triangular cliff in the distance is on the south side of the Colorado River. (J. K. Hillers 888, courtesy of the U.S. Geological Survey Photographic Library.)

B. (September 21, 1968.) The cottonwood trees are gone, replaced with coyote willow, tamarisk, and two young cottonwood trees on the left side of the channel in the midground. (H. G. Stephens AJ-4.)

C. (March 27, 2003.) The channel now appears to be overgrown, and the increase in cottonwood is striking. Tamarisk remains along the creek, although in this reach it appears to be overwhelmed in a sea of native species. (S. Young, Stake 3130a.)

Summary. Kanab and Havasu Creeks enter the Colorado River within a few river miles of each other in western Grand Canyon, with Kanab Creek flowing from the north and Havasu Creek flowing from the south. Kanab Creek formed a substantial arroyo in the period 1882 through 1909, and that arroyo now supports a significant stand of native and nonnative riparian vegetation. Whereas Kanab Creek forms a beautiful but barren canyon, Havasu Canyon is world renowned for its blue, calcium-rich water and spectacular waterfalls. Extremely large floods have occurred in both drainages, and these floods have altered riparian vegetation along both creeks. Kanab Creek has little woody riparian vegetation in its canyon, but Havasu Creek supports lush and highly valued riparian vegetation. Havasu Creek had a series of large floods in the early twentieth century that changed a channel lined mostly with coyote willow into a gallery forest of Frémont cottonwood, Arizona ash, and mesquite.

Both Kanab Creek and Havasu Creek join the Colorado River in Grand Canyon, but they flow from opposite directions. Kanab Creek (fig. 12.1) has its headwaters in southern Utah and flows south to meet the Colorado River at river mile 143.5, whereas Havasu Creek has its headwaters in the high plateaus of northern Arizona near Williams and flows north to meet the river at mile 156.7. Although Kanab Creek has a larger watershed, it has low perennial flow in a nearly barren bedrock canyon. Havasu Creek benefits from high flow rates in its lower reaches, created by discharge from large springs, and the steady flow sustains a lush and highly valued riparian area upstream from the Colorado River.

Kanab Creek

Kanab Creek has its headwaters in the Paunsaugunt and Markagunt Plateaus of southern Utah and drains 2,313 square miles of southern Utah and northern Arizona. The river passes through alternating alluvial and bedrock reaches as it flows down what is locally called the Grand Staircase, a series of cliffs formed in Mesozoic sandstones,[1] upstream from Kanab, Utah. Flowing south across the Utah-Arizona border, Kanab Creek is joined by its principal tributary, Johnson Wash, south of the Arizona town of Fredonia. The creek then carves a meandering canyon through the Paleozoic rocks characteristic of Grand Canyon. Like so many other watercourses in the southwestern United States, Kanab Creek has alternating reaches of perennial and ephemeral flow.

History of Photography

The first photographs of both Kanab Creek and Johnson Wash were taken by members of the Powell Expedition in the winter of 1871–1872.[2] Because E. O. Beaman had recently quit the expedition and James Fennemore had yet to arrive on the scene, the most likely photographer was Clem Powell, assisted by Jack Hillers. Other photographers include Herbert E. Gregory, a prominent geologist of the early twentieth century, and Reed Bailey, a geologist with the U.S. Forest Service who was concerned about overgrazing, range deterioration, and channel erosion. Numerous photographs were taken by photographers on river trips, the earliest being the Kolb brothers (1911) and Eugene C. La Rue (1923). Ninety-nine repeat photographs document changes in riparian vegetation along Kanab Creek and Johnson Wash in their alluvial and bedrock reaches.

Floods, Channel Change, and Flow Regulation

The flood history of Kanab Creek is relatively well known owing to its rich history as one of the primary Mormon settlements in southern Utah.[3] The current arroyo of Kanab Creek at Kanab exposes stratigraphy deposited over the past 5,200 years.[4] Arroyo cutting and filling occurred several times before Mormon settlement in 1870, indicating that human land-use practices were not necessary for arroyo downcutting in this watershed. The most recent episode of cutting and filling occurred about 600 years ago, or roughly in A.D. 1350.

At the time of settlement, the channel of Kanab Creek at Kanab was no more than about 3 feet deep and perhaps 30 feet wide.[5] Between Kanab and the start of Kanab Canyon, the channel was described in the following terms: "This usually dry course of the stream is along a level plain where the sands drift, and sometimes obliterate all traces of the water-course."[6] High groundwater conditions prevailed, leading to dense stands of coyote willow on the floodplain, which gave Kanab its Paiute name (kanab, "willow"). The downcutting of Kanab Creek commenced in 1882 and ended with a flood spawned from the large regional storm in 1909.[7] Numerous floods, notably in 1883, 1884, 1885, 1890, 1896, and 1909, are associated with the downcutting.[8] Widening of the channel within the arroyo walls persisted until the early 1940s, when the arroyo began to store sediments in low floodplain deposits.

The gaging record for Kanab Creek, like that of the Escalante River (fig. 9.2A), is short and completely misses the early period of flooding (fig. 12.2). The annual flood series is pieced

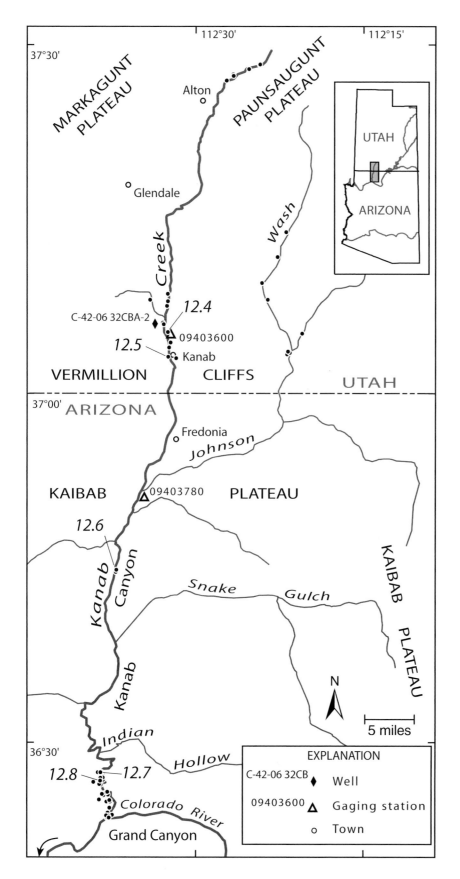

Figure 12.1 Map of Kanab Creek, southern Utah and northern Arizona.

together from the gaging records near Kanab and Fredonia, which are separated by contributions from Johnson Wash (fig. 12.1). The flood of record is 4,630 ft³/s at the Fredonia gage in 1970. Flow in Kanab Creek is regulated by a diversion dam about 5 miles upstream from Kanab. Eleven dams at this site were repeatedly destroyed during the period of arroyo downcutting. Rebuilt after the flood of 1909, the current dam has withstood nearly a century of subsequent floods. Although surface water is used for irrigation in Kanab, groundwater is used as well, and from 1995 through 2002 water levels in one well decreased to far below the rooting depths of most riparian species along Kanab Creek (fig. 12.3).

Changes in Riparian Vegetation

Numerous photographs of Kanab Creek and Johnson Wash show that the channel was essentially devoid of riparian vegetation following the end of the period of arroyo downcutting and widening (fig. 12.4).[9] Tamarisk likely was the earliest woody species to gain a foothold on the low floodplains that developed after 1941. By the 1980s, it formed nearly monospecific stands along these wide terraces. By 2000, however, Frémont cottonwood had germinated and was rapidly overtaking tamarisk for domination of the riparian zone.

At Kanab, the ill-defined and shallowly incised channel viewed by the first Mormon settlers and photographed in 1871 downcut into a deep, incised channel beginning in 1882 (fig. 12.5).[10] Tamarisk was the first colonizer, but, as observed elsewhere, cottonwood and other native species gained a foothold and have steadily increased in size within the channel. Despite declines in groundwater levels (fig. 12.3), riparian vegetation has increased. Russian olive, which is common along other rivers in the region, is a minor component along Kanab Creek except in the vicinity of Fredonia.

Downstream from Fredonia, Kanab Creek gradually incises into the Paleozoic Kaibab Limestone, which dips upstream. Despite confinement within walls of bedrock, the channel behaves like an arroyo, and 50 feet of downcutting occurred in this reach (fig. 12.6). Riparian vegetation, once dominated

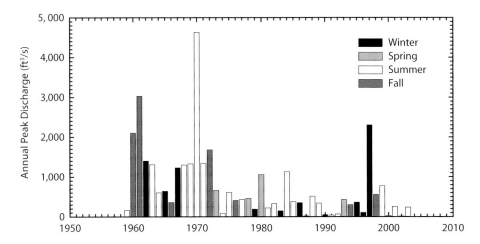

Figure 12.2 Combined annual flood series for Kanab Creek near Kanab, Utah (station 09403600; 1959–1968, 1979–2003), and of Kanab Creek near Fredonia, Arizona (station 09403780; 1969–1978). Johnson Wash is the major tributary entering between the two gaging stations.

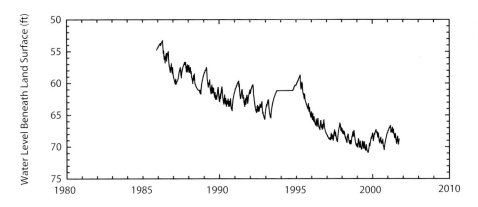

Figure 12.3 Groundwater levels for well C-42-06 32CBA-2 near Kanab, Utah.

by coyote willow, now has a mixture of cottonwood, tamarisk, coyote willow, and Russian olive. These observations support others showing that natives can compete with nonnatives along Kanab Creek.[11]

In the Supai Group of the Paleozoic strata, Kanab Creek is again ephemeral, with scattered water holes and riparian vegetation. Within the Redwall Limestone and Muav Limestone, spring discharge creates perennial flow for the last 9 river miles to the Colorado River. Although alluvial terraces have a few gnarled and stunted cottonwood trees, the most common species present are tamarisk, coyote willow, and seepwillow. At Showerbath Spring, water dripping from the Redwall Limestone creates a hanging garden that appears to be about the same 128 years after Jack Hillers and William Bell photographed it in 1872.[12] A nearby Bell view shows a persistent redbud tree (*Cercis occidentalis*), but also thick growth of tamarisk in an alcove (fig. 12.7). Farther down-

stream, the channel of Kanab Creek looks scoured compared to its appearance in 1872 (fig. 12.8), and deposition of a driftwood log in the foreground suggests the reason: disturbance by large floods.

Kanab Creek, therefore, has a mixed history of change in riparian vegetation. In its alluvial reaches, thick stands of coyote willow were destroyed during floods and arroyo downcutting, and now a mixture of native and nonnative trees is growing in the channel. This net increase in biomass contrasts strongly with change in the bedrock channel, where riparian vegetation was sparse in 1872 and continues to remain so. The combination of low base flow and periodic large floods does not allow significant riparian vegetation to become established in Kanab Canyon.

Havasu Creek

Havasu Creek is a perennial stream that begins at Havasu Springs on

the Havasupai Indian Reservation in northern Arizona (fig. 12.9). This creek is part of a very confusing 3,020-square-mile watershed that has endured numerous name changes since it was viewed by Francisco Garcés in 1776. The Spaniard disregarded any names that the inhabitants might have had for this watercourse and gave it two names—Río Jabesua de San Antonio and Río Cabezua—neither of which was ever used.[13] Lieutenant Joseph Ives visited the canyon in 1858 and named it Cataract Creek; Cataract Creek remains the name for the headwaters watercourse extending northward from the north side of Bill Williams Mountain near present-day Williams, Arizona. John Wesley Powell called it Coanini Creek, but the name of the inhabitants whom Garcés met became the name of the canyon and creek.

Here, we ignore all of the headwaters reaches of Havasu Creek, including Hualapai Canyon and Cataract Creek, and concentrate on changes in

A. (1939.) Herbert Gregory photographed the channel of Kanab Creek at its confluence with Tiny Canyon *(right midground)* as part of his work on the geology of the high plateaus of Utah. In this downstream view, Kanab Creek is wide, with a trapezoidal, denuded channel. Arroyo downcutting had lowered the bed elevation from the high bench shown prominently along Tiny Canyon at left. Downcutting ceased about 1909, and channel widening stopped just after this photograph was taken. This section of channel is just upstream from the water-supply reservoir for Kanab, Utah, but the reservoir probably does not affect the grade of Kanab Creek at this point. (H. E. Gregory 950, courtesy of the U.S. Geological Survey Photographic Library.)

B. (October 9, 1984.) The roadbed from which Gregory took his photograph has been straightened, eliminating the guardrails from the view. Large floods ended about 1941 in southern Utah, and the channels in the region narrowed in response to decreased floods and to the influx of riparian vegetation. A mixture of tamarisk and native riparian vegetation obscures the view of the floodplain, which at this time was 5 to 10 feet above the main channel of Kanab Creek. Some small cottonwood trees appear above the dense tamarisk in the foreground. The channel of Kanab Creek, which is perennial here, appears on the extreme left side of the view. Pumping for domestic use occurs at several sites upstream from here. (R. H. Webb.)

C. (April 27, 2000.) The new grove of cottonwood trees has grown to the extent that much of the far arroyo wall is obscured. The tamarisk trees in the foreground are becoming senescent; some are falling over as a result of the large amount of growth in recent decades. The formerly sharp lines on the alluvial terraces in the midground and background are now rounded, reflecting recent surface erosion. (R. H. Webb, Stake 1227.)

Figure 12.4 Kanab Creek at Tiny Canyon, Utah.

A. (1871.) This eastern view across Kanab Creek at Kanab was taken on an unknown date during the winter of 1871–1872, when the Powell Expedition overwintered in town. The photographer is unknown and could have been Clem Powell, James Fennemore, or Jack Hillers, although the latter photographer is generally credited with images in this part of southern Utah. Kanab Creek flows in a shallow bed, perhaps 6 feet deep and 50 to 100 feet wide. The vegetation present is coyote willow, big sagebrush, and fourwing saltbush (Webb, Smith, and McCord 1991). The word *kanab* means "willow" in Paiute. (J. K. Hillers, from Dutton 1882.)

B. (November 4, 1990.) Beginning in 1882 and ending around 1909, the bed of Kanab Creek dropped about 75 feet at this site during a classic period of arroyo downcutting. Widening continued through the first third of the twentieth century, resulting in a channel 240 to 350 feet wide. Because of the extensive channel change, this camera station may be as much as 20 feet behind the original camera station. After channel widening ceased, deposition of low floodplains began, and riparian vegetation became established. In 1990, that vegetation includes tamarisk, Frémont cottonwood, and coyote willow. (R. H. Webb.)

C. (April 27, 2000.) The low floodplain has increased in height, and riparian vegetation continues to grow with new individuals established in the intervening decade. The species present here include tamarisk and Russian olives intermixed with cottonwood, black willow, coyote willow, and big sagebrush. (R. H. Webb, Stake 2055.)

Figure 12.5 Kanab Creek at Kanab, Utah.

A

B

Figure 12.6 Kanab Creek south of Fredonia, Arizona.

A. (1872.) Kanab Creek was known for its willow-lined channels when it was first explored. As this photograph shows, the transition reach between the alluvial channels upstream and the bedrock channel downstream was once a near-level alluvial fill consisting of high-density coyote willows lining the channel and xerophytic vegetation on the low floodplains. The Wheeler Expedition crossed the canyon at this point. (W. A. Bell 106-WB-590, courtesy of the National Archives.)

B. (2000.) The channel bed has downcut as much as 50 feet in this reach, creating an incised channel confined within the larger bedrock constraints. The now dry floodplain, which once was grazed heavily, supports sparse greasewood, but the channel sustains a dense assemblage of riparian vegetation, obscuring the open water. Cottonwood, Russian olive, tamarisk, and coyote willow are the most common species in this reach. (D. Oldershaw, Stake 3966.)

A   B

Figure 12.7 Photographs of Kanab Creek near Showerbath Spring in Kanab Canyon.

A. (October 1872.) This view, taken by William Bell of the Wheeler Expedition, shows an overhanging wall on Kanab Creek upstream from Showerbath Spring, a landmark in Kanab Canyon. This camera station is not accidental; most glass-plate photographers began their work near permanent water, and perennial flow in Kanab Creek begins at about this site. The tree at lower left is a redbud, a native leguminous tree that is uncommon in riparian settings except in Grand Canyon. (W. A. Bell 106-WB-274, courtesy of the National Archives.)

B. (October 22, 1995.) The redbud is persistent, but tamarisk now dominates the lower part of this view, and a few individuals of Russian olive are present nearby. Longleaf brickellbush is perhaps the most common riparian species in Kanab Canyon, and catclaw, coyote willow, netleaf hackberry, and scattered cottonwood are also present. (D. Oldershaw, Stake 2403.)

riparian vegetation in the reach of perennial flow from upstream of Supai to the Colorado River. Our treatment is not voluntary: few photographs have documented anything in the upstream reaches of Havasu Creek. It is almost as if photographers decided to save their film for the spectacular scenery downstream, including the travertine waterfalls, the riparian vegetation, and the colorful bedrock walls.

History of Photography and Observations on Riparian Vegetation

Havasu Canyon has a substantial history of landscape photography owing to its stunning landscape and vibrant riparian ecosystem.[14] The first photographs of Havasu Creek were taken by Ben Wittick, who visited Diamond Creek in 1883[15] and Havasu Canyon in 1885. Wittick had studied under William Henry Jackson, noted landscape photographer from Denver, and many of Wittick's photographs are incorrectly attributed to Jackson in photograph archives. The earliest river runners missed the mouth of Havasu Canyon, and those who found it did not venture far up the canyon. Other notable photographers include Emery and Ellsworth Kolb, who visited several times in the first two decades of the twentieth century. Ninety-six photographs were matched to document changes in riparian vegetation.

In the winter of 1858, Lieutenant Joseph Ives visited Havasu Canyon briefly during his epic journey upstream on the Colorado River. He had abandoned his boat the *Explorer* months earlier and was traveling over-land to the Hopi villages when he and his party heard word of the existence of a village deep within Grand Canyon. He found Havasu Creek to be 30 to 45 feet wide and lined with cottonwood and willows.[16] In 1881, Frank Cushing visited Havasu Canyon; at Havasu Springs, he found "a grove of fresh green willows and cotton-woods" and described "luxuriant vines and shrubbery; willows, cottonwoods; flags, tules, and other aquatic plants, with a marked absence of grasses, save a kind of cane."[17] "Thick willows" bordered the flowing water, leading Cushing to describe the Havasupai as living in "the Nation of Willows." Cushing was no botanist, but he could have been describing Havasu Canyon in the twenty-first century, except that he did not mention Arizona ash or mesquite in his descriptions.

A

B

C

Figure 12.8 Photographs of Kanab Creek in Kanab Canyon.

A. (September 15, 1872.) This upstream view shows one of the picturesque pinnacles in the perennial reach of Kanab Canyon. Photographer Jack Hillers, under orders from John Wesley Powell, slowly worked his way upstream from the Colorado River after the 1872 river trip ended, systematically photographing Kanab Canyon (Webb, Smith, and McCord 1991). This view shows sparse riparian vegetation, with a grove of what may be coyote willow in the center on the outside of a bend. (J. K. Hillers 632, courtesy of the National Archives.)

B. (September 24, 1968.) The riparian vegetation once present in the foreground has been swept away, along with the low terrace on which it was growing. Channel widening has removed much of the finer-grained substrate on the outside of the channel bend. Evidence of the cause for the changes appears in the foreground. A juniper log, 12 feet long and 2 feet in diameter, is new to the foreground. This driftwood was deposited 20 feet above the base of a channel that is 144 feet wide, indicating that the flood that deposited the log had to be extraordinarily large. Only one riparian plant is readily visible *(center, downstream of large boulder)*, and it cannot be identified at this distance. (H. G. Stephens, AD-2.)

C. (March 23, 1990.) Our match shows little change at this site, other than the shifting of boulders during periodic floods in Kanab Canyon. The log remains across the foreground of the view, and the channel remains barren as well. (R. H. Webb, Stake 1235.)

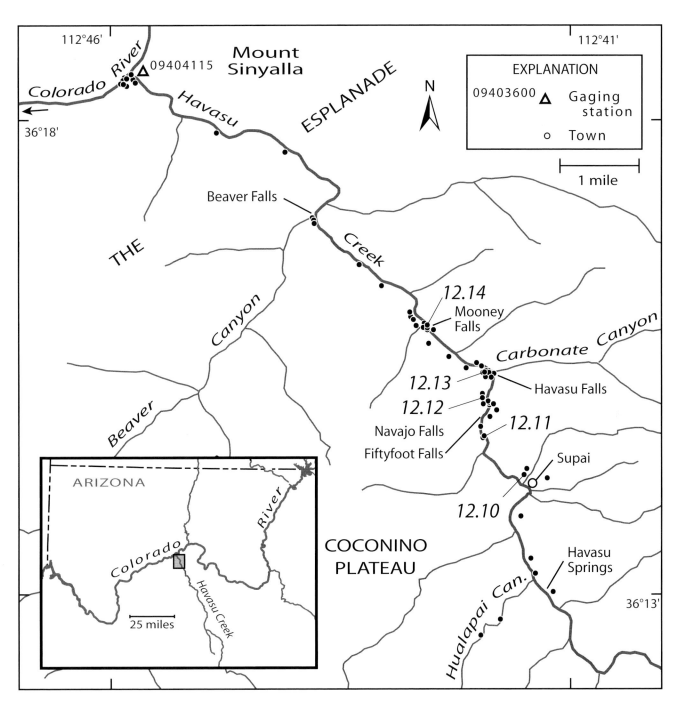

Figure 12.9 Map of Havasu Canyon, northern Arizona.

Floods and Flow Regulation

Havasu Creek has a short gaging record beginning in the fall of 1990 following an extremely large flood. That
discharge, which peaked at 20,300
ft³/s on September 3, 1990, caused significant damage to Supai, changes to
the waterfalls in Havasu Canyon, and
destruction of many riparian trees.[18] A
second flood, in February 1993, peaked
at 13,400 ft³/s and caused less damage.

The largest flood known to have
occurred in Havasu Canyon, beyond

anecdotal stories handed down for
centuries, occurred in January 1910.[19]
This flood, spawned by warm rainfall
on snow, effectively destroyed Supai,
and its "20-foot wall of water" caused
large changes to the waterfalls on
Havasu Creek. This flood followed
a decade of smaller but damaging
floods, such as the events in March and
November 1905 that caused significant
damages elsewhere in the region. Other significant floods occurred in 1920,
1921, 1928, 1935, and 1939. Flooding in Havasu Creek, as with many

other watercourses in the region, was
minimal until the latter third of the
twentieth century, and the 1990 flood
was unusual in its size and effects on
Havasu Canyon only because of the
long lull in big events.

The headwaters channels leading to
Havasu Canyon have small stock tanks
and small earthen dams, including
the Redlands Reservoir and Markham
Dam.[20] These structures, which failed
during the 1993 flood, are primarily
for water storage for livestock operations. Clearly, they were not designed

for flood-control purposes, nor do they significantly influence low-flow discharges in Havasu Canyon.

Changes in Riparian Vegetation

Despite the fact that Havasu Creek is confined within a deep bedrock canyon, it has reaches that function as if they were in deep alluvium. Base level in Havasu Canyon is controlled by the waterfalls, which in turn are controlled by carbonate deposition from the supersaturated waters issuing from Havasu Spring and other springs at about the same elevation. Some waterfalls, such as Fiftyfoot Falls and Beaver Falls, have been destroyed historically, whereas others, such as Navajo, Havasu, and Mooney Falls, are relatively permanent landscape features.

At Supai, photographic evidence shows that historical reports of riparian vegetation in Havasu Canyon were accurate (fig. 12.10A). The channel is lined with dense coyote willow, much as the pre-arroyo channel of Kanab Creek was lined with willows, and only a few scattered cottonwood trees were present along the stream. By 1941, growth of native riparian trees was startling, including the occurrence of Arizona ash, which was not mentioned in historical accounts (fig. 12.10B). The net effect by 2000 was an almost complete conversion of riparian vegetation from a nearly monospecific stand of coyote willow to a mixed stand of cottonwood-ash-willow, with mesquite on the margins (fig. 12.10C). Similar changes occurred downstream from Supai, with agricultural fields overtaken by the encroaching riparian forest (fig. 12.11).

Destruction of Fiftyfoot Falls during the 1910 flood created a headcut that propagated upstream toward Supai.[21] This alluvial section is unstable, and tamarisk is common. Navajo Falls is also unstable, but it withstood the 1910 flood, albeit with major changes (fig. 12.12). This hydraulic feature, which is better described as a series of cascades than as a single waterfall, had a large increase in riparian vegetation that was only slightly damaged during the 1990 flood. Following cessation of floods after the 1993 event, riparian vegetation has increased even more at this site.

Numerous photographs document long-term changes in riparian vegetation at both Havasu Falls and Mooney Falls, the most spectacular waterfalls in Havasu Canyon.[22] Both waterfalls had changes in their precipices resulting from both the 1910 and the 1990 floods, but riparian vegetation above and below both waterfalls recovered quickly. At Havasu Falls, a classic view shows the waterfall after the 1910 flood cut a significant notch in its top, but before the 1990 flood (fig. 12.13). The 1990 flood cut a wider notch at the top of the falls, and riparian vegetation above and below was damaged. However, by 2000, riparian vegetation was increasing at the top of the waterfall, and trees lining the pool downstream had grown considerably (fig. 12.13C). In one of the earliest photographs of Havasu Canyon, Mooney Falls appears to fall over a ledge with little visible channel (fig. 12.14A). Floods have channelized water at the top of the falls, but it is clearly apparent that riparian vegetation remains dense in the vicinity.

Hundreds of riparian trees from Havasu Canyon, mostly Arizona ash, were washed into the Colorado River in western Grand Canyon in September 1990. Dendrochronologic analyses of these dead trees revealed that Arizona ash can achieve diameters approaching one foot in ten years.[23] Although the oldest ash tree was about one hundred years old, most of the trees killed by the 1990 flood were less than forty years old. Given the high growth rates of these trees, it is not surprising that repeat photography shows regrowth in a decade of reaches severely damaged by the 1990 flood.

A. (Before 1910.) This upstream view shows a unique formation in the Supai Formation known on U.S. Geological Survey maps as the Wigleeva, but locally called the Wig-Li-I-Wa Columns. Taken before the devastating flood of 1910, this fuzzy photograph shows the village of Supai on the extreme right side, with only a few buildings visible. Coyote willow is the dominant riparian plant in this view, although isolated cottonwood trees can be seen at various points. Open water is visible in the lower center, showing that at the time of this photograph the river flowed on the valley closest to the camera station. This view shows why the Havasupai were known as the "Nation of Willows." (W. W. Bass 39, courtesy of the Arizona Historical Society, Tucson.)

B. (1941.) This view, taken from a much lower camera station but exhibiting a view similar to photograph A, shows considerable change has occurred in a little more than thirty years. Agricultural fields now dominate the bottomlands, and the creek is hidden behind a wall of Frémont cottonwood and Arizona ash trees. When visited in 2000, this camera station was found to be blocked by catclaw, cottonwood, and netleaf hackberry. (J. Muench B-1871, used with the photographer's permission.)

C. (September 6, 2000.) Owing to several large floods in the preceding ninety years, including the 1910 flood and another of similar magnitude in 1990, the channel of Havasu Creek has shifted across the narrow valley and away from the camera station. The creek is lined with tall cottonwood and ash trees, showing how quickly these species can grow and how fast a riparian ecosystem can rebound from devastating floods provided that water remains available for the plants. The water tank in the foreground provides the domestic water supply for Supai. (T. Brownold, Stake 2175.)

Figure 12.10 Photographs of Havasu Creek at the Wigleeva.

A

B

Figure 12.11 Photographs of Havasu Creek downstream from Supai.

A. (1885.) This view across Havasu Canyon shows two Pai standing in a cornfield with a peach orchard behind and to the right. The camera station is downstream from Supai and upstream from the former location of Fiftyfoot Falls. Riparian vegetation lining the channel in the midground appears to be mostly tall coyote willow; no cottonwood or ash trees are visible. (B. Wittick 16245, courtesy of the Palace of the Governors [MNM/DCA].)

B. (September 11, 2000.) The channel has shifted toward the camera station, and the former agricultural fields are now occupied by lush riparian vegetation, including Frémont cottonwood, black willow, Arizona ash, coyote willow, seepwillow, and a few tamarisk. (T. Brownold, Stake 2173.)

A. (1899.) Navajo Falls is the first main waterfall downstream from the village of Supai, although Fiftyfoot Falls once existed there (Melis and others 1996). This upstream view shows water cascading down a broad slope before encountering vertical falls at the bottom. The riparian vegetation present is a mixture of Frémont cottonwood and Arizona ash; coyote willow is no doubt also present. (H. G. Peabody GrCa 14731, 8984, courtesy of Grand Canyon National Park.)

B. (June 25, 1991.) The 1910 and 1990 floods significantly changed the shape of Navajo Falls and the riparian vegetation. The channel has shifted left, and most of the flow enters this view from the lower right. Although the density and size of both cottonwood and ash have increased dramatically since 1899, dead trees, killed by the September 1990 flood, appear through the view. (T. S. Melis.)

C. (September 8, 2000.) The falls have once again changed, showing the instability of these signature landforms of Havasu Canyon. Although one severely damaged tree appears on the lower left, the riparian zone has responded to the floods of the early 1990s by once again increasing in size and density, and the channel is now more obscured behind this vegetation. Arizona ash forms the wall across the bottom of the view. (T. Brownold, Stake 2156.)

Figure 12.12 Photographs of Havasu Creek at Navajo Falls.

A. (June 1988.) This view across Havasu Falls shows one of the classic waterfalls and plunge pools in Havasu Canyon. A combination of Frémont cottonwood, Arizona ash, and seepwillow forms the riparian vegetation along the channel leading to the waterfall and around the pool downstream. Above the approaching channel, several mesquite trees are present on channel right. A dead cottonwood tree stands on the edge of the pool. (T. Brownold, courtesy of the photographer.)

B. (June 25, 1991.) The 1990 flood has clearly scoured the channel approaching the waterfall, removing much of the woody riparian vegetation present in 1988. Similarly, the trees on the downstream side of the pool have been thinned. (T. Brownold.)

C. (September 8, 2000.) The density of riparian trees has increased, but not to the extent of what was present in 1988. A combination of cottonwood and seepwillow lines the channel approaching the falls. The trees on the downstream side of the pool have grown, but no new individuals are present. Through this period, the mesquites appear to have grown. The cottonwood snag, which died before 1988, has toppled over. (T. Brownold, Stake 2144.)

Figure 12.13 Photographs of Havasu Creek at Havasu Falls.

A   B

Figure 12.14 Photographs of Havasu Creek at Mooney Falls.

A. (1885.) This superbly composed photograph, by master landscape photographer Ben Wittick, shows a man standing at the base of Mooney Falls in 1885. The view is framed by Arizona ash, Frémont cottonwood, and wild grape. (B. Wittick 16105, courtesy of the Palace of the Governors [MNM/DCA].)

B. (September 9, 2000.) The view can still be framed by riparian trees, only they are different ones than were present in 1885. The falls has a decided notch, the combined result of the 1910 and 1990 floods. Riparian species present in this view include cottonwood *(right and top),* Arizona ash *(left),* wild grape *(right),* and seepwillow *(bottom).* If anything, the wild grape is more profuse now than in 1885. (T. Brownold, Stake 2187.)

13   The Virgin River

Summary. The Virgin River is highly valued for its water supply, its scenery, and its habitat for wildlife. Riparian vegetation along the East Fork of the Virgin River through Parunuweap Canyon appears to be similar to that present in 1873 despite large floods and channel changes. Riparian vegetation along the North Fork through Zion National Park has generally increased except in The Narrows, a deeply incised bedrock reach with little room for storage of alluvial terraces upon which riparian vegetation can grow. Downstream from the confluence of the two forks, riparian vegetation has increased as the once-wide channel has narrowed. In northwestern Arizona, the Virgin River flows across a broad alluvial valley to its terminus in Lake Mead, supporting extensive stands of native and nonnative vegetation. Even a large dam-break flood in 1989 did not reverse the trend of increase in riparian vegetation downstream from the Virgin River Gorge.

The Virgin River is the primary drainage of southwestern Utah (fig. 13.1). This river system spans the Colorado Plateau to Great Basin transition, cutting spectacular canyons through Pleistocene lava flows as well as the Mesozoic and Paleozoic strata that define the Grand Canyon region.[1] The North Fork of the Virgin River, which creates Zion Canyon in Zion National Park, flows south in a deep slot canyon through the Markagunt Plateau. The East Fork of the Virgin River parallels Kanab Creek in its headwaters on each side of the Sevier fault, which forms the divide between the Paunsaugunt and Markagunt Plateaus. Southward, the East Fork turns west and carves Parunuweap Canyon, then flows to its confluence with the North Fork

at Springdale, Utah. The combined watercourses flow southwest, past St. George, Utah, through the northern Virgin Mountains in the Virgin River Gorge, and across extreme northwestern Arizona to Lake Mead in Nevada. The Virgin River now creates a delta known more for its tamarisk thicket than for its scenery.

The Virgin River is highly valued for agricultural and domestic water supplies, for its scenery, and for its importance to wildlife and endangered species. Agricultural lands occupy the wide valleys down the length of the river, growing alfalfa, fruit, and produce. Zion National Park is a tourist magnet in southwestern Utah, providing an economic stimulus for this part of the state. The riverine environment supports notable endangered fish species, and the riparian zone, particularly in Zion Canyon, is renowned for its scenic value as well as its wildlife habitat.

Historical Observations of the Virgin River

Most descriptions of the Virgin River were made first by trappers seeking pelts and second by Mormon settlers, who settled in the basin at Ash Creek (fig. 13.1) in 1852.[2] Trappers repeatedly sought beaver in the Virgin River, and river otter were also seen between present-day La Verkin and Washington, Utah.[3] Jedediah Smith traveled down the river through the Virgin River Gorge in 1826 and saw scattered cottonwood except in tributaries such as Beaver Dam Wash, where it was abundant.[4] Antoine Leroux trapped beaver down the Virgin River in 1837. The number of animals removed from the river is unknown, but its proximity to the Old Spanish Trail provided

access to many nineteenth-century travelers, and the population of beaver likely was decimated.

Mormons settled in St. George and the surrounding area in 1861. They coveted the lower elevations of this river for its seemingly abundant water and the relatively warmer winter temperatures, which allowed cultivation of cotton to supply the Utah settlements with clothing. Mormon settlers wrote about the landscape in which they tried to make a living, providing an invaluable record of environmental change.[5] The town of Springdale was established in 1862 near the confluence of the North and East Forks because five springs provided a convenient water supply.[6] Because of its unique wide floodplain within a deep bedrock canyon, Zion Canyon was used extensively for agriculture by the early twentieth century before the establishment of Mukuntuweap National Monument.[7]

Frémont cottonwood, arrowweed, and coyote willow reportedly were common along the Virgin River and its tributaries in 1893, particularly in the vicinity of St. George.[8] Willows were most common along the North Fork of the Virgin River,[9] as they were along Kanab Creek to the east (chapter 12). Settlers used cottonwood logs in particular to build cabins in the valleys[10] before more durable timber from higher elevations became readily available.

Like for many other perennial reaches that had woody riparian vegetation in the nineteenth century, beaver were reported in Zion Canyon.[11] When they disappeared has not been recorded, but in the 1890s floods began to cause arroyo downcutting and channel widening. In order to protect properties valuable to the National

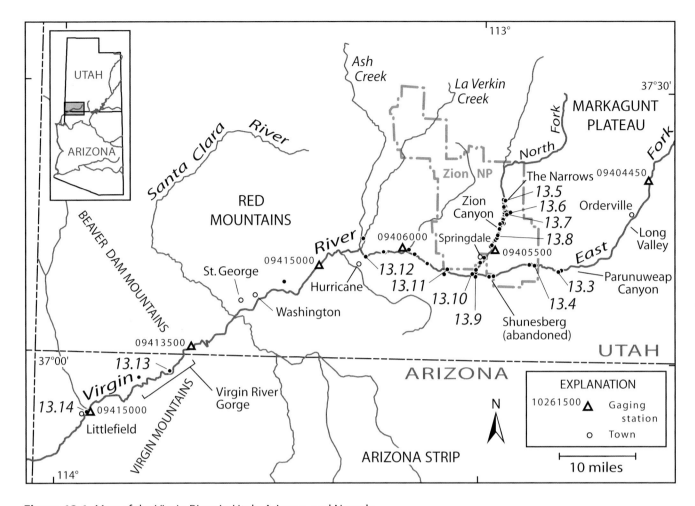

Figure 13.1 Map of the Virgin River in Utah, Arizona, and Nevada.

Park Service, considerable channel-stabilization effort was expended to force the river into a fixed channel.[12] Many thousands of feet of revetment, concrete deflection structures, and grade-control structures remain in this river, which undoubtedly have had an impact on riparian vegetation. On one hand, the stable channel provides a stable floodplain to allow long life for riparian trees, but that stability reduces the amount of floodplain area that might be inundated during floods, limiting the potential for these species' reproduction.

Historical Photography of the Virgin River

The Virgin River has long been a favorite subject of photographers. John Wesley Powell and two companions traversed Parunuweap Canyon to the now-abandoned townsite of Shunesburg in the fall of 1872, and his photographer, John K. Hillers, returned in April 1873. Hillers took at least thirty-six photographs of the East Fork, the North Fork, and surrounding uplands during this trip. These photographs provide a baseline for assessing change of riparian vegetation along the Virgin River. Some of Hillers's photographs of Zion Canyon are considered to be classics in Western landscape photography.[13]

In 1909, Mukuntuweap National Monument, the predecessor to Zion National Park, was established by order of President William Howard Taft.[14] The name was changed to the current one in 1917 because the original name, a Paiute word for the canyon, was difficult to pronounce. The publicity caused photographers to flock to what was at the time one of the few national parks in the country. The National Park Service hired photographers, including George Alexander Grant in 1929 and the legendary Ansel Adams in the 1940s, to provide imagery for its publications. Scientists were attracted by the natural environment, in particular the geology. Herbert E. Gregory, pioneering geologist of the Colorado Plateau, visited the park on numerous occasions in the 1920s and 1930s.[15] As a result, numerous photographs are available to document change in riparian vegetation along the Virgin River.

Two previous repeat-photography efforts documented changes in the Virgin River and other parts of Zion National Park. In the mid-1990s, Richard Hereford analyzed repeat photography to study changes in channel geometry and riparian vegetation related to climatic effects since settlement of the basin;[16] we revisited several of the camera stations that Hereford used (see fig. 13.10B). In 1997, Robert J. Warren matched 161 of G. A. Grant's views of Zion National Park, most of which depict riparian vegetation along the Virgin River or in tributary canyons.[17] We include two of Grant's views here, including one that Warren matched in 1997 (see fig. 13.6). In our study, we matched 125 views of the East Fork, the North Fork, and the main stem of the Virgin River as far downstream as Littlefield, Arizona (fig. 13.1).

Floods, Channel Change, and Flow Regulation

Historic flooding on the Virgin River must be pieced together from several sources. The first and perhaps largest flood known to have occurred on the Virgin River resulted from a large regional storm in January and February 1862.[18] The storm that spawned this flood covered most of the western United States and lasted for forty-four days in southern Utah; arroyo downcutting in the Virgin River basin dates to this flood.[19] In the 1860s, many acres of agricultural lands were washed away between the towns of Virgin and Rockville, Utah,[20] and irrigation diversion dams were destroyed almost as quickly as they were built, particularly in 1883 and 1884.

The peak discharge of the 1862 flood may have been exceeded during flooding in December 1889. On December 7 and 15, floods on the Virgin River had a stage 2 feet higher in a wider and deeper channel than the flood of 1862.[21] A third large flood occurred on September 1, 1909; billed "the greatest flood in the history of the county," this flood caused extensive damages along the Virgin River.[22] The peak discharges for the 1862, 1889, and 1909 floods are unknown but were probably comparable to floods that occurred later in the twentieth century.[23]

The net effect of these disastrous floods was the conversion of fertile tracts of floodplain inhabited by Mormon pioneers into wide, eroded stream channels. Some have blamed the flooding on overgrazing, particularly by sheep,[24] but the period during which these floods occurred was the wettest in three hundred years according to tree-ring records.[25] The extensive damage forced inhabitants to move from the floodplains to the uplands, creating the necessity for longer canals from the Virgin River to supply these upslope sites. One example was establishment of the town of Hurricane, which is high above the Virgin River in its basalt gorge (fig. 13.1).

As was the case for other rivers in

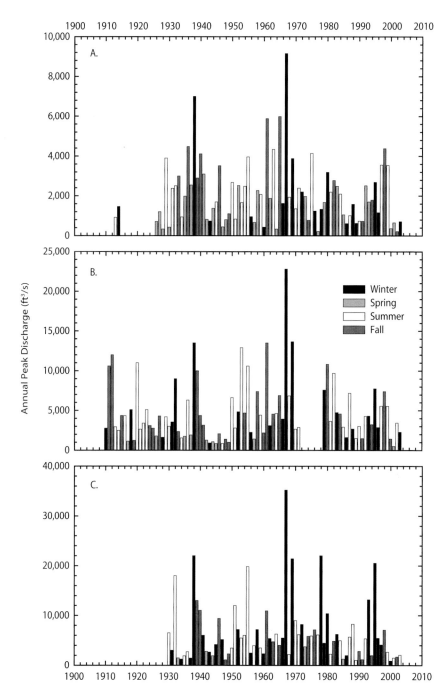

Figure 13.2 Annual flood series for the Virgin River.

A. North Fork Virgin River near Springdale, Utah (station 09405500; 1913–1914, 1926–2003).

B. Virgin River at Virgin, Utah (station 09406000; 1910–1971, 1979–2003).

C. Virgin River near Littlefield, Arizona (station 09415000; 1930–2003).

the region, floods were much smaller from the 1930s (and especially after 1941) through the 1960s, but channels continued to widen during the smaller events. Engineering plans for channelization in Zion National Park were drawn in 1929.[26] A small flood in 1932, possibly related to the storm that caused the largest flood on the Escalante River (chapter 9),

damaged floodplains and structures in Zion Canyon.[27] Although the discharge of this flood at Springdale was not particularly unusual (fig. 13.2A), the damages it caused spurred channelization efforts in Zion National Park. Comparison of tree-ring records with streamflow in the Virgin River indicates that high historic runoff coincides with periods of channel wid-

A

B

Figure 13.3 Photographs of the East Fork Virgin River in Parunuweap Canyon.

A. (April 1873.) This upstream view of the East Fork of the Virgin River in Parunuweap Canyon is in the relatively shallow bedrock canyon upstream from Zion National Park. A vertical bank that may have been eroded during the 1862 flood is in the right midground. The foreground trees appear to be either Arizona ash or black willow (leafless in this early spring view), and coyote willow appears in the center on top of the bank. A skunkbush drapes the cliffs on the left. (J. K. Hillers 57-PL-29, courtesy of the National Archives.)

B. (October 2, 1999.) The channel has narrowed slightly but is not as deep as in 1873. Frémont cottonwood is common in this reach and appears in the left horizon. Two large black willow trees are prominent on river left; box elder is on the slopes well above the channel; and netleaf hackberry is also present. Coyote willow *(foreground)* and tamarisk are also in this reach. (D. Oldershaw, Stake 2390.)

ening and lateral migration, whereas lower runoff is concurrent with channel narrowing.[28]

Unlike rivers farther south or east, the Virgin River had only one large, natural flood in the wet period from 1964 through 1995. With a peak discharge of 35,200 ft³/s, the 1966 flood is the largest natural flood in the period of gaging record, which is from 1930 through 2003 at Littlefield, Arizona (fig. 13.2C).[29] Unusual, but not particularly rare meteorological conditions developed over the western United States in early December 1966 as a cutoff low-pressure system formed in the upper atmosphere over the Pacific Ocean.[30] Although cutoff lows

are common in the winter months, this one had warm moisture that fell on a preexisting snowpack. Flooding was widespread in four western States. Southern California had extensive flooding; large floods occurred on rivers draining the southern Sierra Nevada and the Transverse Ranges south to the Mexican border; and severe flooding occurred in mountain streams near Las Vegas and in central Arizona.[31]

Recent floods on the Virgin River have been relatively small, but one flood that resulted from a dam failure was very large. The Quail Creek Dam, built in a side canyon off the Virgin River upstream from Washington,

failed with little warning on December 31, 1989. The peak discharge of this flood was 61,000 ft³/s at the Littlefield gaging station (not shown in fig. 13.2C). This flood washed out considerable amounts of riparian vegetation, inducing a new cycle of tamarisk establishment on the newly eroded floodplains downstream. As a result, the Virgin River downstream from Washington has a different recent-flood history than the rest of the watershed.

A   **B**

Figure 13.4 Photographs of the East Fork Virgin River in Parunuweap Canyon.

A. (April 1873.) This upstream view of the East Fork of the Virgin River shows typical conditions in the lower part of Parunuweap Canyon. This site is about a mile downstream from a boulder-clogged reach with a high gradient and perhaps 3 miles downstream from a 40-foot waterfall that marks the eastern boundary of Zion National Park. The riparian vegetation is difficult to identify, but several individuals of what appear to be Arizona ash are on river left *(right side)*. (J. K. Hillers 57-PL-14, courtesy of the National Archives.)

B. (October 29, 2002.) The channel has shifted in this narrow canyon, where flood flows are capable of entraining extremely large boulders such as the one now gone from the view. All of the vegetation appears to have changed, and the new trees on river left are now much larger. Riparian vegetation in the vicinity includes cattail, seepwillow, tamarisk, Russian olive, Frémont cottonwood, coyote willow, Arizona ash, box elder, and netleaf hackberry. Parunuweap Canyon has one of the most diverse riparian ecosystems in the Southwest. Tamarisk was being eradicated from the lower part of the canyon at the time of this photograph. (D. Oldershaw, Stake 2377.)

Change in Riparian Vegetation

Parunuweap Canyon (East Fork of the Virgin River)

The headwaters of the East Fork of the Virgin River is a divide between the Markagunt Plateau and the Paunsaugunt Plateau (fig. 13.1). The river flows southwesterly past productive agricultural lands in Long Valley before turning west and entering Parunuweap Canyon. This relatively shallow canyon in Navajo Sandstone gradually deepens as the river flows west to Zion National Park. In the eastern part of this canyon, the floodplain is wide enough to allow creation of terraces that support riparian vegetation (fig. 13.3). How-

ever, once the river enters the extreme eastern part of Zion National Park, it is narrower and confined within a deeper canyon, but riparian vegetation grows in protected zones within the canyon (fig. 13.4). Riparian vegetation—composed mostly of native species—has increased slightly in these reaches.

Parunuweap Canyon in Zion National Park has been closed to general visitation since the mid-1990s, and this reach has become a research natural area. Livestock grazing was eliminated in the 1930s, and tamarisk and Russian olive currently are being eradicated along the river corridor.[32] The lush riparian vegetation, with perhaps twenty species of native trees and shrubs, has

one of the highest diversities of woody riparian vegetation in the Southwest. Repeat photography of fifteen views taken from 1873 through 1936 does not show a large amount of change in riparian vegetation, although native species have increased considerably in the alluvial reaches upstream from the confluence with the North Fork near Springdale.

Zion Canyon (North Fork)

The headwaters of the North Fork of the Virgin River are atop the southern slopes of the Markagunt Plateau, from which the river quickly plunges into the narrow defile of Zion Canyon. In

A
B

Figure 13.5 Photographs of the North Fork Virgin River in The Narrows, Zion National Park.

A. (April 1873.) This upstream view, from midchannel just downstream from the mouth of Mystery Canyon, shows typical conditions in the Zion Narrows. Jack Hillers set up his darkroom tent in the left foreground. Springs flow down the bedrock slopes in the center. The North Fork of the Virgin River is perennial in this reach, but woody riparian vegetation is confined to steep bedrock walls and in sites with some protection against the force of floodwaters. Box elders appear on river right *(left side)* downstream from a bedrock projection. Ponderosa pine and other conifers are higher on the slopes. (J. K. Hillers 631, courtesy of the National Archives.)

B. (September 27, 2002.) Several large boulders persist, providing a frame of reference at this site. Box elders have increased slightly on the left side, and Arizona ash surround the spring conduit at right center. Despite the loss of the Ponderosa pine on the right side, pines have increased on the background slopes. (D. Oldershaw, Stake 4285.)

a reach appropriately called The Narrows, Hillers took several photographs during his trip to Zion in April 1873 (fig. 13.5). Little has changed in riparian vegetation near river level, although tree growth on the slopes above the river has been considerable.

Downstream from The Narrows, the canyon widens enough to allow deposition of the sediments that form high terraces on both sides of the river. Numerous photographs have been taken from Observation Point, high above the canyon floor on the west side of the canyon (fig. 13.6). A time series of the downstream view from this viewpoint shows that the channel of the Virgin River has narrowed considerably and riparian vegetation has increased. Streamside vegetation is considerable and lush, and native

riparian trees have increased in size and density throughout this reach (fig. 13.7). Tamarisk, once common in this reach, is now rare owing to eradication efforts by the National Park Service that began in 1987.[33] Effective eradication was completed by 1991, and tamarisk removal is ongoing because of continued establishment from seed sources outside of the park.

Through the middle of the twentieth century, the Virgin River in Zion Canyon was channelized to minimize flood damages to roads and structures and to constrain meandering. Observers of the riparian forest along the Virgin River have concluded that the Frémont cottonwood are mature individuals that will probably die off in the coming thirty years and are not being replaced.[34] The primary reason given

for this prediction is that channelization limits overbank flooding and consequent germination and establishment of some woody riparian species, notably Frémont cottonwood. Given that many of these trees established on higher terraces in the 1970s and 1980s, cottonwood regeneration seems unlikely without future floods that will overtop those terraces.

The extreme topography in the lower reaches of Zion Canyon is conducive to slope failure that creates large landslides. Downstream from the Court of the Patriarchs, a prehistoric landslide from Sentinel Mountain on the west side of the canyon has periodically dammed the Virgin River, backing up a lake nearly as far upstream as The Narrows. Radiocarbon dating indicates that this lake persisted for

A

B

C

D

E

Figure 13.6 Photographs of the North Fork of the Virgin River from Observation Point in Zion National Park.

A. (1903.) Numerous photographs have been taken of the stunning view downstream from Observation Point, which stands high above the Virgin River downstream from The Narrows. This time series of photographs, taken from slightly different camera stations, shows changes in riparian vegetation in a relatively wide part of Zion Canyon. In 1903, the channel of the Virgin River is wide and barren. Meander propagation patterns are evident on the floodplain, and older groves of Frémont cottonwood remain along the channel. A faint road leading upstream to The Narrows appears on river left. (W. T. Lee 2713, courtesy of the U.S. Geological Survey Photographic Library, Stake 1986.)

B. (September 12, 1929.) Although the channel remains as wide as it was in 1903, woody riparian vegetation is becoming established on the incipient low floodplain. More channel avulsions are apparent at center, probably the result of the 1909 flood, and the road leading to The Narrows has been widened. The groves of cottonwood appear to have increased in height. (G. A. Grant 79-Z-102, courtesy of the National Archives, Stake 2386.)

C. (October 14, 1948.) The position of the low-flow channel appears to have stabilized, in part because of channelization. At this time, the open reach in the bottom half of the view is not channelized, but the reach through the middle third of the view is. Channelization in the 1930s resulted in a straighter and narrower channel. Relatively small floods in the middle of the twentieth century probably contributed to channel stability. The cottonwood groves are no longer distinct, and the road to The Narrows has been paved. (P. T. Reilly 97.46.173.42, courtesy of the Cline Library, Stake 2386.)

D. (1976.) This view is taken from a point along the trail below Observation Point. The reach in the bottom half of the view is not channelized, but the reach downstream is (see photograph C). The channel has narrowed considerably between 1948 and 1976, despite the effects of the 1966 flood. The channelized reach at center is barely visible because of cottonwood growth and increases in other species. Even though the Virgin River in the lower half of the view has not been channelized, the channel has narrowed considerably. (D. Lyngholm 95.55.1741, courtesy of the Cline Library, Stake 2387.)

E. (October 31, 2002.) The Virgin River has narrowed considerably, although the floodplains remain open in parts of the view. This openness likely is the effect of low winter floods, which inhibit germination of species such as cottonwood. Tamarisk-eradication efforts by Zion National Park have reduced but not eliminated this nonnative species. (D. Oldershaw, Stake 2386.)

A

B

Figure 13.7 Photographs of the North Fork of the Virgin River at the Great White Throne in Zion National Park.

A. (Before 1930.) This downstream view on the North Fork of the Virgin River shows a close-up of conditions viewed from Observation Point (fig. 13.6.) This view, with the Great White Throne looming above the river, is an extremely popular photographic subject. The channel remains relatively wide, with only a small amount of riparian vegetation, although slopes support dense riparian species, including bigtooth maple. Shortly after this photograph was taken, the North Fork was channelized in this reach, forcing the river against the west *(right)* bank. (H. Peabody 79-HPA-14-1762, courtesy of the National Archives.)

B. (November 4, 2000.) The channel has narrowed and is nearly obscured from this camera station by riparian vegetation. Netleaf hackberry, Frémont cottonwood, shrub oak, tamarisk, box elder, seepwillow, and coyote willow compose the riparian vegetation in the foreground and along the river. Photographs from 1873 in this reach, however, show that Frémont cottonwood has decreased. The levees built to constrain the channel in the 1930s were washed away in the 1960s, allowing the channel to shift considerably (D. Sharrow, Zion National Park, written communication, 2005). (D. Oldershaw, Stake 1970.)

3,000 to 4,000 years of the Holocene, and its dam eroded to the point that sediments stopped accumulating less than 3,600 years ago.[35] Smaller landslides occurred in this reach in 1923, 1941, and 1995, which were El Niño years. The likeliest failure scenario is that high precipitation during the wet winters lubricated the shales in the Kayenta Formation beneath the Navajo Sandstone and increased the mass of the unconsolidated sediments, inducing movement.[36]

Landslide deposits occur at several points in Zion and Parunuweap Canyons,[37] indicating that this process controls the grade upstream on the Virgin River. A moderate earthquake at St. George on September 2, 1992, caused another landslide on the north side of Springdale to move, destroying three houses and closing the road into Zion National Park.[38] The existence of periodically active landslides underscores the fact that both forks of the Virgin River have been repeatedly dammed, creating depositional environments that are not expected in such deep bedrock canyons.

Downstream, riparian vegetation has also increased along the wider reaches of the North Fork in Zion Canyon (fig. 13.8). An overview photograph from a ridge south of the confluence of the East and North Forks shows considerable increases of riparian vegetation despite agricultural clearing and development (fig. 13.9). Water is diverted from the Virgin River within Zion National Park to supply the agricultural fields, which are of limited extent owing to the relatively narrow bottomlands constrained by the combination of bedrock outcrops and landslide deposits.

Virgin River Main Stem

Just downstream from the confluence, the Virgin River turns westward, and the canyon widens owing to the presence of the Petrified Forest Member of the Chinle Formation at river level and the Moenkopi Formation in low cliffs. As is the case with nearly every alluvial river in the region, the channel was wide and barren following the floods of the 1880s through the 1930s (fig. 13.10). As flows stabilized, riparian vegetation followed, helping

A

B

Figure 13.8 Photographs of the North Fork of the Virgin River in Zion National Park.

A. (April 1873.) This upstream view on the Virgin River shows the canyon walls upstream from Springdale, Utah. The trees, which are leafless in this early spring view, include Frémont cottonwood *(right)*, Arizona ash *(center)*, and Gambel oak *(left)*. The driftwood in the foreground may be residual from the 1862 flood. (J. K. Hillers 594, courtesy of the U.S. Geological Survey Photographic Library.)

B. (October 5, 1999.) The view is now blocked by cottonwood, Utah juniper, and other riparian species. Blockage of historical views by increased riparian vegetation is common in the heavily photographed Zion Canyon. (D. Oldershaw, Stake 3994.)

trap sediments that accumulate on low floodplains. The increase in riparian vegetation apparently accelerated in the wet period of the 1990s (figs. 13.10B and 13.10C). Both native and nonnative species increased, and xerophytic shrubs became established on the floodplain as channel narrowing reduced overbank flooding and the setting became more xeric (fig. 13.11).

Westward, the Virgin River gradually cuts another canyon, this time into the combination of Kaibab Formation (mostly limestone) and basalt upstream from the cities of La Verkin and Hurricane, Utah. Photographs of a narrow bridge at an old hot springs near the mouth of this canyon show that riparian vegetation has increased in this reach as well (fig. 13.12), although, as with most bedrock-lined reaches, the increases are relatively small. The Virgin River crosses the Hurricane fault at this point and immediately enters another canyon cut

through basalt flows. West of this point, the river turns toward the southwest and flows in a wide alluvial valley to Washington and St. George (fig. 13.1). Failure of the Quail Creek Dam in December 1989 destroyed large amounts of riparian vegetation downstream.

Downstream from St. George, the Virgin River enters the Virgin River Gorge, a narrow defile through Paleozoic rocks of the northern Virgin Mountains. Interstate 15 follows the Virgin River through much of the Virgin River Gorge, and freeway construction has caused extensive changes in channel morphology. In addition, the 1989 dam-break flood had serious impacts in this narrow canyon. The peak discharge was 73 percent larger than the biggest known natural flood and destroyed much of the riparian vegetation in the canyon. As a result, tamarisk is the most common species present here (fig. 13.13).

Near Littlefield, Arizona, riparian vegetation—both native and non-native species—has increased considerably (fig. 13.14). The once wide, barren channel is now much narrower.[39] However, downstream from this point, vegetation along the Virgin River gradually becomes a monospecific stand of tamarisk. Near the Arizona-Nevada border, repeat photography shows extensive encroachment of tamarisk.[40] The delta created where the Virgin River empties into Lake Mead has few native species present.

A

B

Figure 13.9 Photographs of the confluence of the East Fork and North Fork of the Virgin River south of Zion National Park.

A. (April 1873.) This view across the confluence of the East Fork and North Fork of the Virgin River looks upstream into Zion National Park. The pioneer town of Springdale is in the midground, and fields east of Rockville are in the left foreground. The East Fork meets the North Fork from the right side of this view and is obscured mostly by the pinyon in the foreground. The river is readily apparent in the center of the view (North Fork) and in the lower left center (Virgin River). (J. K. Hillers 80, courtesy of the U.S. Geological Survey Photographic Library.)

B. (September 25, 2002.) Despite extensive floodplain activities, which include agriculture and establishment of homes and businesses, woody riparian vegetation has increased along the Virgin River throughout this view. From this distance, Frémont cottonwood and tamarisk are the only species that can be identified, although several others are also present. (D. Oldershaw, Stake 2207.)

A. (1937.) This downstream view on the Virgin River west of Rockville shows a wide, barren channel in a meander bend. Frémont cottonwood of various sizes are visible throughout the view, although the trees are mostly scattered. The smaller shrubs cannot be identified, but some in the center along channel right might be young tamarisk. The main road to Zion National Park from the west appears along channel right. (H. E. Gregory 9-2455, courtesy of the Marriott Library.)

B. (1994.) The channel has shifted to the left, abandoning the broad meander bend. The low floodplain that developed is covered with riparian vegetation. Most of the trees appear to be cottonwood, and groves of either arrowweed or coyote willow appear in the left center on the inside of the bend. Utah Route 9 is now paved. (R. Hereford.)

C. (October 31, 2000.) The increase in riparian vegetation is striking, even between 1994 and 2000. Both the size of persistent trees and the overall density of vegetation have increased. Riparian vegetation in this reach is mostly cottonwood, tamarisk, arrowweed, and box elder. (D. Oldershaw, Stake 2814.)

Figure 13.10 Photographs showing the Virgin River at Rockville, Utah.

A

B

Figure 13.11 Photographs showing the Virgin River at Grafton, Utah.

A. (May 1936.) This view south across the Virgin River shows the abandoned town of Grafton. The Virgin River is wide here, although riparian vegetation is beginning to increase. The low shrubs on the floodplain are probably a combination of arrowweed and tamarisk, and small tamarisk trees are on the far bank. Frémont cottonwood tower above the abandoned buildings of Grafton. (L. M. Huey HL-45-150, courtesy of Special Collections, University of Arizona.)

B. (September 21, 2002.) The view of the river is now blocked by dense riparian vegetation, which includes rubber rabbitbrush *(foreground)*, cottonwood, black willow, tamarisk, Russian olive, and coyote willow. (D. Oldershaw, Stake 4327.)

A

B

Figure 13.12 Photographs showing the Virgin River upstream from La Verkin, Utah.

A. (September 6, 1929.) This upstream view shows the Virgin River at the fault plane of the Hurricane fault just upstream of La Verkin. A hot-springs spa appears on the right side. The only riparian vegetation visible is two large clumps of mesquite on channel right *(left center)* and a solitary Frémont cottonwood on river left *(foreground)*. (G. A. Grant 79-ZBV-Z203, courtesy of the National Archives.)

B. (October 31, 2000.) Once the primary road connecting St. George and the towns along the upper Virgin River, the route is now behind the camera station, passing over the Virgin River on a high bridge. Riparian vegetation has increased along the Virgin River upstream and downstream from the original bridge. The species present include mesquite (greatly expanded on channel right), arrow-weed, and tamarisk. The cottonwood trees on the right side are planted along the now paved road and are irrigated. (D. Oldershaw, Stake 2196.)

A

B

Figure 13.13 Photographs showing the Virgin River in the Virgin River Gorge.

A. (1981.) This upstream view of the Virgin River in the Virgin River Gorge provides a rare view of this reach. Tamarisk was the primary riparian species lining the river, and it was flowering when this photograph was taken. (D. Lyngholm 95.55.1781, courtesy of the Cline Library.)

B. (February 20, 2004.) Tamarisk has increased along the Virgin River despite the 1989 dam-break flood. Seepwillow is now present as well. Grazing remains heavy in this reach. (D. Oldershaw, Stake 4749.)

Figure 13.14 Photographs showing the Virgin River at Littlefield, Arizona.

A. (June 4, 1942.) This downstream view from a bluff above the Virgin River shows the gaging station at Littlefield. The channel is relatively wide and free of vegetation; Frémont cottonwood appear along the right side of the view; and mesquite and tamarisk trees are on river left. Although the river is used extensively for irrigation and domestic water supply, no dams have been built across the main channel. A dam built upstream in the Virgin River Gorge in the late 1800s washed out almost immediately after construction. (J. A. Baumgartner 3601, courtesy of the U.S. Geological Survey.)

B. (October 30, 2000.) Despite several large floods in the intervening fifty-eight years, including the dam-break flood of January 1, 1989, riparian vegetation has increased in this reach. Most of the foreground vegetation is native, including cottonwood, arrowweed, mesquite, and catclaw. At lower right, a marsh containing reeds, cattail, and carrizo grass has developed. On river left, native coyote willow lines the bank, and a prominent zone of tamarisk with occasional cottonwood appears behind. The vertical bank at center is on a small floodplain terrace that developed after 1942. (D. Oldershaw, Stake 1727.)

14 Colorado River Reservoirs

Summary. Dams constructed on the Colorado River, beginning with Hoover Dam in 1935 and ending with Glen Canyon Dam in 1963, inundated hundreds of miles of potential riverine habitat, eliminating the riparian vegetation. Although riparian vegetation once was locally lush, especially at major tributary confluences, most of the submerged river corridor was either barren sand or bedrock. This initial loss of the original riparian vegetation has been offset in part by new patches of lush cottonwood and willow habitat, and tamarisk has become established along reservoir margins. Fluctuation in reservoir elevations, combined with high surface-water and groundwater salinity, encourages growth of tamarisk, although native species such as mesquite, black willow, and even cottonwood are present in deltas.

Riparian vegetation in the Southwest has been killed locally by a variety of mechanisms, but by far the largest reductions in potential riverine area are the reaches beneath reservoirs. Dams were built across many rivers in the region, including the Mojave, Verde, Salt, and Gila Rivers, but none of these has had as many free-flowing reaches inundated as the Colorado River (fig. 14.1). Two of the largest reservoirs in the United States—Lakes Mead and Powell—are on this river, as well as several smaller reservoirs in the lower basin. These reservoirs are unique in the region in that they generally are long and narrow, with the exception of parts of Lake Mead, and therefore inundate 404 miles of what once was riverine riparian habitat. How much woody riparian vegetation was actually in this habitat before inundation affects the overall interpretation of regional change in riparian vegetation.

Flood control, regulation of water deliveries, and hydroelectric power generation are the reasons for the existence of these dams and reservoirs. Authorization by Congress for and construction of these projects through the middle third of the twentieth century have their roots in a national desire to "reclaim" the desert—to create productive agricultural lands in an arid environment—as well as to protect diversion structures and towns built in the floodplains of the lower Colorado River following the disastrous events of 1905 (chapter 27). Competing plans for development ranged from small structures and reservoirs with small surface areas that would limit evaporation[1] to large structures strategically located for political and economic reasons.[2] The small-structures scheme would have inundated all free-flowing reaches of the Colorado River, eliminating its riparian vegetation, whereas the adopted plan unintentionally left free-flowing reaches, such as those through Cataract Canyon and Grand Canyon (chapters 7 and 11).

Operations of dams and reservoirs have had several unintended effects on riparian vegetation. Before the construction of dams, the Colorado River delivered prodigious amounts of sediment to the desert reaches and the Gulf of California.[3] Now, this sediment is trapped in the deltas and beds of the reservoirs. Riparian vegetation, in particular early-successional species such as tamarisk, thrive in these deltas, creating habitat for birds. Water-level fluctuations can either drown the newly established plants during wet years or lower groundwater levels beyond maximum rooting depth during drought years. Many deltas in the region appear to be in perpetual succession and are prime habitat for nonnative species.

Lake Powell

Glen Canyon Dam, completed in 1963, impounds Lake Powell, a reservoir with the longest shoreline in the world.[4] The dam is 583 feet above the former river level, allowing backup of Colorado River water not only into 186 river miles of the formerly free-flowing Colorado River, but also up the San Juan, Dirty Devil, and Escalante Rivers and numerous small tributaries. Whatever riparian vegetation was present in these reaches was killed under the rising waters between 1963 and 1980, when Lake Powell filled for the first time.

The new shoreline offered some substrate for new establishment of riparian vegetation, although this substrate was subject to large fluctuations in water level and was mostly bedrock. The upstream ends of the lake are a different story. By 1986, about 0.9 million acre-feet of sediment had accumulated in the reservoir and its delta, including 52,100 acre-feet in the channels of the Colorado and San Juan Rivers.[5] At this rate of accumulation (37,000 acre-feet per year), the reservoir would fill with sediment in about seven hundred years. As the reservoir fluctuated, the delta was exposed, and riparian vegetation—primarily tamarisk—became established.[6]

Compilations of historical photographs suggest that most of the river corridor consisted of water flowing between barren sandstone bluffs.[7] Extensive but isolated areas of lush riparian vegetation were present along the predam Colorado River (fig. 14.2). However, many of these photographs were taken precisely to show how unusual the thick, native riparian vegetation was. Most of the historical photographs taken for scientific documentation—by E. O. Beaman for

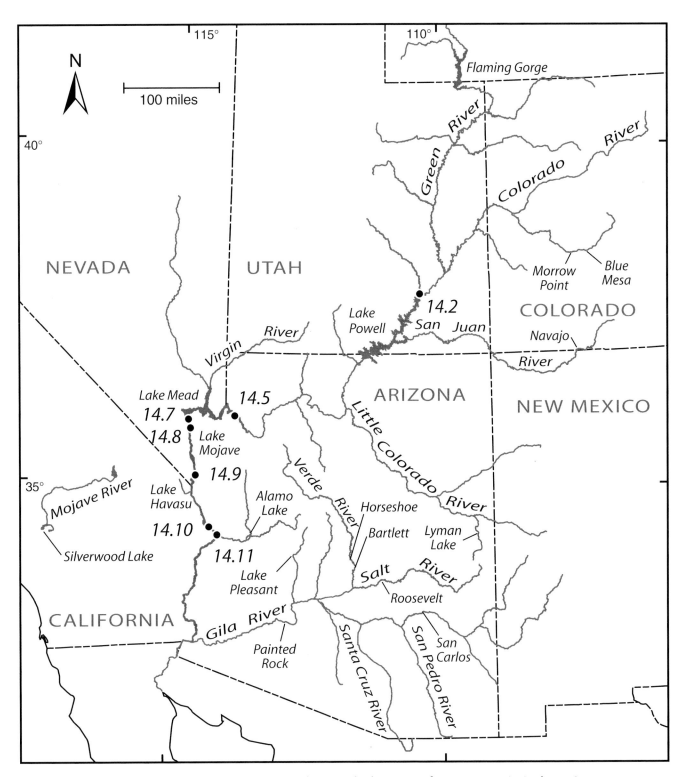

Figure 14.1 Map of the southwestern United States showing the locations of major reservoirs in the region.

the Powell Expedition in 1871 and by Franklin A. Nims for the Brown-Stanton Expedition in 1889—show that in most reaches, riparian vegetation was sparse. Some broad terraces supported dense stands of coyote willow, but Frémont cottonwood generally was found only at tributary mouths or within tributaries. For example, photographs of the lower San Juan River, in the vicinity of the now submerged juncture with Neskahi Wash, show scattered groves of small Frémont cottonwood (fig. 14.3).[8]

The amount of habitat available for riparian vegetation was limited owing to the wide river passing through a narrow canyon (fig. 14.4). Many reaches consisted primarily of a channel lined with bedrock on both sides, particularly where Navajo Sandstone was at river level. Broad terraces similar to those that remain downstream of the dam (chapter 11) occurred along one or both banks, most notably in Wingate Sandstone or Chinle Formation. Many of these banks were covered with

Figure 14.2 Photographs of the confluence of the Dirty Devil and Colorado Rivers.

A. (Late June 1889.) This upstream view on the Dirty Devil River at the Colorado River was taken from the downstream side of the confluence. A continuous row of shrubs lines the left bank of the Dirty Devil and the right bank of the Colorado. These shrubs likely include coyote willow and desert olive. (F. A. Nims 100, courtesy of the National Archives.)

B. (September 1, 2000.) The mouth of the Dirty Devil River is now submerged more than 250 feet beneath the surface of Lake Powell at full-pool elevations. By October 2003, ongoing drought depletes the storage of Lake Powell, the Colorado River is channelized through terraces in the delta, and tamarisk in particular is colonizing the newly exposed substrate. (R. H. Webb, Stake 3053.)

Figure 14.3 (January 1894.) Photograph of the San Juan River downstream from Paiute Farms. This unique photograph, from river right on the San Juan River, shows an open grove of Frémont cottonwood on river left upstream from a bend. On the basis of the geology, this site likely is downstream from Paiute Farms, a now abandoned farming area along the San Juan River. (C. Goodman P0068:039, courtesy of the Marriott Library.)

xerophytic vegetation, and obligate riparian vegetation occurred mostly on streamside cutbanks. Otherwise, both the point bars on the inside of bends and the wide reaches associated with tributary mouths sustained some level of riverine riparian vegetation.

As part of the Glen Canyon Salvage Program, riparian vegetation beneath the future Lake Powell was estimated using a combination of field measurements and analysis of aerial photography.[9] These measurements indicated that 3,421 acres of riparian vegetation was present,[10] or only 13 acres per river mile of the inundated Colorado and San Juan Rivers. This calculation suggests that the combined width of the floodplain on both sides of these rivers supporting riparian vegetation was only 54 feet. Coyote willow was the most common native species along the river in the late 1950s, and black willow was the most common riparian tree after nonnative tamarisk. In contrast, cottonwood was not common along the main-stem Colorado and San Juan Rivers but was locally dense in tributaries.[11] Tamarisk was established to such an extent as to be noted in Glen Canyon in the mid-1930s[12] and is clearly visible in 1950s photographs.[13]

Lake Mead

Hoover Dam, completed in 1935, is one of the engineering marvels of the twentieth century.[14] Standing 726 feet above the former bed of the Colorado River, this massive plug of concrete impounds Lake Mead, which stores 28.5 million acre-feet of water at full-pool elevation and backs water 115 river miles up to mile 239 in western Grand Canyon.[15] Sediments delivered to Lake Mead from the Colorado River and to a lesser extent from the Virgin River created deltas at the head of the lake. By 1949, about 5 percent of the reservoir's capacity was filled with sediment, and studies suggested that the reservoir would completely fill in about four hundred years.[16]

Construction of Hoover Dam occurred in the era before environmental-impact statements, and no studies documented the resources submerged beneath Lake Mead. The only information on the type and amount of riparian vegetation that once was along the Colorado River comes from anecdotal observations and photographs. Few written observations are available concerning this reach on account of its isolation. In 1837, Antoine Leroux built skin canoes at the mouth of the Virgin River and floated downstream.[17]

He found "timber," built seven wooden canoes, and paddled to the Gulf of California, seeking beaver pelts. He observed "a few scattered" Frémont cottonwood 40 miles downstream from the mouth of the Virgin River. In 1893, C. Hart Merriam described woody vegetation in the Mojave and Great Basin Deserts, including riparian vegetation along the Colorado River.[18] Although coyote willow, arrowweed, and seepwillow[19] were common on both sides of the river in Great Bend, just upstream of the present location of Hoover Dam, neither Frémont cottonwood nor black willow is listed as a species along the Colorado River.

Photography confirms that woody riparian vegetation was relatively sparse, particularly in narrow reaches, and that few trees were present.[20] Numerous photographs downstream of mile 239 in western Grand Canyon show a barren river corridor with dense mesquite in the high-water line (see chapter 11) and no cottonwood or black willow (fig. 14.5). Creation of Lake Mead caused deposition of a delta in this reach, with a moonscape appearance when it emerged from beneath the lake. As was the case with Lake Powell, riparian vegetation, primarily tamarisk, became established in dense, monospecific stands.

Figure 14.4 (December 21, 1889.) Photograph of the Colorado River in Glen Canyon. The Stanton Expedition systematically photographed the Colorado River through Glen Canyon. This site, just upstream from present-day Glen Canyon Dam, is now submerged beneath about 568 feet of water at the full-pool elevation of Lake Powell. Essentially no riparian vegetation was present in this view in 1889. (F. A. Nims 234, courtesy of the National Archives.)

Tamarisk was not the only species to colonize the exposed parts of the delta. By the late 1990s, the delta within the western part of Grand Canyon locally supported dense stands of black willow. These stands, which apparently established following the high-flow period of the mid-1980s, were decimated by rising lake levels in the early 1990s and then further damaged by falling reservoir levels associated with the early-twenty-first-century drought. Tributary confluences did have woody riparian vegetation near river level, including black willow and tamarisk (fig. 14.6). In general, wider reaches of the Colorado River had high terraces containing mesquite and other low riparian species, with few and scattered individuals of cottonwood and black willow (fig. 14.7).

Lake Mohave

Davis Dam, 200 feet high and completed in 1950, impounds Lake Mohave, a narrow, 59-mile-long reservoir that extends upstream almost to the base of Hoover Dam (fig. 14.1). Davis Dam serves two primary purposes: it provides additional water storage for delivery to the lower basin and produces hydroelectric power. Its isolation in Black Canyon and the wider valley downstream between desert mountains makes significant water transfers from this reservoir difficult. Because Davis Dam was built well after completion of Hoover Dam, no delta has formed at its upstream end.

The steep walls of Black Canyon provided little room for establishment of woody riparian vegetation (fig. 14.8). Although mesquite and arrowweed were common along the wider reaches, watercolors made in 1857 by Heinrich Balduin Möllhausen[21] and photographs taken in 1871 by Timothy O'Sullivan[22] show a barren canyon with little riparian vegetation. Several relatively lush areas are exceptions; Cottonwood Cove, the mouth of El Dorado Canyon, and "Cottonwood Island" (chapter 27) had significant amounts of cottonwood that were utilized by miners and steamboat operators. The banks of the Colorado River were mostly either rock outcrops or sandy coves with small amounts of woody riparian vegetation above the high-water line (fig. 14.9).

Lake Havasu

Lake Havasu is, in one sense, the most important of the reservoirs on the Colorado River. The intakes and pumping stations for the Colorado River Aqueduct, which supplies water to southern California, and for the Central Arizona Project (CAP), which supplies agricultural and domestic water to central and southern Arizona, are on opposite sides of the lake just upstream from Parker Dam. The dam, 320 feet high and completed in 1938, was built in a narrow bedrock constriction between the Whipple Mountains in California and the Bill Williams Mountains of Arizona. The lake and delta extend 44 river miles upstream into the Topock Gorge.

The delta of Lake Havasu forms the basis of much of the Lake Havasu NWR (see chapter 27). Downstream, the Topock Gorge, formed in volcanic terrane, supported little riparian vegetation before construction of Parker Dam extended the delta into this reach. At the north end of the Chemehuevi Valley, the delta is wide, and extensive stands of tamarisk and cattail occur within other native vegetation; Lake Havasu NWR is a magnet for waterbirds and migratory songbirds. The lake is relatively wide in the Chemehuevi Valley, where extensive mesquite bosques once grew and small groves of Frémont cottonwood and black willow were present (fig. 14.10). The mouth of the Bill Williams River, just upstream from Parker Dam, supported considerable riparian vegetation, including what appeared to be a continuous but thin band of cottonwood-willow vegetation (fig. 14.11).

A

B

Figure 14.5 Photographs of the Colorado River in western Grand Canyon.

A. (November 1909.) This upstream view on the Colorado River at mile 270.3 shows a barren debris fan and the typically barren banks of the river corridor. Patchy mesquite appears on both banks in discontinuous lines; otherwise, no riparian vegetation exists. (R. A. Cogswell 1989.022:1088-ffALB vol. 2, courtesy of the Bancroft Library.)

B. (March 12, 1998.) Comparison of the black-and-white-striped bedrock layers in photographs A and B shows that Lake Mead has raised the water level here by perhaps 75 feet. A stand of black willow has become established on deltaic terraces, but rising waters of Lake Mead during the 1997–1998 El Niño period submerged the bases of these trees. Tamarisk is the more common species present on deltaic deposits in this reach (Turner and Karpiscak 1980). (D. Oldershaw, Stake 3514.)

Figure 14.6 (September 1923.) Photograph of the Colorado River beneath Lake Mead. The feathery trees on the far bank in the center of the view are probably tamarisk. (E. C. La Rue 790, courtesy of the U.S. Geological Survey Photographic Library.)

A. (June 23, 1929.) This view across Big Bend and upstream on the Colorado River, taken before construction began on Hoover Dam several miles downstream, shows the annual spring flood in progress. Higher floodplains support riparian vegetation, likely mesquite, and patchy vegetation on the lower floodplain is likely a combination of coyote willow, seepwillow, and arrowweed. What may be Frémont cottonwood or black willow trees or both appear singly along the nearest bank. (J. E. Stimson BCP311A, courtesy of the Bureau of Reclamation, Washington, D.C.)

B. (February 19, 2004.) The camera station is from a place now known as the Hemenway Wall on U.S. Route 93, which passes over Hoover Dam. The persistence of several creosote bushes in the foreground starkly contrasts with the complete change in the background. The white shoreline shows the effect of the early 2000s drought on water levels in Lake Mead. (D. Oldershaw, Stake 4747.)

Figure 14.7 Photographs of the Colorado River at the head of Black Canyon.

Figure 14.8 Photographs showing the Colorado River above Willow Beach.

A. (May 16, 1938.) After completion of Hoover Dam in 1935, the U.S. Geological Survey established a gaging station in Black Canyon to document dam releases. This downstream view from near the former stream-hydrographer residence shows a small riffle formed at a tributary mouth. The car at lower left traveled a road down this tributary to reach the site. A line of arrowweed appears on the far side of the river, growing on a low terrace deposit, with taller mesquite trees behind. What appears to be tamarisk is present on river left and at lower left in the view. (W. L. Heckler 2346, courtesy of the U.S. Geological Survey.)

B. (December 17, 2002.) This reach of Lake Mohave is slightly wider than most of Black Canyon, and back-flooded tributaries still support arrowweed and mesquite. The difference in lake and river level between the two views is about 20 feet. Tamarisk still grows at this site along the rocky shoreline. (D. Oldershaw, Stake 4496.)

Figure 14.9 Photographs of the Colorado River at Davis Dam and Lake Mohave.

A. (No date, probably early 1920s.) The camera station is on the left bank of the Colorado River upstream from the current towns of Laughlin, Nevada, and Bullhead City, Arizona. In the 1920s, the Colorado River was unregulated and subject to large annual floods. That flood regime is abundantly apparent in this upstream view of the river. A sparse mesquite bosque appears at left center; otherwise, the river channel is lined with massive sandbars. Barren, desert hillsides loom above the river in the midground. (E. C. La Rue 848, courtesy of the U.S. Geological Survey Photographic Library.)

B. (October 27, 2000.) Davis Dam was built in 1950 for additional reservoir storage and hydroelectric-power generation. The former riparian ecosystem is now dead and under about 190 feet of water at full-pool elevation. The white stripe across the hillsides in the right midground shows that the reservoir is drawn down from its normal-pool elevation at the time of this photograph. The formerly arid hilltop is now just emergent above the reservoir and supports a dense thicket of tamarisk and other species. (D. Oldershaw, Stake 2123.)

A

B

Figure 14.10 Photographs of the Colorado River at Pittsburg Flat (Lake Havasu).

A. (November 8, 1922.) This generally upstream view of the Colorado River, the middle section of a wider, panoramic view, shows a wide section of river corridor once called Pittsburg Flat. Scattered groves of cottonwood and black willow grow in backwaters along the river; overall, most of the channel margin is either barren sand or dense groves of arrowweed. A small grove of mesquite and catclaw, well above the influence of the Colorado River, appears in the center. (E. C. La Rue 1579, courtesy of the U.S. Geological Survey Photographic Library.)

B. (February 19, 2003.) Lake Havasu now inundates this reach, and vegetation consisting of tamarisk, mesquite, and arrowweed demarcates the former high lake shoreline. The near-shore part of Lake Havasu City appears in the center. (R. H. Webb, Stake 4466.)

A

B

Figure 14.11 Photographs of the confluence of the Bill Williams and Colorado Rivers.

A. (March 3, 1938.) This view across the Colorado River *(foreground)* and upstream on the Bill Williams River *(background)* shows interaction of both rivers' waters during a flood on the Bill Williams River. Parker Dam was completed shortly after this photograph was taken. The Bill Williams River has mostly shrubby riparian vegetation, probably including coyote willow, seepwillow, and arrowweed, and a line of Frémont cottonwood appears on the left side *(channel right of the Bill Williams)*. The taller dark trees in the center likely are mesquite. (Unknown photographer, courtesy of the Bureau of Reclamation, Washington, D.C.)

B. (December 16, 2002.) The original camera station is under more than 300 feet of water offshore from the pumping station for the Colorado River Aqueduct and about a half-mile upstream from Parker Dam. A marina and trailer park appear on the right side, and the western end of Bill Williams NWR is visible in the center. (D. Oldershaw, Stake 4391.)

15 The Bill Williams River and Major Tributaries

Summary. The Bill Williams River watershed, the principal drainage of the low-desert landscapes of west-central Arizona, sustains significant amounts of native and nonnative vegetation in alternating reaches of ephemeral and perennial flow.[1] With headwaters in the basin-and-range country of western Arizona, its two major tributaries—the Big Sandy and Santa Maria Rivers—join just upstream of Alamo Dam. Date Creek is a small tributary of the Santa Maria River with a puzzling anecdotal history of riparian vegetation change. In repeat photographs, riparian vegetation has increased along most of the reaches, although the increases are slight in the ephemeral reaches. Downstream from Alamo Dam, the Bill Williams River flows through alternating bedrock canyons separated by small alluvial valleys. The Bill Williams River NWR, extending upstream from Lake Havasu about 12 miles into the watershed, is an important bird and animal sanctuary supported by lush and extensive riparian vegetation. Riparian vegetation, especially Frémont cottonwood, now forms a completely different riparian ecosystem than existed historically downstream from the dam.

The Bill Williams River system is the principal drainage network in west-central Arizona, extending from the Basin and Range mountains in the western part of Arizona to the Colorado River (fig. 15.1). This river is named for William Sherley "Old Bill" Williams, who in 1837 trapped beaver along the watercourse that bears his name.[2] However, the part of the drainage basin that bears Williams's name is relatively short, extending downstream from the triple confluence of the Big Sandy and Santa Maria Rivers and Date Creek. The longer Big Sandy River basin contains most of the drainage area and produces the largest floods.

The Bill Williams River and its major tributaries flow through broad Basin and Range valleys with extensive rangelands used for livestock production. Its headwaters on the western side of the central Arizona highlands are positioned to intercept moisture from both warm-winter storms and dissipating tropical cyclones, and as a result the river has produced some of the largest floods in Arizona on the basis of drainage area. Although some tributaries are perennial, such as Burro Creek, most reaches are intermittent or ephemeral, with generally shallow water tables in the alluvial aquifers. Groundwater development is not significant except on the lower Bill Williams River at Planet Ranch, where alfalfa was irrigated from the alluvial aquifer. Flow regulation on the Bill Williams River has led to extensive establishment of riverine riparian vegetation, which forms the basis of the Bill Williams River NWR at the river's mouth.[3]

The Big Sandy River

The Big Sandy River has its headwaters in the Hualapai and Aquarius Mountains and drains about 3,200 square miles of west-central Arizona (fig. 15.1). This unregulated watercourse, the longest in the Bill Williams system, has several large tributaries, notably Willow Creek, on the Hualapai Indian Reservation of northwestern Arizona, and Burro Creek on its western side.[4] Through much of its alluvial reaches, the Big Sandy River has a braided channel that is shallowly incised into alluvium.[5] Because this watershed is isolated, with only the small town of Wikieup in the basin, little historical information is available on long-term changes in riparian vegetation.

Floods and Channel Change

As is characteristic of the entire watershed, years with extremely large floods follow years with little or no flow (fig. 15.2). The flood of record from 1966 through 2002, 68,700 ft³/s in 1993, contrasts with several years of peak flow less than 1,000 ft³/s. Reportedly, the flood on September 6, 1939, had a peak discharge of 100,000 ft³/s near Signal, Arizona.[6] These extreme fluctuations in flood discharges present opportunity to riparian vegetation, which can become established following large floods and have sufficient time before the next event to grow to a flood-resistant size. However, an eyewitness account of a December 1908 flood described vegetation stripped from the channel and old cottonwood eroded from the banks.[7]

The Big Sandy River has a wide floodplain with banks, where present, of less than 20 feet. Downcutting of 5 to 10 feet occurred at an unknown time in the last half of the nineteenth century, but large floods reportedly occurred from the 1880s through the early 1900s.[8] In the reach near Wikieup, the channel shifted considerably during floods, in particular those in 1978, 1990, and 1991.[9] Downstream, a large flash flood from Bronco Creek deposited considerable sediment in the Big Sandy River, and its channel subsequently adjusted by moving away from the tributary in a narrower channel.[10]

Change in Riparian Vegetation

In the late 1960s, a section of Willow Creek was used for experimentation regarding the effect of riparian vegetation removal on evapotranspiration losses.[11] In 1960, measurements along 4.1 miles of channel showed that the riparian ecosystem was dominated by

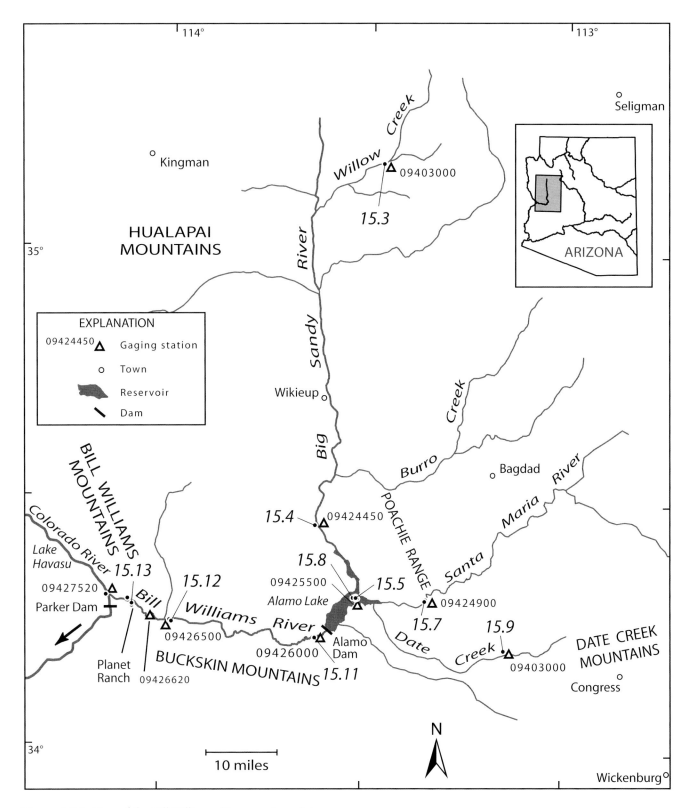

Figure 15.1 Map of the Bill Williams River, western Arizona.

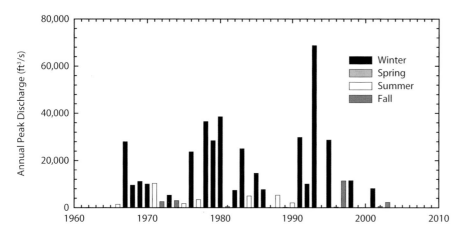

Figure 15.2 Annual flood series for the Big Sandy River near Wikieup, Arizona (station 09424450; 1966–2003).

red willow, Frémont cottonwood, Arizona ash, and seepwillow, with smaller amounts of netleaf hackberry, tamarisk, coyote willow, and desert willow at an average cover of 24 percent.[12] The trees that were removed ranged in age from ten to nineteen years. At the lower gaging-station site, riparian vegetation has increased (fig. 15.3).

The floodplain of the Big Sandy River in its reaches near and upstream from Wikieup (fig. 15.1) sustains dense riparian vegetation, especially tamarisk, but also Frémont cottonwood in some reaches and most of the other native riparian species present in the region. Downstream from Wikieup, the river passes from the wide alluvial valleys for which the watercourse is named into several narrow reaches partially or totally confined by bedrock. Only four repeat photographs document changes in riparian vegetation in these reaches. Downstream from Burro Creek, the gaging station was built in a reach partially controlled by bedrock; both native vegetation and nonnative vegetation have increased here (fig. 15.4). At the confluence of the Big Sandy and Santa Maria Rivers, where the Bill Williams River begins, the Big Sandy River looked as if it were well named in 1939 (fig. 15.5). Now it is difficult to tell where the channel is in this reach, despite the fact that it is occasionally inundated by Alamo Lake.

The Santa Maria River

The Santa Maria River drains 1,439 square miles in the southern part of the Bill Williams watershed (fig. 15.1).

In 1604, the Spanish explorer Juan de Oñate named the entire Bill Williams watershed the Río Santa María and reserved the name "Río de San Andrés" for this particular watercourse.[13] The Santa Maria River is unregulated, and agriculture in the upper parts of the drainage basin is watered using groundwater, not surface-water diversion. Mining operations are locally important in the upper part of the watershed near Bagdad (fig. 15.1). From the gaging station to the river's confluence, the Santa Maria is similar to the Big Sandy, with its braided, sandy channel.[14]

Floods and Channel Change

As shown in figure 15.6, floods are large on the Santa Maria River, but not as large as those that have occurred in the larger Big Sandy River watershed, which has twice the drainage area. The largest flood on the Santa Maria River was 33,600 ft³/s during the dissipating tropical cyclone of August 1951. Because this river is not regulated and floods regularly occur, channel narrowing as noted in our photography is not as significant on the Santa Maria River as it is downstream on the Bill Williams River.[15] From 1953 to 1979, the channel did narrow during a period of small floods.[16] From 1987 through 1992, the channel widened[17] in response to moderate flooding (fig. 15.6).

Change in Riparian Vegetation

Fifteen sets of repeat photographs document changes in riparian vegeta-

tion on the Santa Maria River, mostly in the vicinity of its confluence with the Big Sandy River. Two sets of repeat photographs at the gaging station near Bagdad show some changes in riparian vegetation (fig. 15.7). Wide ephemeral channels such as the Santa Maria River, especially those periodically scoured by floods, typically cannot sustain large amounts of obligate riparian species. This reach was dry during the droughts in both 1996 and 2002. At the Santa Maria's confluence with the Big Sandy River, riparian vegetation increases again (fig. 15.8), encouraged by the ephemeral delta of Alamo Lake as well as what likely are elevated water levels in the alluvial aquifer. Although tamarisk is present in large amounts along both the Big Sandy and the Santa Maria, the increases are more striking in the native species, in particular Frémont cottonwood and black willow. As a result, Bald Eagles are known to nest in the upper parts of Alamo Lake.[18]

Date Creek

Date Creek is an unregulated tributary of the Santa Maria River, with headwaters in the relatively low Date Creek Mountains north of Congress. This creek drains an area of about 300 square miles. This watercourse, despite its isolated location in west-central Arizona, has a long history on account of what once was a verdant and fertile setting that attracted both settlers and Native Americans. The latter called it Ah-ha Carsona (Pretty Water), and the current name refers to the abundance

Figure 15.3 Photographs of Willow Creek.

A. (September 3, 1969.) This downstream view on Willow Creek is 1.5 miles downstream from Tuckayou Wash in the Cottonwood Mountains of northwestern Arizona. A short-lived gaging station appears in the right midground. Several Frémont cottonwood and black willows appear in the view. (Unknown photographer 9-4242, courtesy of the U.S. Geological Survey.)

B. (November 21, 2001.) Most of the view of the channel is now blocked by young cottonwood, with Arizona ash and seepwillow growing along the channel. Cattle grazing continues in this reach. (D. Oldershaw, Stake 4308.)

A

B

Figure 15.4 Photographs showing the Big Sandy River downstream from Burro Creek.

A. (September 2, 1969.) This downstream view on the Big Sandy River shows the reach approaching the primary gaging station downstream from Wikieup. This perennial reach has seepwillow *(left side),* coyote willow, and black willow *(left midground);* Frémont cottonwood appear in the distance *(center).* Tamarisk is not obvious in this view. (Unknown photographer 9-4244.5, courtesy of the U.S. Geological Survey.)

B. (May 24, 2001.) Flood damage, possibly during the 1993 flood, has widened the channel here, although woody riparian vegetation has increased. Tamarisk is clearly visible in the right foreground, but the amount of black willow, in particular, has increased in this reach. Other species present include mesquite, cottonwood, coyote willow, and seepwillow, as well as a variety of aquatic species such as sedges. Cattle grazing is intensive here, and although many cottonwood seedlings are present, grazing has damaged them as well as the sedges. (D. Oldershaw, Stake 2457.)

Figure 15.5 The Big Sandy River at its confluence with the Santa Maria River.

A. (September 1939.) This upstream view on the Big Sandy River, its confluence with the Santa Maria River in the foreground, shows a largely barren channel and obvious flood damage following the severe floods associated with a series of dissipating tropical cyclones in September 1939. A foothill palo verde appears in the left foreground, and tamarisk lines the channel of the Santa Maria River on the right side. Other trees along the Big Sandy River are difficult to recognize from this distance, but probably are black willow. (J. A. Baumgartner 2837, courtesy of the U.S. Geological Survey.)

B. (November 2, 2002.) This reach is now occasionally inundated by Alamo Lake, which last reached this area in 1993. The growth of riparian vegetation is extensive, and the species present are Frémont cottonwood, black willow, seepwillow, and tamarisk, with desert broom in drier settings. Burro grazing is clearly apparent at this site. (D. Oldershaw, Stake 2392.)

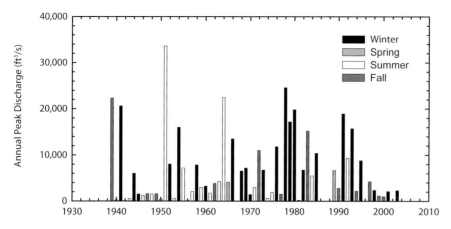

Figure 15.6 Combined annual flood series for the Santa Maria River near Alamo (station 09425500; 1939–1965) and for the Santa Maria River near Bagdad (station 09424900; 1966–1985, 1989–2003).

of yucca (*yucca* means "wild dates") in the countryside nearby.[19] Observations at an unspecified location in 1863 (likely near the headwaters) reported that a beautiful stream of clear water ran through a meadow lined with scattered cottonwood trees. Photographs at the former gaging station, which operated for a short period in the early 1940s, show nothing resembling this description (fig. 15.9). Since the 1940s, this ephemeral reach has had few changes in riparian vegetation; the four pairs of photographs in this minimal record are all near this former gaging site.

The Bill Williams River

Within a relatively short distance, the Big Sandy River, the Santa Maria River, and Date Creek join to form the Bill Williams River (fig. 15.1). This relatively short watercourse flows west through a series of bedrock canyons and alluvial valleys to join the Colorado River at Lake Havasu.[20] Since 1968, the Bill Williams River has been fully regulated by Alamo Dam, a flood-control structure operated by the U.S. Army Corps of Engineers.

The Bill Williams River NWR manages the lower third of the Bill Williams River, roughly 12 miles in length.[21] This reach of the river is strongly influenced by its delta in Lake Havasu. Originally part of the Lake Havasu NWR, established in 1941, the Bill Williams River NWR was separated owing to its geographic isolation and its uniqueness, notably

for attracting birds and other wildlife to the riparian zone of the lower Bill Williams River. Its 2,300 acres of riparian vegetation include a cottonwood-willow forest and 500 acres of cattail marshes. This riparian ecosystem adjacent to surrounding desert uplands is the reason that more than 343 species of birds, 56 species of mammals, and 28 species of reptiles and amphibians have been observed in the refuge.[22] Southwestern Willow Flycatcher and Yuma Clapper Rail, two endangered species, are among the birds that use the refuge.

Early Observations of the Bill Williams River

The first European explorers of the Colorado River (see chapter 27) visited the lower Bill Williams River. In 1775, Francisco Garcés found "every sort of riverland tree" in the canyon.[23] Bill Williams himself was attracted to the abundance of beaver in this isolated canyon.[24] Before accompanying Lieutenant Joseph Ives on his epic steamboat journey, Heinrich Balduin Möllhausen[25] traveled with Lieutenant Amiel Whipple in 1885 to the Bill Williams River, where they were extremely impressed with the extent of beaver dams. Later that year, Ives noted the presence of willow thickets (probably coyote willow) at the mouth.

Floods and Flow Regulation

The Bill Williams River has produced some of the largest floods, in terms of

peak discharge per unit drainage area, in Arizona history (fig. 15.10A). These floods are little known because most of the watershed, in particular the reach downstream from the Big Sandy–Santa Maria confluence, has not had a significant human population. The flood in February 1891 had a peak discharge of 200,000 ft³/s, larger than any measured on either the Verde River or the Salt River (chapters 23 and 24) and comparable to the largest known Colorado River floods (chapter 11). Flood peaks in January 1916 and February 1927 were 175,000 and 125,000 ft³/s, respectively (fig. 15.10A).

Alamo Dam completely regulates flow in the Bill Williams River. Built 39 miles upstream from the confluence of the Bill Williams and Colorado Rivers, the dam is 283 feet high and can store nearly one million acre-feet of water.[26] The dam and reservoir are designed solely for flood-control purposes, although there are no municipalities downstream to protect from floods. The structure can accommodate 820,000 ft³/s of inflow, or four times the peak discharge of the February 1891 flood, and the spillway is designed to release 375,000 ft³/s. The dam can release only about 7,000 ft³/s through its base. In essence, this reservoir was designed as extra storage capacity for Lake Havasu to prevent rapid rises in lake level or unanticipated large discharges downstream.

Operations of Alamo Dam are very effective at controlling floods on the Bill Williams River.[27] We used reservoir inflow data[28] to calculate the annual flood series for the Bill Williams River in the absence of Alamo Dam (fig. 15.10B). This analysis required the assumption that the maximum hourly inflow rate during a year is equivalent to the instantaneous peak discharge, which is reasonable for larger floods but questionable for small events with flashy peak discharges. After 1968, floods increased on the Bill Williams River in a manner resembling the annual flood series for the Gila, San Francisco, and Santa Cruz Rivers (chapters 17, 18, and 21). Flood-control operations are extremely effective; for example, the 1993 flood inflow to Alamo Lake was 105,000 ft³/s, but the dam released only 6,980 ft³/s (fig. 15.10).

Figure 15.7 Photographs of the Santa Maria River at the gaging station near Bagdad.

A. (September 2, 1969.) This downstream view of the gaging station reach on the Santa Maria River shows a wide, barren channel with riparian vegetation densely packed on low terraces. This site is on the edge of the Arrastra Mountain Wilderness Area in the foothills of the Poachie Range in west-central Arizona. The sparse riparian vegetation cannot be identified from this distance. (Unknown photographer 9-4249, courtesy of the U.S. Geological Survey.)

B. (November 20, 2002.) The channel has widened slightly and downcut about 6 feet at this site during floods in the intervening thirty-three years. Riparian vegetation remains sparse, although burrobrush is now common in the channel; a few seepwillow are present downstream from the gaging station, and tamarisk is behind the camera station and scattered through this reach. A solitary black willow appears to the left of the stilling well, and vegetation on the low terrace is foothill palo verde. (D. Oldershaw, Stake 4375.)

Figure 15.8 Photographs of the Santa Maria River near its confluence with the Big Sandy River.

A. (December 20, 1939.) This view across the Santa Maria River from river left shows the former gaging station near the confluence with the Big Sandy River. The channel is wide and sandy with few riparian species visible. One Frémont cottonwood is in the center of the view, and another appears to the right of the vertical cliff profile above the gaging station. The foreground plants are xerophytic vegetation characteristic of the lower Sonoran Desert, including a single saguaro in front of a foothills palo verde at left, creosote bush throughout the view, and what appears to be greythorn on the right. (H. S. Hunter 2861, courtesy of the U.S. Geological Survey.)

B. (November 3, 2002.) This reach is now periodically inundated by Alamo Lake. The increase in density of riparian vegetation is striking. Tamarisk is extremely dense in the midground, but a mixture of native species—including black willow, Frémont cottonwood, seepwillow, mesquite, and arrowweed—appears closest to the channel in the midground. A solitary creosote bush appears in the left foreground. This reach is grazed heavily by cattle and burros. (D. Oldershaw, Stake 4347.)

A

B

Figure 15.9 Photographs of Date Creek.

A. (January 18, 1940.) This upstream view shows a short-lived gaging station on Date Creek about 20 miles west of Congress, Arizona. Despite the presence of bedrock walls, Date Creek is ephemeral at this site, although springs once provided sufficient water for livestock nearby. Besides the desert vegetation, a mesquite tree appears on channel right *(left side)*. (H. S. Hunter 2848, courtesy of the U.S. Geological Survey.)

B. (September 28, 2002.) A small amount of aggradation has occurred here, obscuring part of the large boulder in the channel. There has been only a slight increase in riparian vegetation, mostly in desert willow *(on low terrace)* and burrobrush *(edge of channel)*. Other species in the vicinity include desert willow and desert broom, with mesquite behind them. (D. Oldershaw, Stake 4292.)

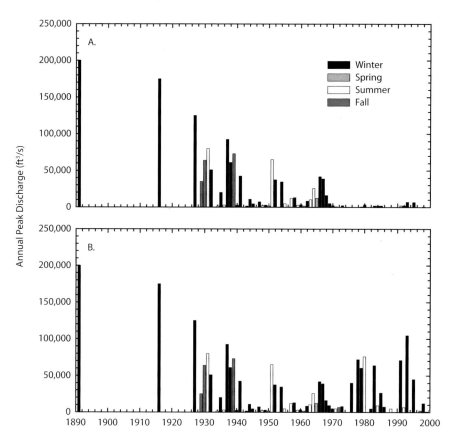

A. The combined records of the Bill Williams River near Planet Ranch (station 09426500; 1891–1939) and the Bill Williams River downstream from Alamo Dam (station 09426000; 1940–2000).

B. The combined records of the Bill Williams River shown in graph A before 1965, and the flood series developed from records of inflows to Alamo Lake from 1966 through 2000.

Figure 15.10 Annual flood series for the Bill Williams River.

Base flow in the Bill Williams River decreased following completion of Alamo Dam.[29] This decrease coincided with several hydrologic factors that affect the water budget on the Bill Williams River NWR: withholding of flows to fill Alamo Reservoir, evaporative losses from the reservoir surface, and low-runoff conditions associated with the end of the midcentury drought in the 1970s. At Planet Ranch, the additional factors of increased groundwater extraction and recharge of the alluvial aquifer affected flow entering the refuge. These factors created a difference of roughly 15,000 acre-feet per year in what was released downstream from Alamo Dam in the 1970s and 1980s and what was measured at the Planet Ranch gaging station from 1929 through 1946.

Flow regulation has both hurt and protected riparian vegetation in the reach downstream from Alamo Dam. Initially, filling of Alamo Reservoir, combined with heavy groundwater use at Planet Ranch (see the next section),

killed mature Frémont cottonwood trees in the delta at Lake Havasu.[30] Despite floods accidentally timed for germination of native plants in 1983, native riparian vegetation did not become established in significant amounts, again owing to groundwater withdrawals. Early spring flood releases in 1993 and 1995 and subsequent low-flow releases from the dam caused germination, establishment, and fast growth of native riparian trees.[31] Overall, the presence of Alamo Dam prevented flood damage, which would have been severe (fig. 15.10B), and allowed significant growth of riparian vegetation.[32] Within the refuge, the difference between events in the 1980s and 1990s is related to the history of groundwater extraction upstream at Planet Ranch.

Groundwater Development

Planet Ranch includes the former mining town of Planet, which was abandoned in 1921. Copper depos-

its were discovered in the vicinity in 1864,[33] and mining operations likely used wood harvested from the riparian zone, in particular mesquite. Ranching supplanted mining, and surface water and groundwater were used to irrigate agricultural fields. Planet Ranch became a major ranching operation with extensive fields; water for irrigation was pumped mostly from the alluvial aquifer, with lesser amounts diverted from the Bill Williams River.[34] The alluvial aquifer underlying Planet Ranch, locally referred to as the Planet basin, is the most significant groundwater source downstream from Alamo Dam.[35]

In 1984, the city of Scottsdale purchased the 3,400-acre property from Arizona Ranch and Metals Company. The city's goal was to "water farm," with the intention of either transferring or trading its water rights to increase its potential water supply. The city cleared the bottomlands of remaining vegetation and continued irrigation of about 2,000 acres of alfalfa, continuing the practice of irrigation with groundwater. The water right of 15,500 acre-feet per year was fully used in the late 1980s and early 1990s, and surface water rarely crossed Planet Ranch, instead infiltrating into the drawn-down alluvial aquifer. The heavy use of surface water and groundwater at the eastern end of the Bill Williams River NWR, combined with the severe drought that began in 1989, killed an estimated 85 percent of the trees,[36] a statistic that is not verified in analyses of aerial photography.[37] Releases in 1993 and 1995, which had peak discharges of less than 7,000 ft³/s (fig. 15.10A), damaged the irrigation ditches and field, causing agriculture to be scaled back at Planet Ranch. In 2004, no crops were irrigated in the Planet basin.

Changes in Riparian Vegetation

Extensive studies of germination, establishment, and response of young plants to groundwater fluctuations have been conducted in the reach between Lake Havasu and Alamo Dam.[38] Eighteen repeat photographs document changes in the latter two-thirds of the twentieth century in this 47-mile reach, both upstream from and within the Bill Williams River NWR. Just below Alamo Dam, a photograph taken to document conditions at the proposed dam site shows a largely barren channel in 1923 (fig. 15.11). Now, channel-bed degradation has occurred, as is common downstream from dams,[39] and the channel is lined with cobble bars that provide poor habitat for growth of riparian vegetation. Nonetheless, tamarisk and a variety of native riparian species have increased in this reach. Beaver dams are now common here, as they apparently were in the 1800s.

At the former gaging station downstream from Planet Ranch, riparian vegetation clogs a once-barren channel (fig. 15.12).[40] Perennial flow begins at the downstream end of the alluvial valley occupied by Planet Ranch. Riparian vegetation on the ranch was mapped using 1956 aerial photography and field checked in 1962.[41] Vegetation encroached on the channel only five years after the last major flood in August 1951 (fig. 15.10), during which riparian vegetation upstream from the present site of Alamo Dam was partially destroyed.[42] In 1962, the number of acres of riparian vegetation totaled 1,680, and 1,138 of these acres had 100 percent density of both native and nonnative species. Tamarisk occupied the largest amount of area (38 percent), followed by mesquite (18 percent), cottonwood-willow forests (17 percent), and mixed shrubs (14 percent), which include seepwillow, arrowweed, and burrobrush. Therefore, even before operations of Alamo Dam began, and with a regime of extremely large floods, riparian vegetation was increasing on the Bill Williams River. Downstream from Planet Ranch, riparian vegetation is lush (fig. 15.12B), and more than forty beaver dams are present in the channel.[43]

One might be tempted to conclude that the vegetation on the Bill Williams River NWR is a mature stand of trees, given its density and stature, and that some older trees are present in this reach.[44] However, in July 1989, a lightning strike sparked a 300-acre fire on the Planet Ranch that swept downstream, scorching 590 acres of riparian vegetation in the refuge, including all the area shown in figure 15.12.[45] Of the riparian vegetation burned, 325 acres were pure tamarisk, and 225 acres were either Frémont cottonwood or mixtures of cottonwood, willow, and tamarisk. Nearly all of the cottonwood affected by the fire died, making the changes shown in figure 15.12 and in other repeat photographs in this reach more remarkable. The repeat photographs mirror changes in percentage of channel area occupied by riparian vegetation as interpreted from aerial photography; the largest changes occurred between 1980 and 1996.[46]

Other fires in the riparian zone between Planet Ranch and Lake Havasu occurred in September 1990 (548 acres), August 1994 (80 acres), and April 1995 (95 acres).[47] The dam releases of 1993 swept the charred trees downstream, where they accumulated in a former channel splay in the refuge. Clearly, riparian vegetation, once severely damaged by groundwater withdrawals and fire, has quickly rebounded, although the evidence of hydrologic drought and fire remain on the landscape. The presence of tamarisk increases the possibility of future fires, which might reverse the trend of increases in native species.

In the delta of the Bill Williams River, riparian vegetation has increased considerably, blocking several views upstream from Lake Havasu (fig. 15.13). Despite the large, natural increases in native riparian species in this watershed, revegetation was a priority of the staff of Bill Williams River NWR in the late 1980s and early 1990s. Extensive revegetation projects were conducted following the fires in the riparian zone downstream from Planet Ranch.[48] The concern apparently centered on self-replacement of tamarisk, which consistently becomes the species first established following disturbances, in particular fire. Now, shading by cottonwood and black willow appears to be causing declines in tamarisk, which reproduces poorly in partial or total shade (chapter 4). Given the extremely fast growth rates of cottonwood and black willow, which grow most of the year along the Bill Williams River, tamarisk may become less significant in this riparian ecosystem over time. The ever-present possibility of another fire, the potential for renewed heavy groundwater use at Planet Ranch, and reductions in streamflow caused by the ongoing drought remain potential threats to the Bill Williams River NWR.

A

B

Figure 15.11 Photographs of the Bill Williams River downstream from Alamo Dam.

A. (April 20, 1923.) This view represents the center of a panoramic photograph of the Bill Williams River at the Alamo Dam site, as proposed by Eugene C. La Rue, the photographer. The channel has perennial flow, but riparian vegetation is in scattered groups. Solitary Frémont cottonwood appear in the right foreground and in the center midground, and what may be tamarisk appears along the left bank *(right side)*. (E. C. La Rue 1623, courtesy of the U.S. Geological Survey Photographic Library.)

B. (November 3, 2002.) Alamo Dam was completed in 1968, but it was not built at the site La Rue identified. The crest of the dam is just visible in the center distance. Operations of this dam have caused significant channel downcutting in this reach, and the bed material is obviously coarser now than in 1923. Riparian vegetation clearly has increased in this reach, notably tamarisk, black willow *(right foreground)*, cottonwood, seepwillow, arrowweed, cattail, burrobrush, and mesquite. Burros graze in this canyon, and beaver are widespread, building dams out of available materials, including cattail leaves. (D. Oldershaw, Stake 4349b.)

Figure 15.12 Photographs of the Bill Williams River near Planet Ranch.

A. (September 14, 1939.) This downstream view on the Bill Williams River shows the river in flood downstream from the former gaging station at Planet Ranch. This runoff is the cumulative result of precipitation from two dissipating tropical cyclones in early September 1939; the peak discharge in this month was 86,000 ft^3/s on September 11, and the flow clearly has dropped in this view. A grove of Frémont cottonwood and black willow appears on channel left in the foreground; more of the cottonwood-willow vegetation appears on river right *(right side)* and in the left distance. Flood damage to the riparian vegetation is subtle in the left midground. (Unknown photographer 4393, courtesy of the U.S. Geological Survey.)

B. (November 21, 2002.) The increase in riparian vegetation at this site completely obscures the view of the once wide and barren channel. The species are cottonwood (nearly leafless on this late November date), black willow (dark leaves), mesquite, tamarisk, arrowweed, catclaw, and two species of saltbush. In July 1989, riparian vegetation completely burned in this reach, and what appears in this view is thirteen years of germination, establishment, and growth. (D. Oldershaw, Stake 4377.)

Figure 15.13 Photographs of the Bill Williams River about 4.5 miles upstream of its confluence with the Colorado River.

A. (September 6, 1939.) This view across the Bill Williams River, apparently taken from a boat, shows a flood that resulted from the first of three dissipating tropical cyclones to affect the basin in September 1939. Black willow forms a dense band at left and right; arrowweed appears to be the lighter-colored, taller shrub that is closer to the channel in the midground; the low shrubs partially submerged in the flow may be sweet bebbia; and at least one Frémont cottonwood appears on the right. The black mass in the river at left center may be a driftwood log. (Unknown photographer 4395, courtesy of the U.S. Geological Survey.)

B. (December 16, 2002.) Our match is not even close to the original camera station, access to which is totally blocked by the large increase in riparian vegetation. Instead, we took a photograph from a boat at the extreme upstream end of navigation on Lake Havasu in the Bill Williams delta. Tamarisk forms a wall across the foreground, and the taller trees behind the tamarisk are black willow. Cottonwood trees are dense in this reach but occur on higher terraces that are not visible from this location. (D. Oldershaw, Stake 4393a.)

Summary. Several small watersheds sustain important stands of native and nonnative riparian vegetation in southeastern Arizona. The San Simon River, heavily impacted by land-use practices, is a tributary of the Gila River. Whitewater Draw and Guadalupe Creek near Douglas are tributaries of the Río Yaqui in Mexico. Leslie Creek, an important tributary of Whitewater Draw, has an NWR to protect native fish and a thriving riparian ecosystem. Animas Creek, a small ephemeral stream in southwestern New Mexico, arises in the Peloncillo and Animas Mountains and flows northward into the Animas Playa. The flood histories of these watersheds differ markedly from others in southern Arizona and probably reflect the regional climatic influences operating in the Chihuahuan Desert instead of conditions farther west. Except for the single matched pair from Guadalupe Creek, the original photography was taken after arroyo downcutting had removed riparian vegetation, which is particularly significant in the case of the San Simon River. Riparian vegetation—primarily tamarisk—has increased along the channels of the San Simon River and Whitewater Draw, but there also are notable increases in mesquite and native obligate riparian species. Native riparian vegetation has increased along Guadalupe Canyon and the Animas River.

With the exception of the Gila and San Pedro Rivers, no perennial streams flow across the broad alluvial valleys of southeastern Arizona and adjacent New Mexico. Instead, ephemeral streams cross these valleys in deeply incised arroyos. Several important mountain ranges—notably the Chiricahua and Dos Cabezas Mountains of Arizona, the Peloncillo Mountains of both Arizona and New Mexico, and the Animas Mountains of New Mexico—serve as headwaters for these streams. Once the Arizona watercourses were an impor-

tant source of irrigation and domestic water, but now groundwater provides most of the irrigation for southeastern Arizona. Channel downcutting, which damaged surface-water irrigation structures, induced this change; consequently, arroyo formation had a major impact on the region's rural economy and use of surface water.

San Simon River

The headwaters of the San Simon River are in the Chiricahua and Peloncillo Mountains of Arizona and New Mexico (fig. 17.1). After its tributaries exit these mountains, the river flows northwesterly to its confluence with the Gila River just upstream from Safford, Arizona. The area of this watershed is 2,192 square miles. The broad valley traversed by the San Simon River is a major agricultural and ranching area in southeastern Arizona. Much of the watershed is covered with low trees and shrubs characteristic of the interface between the Sonoran and Chihuahuan Deserts, such as xerophytic mesquite, creosote bush, and numerous species of shrubs and cacti.

Early Observations of the San Simon River

When the Spaniards first saw the San Simon River in the sixteenth century, they named it the Río San Domingo. Cottonwood and "willow thickets" lined the banks of what was also called Valle de Sauz (Valley of Willows), Río de Sauz, and Ciénega de Sauz. A visitor in 1849 named the river Welcome Creek, but by 1879 the name "San Simon River" appeared on maps. Although the early names of the watercourse are confusing, they show that surface water was present, at least in local reaches. Settlement of the San Simon Valley began in 1859 with the

establishment of the town of San Simon in a reach of perennial water.[1] However, it is extremely unlikely that the river ever flowed continuously from its headwaters to the Gila River except during rare floods.[2]

Floods, Channel Change, and Groundwater

In 1882, the channel had essentially no banks except near the confluence with the Gila River, where banks were about 3 feet high and the channel was 20 feet wide.[3] South of the present-day town of San Simon (fig. 17.1), the San Simon Ciénega forms from high groundwater levels over a 5-mile reach that contained 1,200 acres of riparian vegetation in 1913.[4] The ciénega serves as a starting point for the San Simon River; the valley upstream is broad with a low gradient, and its slope may be controlled by deposition in the vicinity of the ciénega.

As occurred in nearly all of the watersheds in the region, a series of floods in the late nineteenth and early twentieth centuries carved a deep arroyo channel.[5] Prehistoric canals in this reach indicate that Native Americans once diverted water for irrigation. The downcutting began in 1883 as a result of the efforts of settlers to channelize the San Simon River near Solomon (fig. 17.1). The newly modified channel was originally 20 feet wide and 4 feet deep. This new channel headcut upstream, carving a new entrenchment that locally was up to 30 feet deep and 600 to 800 feet wide.[6] Because channel downcutting preceded significant stocking of livestock, channelization and road construction are thought to be the primary human influences causing floodplain modification. The downcutting drained the alluvial aquifer, eliminating most of the existing riparian trees.

The gaging record of the San Simon River (fig. 16.1), which begins with the flood of record (27,500 ft³/s) in 1931, shows the end of this period of high runoff. Unlike the Gila River, flood frequency on the San Simon River did not increase in the late 1970s. The gaging station was discontinued in 1980, which prevents evaluation of the effects of regional storms in 1982–1983 and 1993 that spawned large floods elsewhere in central and southern Arizona.

Surficial erosion in the 1930s, induced by the interrelated factors of overgrazing and arroyo downcutting, prompted installation of channel-stabilizing structures along the San Simon River. These small-scale erosion-control structures, such as earthen dikes, were built to stem downcutting and channel widening. In 1953, the first large grade-control structure was built in an attempt to fill the arroyo to its presettlement level;[7] another dam, built in 1980, raised the total number of erosion-control structures to eighteen. Channel-stabilization efforts, coordinated by the Natural Resources Conservation Service and the BLM, continue throughout the San Simon River watershed.[8]

Land-use practices within the San Simon River watershed changed around 1910 in response to the discovery of an artesian aquifer near the town of San Simon.[9] Before that time, widely dispersed ranching was the only economic development in the basin because of the sparse surface-water supplies. Groundwater development started a new agricultural era, with irrigation supplied from the flowing wells.

Changes in Riparian Vegetation

We have matched fifteen photographs of the San Simon River at former gaging stations near the towns of San Simon and Solomon, Arizona (fig. 17.1). At the upstream site near San Simon, channels were mostly barren in the 1930s (fig. 16.2). Riparian vegetation, primarily tamarisk, has increased, but mesquite has increased on channel banks as well. The changes in mesquite are likely related to the overall increase in this species on adjacent rangelands[10] and do not reflect changes in mesquite bosques associated with the river channel. Whether in bosques or on range-

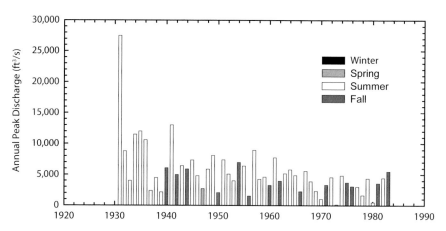

Figure 16.1 Annual flood series for San Simon River near Solomon, Arizona (station 09457000; 1931–1983).

land, mesquite is the most common species along the San Simon River.[11]

As the San Simon River approaches its confluence with the Gila River, its channel changes from ephemeral to at least intermittent. This reach is regulated by the numerous grade-control structures in the watershed, but these structures are unlikely to affect flood magnitudes. The gaging station has been moved several times; at the upstream site (fig. 16.3), what was a barren channel in the 1930s became lined with tamarisk in the 1960s, and this tamarisk had died back before the 2001 photographs were taken. Downstream, depth to groundwater is lower (e.g., fig. 2.3), and native riparian trees are increasing along the San Simon River (fig. 16.4). This mixture of tamarisk, Frémont cottonwood, and black willow is common in the Safford Valley.

Headwaters of the Río Yaqui

Several tributaries of the Río Yaqui of Sonora, Mexico, have headwaters in southern Arizona and extreme southwestern New Mexico. Whitewater Draw, a major tributary of the Río Yaqui, originates from the southwestern part of the Chiricahua Mountains (fig. 19.1). Leslie Creek is a major tributary with headwaters in the southwestern parts of the Chiricahuas. Finally, Guadalupe Canyon heads in New Mexico's Guadalupe Mountains at the southern end of the Peloncillo Range. This watercourse crosses into Arizona as it drains southwesterly, adding minor Arizona drainages before it crosses into Mexico at Rancho Puerto Blanco at 4,200 feet.

The name "Whitewater Draw" refers to salt deposits along the channel, not to rapids.[12] The watershed has a drainage area of 1,023 square miles, making it the largest in the southeastern corner of Arizona. The main channel issues from Rucker Canyon, one of the major canyons in the southern Chiricahuas, and flows westerly before turning south. The watershed increases with the addition of numerous ephemeral channels, such as Double Adobe Wash and Leslie Creek, before the channel crosses the U.S.–Mexico border west of Douglas, Arizona.

As with nearly all of the other alluvial channels in the Southwest, Whitewater Draw has a long history of arroyo cutting and filling.[13] At least eight episodes of arroyo cutting and filling occurred between 3,500 and 8,000 years ago, followed by numerous events in the past 3,500 years. Between around 1884 and 1910, the modern arroyo downcut, although there is some indication that this arroyo may have formed only in the early twentieth century.[14] By the 1950s, the arroyo had a maximum depth of 25 feet near Douglas.[15]

The first observations of riparian vegetation were made by members of the Mormon Battalion in 1846.[16] On December 7, they camped next to what they called "Little Ash Creek," which was lined with Arizona ash, Arizona walnut, and oak, with dense mesquite nearby. In 1892 and 1893, Edgar Mearns visited Whitewater Draw (which he called the San Bernardino River).[17] The watercourse was "wooded" with both black and yewleaf willow, Frémont cottonwood,

Figure 16.2 Photographs of San Simon River near San Simon, Arizona.

A. (May 25, 1939.) This upstream view on the San Simon River, from a low bridge on a county road, shows an ephemeral-channel reach upstream from San Simon, with the Peloncillo Mountains of New Mexico in the background. The channel is wide and barren; creosote bush, a native and dominant species in the Chihuahuan Desert, is on both banks, along with other xerophytic shrubs. What may be the edge of a solitary mesquite tree appears on the extreme right edge of the view. (R. H. Monroe 2537, courtesy of the U.S. Geological Survey.)

B. (December 6, 2001.) Although the bridge has been replaced, this match shows dramatically different conditions in riverine riparian vegetation along the San Simon River. The once-barren channel is now filled with nonnative tamarisk and native seepwillow. Mesquite is now common on the high terrace, although creosote bush remains on the floodplain, and alkali sacaton is present in significant amounts. (D. Oldershaw, Stake 4316.)

A. (August 20, 1931.) This downstream view on the San Simon River shows low summer runoff in an otherwise ephemeral reach. The channel is mostly barren and shows conditions of this well-known arroyo system near the end of the arroyo downcutting phase. No nonnative species are present. (Unknown photographer 890, courtesy of the U.S. Geological Survey.)

B. (June 10, 1964.) By the 1960s, downcutting had ceased, the channel had widened, and tamarisk had become established in this reach. The original camera station was destroyed, so that the match, taken from the bed of the channel, gives a different perspective from the original view. (R. M. Turner.)

C. (October 5, 2001.) It is unclear, owing to an obstructed background, whether the channel downcut between 1964 and 2001, or the active channel narrowed owing to deposition of low-floodplain deposits. The perspective on the bases of the tamarisk trees suggests that downcutting has occurred, and many of the tamarisk have dead branches and stems that might indicate a recent lowering of water levels in the alluvial aquifer. Nearby mesquites appear to be healthy. (D. Oldershaw, Stake 336a.)

Figure 16.3 Photographs of the San Simon River near Solomon.

A. (August 25, 1933.) The gaging station on the San Simon River, abandoned in 1983, was on a bridge on Solomon Road southeast of Solomon. This downstream view from the channel shows the stilling well on the right side. Mesquite appears to be the dominant species on top of the channel banks, and what may be a solitary seepwillow is just beyond and left of the stilling well. (Unknown photographer 894, courtesy of the U.S. Geological Survey.)

B. (June 10, 1964.) The channel banks are mostly obscured by a combination of black willow and tamarisk trees. The shadow of the bridge overhead crosses the foreground, and what appears to be out-of-focus black willow leaves are at lower right. (R. M. Turner.)

C. (October 5, 2001.) Clearly, riparian vegetation has increased here to the point of blocking the view. Because of channel filling, the camera station is about 6 feet too high and too close to the old stilling well, and the old bridge has been replaced. Seepwillow is in the immediate foreground; mesquite is directly above the camera station; and black willow blocks any view of the channel. A large cottonwood tree is just out of this view and likely was present in 1933, and tamarisk is present directly behind the camera station. (D. Oldershaw, Stake 339.)

Figure 16.4 Photographs of San Simon River near Solomon, Arizona.

box elder, Arizona ash, and mesquite, with sycamore and hackberry observed nearby. Springs near the river had "cane" (carrizo grass?) in 1846, but none was present in 1892. Despite the abundant woody riparian vegetation, he saw no beaver.[18]

Water development in the Sulfur Springs Valley may have exacerbated the negative effects of arroyo downcutting on riparian vegetation; between 1933 and 1965, base flow in Whitewater Draw was gradually depleted, because of either increased water development or the mid-twentieth-century drought.[19] The first irrigation wells were drilled in 1910,[20] and a copper smelter was built adjacent to Whitewater Draw to extract the shallow groundwater easily. This smelter, now closed, used groundwater from the alluvial aquifer and discharged the effluent back into the channel at the international border downstream from the gaging station. However, the period of base-flow decreases also encompassed the midcentury drought, which may also have played a role in decreasing surface water.

The annual flood series for Whitewater Draw (fig. 16.5) is different from most of the gaging records in the region. Floods were highest during the midcentury drought, with a peak discharge of 5,060 ft³/s during the summer of 1955, and floods during the last two decades of the twentieth century were notably small, considering that other watersheds in the region were experiencing large floods. This contrast underscores the spatially discontinuous nature of regional climatic fluctuations. Summer moisture enters this watershed before reaching any others in the region, thereby providing more reliable summer precipitation, and dissipating tropical cyclones and winter storms tend to affect watersheds farther north and west. The increase in southern Arizona floods in the 1980s and 1990s resulted primarily from fall and winter storms.

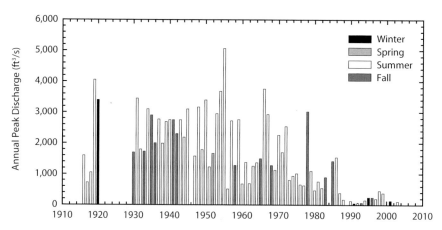

Figure 16.5 Annual flood series for Whitewater Draw near Douglas, Arizona (station 09537500; 1916–2003).

Change in Riparian Vegetation— Leslie Creek

Headwater tributaries of Whitewater Draw in the Chiricahua Mountains are lush with riparian vegetation, typically at elevations higher than considered in this book. Leslie Creek (fig. 19.1) is a lower-elevation tributary with a short perennial reach that sustains a highly valued riparian ecosystem. This influent-effluent reach, within Leslie Canyon, is only three-quarters of a mile long. The Leslie Canyon NWR, established in 1988, is in the Swisshelm Mountains 16 miles north of Douglas, Arizona.[21] The riparian zone consists of a cottonwood-willow gallery forest, and the stream supports a variety of native fish species, including the Yaqui topminnow (*Poeciliopsis occidentalis sonoriensis*) and Yaqui chub (*Gila purpurae*), both endangered species.[22]

Leslie Creek has a gaging station with a short record (1969–2003) that documents an average daily discharge of 1.4 ft³/s and a flood of record of 5,200 ft³/s on September 1, 1994. The watershed area is 79.1 square miles. Arroyo cutting upstream from Leslie Canyon threatened alkali sacaton grassland with gully erosion, and an erosion-control dam was built across Leslie Creek by the Civilian Conservation Corps in the 1930s.[23] Repeat photography of views showing this dam (fig. 16.6) at the upstream end of Leslie Canyon documents increases in riparian vegetation, notably Frémont cottonwood.

Change in Riparian Vegetation— Whitewater Draw West of Douglas

We matched sixteen historical photographs at the gaging station 2 miles west of Douglas, Arizona, and examined several more. A 1918 photograph of this reach shows a freshly downcut arroyo, about 10 feet deep, with no riparian vegetation along the channel or on the banks.[24] Probably owing to the decrease in recent floods, tamarisk has completely clogged the channel of Whitewater Draw in the vicinity of the gaging station (fig. 16.7). This reach, heavily disturbed by construction upstream and smelter operations downstream, is probably representative of conditions along the main channel in this broad alluvial valley.

Our repeat photographs do not document the fate of extensive stands of mesquite along Whitewater Draw, which reportedly became established after 1910.[25] These dense stands probably fit the definition of the term *bosque* and were not part of the general increase in mesquite on uplands in southern Arizona that occurred in the early twentieth century.[26] These stands are threatened by the combination of water withdrawals from the alluvial aquifer and clearing for agriculture. Whitewater Draw represents another case of change in channel geometry and riparian vegetation that likely was caused by a combination of climatic fluctuations, livestock grazing, and agricultural development.

Change in Riparian Vegetation— Guadalupe Canyon

Guadalupe Canyon has headwaters in the Guadalupe Mountains of extreme southwestern New Mexico at elevations mostly below a maximum of about 6,500 feet. Most of the terrain drained by this system is former grassland, now dominated by shrubs. A

A. (October 1941.) This upstream view shows Leslie Creek at an erosion-control dam built by the Civilian Conservation Corps in the 1930s. The stream is in flood, and Frémont cottonwood trees upstream from the dam are in standing water. The channel through the wide valley, which sustains an alkali sacaton grassland, is relatively wide and barren, reflecting conditions at the end of the period of arroyo widening. (B. Brixner Ariz. 4257, courtesy of the Natural Resources Conservation Service.)

B. (April 18, 1993.) The camera position is exact; the road segment at the left center has been widened and moved right, giving the impression of an imperfect match. Riparian vegetation has increased considerably, obscuring most of the view of the dam. The channel upstream from the dam and its small reservoir remains relatively wide, but riparian vegetation lines its banks. (R. M. Turner.)

C. (May 27, 2004.) Few new trees are present in the view, but many have increased in size. The species present are mostly black willow with significant amounts of Arizona ash, wild grape, Arizona walnut, and netleaf hackberry. Only one Frémont cottonwood is present in the midground *(right side),* although others are visible in the distance. Perennial water remains on the downstream side of the dam. Just downstream from the right edge of this view, cottonwood dominates, with black willow, burrobrush, seepwillow, and desert broom. A large mesquite bosque is also well developed on the terrace above the channel. (D. Oldershaw, Stake 1870.)

Figure 16.6 Photographs of Leslie Creek.

Figure 16.7 Photographs of Whitewater Draw near Douglas, Arizona.

A. (January 31, 1963.) This upstream view of Whitewater Draw, taken from the Arizona Highway 80 Bridge about 2 miles west of Douglas, shows a seriously disturbed watercourse, as the fifty-five-gallon drum lying in the water attests. Tumbleweeds, a nonnative xerophytic annual, clog the pool, and young tamarisk individuals are becoming established in the floodplain. The tussocks of grass that dominate the floodplain on channel left *(right side of view)* appear to be alkali sacaton. The hills in the right background are part of the Swisshelm Mountains. (F. S. Anderson 4806, courtesy of the U.S. Geological Survey.)

B. (October 19, 2001.) Tamarisk now dominates the vegetation on the floodplain, and the taller, broadleaf tree at right center could not be accessed for identification, but appears to be a crabapple tree, another nonnative species. The herbs in the left foreground are a native sunflower, and nonnative tamarisk is now present in the floodplain and on the bridge abutments. (D. Oldershaw, Stake 3141.)

favorite trail for early travelers as they followed the southern route across New Mexico Territory, Guadalupe Pass is at the summit of one of the watercourses draining into Guadalupe Canyon. Unfortunately, no gaging stations have recorded flow in Guadalupe Canyon.

An intriguing photograph of Guadalupe Canyon, taken by prominent early-day photographer C. S. Fly in the mid-1880s,[27] shows a U.S. military encampment semipermanently transgressing a short distance into Mexico. When the troop was stationed at this strategic water source to prevent use by the Apache, most of the soldiers were sent in pursuit of a recently escaped Apache band that had headed into Mexico. The seven soldiers left behind to protect the spring and a military supply train were attacked by those Apache. Before the end of the fighting, four ambulances and several wagons containing forty days' rations and ten thousand rounds of ammunition exploded or burned, and two of the soldiers were killed.

Besides the fascinating Indian-soldier undertones of this view, much can be learned from the landscape details captured here, in particular the increases in native riparian vegetation (fig. 16.8). In 2002, the once prominent sycamores shown in Fly's photograph are partially hidden by a large increase in native species, including black willow, Frémont cottonwood, and velvet mesquite. Tamarisk is also present at this site in low density.

Animas Creek

Animas Creek has the only watershed wholly within New Mexico that we consider in this book for changes in riparian vegetation. This watercourse drains northward into the closed basin of Lake Animas, ordinarily a dry playa. Although the watershed extends to 8,800 feet in the Animas Mountains and above 5,000 feet in the less distinct Peloncillo Mountains, most of its area is grassland, either relatively pristine or degraded with mesquite encroachment. Ranching is the primary land use in this sparsely settled watershed. The channel flows through deep al-luvium along most of its course, interrupted by a short reach, called "the Box," which passes through a narrow canyon. Ciénegas once marked the course of Animas Creek on the Diamond A Ranch (formerly the Gray Ranch),[28] a historic property occupying the largest part of the watershed, and groundwater levels remain high in the upper part of the watershed.

A gaging station at a site downstream from the Box provides an incomplete record of peak flows on Animas Creek from 1959 through 1999. The largest flood in the record had a peak discharge of only 3,400 ft³/s on October 13, 1974. Change in woody riparian vegetation can be documented using only one set of repeat photographs (fig. 16.9). The initial view documents what appear to be pre-arroyo conditions in this part of the watershed, and the match shows the extent of downcutting and channel movement in this reach. Woody riparian vegetation, notably Frémont cottonwood and black willow, has increased in the view.

Figure 16.8 Guadalupe Canyon, Sonora, Mexico.

A. (Ca. 1885.) This view shows a U.S. military encampment on Guadalupe Creek, Sonora, Mexico, at the time of Geronimo's surrender. This semipermanent encampment was positioned here to prevent access to water by Mexico-bound Apache who had escaped from their guards at a camp near Fort Apache, Arizona Territory. The large white-barked trees are sycamores. The scattered dark shrubs on the midground hill are one-seed junipers. (C. S. Fly 9928, courtesy of the Arizona Historical Society, Tucson.)

B. (November 16, 2002.) The sycamores at left have been eclipsed by large black willows; a Frémont cottonwood tree extends into view at right foreground; velvet mesquite partially blocks the view at left foreground. Although not discernable, considerable downcutting has occurred since the first photograph was taken. Besides increased riparian vegetation, the photograph shows that the uplands support a heavier growth of mesquite and acacia. Many woody riparian species are present, including Frémont cottonwood, black willow, sycamore, Arizona ash, and velvet mesquite; no tamarisk is present. (R. M. Turner, Stake 3221.)

Figure 16.9 Animas Creek on the Gray Ranch, southwestern New Mexico.

A. (November 20, 1913.) This view upstream and across Animas Creek shows the mouth of Indian Creek on the left side and the Animas Mountains in the background. A series of terraces, cut by Animas Creek, is evident beyond the creek's bed. No riparian forest is found along Animas Creek, the banks of which appear newly eroded; instead, a few scattered shrubs occupy the grassland on the left side. Little woody riparian vegetation and no trees are visible along Animas Creek, although a small clump of chittamwood can be seen at far right in the stream channel. (A. T. Schwennesen 23, courtesy of the U.S. Geological Survey Photographic Library.)

B. (November 9, 1994.) Season and time of day in our match are approximately replicated to facilitate comparison of landforms and to interpret changes. The terrace slopes next to the channel, which appear to be freshly eroded in 1913, have healed. Woody plants have increased throughout the grassy uplands, especially at left. These shrubs, mainly honey mesquite and soaptree yucca, have advanced from areas at lower elevation *(out of view to the left)*. Woody riparian vegetation has increased, and the channel of Animas Creek is greatly altered. The new riparian species present are Frémont cottonwood and black willow, and the small clump of chittamwood has grown slightly in the eighty-one years between the photographs. (R. M. Turner, Stake 1615.)

17 The Upper Gila River

Summary. The Gila River is the primary drainage for most of Arizona. With its headwaters in New Mexico, the Gila River upstream from Safford, Arizona, is largely unregulated, although flow diversions affect base-flow discharges at least locally. Below Coolidge Dam, the Gila River becomes fully regulated until the entire surface-water flow is diverted from the channel upstream of Florence, Arizona. Historically, the upper Gila River generated large, damaging floods while providing irrigation waters for fertile agricultural lands downstream. This river appears to respond strongly to regional climatic fluctuations, and flood frequency has been inferred to be low before 1891, high between 1891 and 1920, low through the middle part of the twentieth century, and high again after 1964. Despite considerable agriculture, groundwater levels near the river remain high. Cottonwood gallery forests once present in wide reaches were swept away in channel-widening floods in the early decades of the twentieth century. Cottonwood has increased in narrow reaches and bedrock canyons but has decreased in the wide valleys where it once was common. Channel narrowing during the middle of the twentieth century allowed establishment of dense tamarisk, but native species, such as mesquite, increased as well. Extremely large floods in the last third of the twentieth century failed to reduce tamarisk significantly, and tamarisk eradication with no follow-up maintenance resulted in reestablishment.

We discuss the Gila River in two chapters because of its importance to Arizona hydrology, its length, and changes along its course. We present changes in the reach from the Arizona–New Mexico border to the Ashhurst-Hayden Diversion Dam upstream from Florence in this chapter, and we consider the reach downstream to the confluence with the Colorado River near Yuma in chapter 26. The intervening chapters discuss changes in important tributaries, including the San Francisco and San Carlos Rivers (chapter 18); the San Pedro River and its tributaries (chapters 19 and 20); the Santa Cruz River and its tributaries (chapters 21 and 22); and the Verde and Salt Rivers and their tributaries in central Arizona (chapters 23 through 25).

The Gila River has its headwaters in the Gila Wilderness of western New Mexico (fig. 17.1). Several forks of the river combine upstream from Cliff, New Mexico, and the river plunges through a bedrock canyon of granite, locally called the Connor Box, before reaching the Arizona border. Beginning at Duncan, the river flows northwesterly through a relatively wide alluvial valley supporting agriculture—primarily hay production—before entering a canyon developed in volcanic rocks called the Middle Box. Upstream from Safford, Arizona, the river enters a broad valley with extensive agricultural production and flows northwesterly to San Carlos Reservoir. Downstream from Coolidge Dam, the river enters a series of short bedrock canyons through the Galiuro Mountains to emerge at Winkelman, Arizona. A short and relatively narrow segment through alluvium ends near the small town of Kelvin and enters yet another bedrock canyon, through which the river flows for 18 miles to emerge at Ashhurst-Hayden Dam upstream from Florence, Arizona.

The Gila River is an extremely important source of irrigation waters in eastern and central Arizona, and small diversion structures are common along its length. This river has also had extremely large floods in its history, resulting in construction of Coolidge Dam, which serves as both a flood-control structure and a flow-regulation dam for surface-water irrigation projects downstream of this reach.

Upstream from San Carlos Reservoir, the Gila River is a free-flowing watercourse that remains subject to large floods; downstream of Coolidge Dam, the river is operated as an irrigation canal with the exception of certain periods, such as January 1993, when headwater floods overwhelmed the dam and reservoir's ability to contain all the runoff.

Early Observations of the Upper Gila River

Early Spanish explorers, such as Francisco Coronado, crossed the upper Gila River but made no mention of its resources. The first description of the Gila appears in the narrative account of James Ohio Pattie, who ranged widely through the western United States between 1824 and 1830. Pattie and his party of trappers first reached the Gila River upstream from the present-day Arizona–New Mexico border on December 14, 1824, and immediately caught thirty beaver.[1] He described downstream travel through one bedrock-confined canyon with tangled growth of wild grape and shrubs and another reach—probably near present-day Cliff, New Mexico, with "banks covered with tall cottonwoods and willows."[2]

Pattie's group found that beaver had already been trapped from the river just downstream from the present-day Arizona–New Mexico border, so they continued downstream. Pattie described the river bottomlands near present-day Safford as thickly vegetated with mesquite; he made no mention of cottonwood. After a brief excursion up the San Pedro River, the trapping party regained the Gila River upstream from present-day Safford, where Pattie trapped and killed an otter.[3]

Pattie returned to the Gila River several times, always trapping for

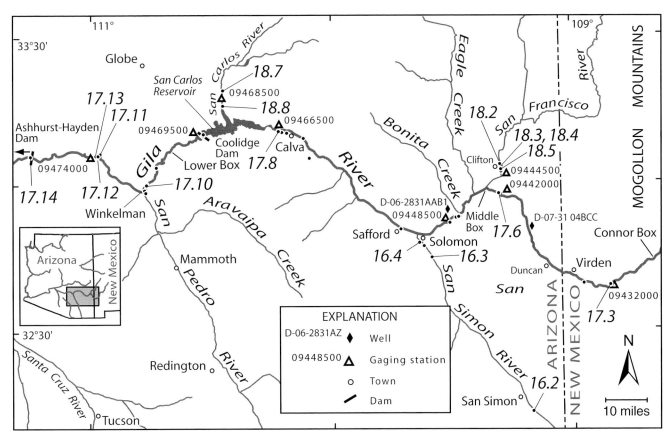

Figure 17.1 Map of the Gila River from the Arizona–New Mexico border to Florence, Arizona.

beaver. In January 1826, his group traveled down the Gila and observed that "there is here little timber, beside musqueto-wood, which stands thick" near the site of present-day Coolidge Dam.[4] In September 1829, Pattie made his final trip to the Gila and found that most of the beaver had been trapped out upstream from the San Pedro River.[5] As Pattie's account shows, the Gila River was a frequent destination for all the pre-1850 trappers in the West.[6] As late as 1884, beaver were reported to be abundant on the Gila River near the confluence with the San Carlos River.[7]

The 1846 war with Mexico caused the next incursion of Gila River observers. The Army of the West, guided by Kit Carson and led by Lieutenant Colonel W. H. Emory, came to the upper Gila River near the present-day Arizona–New Mexico border and traveled along its channel en route to California. Near the Arizona–New Mexico border, Emory observed that "[t]he growth of trees and weeds was very luxuriant; the trees chiefly cottonwood, a new sycamore, mesquite." At the mouth of the San Pedro River, the Gila River had a "bottom three miles wide . . . principally of deep dust and sand, over grown with cotton-wood, mezquite, chamiza, willow, and the black willow."[8] Dr. John S. Griffith, a physician and naturalist accompanying the expedition, described the riverine setting as a river filled with fish and lined with coyote willow and cottonwood trees.[9] Other descriptions noted the occurrence of mesquite bosques along the river.[10]

Groundwater conditions were high along the river, and malaria forced abandonment of several early towns.[11] Settlers affected riparian vegetation by clearing floodplains—mainly mesquite—for fields and using the wood for construction and fuel. Safford was established in 1872 by Mormon farmers, some of whom soon began to cut the bosques of mesquite just upstream to fuel smelters associated with local mines.[12] With the discovery of copper locally in the mountains near Safford and on a large scale near Clifton, woodcutting accelerated to fuel those smelters as well.[13] Undoubtedly, the same pattern was repeated between Winkelman and Kelvin with the discovery of extensive copper ore bodies.

Floods, Flow Regulation, and Channel Change

Because of the Gila River's importance for water supply and its propensity for generating large floods, six streamflow gaging stations record flow on the river from just upstream of the Arizona–New Mexico border to Ashhurst-Hayden Dam (fig. 17.1). The gaging stations below Blue Creek and near Clifton record average daily discharges of 215 and 197 ft³/s, respectively. The gaging stations at the head of Safford Valley and near Calva, both downstream from the San Francisco River (fig. 17.1), have average daily discharges of 512 and 376 ft³/s, respectively. The gaging station below Coolidge Dam has an average daily discharge of 401 ft³/s, and the gaging station at Kelvin, which is below the confluence with the San Pedro River, has an average daily discharge of 542 ft³/s.

The upper Gila River is subject to extremely large floods (figs. 2.2, 17.2), most of which occur during warm-winter storms or fall storms, the latter typically spawned from dissipating tropical cyclones. The earliest known floods occurred before settlement of the Safford Valley and are discussed in chapter 26 in reference to the lower Gila River. The February 1891 flood, which peaked at around 100,000 ft³/s at Florence, is one example of a large nineteenth-century flood on the upper Gila River.[14] The November 1905 flood reportedly peaked at 150,000 ft³/s at San Carlos (now beneath San Carlos Reservoir) and 190,000 ft³/s at Florence.[15] Other large floods occurred in 1906, 1915, 1916, and 1941.[16]

Cadastral surveys made from 1875 through 1894 show that the Gila River in the Safford Valley was relatively narrow and lined with willow, cottonwood, and mesquite.[17] Most of this vegetation was destroyed during channel widening associated with large floods between 1905 and 1916. Over a fifty-year period during the middle part of the twentieth century, the channel narrowed appreciably to about its pre-1905 conditions,[18] in part because of low peak discharges during floods. Near Safford, the channel was widest in 1935.[19] Riparian vegetation took advantage of the new openings on the margins of the channel and encroached on low floodplains.

Relatively small annual floods occurred in the middle part of the twentieth century, notably in the 1950s, but large floods occurred in the 1970s through 1990s, including the 1983 flood,[20] which peaked at 132,000 ft³/s at the head of the Safford Valley gaging station (fig. 2.2) and is the largest recorded flood on the upper Gila River. In January and February 1993, three flood peaks exceeding 100,000 ft³/s at the Calva gaging station coursed through the Safford Valley, filling San Carlos Reservoir and causing major flow releases downstream.

The annual flood series shows that floods on the upper Gila River are nonstationary in time and follow the same pattern of climatic fluctuations described for other sites in southern Arizona.[21] Even the lesser floods of the 1930s and 1940s caused significant damage. The 1941 flood reportedly inundated Duncan, Arizona, with 4 feet of water. The bridge over the Gila River at Duncan was destroyed during the January 1949 flood, which also forced evacuation in all the major

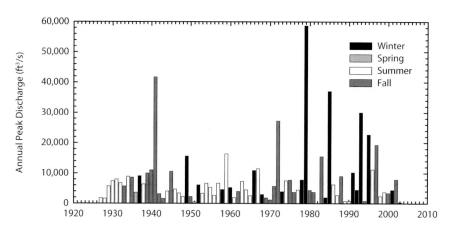

Figure 17.2 Annual flood series for the Gila River below Blue Creek, New Mexico (station 09432000; 1927–2003).

towns downstream to San Carlos Reservoir.[22]

As with most nineteenth-century flow-regulation structures, the first low dams across the Gila River failed during the first flood after construction was completed. The first permanent structure built on the Gila River was the Ashhurst-Hayden Dam, a low-head dam completed in 1922 by the U.S. Indian Service.[23] Located about 20 miles upstream from Coolidge, Arizona, this dam diverts base flow into canals that supply the extensive agricultural development of the Eloy basin. The Ashhurst-Hayden Dam did nothing to control floods, which were common in the first three decades of the twentieth century. Smaller diversion dams are common along the river from the Arizona–New Mexico border to San Carlos Reservoir. Coolidge Dam, completed in 1929, impounds San Carlos Reservoir, which has highly variable water levels. This structure has the dual purpose of flood control and delivery of irrigation water downstream to extensive agricultural areas in the Eloy basin.

Riparian Vegetation Change on the Upper Gila River

Much of the upper Gila River flows in remote bedrock canyons that do not have a recorded history of riparian vegetation. Moreover, either few people ventured into these canyons to take photographs, with the exception of U.S. Geological Survey hydrographers, or those photographs were not pre-

served. Nonetheless, 127 repeat photographs document vegetation change on the Gila River between Blue Creek (just upstream of the Arizona–New Mexico border) and the Ashhurst-Hayden Dam.

Arizona–New Mexico Border to Safford

Just upstream from the Arizona–New Mexico border, the Gila River passes through a series of short bedrock canyons and high alluvial terraces. Photographs associated with the gaging station document change in riparian vegetation in a wide reach upstream from one of these short canyons below Blue Creek, a major tributary (fig. 17.3). This wide reach shows considerable channel shifting that occurred during large floods in the last third of the twentieth century, resulting in removal of a band of Frémont cottonwood, which were replaced with tamarisk and coyote willow. Within the bedrock-confined reach just downstream, riparian native vegetation—primarily in the form of mesquite bosques—is locally dense, and tamarisk is less common. Bosques in this reach are notable for their wildlife, including javelina, turkey, coatimundi, deer, and other large vertebrates. Twenty-three photographs show that native riparian vegetation has increased in this reach despite repeated channel shifting across the wide floodplain. Tamarisk is not widespread here.

In the vicinity of Duncan, the corridor opens into a relatively wide

A. (July 23, 1931.) The Gila River upstream from its juncture with Blue Creek near the Arizona–New Mexico border drains 3,203 square miles of rangeland and the Gila Wilderness. Just downstream from Blue Creek, the channel is somewhat confined within a bedrock canyon, but just upstream, as this view shows, the valley is relatively wide. (J. Baumgartner 1501, courtesy of the U.S. Geological Survey.)

B. (June 11, 1964.) By 1964, cottonwood trees had grown up along the banks of the river behind a lower ribbon of tamarisk trees. The channel position is the same despite a flood of 41,700 ft^3/s in 1941. (R. M. Turner.)

C. (October 5, 2000.) The channel has shifted to the left, probably during one of the four floods that exceeded 27,000 ft^3/s between 1964 and 2000. The cottonwood gallery has been destroyed, and the channel is lined with tamarisk, coyote willows, and some scattered brickellbush. (D. Oldershaw, Stake 346a.)

Figure 17.3 Photographs of the Gila River below Blue Creek.

bottomland supporting agriculture. Despite use of groundwater for agriculture, long-term records of water level downstream from Duncan (fig. 17.4) show that no significant trends have occurred other than a general decline during the midcentury drought and rises of 11 and 4 feet following the 1970 and 1993 floods, respectively. Several historical photographs, notably ones that document a bridge that repeatedly failed during floods in this reach, also show increases in riparian vegetation, in particular cottonwood. Analysis of aerial photography taken between 1935 and 1997 shows the combined effects of initial channel narrowing, agricultural clearing of floodplain vegetation, and increases in riparian vegetation, followed by channel widening during the 1993 flood.[24] The amount of tamarisk along the Gila River increases substantially downstream from Duncan.

Between Guthrie and Safford, the river flows through the Gila Box Riparian National Conservation Area, where grazing has ceased but off-road vehicle use and hunting persist.[25] At the upstream reach, a U.S. Geological Survey gaging station has recorded flows since the early twentieth century (fig. 17.5). Twenty-one photographs associated with the gaging operation have been taken at this site. In general, these photographs show a steady increase in riverine riparian vegetation despite the recent occurrence of large floods (fig. 17.6).

Downstream from Bonita Creek (fig. 17.1), the Gila River exits the Middle Box and enters the wide Safford Valley. Because the Gila River has been

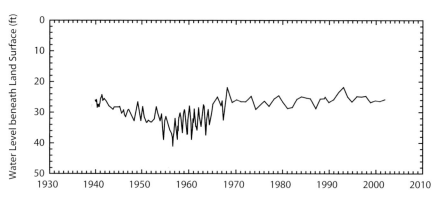

Figure 17.4 Water levels for well D-07-31 04BCC near Duncan, Arizona.

subjected to climatically induced fluctuations in flood frequency (fig. 17.2), damage to floodplains and bosques at the start of this reach was frequent in the 1970s through the 1990s.[26] Overall, however, eleven photographs upstream from Safford show increases in native species and nonnative tamarisk. At Safford, another ten photographs that have been matched show increases as well, although the relative proportion of the increase in nonnative vegetation, primarily tamarisk, is higher.

Reaches Affected by San Carlos Reservoir

The reach of the Gila River upstream from San Carlos Reservoir to the eastern edge of the San Carlos Apache Indian Reservation has long been a place where elimination of riparian vegetation once was considered a viable means of increasing available water.[27] The reach near Calva, Arizona, was used in a demonstration study called the Gila River Phreatophyte Project

to determine the effect of tamarisk removal on reducing evapotranspiration losses in the 1960s and early 1970s.[28] In the late 1940s, about 9,300 acres of riparian vegetation were present between Thatcher and Calva, and tamarisk used an estimated 75 percent of the 23,000 acre-feet per year of water consumed by riparian vegetation in this area.[29]

Groundwater in this reach moves between basin-fill deposits and the alluvial aquifer, and water levels are highly influenced by flows and floods in the Gila River.[30] As discussed in chapter 2, groundwater levels under these conditions are generally shallow, vary seasonally in accord with streamflow variations, and may show diurnal fluctuations owing to withdrawals by riparian vegetation (fig. 2.3).

The San Francisco River (chapter 18) is the largest tributary of the upper Gila River, and this tributary strongly affects flood magnitudes in this reach (figs. 2.2, 17.7). The alluvial channel of the Gila River has widened historically and shifted in this reach, in some places by large distances, thus affecting the riparian vegetation.[31] The change in riparian vegetation has been substantial; where closed gallery forests of cottonwood once grew, tamarisk has become the primary species lining the Gila River.[32] Although floods and channel change are directly involved in causing loss of the cottonwood gallery forests along the upper Gila River, livestock grazing is not directly implicated.[33]

In the mapped area of the Gila River Phreatophyte Project, change in the area occupied by several species of riparian vegetation can be determined

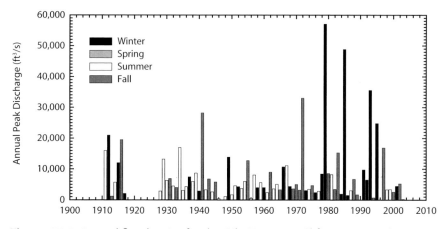

Figure 17.5 Annual flood series for the Gila River near Clifton, Arizona (station 09442000; 1911–1917, 1928–2003).

from 1914 to 1994.[34] Not surprisingly, the area occupied by tamarisk increased from zero percent in 1914 to 7 percent in 1937, 26 percent in 1944, 44 percent in 1964, and 61 percent in 1994. Cottonwood declined precipitously, from about 1,023 mapped trees in 1914 to 49 trees in 1994, although young trees that germinated during the 1993 floods are not included. The loss of cottonwood probably resulted from inundation during record high water levels at the San Carlos Reservoir.[35] Besides cottonwood, seepwillow lost the most area of the mapped native species; its loss might be a reflection of stability as the larger trees assume dominance of a more stable riparian ecosystem.

Changes in the area occupied by mesquite are interesting, given the divergent claims that mesquite bosques have been destroyed along the Gila River[36] and that mesquite has increased throughout southern Arizona.[37] Mesquite at all density levels occupied 21 percent of the area in 1914 and increased to 24 percent in 1937, 39 percent in 1944, and 44 percent in 1964, finally decreasing to 34 percent in 1994. The decrease between 1964 and 1994 is related to a renewed channel widening associated with floods from 1972 through 1993. Particularly in light of the later event, which consisted of multiple flood pulses, the increase in tamarisk and decrease in mesquite do not suggest that floods, whether artificially released from dams or naturally occurring, would help reduce the prevalence of nonnative vegetation in this setting.

Repeat photography of this area from the Calva railroad bridge (twelve views matched; e.g., fig. 17.8) shows another problem with restoration of floodplains, particularly with respect to eradication of tamarisk. In 1932, cottonwood and willow were still obvious on the floodplain. By 1964, tamarisk was ubiquitous, and native species could not be seen. All woody riparian plants, mostly tamarisk, were removed by 1974, but no additional channel maintenance was performed after the Gila River Phreatophyte Project ended. As a result, tamarisk steadily encroached onto the floodplain, despite large floods and attempts at eradication.

Downstream from Coolidge Dam, the river enters a series of bedrock canyons en route to Winkelman. Coolidge Dam is operated as a flood-control dam, and, except in 1993, this operation has been highly successful (fig. 17.9A). Flow releases from Coolidge Dam are generally highest during the summer months and low during the winter, which creates prime conditions for the growth of riparian vegetation. In the 1980s, reports from river runners downstream from the dam described a dense tamarisk thicket with water running through groves of trees.[38]

Nine historical photographs document changes in this reach, although most of the views show channel changes associated with construction activities at the base of the dam. Several photographs show conditions downstream from the disturbed areas, and cottonwood, black willow, and tamarisk have increased in this reach despite destruction of riparian vegetation during the 1993 high releases (fig. 17.10). Cottonwood trees along this reach provide nesting habitat for Bald Eagles.

Winkelman to Florence

The San Pedro River (chapter 19) enters the Gila River downstream from Winkelman and partially negates the flood-control effectiveness of Coolidge Dam (fig. 17.9B). Floods in 1983 and 1993 were of approximately the same magnitude as the 1891, 1916, and 1926 events, the latter also originating from the San Pedro River. The wide, denuded reach that existed before completion of Coolidge Dam has narrowed significantly (figs. 17.11, 17.12). During the midcentury drought and persisting through 1992, tamarisk created a closed canopy over much of the river in this reach, but at least part of this dense tamarisk was swept away during the 1993 releases from Coolidge Dam. However, as shown in eighteen sets of repeat photographs, cottonwood, mesquite, and tamarisk have increased near the gaging station at Kelvin (fig. 17.13).

The Ashhurst-Hayden Dam, which diverts most of the surface flow into irrigation canals, provides a convenient place to divide the upper from the lower Gila River. This small structure allows the passage of large floods while completely shifting small flows out of the channel. A small reservoir upstream from the dam has silted in and periodically dries, allowing encroachment of riparian vegetation (fig. 17.14). Nonnative vegetation, including both tamarisk and Athel tamarisk, dominates the banks of the river at this point, although dense mesquite also has been increasing.

A

B

Figure 17.6 Photographs of the Gila River near Clifton, Arizona.

A. (August 20, 1930.) The bridge on the right side of this photograph spans the Gila River between Clifton and Safford. This bridge formerly was the main route between the two cities. The Gila River flows right to left in this view, and downstream from this point it enters a narrow canyon. Isolated cottonwood trees are present on the near side of the channel, behind small mesquite trees that are established above the floodplain. Native coyote willows and brickellbush line the far side of the channel. (H. Pritchett 1136, courtesy of the U.S. Geological Survey.)

B. (June 24, 1964.) Few large floods occurred on the Gila River in the middle part of the twentieth century, and the riparian vegetation responded by growing in dense assemblages. Cottonwood trees in particular have benefited, although the largest increase may be in the mesquite trees in the foreground. The perennial flow in the Gila can be seen only at far left, owing to the increased stature of the plants. (R. M. Turner.)

C

D

C. (May 18, 1980.) Large floods in the 1970s scoured much of the floodplain, but most of the larger riparian plants remain and have grown. The largest flood, in 1979, had a discharge of 57,000 ft³/s, which is the flood of record at this station. If the trees were not destroyed or damaged during the floods, they benefited from the increased water infiltrating into the floodplain. The gaging station, which was moved from a site nearby, now appears on the upstream side of the bridge at far right. (R. M. Turner.)

D. (October 4, 2000.) Four floods larger than 15,000 ft³/s occurred at this site after the 1980 photograph. Despite these floods, riparian vegetation increased to the point of completely obscuring the channel. Even the opening under the bridge is nearly blocked by the vegetation, which includes cottonwood, mesquite, coyote willow, black willow, and brickellbush. Tamarisk is also present at this site, but it is a minor component of the riparian ecosystem. (D. Oldershaw, Stake 358.)

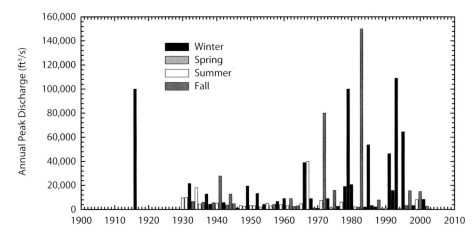

Figure 17.7 Annual flood series for the Gila River near Calva, Arizona (station 09466500; 1916, 1930–2003).

A

B

Figure 17.8 Photographs of the Gila River at Calva, Arizona.

A. (March 6, 1932.) The Gila River was ravaged by floods in the 1910s and 1920s, including one flood estimated at 100,000 ft³/s in 1916. This upstream view from the railroad bridge at Calva shows a wide, braided channel. An open cottonwood gallery forest appears in the midground at left *(channel right),* and the low shrubs at right appear to be native willows. (W. E. Dickinson 1053, courtesy of the U.S. Geological Survey.)

B. (June 18, 1964.) In the intervening thirty-two years, the highest discharge through this reach was 27,900 ft³/s on October 1, 1941. Dense tamarisk has become established, creating a floodplain where the main channel once was. The channel has shifted to the right and is much smaller than it was in 1932. (R. M. Turner.)

C. (October 17, 1973.) To reduce evapotranspiration, phreatophytes (mostly tamarisk) were removed from this reach in 1970, and the floodplain was reseeded to native grasses. Those grasses are mostly gone at the time of this photograph, one year after a flood of 80,000 ft^3/s passed through this reach. The small shrubs throughout the view are mostly tamarisk that are reestablishing in the reach. (R. M. Turner.)

D. (May 21, 1984.) In October 1983, a flood of 150,000 ft^3/s passed through this reach in the wake of Tropical Storm Octave. This flood followed one that peaked at 100,000 ft^3/s in 1978. These floods shifted the channel back into the view, and driftwood racks appear throughout the foreground. Despite this flood, dense tamarisk appears on both floodplains. (R. M. Turner.)

E. (October 6, 2000.) Except for withdrawals of domestic and irrigation water at low-head diversion dams, the Gila River is unregulated upstream from Calva. In January and February 1993, three floods exceeding 100,000 ft^3/s passed through this reach. Despite these floods, tamarisk has grown considerably, blocking the view of the river channel from this camera station. Despite the enormous effort at tamarisk removal and river restoration, the tamarisk has attained a higher biomass than it possessed in 1964. The channel has shifted from the right side to the left side of this view. (D. Oldershaw, Stake 331a.)

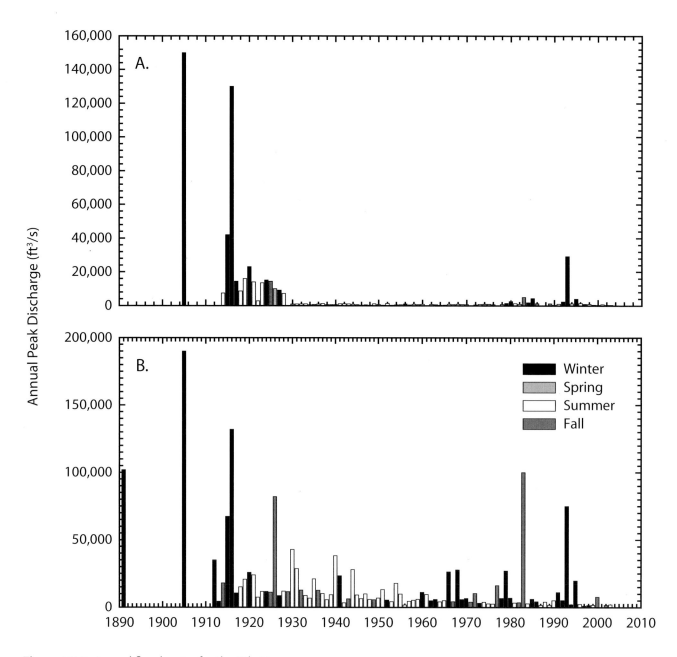

Figure 17.9 Annual flood series for the Gila River.

A. The Gila River below Coolidge Dam, Arizona (station 09469500; 1905, 1914–2003).

B. The Gila River at Kelvin, Arizona (station 09474000; 1891, 1905, 1912–2003).

Figure 17.10 Photographs of the Gila River upstream from Winkelman, Arizona.

A. (February 1928.) Downstream from Coolidge Dam, the Gila River flows through short segments of bedrock canyon on its 30-mile course to Winkelman. Downstream of Dripping Springs Wash, the river flows through a canyon of tilted limestone, and the railroad spur line to the Christmas Mine appears in the foreground and at left in this upstream view. The channel of the Gila River is mostly denuded, with patchy, low clumps of mesquite at several points along the right bank. (C. A. Amsden PO5192, courtesy of the Southwest Museum.)

B. (1944.) Edwin McKee, a prominent geologist, visited this section of the Gila River to examine the extensive copper-mining operations both upstream and downstream. His view, from slightly upstream of the original camera station, shows much of the same reach. Owing to flood-control operations of Coolidge Dam, completed one year after the original view was taken, the channel has narrowed, and nonnative tamarisk lines its banks. Mesquite is becoming established on the back side of the floodplain. (E. D. McKee 95.48.1534, courtesy of Northern Arizona University, Flagstaff.)

C. (February 10, 1994.) The main road, Arizona Highway 77, between Winkelman and Globe now occupies the former railroad grade. The 1993 flood scoured the channel free of nonnative tamarisk, depositing coarse-grained sediments on the floodplain in the same place as is apparent in the 1928 view (photograph A). A dense mesquite bosque has developed on the point bar in the midground, and Frémont cottonwood is now common in this reach, particularly downstream. (R. H. Webb, Stake 3108.)

A

B

Figure 17.11 Photographs of the Gila River near Kelvin, Arizona.

A. (September 2, 1915.) This upstream view from a railroad grade shows a mill associated with copper-mining operations along the Gila River in this reach. Wooden A-frame structures suspend a pipeline across the river in the midground. The channel is wide and barren, and a nearly continuous, although thin, line of mesquite is present on the left bank. (Photographer unknown, courtesy of the U.S. Geological Survey.)

B. (October 2, 2000.) The mill has been abandoned, but the railroad remains active. The channel has narrowed substantially owing to flow regulation and encroachment of riparian vegetation. Creosote bush appears in the immediate foreground, and mesquite and tamarisk form the wall of vegetation blocking most of the view of the channel. The upper branches of Frémont cottonwood, which are common in this reach, appear at right. (D. Oldershaw, Stake 433.)

Figure 17.12 Photographs of the Gila River near Kelvin, Arizona.

A. (Ca. 1908.) This view, the right *(upstream)* quarter of a panorama, shows the Gila River in the vicinity of Kelvin. The wide channel is denuded, and low terraces sustain dense stands of mesquite. No Frémont cottonwood trees are obvious. (Photographer unknown, PAN US GEOG-Arizona No. 3, courtesy of the Library of Congress, Washington, D.C.)

B. (October 1, 2004.) The once-wide channel has narrowed considerably in response to regulation by Coolidge Dam, although occasional large floods emanate from the San Pedro River, a major tributary that joins the Gila River in the distance. Riparian vegetation consists of well-established Frémont cottonwood, tamarisk, black willow, catclaw, and mesquite. (D. Oldershaw, Stake 1429d.)

Figure 17.13 Photographs of the Gila River at Kelvin, Arizona.

A. (May 21, 1945.) This view, from a hill on the south side of the Gila River, shows the bridge at Kelvin from which discharge measurements are made at high flows. The river flows from right to left in this view and is regulated by Coolidge Dam, completed in 1928. The largest historic flood at this station to this date is 132,000 ft³/s in 1916, and three other annual flood peaks exceeded 40,000 ft³/s before 1945. Floods averaged less than 40,000 ft³/s for fifteen years prior to this photograph. In this view, tamarisk is becoming established in the foreground, and mesquite and cottonwood trees appear on the far bank. Mineral Creek, spanned by a railroad bridge, enters the Gila River at left. (W. L. Heckler 3709, courtesy of the U.S. Geological Survey.)

B. (October 2, 2000.) The view is blocked by cottonwood trees, particularly at left and right, with small trees in the foreground, and by large tamarisk trees. This increase in riparian vegetation has occurred despite the fact that floods of 100,000 and 74,900 ft³/s in 1983 and 1993, respectively, passed through this reach. (D. Oldershaw, Stake 430.)

A. (April 3, 1939.) Ashhurst-Hayden Dam, built as an irrigation diversion structure in 1922, channels all low flow of the Gila River into nearby canals for delivery to the agricultural areas around Florence, Eloy, and Casa Grande. What may be nonnative Athel tamarisk trees, well pruned, appear at lower right. (Photographer unknown 1782, courtesy of the U.S. Geological Survey.)

B. (November 7, 1980.) Active clearing of the small reservoir area upstream from the dam has kept the channel clear of any riparian vegetation. Tamarisk lines the far bank upstream of the standing water. Athel tamarisk appears where the well-pruned trees were in 1939. (R. M. Turner.)

C. (November 28, 2001.) Riparian vegetation has increased substantially at this site, although most of it is nonnative. Seepwillow is present in the channel, but there is more tamarisk, much of which is dead. Mesquite is present at various places in the view, but it is not dense here. The Athel tamarisk appear to have increased significantly in size at lower right. (D. Oldershaw, Stake 1006.)

Figure 17.14 Photographs of the Gila River at Ashhurst-Hayden Diversion Dam.

Summary. The San Francisco and San Carlos Rivers, which flow southwesterly from the higher elevations of western New Mexico and eastern Arizona, are important perennial tributaries of the Gila River. Neither river has significant flow regulation, and they have different flood histories: the San Francisco River is highly affected by regional climate, and floods in the San Carlos appear to be random in time. Both rivers have had increases in riverine riparian vegetation, although mostly tamarisk has increased on the San Francisco River near Clifton. Most of the increases along the San Carlos River are in native species, especially Frémont cottonwood and black willow, although tamarisk currently forms a dense band closest to perennial flow.

Several tributaries enter the upper Gila River from the north (fig. 17.1). The two largest—the San Francisco River and the San Carlos River—are the subject of this chapter. Several other important tributaries, such as Eagle and Bonita Creeks, have insufficient photographic evidence to document change in riparian vegetation. The San Francisco River, in particular, is extremely important to the hydrology of the upper Gila River because of its production of remarkably large floods and because its annual flood series mirrors that of the main-stem Gila River. The San Carlos River seemingly has little in common with either the San Francisco River or the Gila River except that riparian vegetation has increased along all three watercourses.

The San Francisco River

The San Francisco River drains 2,766 square miles of eastern Arizona and western New Mexico (fig. 17.1), flowing mostly in bedrock canyons with shallow alluvial fill. With headwaters

in the northern Mogollon Mountains, the San Francisco River flows mainly through wildlands and rangeland until it reaches Clifton, Arizona, the only significant town in the drainage basin. Its major tributary, the Blue River, drains the Mogollon Rim and White Mountains of east-central Arizona and flows south to its confluence with the San Francisco River about 20 miles upstream from Clifton. This town, established in 1872 in support of nearby rich copper deposits,[1] was built on the narrow floodplain of the San Francisco River, and it has suffered more flood damage in its 130 years of existence than any other town in Arizona.

Early Observations of the San Francisco River

Few nineteenth-century scientific observers traveled along the San Francisco River because of its isolation. The omnipresent James Ohio Pattie and his group of trappers reached the mouth of this river on January 1, 1825, and sought beaver about 4 miles upstream. They caught 37 in one night, spurring them to investigate the drainage up into New Mexico. He reported a take of 250 beaver in two weeks of trapping on the river.[2] Although no reports document the amount or condition of woody riparian vegetation, it must have been extensive to sustain this many beaver.

Floods

The San Francisco River has a long history of flooding documented by reports of damages to Clifton and by one of the longest gaging records in Arizona (fig. 18.1). The average daily flow rate at Clifton is 224 ft³/s. Flood damages here were extensive during February 1891, when a discharge of

65,000 ft³/s passed through town. During the winters of 1905 and 1906, a period of sustained flooding throughout Arizona, peak discharges of 60,000 and 65,000 ft³/s occurred on the San Francisco River during two separate events. On December 4, 1906, flooding followed a severe winter storm, which lasted thirty hours and created a peak discharge of 70,000 ft³/s (fig. 18.1). The 1906 flood took eighteen lives and destroyed considerable amounts of property in Clifton.[3] Another flood in October 1916 peaked at 60,000 ft³/s.

A sustained period of small floods in the middle of the twentieth century (fig. 18.1) and the economic growth of Clifton thanks to the riches of its nearby copper mines may have lulled the residents into complacency. Heavier flooding resumed, however, with another large event spawned by Hurricane Joanne in October 1972. This flood, which peaked at 64,000 ft³/s, caused extensive damage in Clifton as well as in the vicinity of Safford.[4] The largest historical flood on the San Francisco River occurred on October 2, 1983, as a result of rainfall from Tropical Storm Octave (see chapter 21). The peak discharge of this flood, 90,000 ft³/s, caused substantial damage to Clifton, where approximately 130 homes and businesses were damaged or destroyed.[5]

As occurred during the first two decades of the twentieth century, flooding in the late twentieth century declined after the 1983 flood (fig. 18.1). Smaller floods occurred in January 1993, January 1995, and October 2000 during regional storms. Damages in Clifton were small because the 1983 flood had spurred extensive changes in floodplain protection and management, reducing the potential for damages. The annual flood series for the San Francisco River (fig. 18.1)

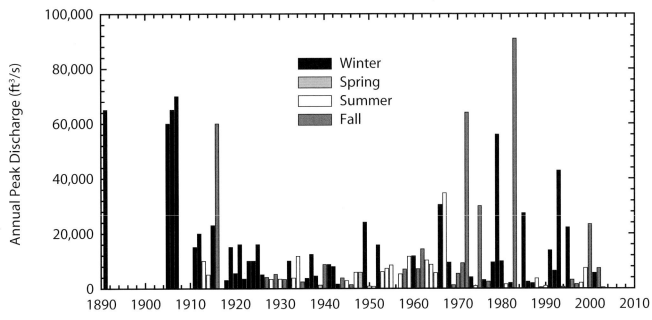

Figure 18.1 Annual flood series for the San Francisco River at Clifton, Arizona (station 09444500; 1891, 1905–1907, 1911–1912, 1915–1916, 1918–2003).

appears to be nonstationary, a conclusion also determined through analysis of tree-ring specimens collected from the basin.[6]

Despite its extensive flood history, the San Francisco River remains unregulated except for small diversion structures for local irrigation supply. The amount of agricultural land—approximately 2,700 acres upstream from the gaging station—is highest in New Mexico, although ranches and a diversion structure are present upstream from Clifton near the confluence with the Blue River.

Changes in Riparian Vegetation

Abundant historical photography documents changes in riparian vegetation at Clifton, but at no other point on the San Francisco River. At a bridge upstream from the main part of town (fig. 18.2), riparian vegetation—both native and nonnative species—has increased on the river's banks. Through Clifton (figs. 18.3, 18.4, and 18.5), the earliest photographs, dating to the 1891 flood (not shown), show a channel devoid of riparian vegetation. The reasons for the barren channel are likely related to the presence of a community along a bedrock-confined river; clearing for firewood, construction of

flood protection, and the occurrence of large floods likely stripped from the channel what riparian vegetation may have been present. Currently, riparian vegetation in the channel consists of tamarisk and mesquite. These camera stations, all showing views of heavily disturbed channel reaches, may not reflect the broader conditions along the river.

The San Carlos River

The San Carlos River drains 1,026 square miles of lands entirely on the San Carlos Indian Reservation of east-central Arizona (fig. 17.1). It was originally called the Río San Carlos, named by Francisco Garcés on November 4, 1775,[7] making it one of the oldest-named streams in the region. This watercourse and its alluvial aquifer provide water supplies and irrigation water for agriculture in the vicinity of San Carlos, Arizona.

Early Impressions of the San Carlos River

On January 31, 1825, Pattie's party of trappers reached the mouth of the San Carlos River. Although the floodplain was "plentifully timbered with trees, and the land fine for cultivation," they

found no beaver.[8] After returning to the Gila, on February 13, he decided to name this reach the "Deserted Fork" because of the lack of beaver in its waters. After establishment of the San Carlos Apache Indian Reservation in 1871, Fort San Carlos was created along the river about 3 miles upstream from its confluence with the Gila River.[9] Numerous photographs of this fort, which cannot be matched owing to submergence of the site under San Carlos Reservoir, show scattered groves of cottonwood along the channel.

Floods and Flow Regulation

The San Carlos River has a flood history that is more like that of the Salt River (chapter 24) than that of the Gila River (chapter 17), despite its status as a tributary of the latter. This river has an average daily flow rate of 60 ft[3]/s. The similarity between the San Carlos River and the Salt River might reflect the fact that they share headwaters in central Arizona. Although flooding on both the San Francisco and the Gila Rivers, with headwaters in the Mogollon Mountains of New Mexico, was low in the middle of the twentieth century, flood frequency was high on the San Carlos River (fig. 18.6). The largest flood on this river, which

A. (May 26, 1939.) This downstream view across the San Francisco River is at a road bridge upstream from Clifton. Children are swimming in the river, which is lined with sparse riparian vegetation (probably mostly seepwillow). (R. H. Monroe 2521, courtesy of the U.S. Geological Survey.)

B. (June 24, 1964.) Riparian vegetation remains sparse along this channel bank, but a single mesquite tree is now present on the extreme left side, and tamarisk has become established. (R. M. Turner.)

C. (October 5, 2000.) Riparian vegetation has clearly increased within the view. The species present include tamarisk, seepwillow, cottonwood, black willow, and coyote willow. (D. Oldershaw, Stake 360.)

Figure 18.2 Photographs of the San Francisco River upstream from Clifton, Arizona.

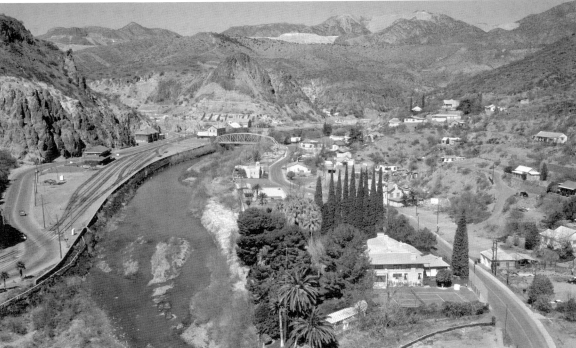

Figure 18.3 Photographs of the San Francisco River at Clifton, Arizona.

A. (Ca. March 1915.) This upstream view of the San Francisco River shows its confluence with Chase Creek *(left center)* and the center of Clifton. Bank protection is extensive along the river owing to flood damage that occurred during the first decade of the twentieth century. Floods have swept the channel clear of riparian vegetation. (O. A. Risdon 6, courtesy of the Arizona Historical Society, Tucson.)

B. (March 11, 2003.) Most of the buildings and the bridge remain in place, allowing many comparisons of changes in the intervening eighty-eight years. Several new tailings piles are highly visible on the distant hills. The bank protection present in 1915 remains, although some flood walls have been extended *(left side)*. Trees planted near buildings before 1915 are still alive, and palms are now abundant in this view. The riparian vegetation in the channel includes mesquite, tamarisk, and seepwillow. (D. Oldershaw, Stake 4607.)

Figure 18.4 Photographs of the San Francisco River at Clifton, Arizona.

A. (Ca. March 1915.) This downstream view, from a position on a hillside upstream from the camera station for figure 18.3, shows the lower part of Clifton as well as the mining operations closest to town. Bank protection consists of a flood wall with a dike behind in the foreground and a high dike system extending downstream on both banks to the tailings pile. The channel is completely devoid of riparian vegetation, but a small grove of cottonwood appears on river left at the bridge abutment; these plants may have been present before creation of the town. (O. A. Risdon, courtesy of the U.S. Geological Survey.)

B. (March 11, 2003.) Although the original bridge remains, a new bridge for U.S. Highway 161 spans the San Francisco River, partially blocked from view by the old bridge. Most of the buildings in this part of town have been removed. Riparian vegetation has clearly increased, although tamarisk is the most common species present. Other species in the channel include mesquite and seepwillow. (D. Oldershaw, Stake 4611.)

Figure 18.5 Photographs of the San Francisco River at Clifton, Arizona.

A. (Ca. 1927.) This upstream view, from a tailings pile visible in figure 18.4, shows the San Francisco River in the downstream part of Clifton. An arched-rail highway bridge appears next to the older steel-girder bridge, suggesting a date later than 1915 for this view. The long-term gaging station, established in 1927, is visible on the pier closest to the right bank *(left side of bridge)*. Bank protection on channel right is slag from the copper smelter out of view behind the camera station. Riparian vegetation is becoming established on what appears to be a recently deposited low floodplain. The Frémont cottonwood in the right center of the view *(channel left)* may be trees that predate settlement of Clifton. (Unknown photographer 1337, courtesy of U.S. Geological Survey.)

B. (October 5, 2000.) The arched-rail highway bridge has been replaced because of flood damage, and new bank protection appears on channel left *(right side)*. The channel is filled with woody riparian vegetation, mostly tamarisk and mesquite. (D. Oldershaw, Stake 1940.)

peaked at 54,800 ft³/s, occurred during the regional flooding of January 1993; however, the second-largest flood had a peak discharge of 39,200 ft³/s and occurred on January 18, 1952. No other river in the region had a significant flood at that time.

Because the river's wide, alluvial channel has shifted during floods, the gaging station near Peridot has been moved repeatedly. At various times, the gaging station was on a narrow, wooden highway bridge on Reservation Route 810 and on a railroad bridge, all within a two-mile reach. In 1980, the gaging station was moved to a pier of the U.S. Highway 70 bridge about a mile south of Peridot. Talkalai Lake, built in 1979 on the main stem of the San Carlos River about 6 miles upstream from the town of San Carlos, partially regulates flow and provides irrigation water but does not provide significant flood control.

Changes in Riparian Vegetation

Other than historical photography of sites now submerged beneath San Carlos Reservoir, the only photographic evidence for change in riparian vegetation is at the former gaging stations. Photographs from one station, on the Reservation Route 810 bridge, show a wide, barren channel in 1941 (fig. 18.7). Photographs of approximately the same view, taken in 1956, 1964, and 2000, show a steady increase in riparian vegetation, composed mostly of native species. Photographs from the gaging station on the railroad bridge about a mile downstream show similar increases in riparian vegetation (fig. 18.8). Despite channel changes and concurrent losses in established cottonwood trees, possibly associated with the 1993 flood, native riparian vegetation has increased at this site, although tamarisk also is now well established along the channel. In addition, livestock grazing has occurred continuously in this reach.

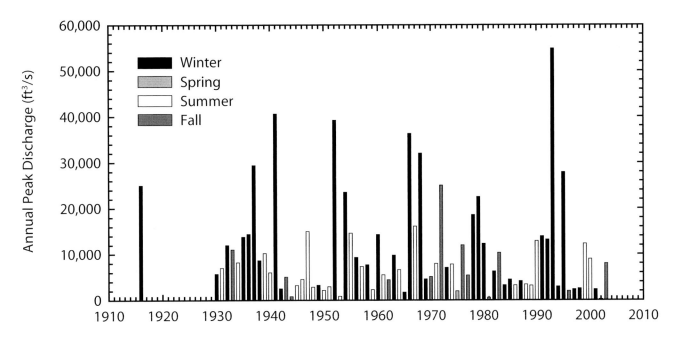

Figure 18.6 Annual flood series for the San Carlos River near Peridot, Arizona (station 09468500; 1916, 1930–2003).

Figure 18.7 Photographs of the San Carlos River near Peridot, Arizona.

A. (October 21, 1941.) In this upstream view from the old bridge just south of San Carlos, Arizona, native shrubs and scattered cottonwood line the wide channel. The San Carlos River near Peridot is a very difficult river to gage accurately because of a high sediment load and a shifting channel. A flood of 40,600 ft^3/s passed through this reach only seven months before this photograph was taken, contributing to the scoured appearance of the channel. (A. J. Hanson 3550, courtesy of the U.S. Geological Survey.)

B. (April 19, 1956.) The channel has narrowed considerably in the intervening fifteen years despite the occurrence of two floods exceeding 23,000 ft^3/s after photograph A was taken. The obvious bulldozer work in the foreground indicates that the channel is actively manipulated at this time. One reason for the channel narrowing may be the increase in nonnative tamarisk trees, which appear along river left *(right center)*. (J. G. R., courtesy of the U.S. Geological Survey.)

C. (June 4, 1964.) The channel has narrowed even more than as indicated in 1956. Native plants have also increased, including the coyote willows at left. The San Carlos River was unregulated at this time, except for diversions for irrigation. (R. M. Turner.)

D. (October 6, 2000.) The San Carlos River became partially regulated by Talkalai Lake in 1979. The channel has shifted out of the view to the right and is narrower than it was in 1964. Most of the plants in the foreground, which is now a stable floodplain, are natives, including both black and coyote willow, seepwillow, and scattered mesquite and catclaw. Riparian vegetation has increased, and the channel remains narrow despite a flood of 54,800 ft³/s in 1993, which is the flood of record for this gaging station. (D. Oldershaw, Stake 333c.)

A. (March 21, 1940.) In 1940, the gaging station for the San Carlos River was on a railroad bridge south of the little town of Peridot. This downstream view from that bridge shows the gage house on the extreme left side and a safety railing next to it. The channel is wide following a flood, and scattered Frémont cottonwood trees are along channel right. The dense shrubs on the low terrace on the right bank are unidentified but might be coyote willow. (R. H. Monroe 3451, courtesy of the U.S. Geological Survey.)

B. (June 4, 1964.) Riparian vegetation has increased to the extent of blocking the downstream view. The channel appears to have shifted to the right, and the terrace is now occupied by cottonwood, black willow, and tamarisk. (R. M. Turner.)

C. (October 6, 2000.) The gaging station has been removed, but the approximate position of the original camera can be established from the background mountains. The channel has shifted back to the left, eliminating the cottonwood trees present in 1964, but black willow is obvious in the foreground, cottonwood seedlings and mature trees appear throughout the view, and tamarisk is adjacent to the river channel. (D. Oldershaw, Stake 334a.)

Figure 18.8 Photographs of the San Carlos River near Peridot, Arizona.

Summary. The riparian ecosystem along the unregulated San Pedro River is one of the most valuable in the Southwest, particularly for birds. This watercourse has alternating reaches of perennial and ephemeral flow, with no flow regulation and minimal surface-water diversions. The closed gallery cottonwood forests along the upper San Pedro River are considered to be threatened by urbanization in and around Sierra Vista, Arizona. The San Pedro River between Benson and San Manuel is rural, with scattered agricultural areas; riparian vegetation in this reach is discontinuous but locally lush. The lower reach of the San Pedro supports a gallery forest near the confluence with the Gila River. Riparian vegetation has generally increased along the river north of the U.S.–Mexico border, although at least one ephemeral reach shows little change. Following channel widening that ceased in the early 1940s, low alluvial terraces were deposited that allowed colonization by woody riparian vegetation in the effluent-influent reaches. Riparian vegetation has steadily increased despite episodic flooding, notably in 1983 and 1993, which affected the northern half of the watershed.

The San Pedro River in southern Arizona has riparian vegetation and biological diversity that qualify it as one of the "Last Great Places" according to The Nature Conservancy.[1] The upper San Pedro River is considered to be an extremely important, year-round habitat for neotropical species, and it also functions as one of several corridors for neotropical birds migrating between Mexico and the United States.[2] The San Pedro Riparian National Conservation Area, established in 1988,[3] is the subject of considerable research on the effects of groundwater development on riparian ecosystems. At least 220 species of birds use this reach. The lower half of the San Pedro River has significant riparian resources, including the Bingham Cienega and a cottonwood gallery forest near the confluence with the Gila River. The lower San Pedro River has been designated as critical habitat for the Southwestern Willow Flycatcher.[4]

With its headwaters in the Mexican state of Sonora (not shown in fig. 19.1), the river enters the United States and flows north to its confluence with the Gila River near Winkelman, Arizona. Although most of the interest in the San Pedro River centers on a reach between the border and near St. David, riparian vegetation closely follows the alternating pattern of perennial-ephemeral flow that characterizes this watercourse along its greater than 150-mile length in Arizona. Draining 4,453 square miles, including 696 square miles in Mexico, this river behaves as if it were two watersheds in one, with a southern half that has a flood record with low variability and a northern half that produced extremely large floods in the last third of the twentieth century. The population of Sierra Vista and its surrounding area is expanding rapidly, creating demands for more water usage from the regional aquifer system, which are thought to pose a severe future threat to the riparian ecosystem.[5]

Ultimately, the case of riparian vegetation change on the San Pedro River represents one of the largest increases in woody riparian vegetation in the Southwest. Many researchers have noted that this river, once swampy, now sustains a verdant forest.[6] If cattle grazing caused arroyos to downcut, then it created this forest because growth and establishment of most of these woody trees required dewatering of the upper few feet of a once saturated alluvial aquifer and disturbance. The current intense interest in the San Pedro River centers on the future effects of water development,[7] not on the rich history of its past.

Early Observations of the San Pedro River

Photographs taken in the nineteenth century and historical accounts indicate a complete change in hydrologic and ecological conditions along the San Pedro River. When James Ohio Pattie first viewed the San Pedro River on March 3, 1825, he and his companions named it "Beaver River" because of the large quantity of these animals present there.[8] Near its confluence with the Gila River, they trapped two hundred animals in one week. They traveled upstream, then left the San Pedro River near present-day Benson to go northeastward and return to the upper Gila River. Before they left the San Pedro, Pattie observed that "its banks are still plentifully timbered with cotton-wood and willow." In September 1829, Pattie and his father returned to the San Pedro River, again trapping beaver in "considerable numbers."[9]

After independence from Spain in 1821, the new country of Mexico looked northward to develop lands south of the Gila River. In 1832 and 1833, land grants totaling about 40,000 acres were awarded in the upper San Pedro River, and large herds of cattle were introduced.[10] Ongoing hostilities with the Apache caused abandonment of these large ranches in around 1840, leaving large herds of wild cattle, likely near the San Pedro River. The Apache selectively culled cows, and by 1846 most of the remaining cattle were bulls, as encountered by the Mormon Battalion.[11] These large herds, which preceded downcutting of the arroyo in this reach by about forty to fifty years, were gone by the 1850s.[12]

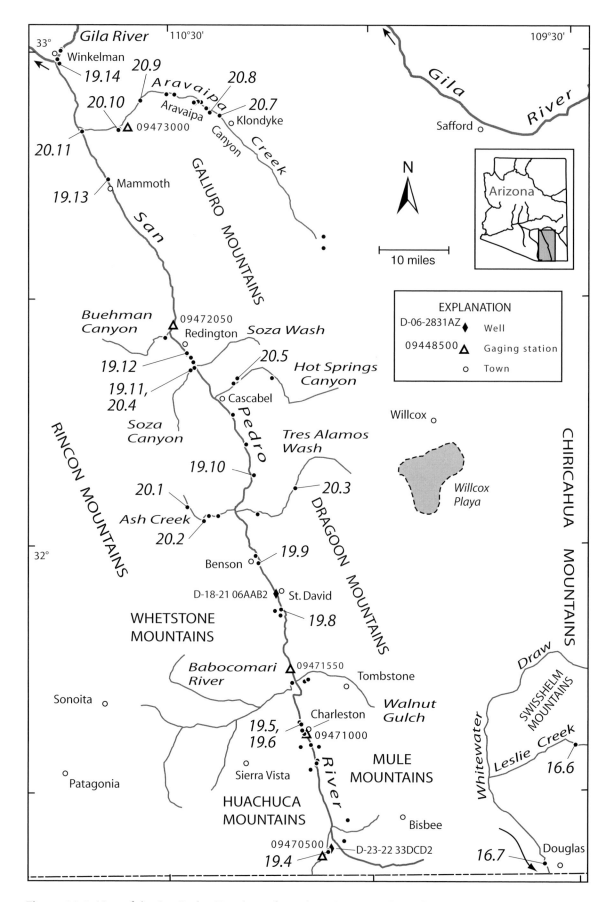

Figure 19.1 Map of the San Pedro River basin from the U.S.–Mexico boundary to the Gila River in central Arizona.

Despite descriptions of swampy conditions indicative of low flow, fish—including Colorado River pikeminnow—were abundant in the San Pedro River.[13] Most observers believe the river was perennial over its full length from the Mexican border to the Gila River, although flow may have been low much of the time.[14] Before development of the coalesced arroyo system in the 1880s, groundwater levels along the lower San Pedro River were high enough to sustain marshes along much of its length.[15] Floodplains were saturated, and the marshy conditions led to mortality in settlements and army encampments.[16] For example, Mormon settlers established St. David in 1877, and by 1878 they were dying of malaria. The floodplains in this reach supported alkali and Wright sacaton grasslands with scattered stands of woody vegetation, including isolated and small groves of cottonwood and willow trees.[17] Part of the reason for high water tables in the alluvial aquifers might have been the small dams built by the thriving beaver population.[18] In fact, the type locality for *Castor canadensis frondator*, once considered a viable subspecies, is the San Pedro River at the U.S.–Mexico border.[19]

In 1892 and 1893, Edgar Mearns visited the San Pedro River at the border between the United States and Mexico (fig. 19.1). At this time, after the arroyo had downcut (see the next section), he observed fish and turtles in the "good-sized stream," but no marshes, and found that trees were limited to the edge of the channel. In addition to Frémont cottonwood, ash, sycamore, box elder, and mesquite, Mearns noted black willow and yewleaf willow but made no mention of forests. Indeed, a photograph accompanying his text shows scattered trees along the river corridor.[20] The only documented nineteenth-century gallery forests along this river were downstream near the confluence of the San Pedro and Gila Rivers,[21] which is what Pattie observed.

Floods, Channel Change, and Groundwater

Both the Santa Cruz and the San Pedro Rivers, as well as many other ephemeral and perennial rivers in the region, developed arroyos as early as eight thousand years ago, with well-preserved sequences of cutting and filling common in the past four thousand years of riverbank stratigraphy.[22] From a geologic-time perspective, arroyos are reasonably synchronous regionally and represent a geomorphic record of climatic change.[23] The prehistoric cutting and filling of arroyos in southeastern Arizona shows that livestock grazing could not have been the only reason for channel change on the San Pedro River.

Arroyo cutting on the San Pedro River likely began in the late 1870s, but the channel was entrenched near the Gila River at the time that Pattie viewed it in the 1820s. By 1883, a welldeveloped arroyo was present near the confluence with the Gila River, and by the mid-1890s an arroyo spanned the length of the San Pedro River. Considerable controversy arose as to just how the arroyo had coalesced: whether a headcut propagated from the confluence upstream by headward erosion[24] or a series of headcuts formed locally at tributary junctures and floods connected these junctures together into the continuous arroyo.[25] The question of causality also arises: Did arroyo formation on the San Pedro River result from overgrazing, did modification of the floodplains for agriculture and roads create gullies that enlarged during floods, or did a period of unusual climate spawn large floods that would have created an arroyo without human influence? All of these causes likely influenced channel change on the San Pedro River, although livestock were present in the basin well before the period of arroyo formation.[26]

Historical channel changes on the San Pedro River are relatively well known.[27] In the mid-1850s, the channel near Tres Alamos Wash had banks about 9 to 12 feet high,[28] and Mormon settlers of St. David in the 1870s reported that the channel was a deep gully with banks up to 20 feet high.[29] The 1887 earthquake, which was centered in the San Bernardino Valley southeast of Agua Prieta, Sonora, and Douglas, Arizona, affected spring flow in the San Pedro River basin, particularly in the vicinity of St. David.[30] Some have suggested that these hydrologic changes may have contributed to arroyo entrenchment, but the arroyo was mostly formed by 1887. Entrenchment at Benson was completed by 1928,[31] followed by a period of channel widening.[32] Channel dimensions stabilized by 1941, and then low floodplains developed within the arroyo banks over a long period extending through 2002.[33]

Three long-term gaging stations have documented flow on the San Pedro River (fig. 19.2). Annual flow in the San Pedro River at the long-term station at Charleston is 56 ft³/s, and the annual flood series (fig. 19.2B) consists of an extremely large flood in 1926 (98,000 ft³/s) followed by a number of relatively small floods extending to 2001. In September 1926, a severe storm in southeastern Arizona caused some of the greatest damage recorded in Arizona history,[34] and the peak discharge entering the Gila River at Winkelman was 85,000 ft³/s. Following this flood, annual flood peaks were mostly small, although the seasonality of flooding has changed from floods caused by summer thunderstorms to ones caused by regional storms in fall and winter (fig. 19.2).

Base flow at Charleston has decreased by about 2 ft³/s during June through October, and annual runoff passing Charleston decreased from greater than 45,000 acre-feet per year from 1928 through 1935 to 20,000 acre-feet per year in the mid-1990s.[35] Water levels in wells fluctuated more than 10 feet, but declined only 0.2 to 0.5 foot per year between 1940 and the mid-1980s (fig. 19.3). Rates of groundwater decline are greater in other wells along the river.[36] Groundwater and surface-water development associated with mining operations in Mexico and with agricultural and urban development in the United States might be affecting the overall aquifer system that creates baseflow conditions in this reach. Summer rainfall has become more erratic,[37] which may be affecting the river. Climatic fluctuations and increased woody vegetation on uplands[38] might be reducing recharge to the alluvial aquifer. Another possible reason for the baseflow declines is the increase in riparian vegetation (see the next section) and its increasing water use in the latter half of the twentieth century.

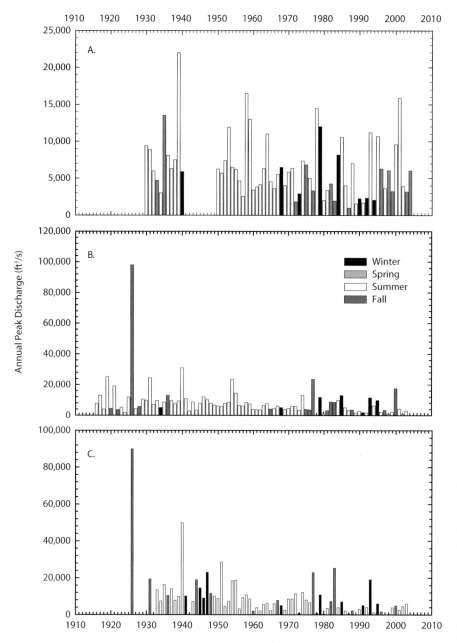

Figure 19.2 Annual flood series for the San Pedro River.

A. San Pedro River at Palominas, Arizona (station 09470500; 1926, 1930–1940, 1950–2003).

B. San Pedro River at Charleston, Arizona (station 09471000; 1916–2003).

C. San Pedro River near Redington, Arizona (station 09472000; 1926, 1931–1998), combined with the San Pedro River at Redington Bridge near Redington, Arizona (station 09472050; 1999–2003).

The 1983 flood on the San Pedro River affected mostly the drainage basin downstream from Redington (fig. 19.2C), particularly from Aravaipa Creek downstream to the confluence with the Gila River. The peak discharge at Charleston was only 8,500 ft³/s, which does not rank within the top-ten floods in this reach. However, tributaries originating in the Santa Catalina and Galiuro Mountains (chapter 20) contributed large volumes of water to the lower San Pedro River. Near Winkelman, the peak discharge of the San Pedro River was 135,000 ft³/s on October 1, 1983.[39] Large amounts of riparian vegetation were swept away during this flood.

In the effluent reach between Palominas and Charleston, Arizona, the alluvial and regional aquifers are disconnected locally owing to the presence of significant amounts of silt and clay lenses in the subsurface.[40] The water-level declines in the alluvial aquifer were probably of the same order of magnitude as the depth of channel downcutting,[41] which was 16 to 33 feet in the vicinity of Charleston and up to 16 feet in the vicinity of Palominas.[42] Decreases in groundwater pumping for agricultural areas, combined with above-average runoff associated with a wet period, allowed a rebound in the alluvial aquifer in the mid-1980s, with as much as an 8-foot rise on one well.[43]

Changes in Riparian Vegetation

Ninety photograph matches document changes in riparian vegetation along the length of the San Pedro River from 1880 through 2003. Previous studies used repeat photography to document the magnitude of changes in riparian vegetation between Fairbank and the U.S.–Mexico border.[44] In the nineteenth century, the river corridor in this reach either was barren or was vegetated with alkali sacaton grasslands as well as scattered and open gallery stands of cottonwood trees and mesquite. Although woodcutting has been suggested as a reason for the open gallery stands,[45] no stumps or other evidence of woodcutting can be seen in photographs of the floodplain. Woody riparian vegetation began expanding with the cessation of channel widening in the 1940s, and the expansion of this community was unchecked from about 1960 to the present, with particular increases in the late 1970s and early 1980s associated with winter floods that enhanced the potential for germination and establishment.

The increases in riparian vegetation are so large that they have been documented with satellite imagery.[46] Multispectal data are particularly well suited for detection of dense trees in contrast to desert vegetation, and the riparian corridor is easily distinguished from adjacent xerophytic vegetation, although the data are in blocks (pixels) of about 90 feet square. For the San Pedro River from its headwaters in Mexico to about Redington, Arizona (fig. 19.1), the total area of riparian vegetation detected on satellite imagery from 1973 is 21,400 acres, or about 1.1 percent of the drainage

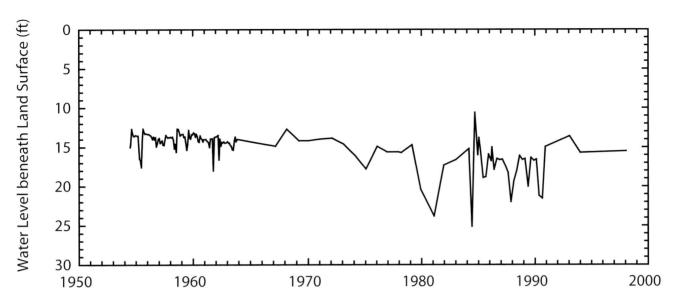

Figure 19.3 Water levels for well D-23-22 33DCD2 near the San Pedro River at Palominas, Arizona.

area.[47] The amount decreased in the 1980s,[48] but by 1997 that area had increased slightly to 22,800 acres, despite livestock grazing and agricultural clearing. Mesquite-dominated vegetation, presumably including mesquite bosques, was distinguished from riparian vegetation, and the 14,200-acre increase in agricultural area from 1973 to 1997 presumably was at the expense of mesquite near the river.

Historical photographs show that riparian trees generally were not abundant before arroyo downcutting.[49] For example, photography of monuments at the U.S.–Mexico border, originally taken in 1892 and matched in 1969 and 1976, shows considerable increases in riparian vegetation along the San Pedro River, primarily Frémont cottonwood and mesquite.[50] Another set of repeat photographs, with the originals taken in 1890 and matches taken in 1962 and 1994, shows substantial increases in woody riparian vegetation at the confluence of the San Pedro River and Babocomari River.[51] Using aerial photography, one researcher concluded that the riparian forest did not begin to develop until after the late 1930s.[52] In 1972 and 1973, one study using aerial photography concluded that mesquite was the most common type of riparian vegetation along the upper reaches of the river.[53] Clearly, the linkage between channel change and establishment of woody riparian vegetation is strong.

The U.S.–Mexico Border to St. David

The San Pedro River crosses the U.S.– Mexico boundary about 5 miles upstream from Palominas, Arizona. The river becomes perennial with effluent discharge from the alluvial aquifer near the border, and groundwater levels near Palominas (fig. 19.3) remain within reach of obligate riparian trees despite fluctuations that likely resulted from agriculture, which increased in the 1970s and 1980s. In the mid- to late 1930s, at the Palominas gaging station on a bridge on Arizona Route 92, the San Pedro River had a barren channel with no stable floodplains within the arroyo walls (19.4A). A few cottonwood were present at this time and probably were established in the 1920s to early 1930s.[54]

After 1941, low floodplains developed at the Palominas gaging station owing to a summer-dominated flood regime (fig. 19.2A), allowing establishment of mostly native woody species (fig. 19.4). Beginning in the mid-1960s, the seasonality of flooding shifted to a fall- and winter-dominated pattern, and woody riparian vegetation steadily increased through the wet period of the late 1970s through mid-1990s. The occurrence of a ten-year flood in October 2000 did not change the view upstream from this bridge. Tamarisk is present in this reach but is diminutive and sparse under the closed canopy of native species.

At the gaging station at Charleston, riparian vegetation has clearly increased along the San Pedro River (figs. 19.5 and 19.6). The amount of tamarisk has been declining in this reach, whereas cottonwood has been increasing.[55] Tamarisk may be at low density in the upper part of the San Pedro River owing to the combination of low floodplain salinity, occurrence of severe freezing, and fast growth rate of cottonwood.[56] Few cottonwood predate the 1926 flood upstream from Fairbank, and no cottonwood predate the 1926 flood near Cascabel.[57] Photographs at the former town of Contention, about 4 miles downstream from Fairbank (fig. 19.1), show increases in cottonwood and other riparian trees from 1882, when the channel is not visible, through 1989.[58]

Some researchers have suggested that the combination of fire, beaver, and high groundwater levels in the nineteenth century may have been the reason for low amounts of woody riparian vegetation along the San Pedro River.[59] Fires purposefully started by Native Americans or naturally sparked by lightning strikes likely swept out of the grasslands of the San Pedro River valley and through the riparian area; now, heavy human visitation and continued lightning strikes raise the possibility of increased fire frequency in a riparian zone heavily laden with fuel loads. On May 31, 1998, a fire caused by a carelessly thrown cigarette

Figure 19.4 Photographs of the San Pedro River at Palominas, Arizona.

A. (May 24, 1939.) This upstream view of the San Pedro River, from the bridge east of Palominas, looks over an open grassland toward mountains across the border in Mexico. Scattered cottonwood trees line the shallowly incised channel; vertical banks about 5 feet high appear in the midground at right. The 1926 flood was not measured at this site, but it had a discharge of 98,000 ft^3/s downstream at Charleston. (R. H. Monroe 2503, courtesy of the U.S. Geological Survey.)

B. (January 23, 1981.) This winter view shows small, defoliated cottonwood and willows that block out most of the background. The channel is deeper, but the floodplain remains relatively free of woody plants. A flood of 22,000 ft^3/s occurred in 1940, a flood of 16,500 ft^3/s in 1958, and another flood of 14,500 ft^3/s on October 9, 1977. Despite these floods, riparian vegetation has increased since 1939. (R. M. Turner.)

C

D

C. (February 7, 1995.) The main channel of the San Pedro River has narrowed, possibly in response to riparian vegetation or earthwork or both. A dense thicket of trees—mostly cottonwood—lines its banks, and older trees have grown up to the right. The plants are leafless in this wintertime photograph. (D. Oldershaw.)

D. (October 8, 2000.) The cottonwood and willows completely block the view. Despite a flood of 14,800 ft³/s in late October 2000, no changes occurred in the channel or riparian community. No woody nonnative species are present in this reach. (D. Oldershaw, Stake 1009.)

A   B

Figure 19.5 Photographs of the San Pedro River at Charleston, Arizona.

A. (May 4, 1954.) The gaging station for the San Pedro River at Charleston has one of the longest records in Arizona. In 1954, the gaging station was at this site, about a quarter-mile downstream from the current station on the highway bridge. This view shows the cableway cross section used for discharge measurements. The view is to the east, and the river flows from right to left. The trees along the far bank are cottonwood. (C. A. Baker 4251, courtesy of the U.S. Geological Survey.)

B. (July 29, 2000.) Cottonwood and willows, with scattered tamarisk, now block most of the view. The railroad on the opposite bank remained in operation when this photograph was taken. (D. Oldershaw, Stake 296.)

Figure 19.6 Photographs of the San Pedro River at Charleston, Arizona.

A. (April 17, 1930.) This downstream view shows channel conditions at the San Pedro River gaging station at Charleston (also see Hereford 1993, p. 20). The gaging station is just upstream from a little granite narrows, which provides a stable-channel cross section. The channel is wide, with little low floodplain development, and the only riparian vegetation present near the bottom of the channel is low shrubs. Most of the riparian plants in this view are mesquite, although a cottonwood is present in the left foreground. (W. E. W. 1441, courtesy of the U.S. Geological Survey.)

B. (July 29, 2000.) Cottonwood block any view of the channel at this site. In the past fifty years, riparian vegetation along the San Pedro River has become extremely dense and is a magnet for neotropical species, in particular birds, but also amphibians, reptiles, and mammals. As a result, this vegetation corridor is highly valued by the Bureau of Land Management, and the area has been converted into the San Pedro Riparian National Conservation Area. Views from many camera stations for the two gaging stations in this reach—at Palominas and at Charleston—are blocked by riparian vegetation, which is significant because the original photographs were taken to document channel conditions as part of gaging-station operation. (D. Oldershaw, Stake 293.)

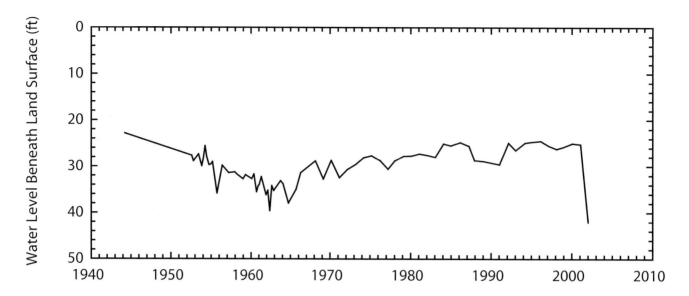

Figure 19.7 Water levels for well D-18-21 06AAB2 near the San Pedro River at St. David, Arizona. The first water level is a single reading for well D-18-21 06AAB3 a short distance away. The final reading shows a significant drop that we cannot substantiate; it may be a measurement error.

burned 750 acres of the San Pedro Riparian National Conservation Area.[60] Although many large cottonwood trees were destroyed,[61] postfire surveys indicated an increase in the number of migratory birds using the burned reach, and alkali sacaton was increasing after the burn.

Beginning with Pattie, trappers repeatedly sought beaver on the San Pedro River in the late nineteenth century until they completely decimated the species here. By the middle of the 1890s, trapping and the drought, compounded by increasing surface-water diversion, eliminated beaver from the river.[62] Beginning in the 1940s, the Arizona Game and Fish Department began reintroducing beaver in rivers throughout Arizona, including the San Pedro.[63] They did not survive in the San Pedro River, despite thriving elsewhere, and more were introduced to the river in 1999. Reintroduction of beaver likely will have the effect of thinning woody vegetation, although the potential for increase owing to resprouting remains a possibility.

Near St. David, development of the once-artesian aquifer has resulted in at least small declines in the alluvial water table. Water levels declined through the middle of the twentieth century, probably as a joint response to water extraction and the midcentury drought (fig. 19.7). Water levels

rebounded near St. David by about 12 feet during the wet period from the late 1970s through the mid-1990s. The San Pedro River in this reach, once with a barely recognizable channel, now has an arroyo that supports a closed gallery forest of Frémont cottonwood and a variety of other species, including tamarisk (fig. 19.8).

St. David to Redington

North of Benson, Arizona, the San Pedro River flows through a broad alluvial valley in a deeply incised arroyo. Riparian vegetation has increased within the arroyo walls, even though the river is at best intermittent in this reach (fig. 19.9). The river channel narrows between the Rincon Mountains to the west and the Winchester Mountains to the east, forming a narrows (called "The Narrows") where groundwater levels approach the level of the channel bed. Several important tributaries enter the San Pedro River in this reach[64] and are discussed in chapter 20.

Downstream from The Narrows, the river flows to the Gila River through a meandering arroyo system with no bedrock control of channel width. Between The Narrows and Redington (fig. 19.1), perennial flow arises in two reaches.[65] The first reach, immediately north of The Narrows, is only 1 to 2 miles long; we have no

repeat photography of this reach. The second perennial reach extends about 5 river miles in the vicinity of Cascabel and includes the mouth of Hot Springs Canyon.

The 1983 flood and other smaller ones periodically shifted the channel in the reach upstream from Cascabel, leading to a riparian assemblage consisting of tamarisk and native shrubs at sites with shallow water levels and facultative riparian species, in particular mesquite, that formed dense stands along the intermediate and high terraces (fig. 19.10). In November 1964, the amount of tamarisk in the reach between Tres Alamos Wash and Redington (fig. 19.1) was estimated to be 451 acres.[66] Tree tobacco is another nonnative species commonly observed in this reach. Downstream from Cascabel, at the joint confluence of Soza Canyon (river left) and Soza Wash (river right), obligate riparian species form a dense stand along perennial water for several miles, and Frémont cottonwood and black willow in particular have increased since the 1960s (fig. 19.11).[67]

Historically, the San Pedro River did not have significant amounts of woody riparian vegetation at Redington (fig. 19.1).[68] This condition continues (fig. 19.12) and is unlikely to change because the river is ephemeral, the channel is unstable, and ground-

Figure 19.8 Photographs of the San Pedro River at St. David, Arizona.

A. (1890.) This view, from a hill overlooking the west side of the San Pedro River, is to the east-northeast, with the Dragoon Mountains in the background. The river channel is barely discernible in the midground and is lined with isolated small trees and shrubs growing in what appears to be an alkali sacaton grassland. Mesquite dominates the foreground slopes. (G. Roskruge BQ-7, courtesy of the Arizona Historical Society, Tucson.)

B. (November 20, 2003.) The camera station is lower owing to blading of the hilltop. The riparian vegetation growing along the San Pedro has greatly increased in size, density, and species composition. Cottonwood is the dominant tree, with interspersed black (Goodding) willow, ash, fourwing saltbush, and tamarisk. The foreground vegetation has also increased, including mesquite (at least one of which still persists from 1890), catclaw, greythorn, crucifixion thorn, Lehmann's lovegrass (a nonnative), and tumbleweed. The road in the foreground leads into a recent housing development. (R. H. Webb, Stake 3734.)

Figure 19.9 Photographs of the San Pedro River at Benson, Arizona.

A. (August 27, 1936.) This downstream view from the main, two-lane bridge of Arizona Highway 86 east of Benson shows the San Pedro River in an intermittent reach. Small mesquite bushes are on top of the bank, and what appear to be xerophytic shrubs—possibly creosote bush and other common species of the Sonoran Desert—are on the point bar at right. (McDonald ARIZ-749, courtesy of the Natural Resources Conservation Service.)

B. (November 28, 2000.) Four bridges now cross the San Pedro River at this point: the old highway bridge, which appears to be intact; two bridges for eastbound and westbound lanes of Interstate 10 *(midground);* and a railroad bridge behind the camera station. The channel bend remains, but the channel appears to be wider now, although that change is difficult to determine accurately owing to the extreme growth of riparian vegetation in the foreground and midground. Tamarisk is now abundant, but cottonwood, mesquite, greythorn, and fourwing saltbush are also present in large numbers. (D. Oldershaw, Stake 2218.)

A. (December 1963.) This downstream view shows the San Pedro River at the upstream end of an effluent reach upstream from Cascabel. Mesquite and catclaw appear on the high terrace on the right side, and small tamarisk trees are on the low terrace on the left. A solitary mesquite tree protrudes from the vegetation at left center. (R. C. Zimmerman.)

B. (February 29, 1984.) According to R. C. Zimmerman's notes, the banks of the San Pedro River retreated 10 to 15 feet in the period between 1963 and 1984. The mesquite and catclaw on top of the terrace either have died or have died back considerably. The 1983 flood, which was relatively large at this site, removed the low terrace that supported tamarisk in 1963, but more tamarisk appears on river right in the distance. The solitary mesquite at left center has a significant amount of mistletoe in its crown. (R. C. Zimmerman.)

C. (August 15, 2003.) Bank retreat continues on the outside of this bend, although the trees on top of the high terrace appear healthier in this midsummer view. Small black willow seedlings appear in the foreground, and tamarisk has increased considerably on both banks. The mesquite remains on the left side, and a black willow tree is just apparent on its downstream side. (D. Oldershaw, Stake 1367.)

Figure 19.10 Photographs of the San Pedro River south of Cascabel, Arizona.

Figure 19.11 Photographs of the San Pedro River north of Cascabel, Arizona.

A. (1963.) This view across and downstream on the San Pedro River near the mouth of Soza Canyon (chapter 20) shows a continuous line of cottonwood and sycamore trees that appears to be about one tree-width along the relatively flat and wide channel. A line of low shrubs is in front of the cottonwood and appears to be mostly seepwillow. A single tamarisk tree is located downstream on the far right. (R. C. Zimmerman.)

B. (September 3, 2003.) The channel has narrowed considerably, and the density of riverine riparian vegetation has increased considerably. The cottonwood-willow forest has increased in height and density, and the forest is now many trees wide. Seepwillow and mesquite are also present along the channel, and tamarisk is more abundant in this view than in 1963. (D. Oldershaw, Stake 4690.)

water is used to irrigate alfalfa fields along the river. Clearing of native vegetation for agriculture, diversion of flows, and installation of bank protection to minimize flood damages to agricultural fields have caused considerable changes in this reach. The 1926 flood caused channel downcutting in the reach between Cascabel and Redington.[69] The 1983 flood had its highest peak discharge in this reach and caused additional channel changes. After 1935, lateral channel migration commonly occurred during floods in the reach downstream from the gaging station near Redington.[70]

Redington to the Confluence with the Gila River

At Redington, the San Pedro River occupies a wide channel–floodplain system, and historically much of the floodplain has been cleared for agriculture. The river, channelized by arroyo downcutting, has bank protection installed locally to protect agricultural lands against lateral channel migration. Downstream from Redington, the river skirts the eastern foothills of the Santa Catalina Mountains, and tributaries contribute groundwater to the basin. The Bingham Cienega supports a dense stand of Arizona ash, a species not documented to be present here in forests during the nineteenth century.[71] The channel returns to ephemeral status upstream from Mammoth, where a once wide, barren reach remains in that condition (fig. 19.13).

Between 1877 and 1911, the channel near the confluence with the Gila River shifted up to 1,000 feet.[72] Channel avulsions undoubtedly have killed riparian vegetation within the cottonwood gallery forest that has historically dominated this reach, but the riparian area here is as much as a mile wide. Our repeat photography shows that riparian vegetation, primarily native species, increased considerably in this reach during the last third of the twentieth century (fig. 19.14).

Much of the cottonwood gallery forest along the San Pedro River just upstream from its confluence with the Gila River is part of a multiagency management plan. To mitigate the flooding of black willow habitat upstream from Theodore Roosevelt Lake on the Salt River, owing to an increase in the height of Roosevelt Dam (chapter 24), the Bureau of Reclamation joined with The Nature Conservancy to purchase large amounts of land in this reach in 1994.[73]

A

B

Figure 19.12 Photographs of the San Pedro River near Redington, Arizona.

A. (August 14, 1946.) This upstream view on the San Pedro River at the gaging station near Redington shows flow in a reach that normally is ephemeral. Perennial water and dense riparian vegetation occur only about one mile upstream from this point, showing how quickly conditions can change along this river. Mesquite, the only riparian species present, is on top of the channel banks at right. Several other photographs that approximately match this view and show similar channel conditions were taken in 1958 and 1962. (Unknown photographer 456A, courtesy of the U.S. Geological Survey.)

B. (November 8, 2002.) Both channel conditions and the amount of riparian vegetation appear to be similar despite the passage of fifty-six years. Minor channel changes are apparent, including deposition of a cobble floodplain at left and bank erosion on the right, but riparian vegetation remains confined to mesquite, burrobrush, and xerophytic shrubs. (D. Oldershaw, Stake 4356.)

Figure 19.13 Photographs of the San Pedro River at Mammoth, Arizona.

A. (August 14, 1940.) This downstream view from the old Arizona Highway 77 Bridge at Mammoth shows the San Pedro River in flood. The vegetation on either side of the wide channel appears to be mesquite, with no other common riparian species present. (A. A. Fischback 3100, courtesy of the U.S. Geological Survey.)

B. (November 24, 2000.) The old bridge is gone and has been replaced by a new concrete bridge downstream, and as a result the camera station is too low. Tamarisk is now relatively common at this site, and mesquite still lines the channel banks. (D. Oldershaw, Stake 2895.)

A

B

Figure 19.14 Photographs of the San Pedro River at Winkelman, Arizona.

A. (February 20, 1968.) This downstream view on the San Pedro River is taken from near the junction of the Gila River Road and Arizona Highway 77 south of Winkelman. The photo documents channel conditions at the former gaging station just after a relatively minor flood. Tamarisk lines both channel banks, with mesquite on the floodplain behind them and a single cottonwood tree just behind the gaging station. However, a gallery forest of cottonwood is clearly visible in the distance along the San Pedro River just upstream of its confluence with the Gila River. (Unknown photographer, courtesy of the U.S. Geological Survey.)

B. (October 26, 2001.) The channel has shifted to the west, leading to both the abandonment of the gaging station and the growth of riparian vegetation. The increase in woody riparian vegetation is mostly mesquite and cottonwood, but a considerable amount of tamarisk also is present. The increased woody vegetation obscures not only the channel, but also most of the background hills. (D. Oldershaw, Stake 3017.)

Summary. Numerous small tributaries enter the San Pedro River between the U.S.–Mexico border and the Gila River. Riparian vegetation—in particular cottonwood, sycamore, and Arizona ash—has been increasing in reaches with permanent or intermittent water, and tamarisk does not have a substantial presence in these tributaries. Aravaipa Canyon, in particular, is highly valued for its riverine riparian habitat, and matches with some of the oldest photographs of the state of Arizona show that vegetation along this watercourse has increased despite floods, in particular the 1983 event. Other tributaries, such as Ash Creek and Soza Canyon, contain significant amounts of woody riparian vegetation, which increased after 1963. These changes occurred despite continued livestock grazing in Ash Creek, Soza Canyon, and the eastern and western ends of Aravaipa Canyon.

The San Pedro River has numerous tributaries arising from the mountainous terrain on its east and west sides (fig. 19.1). Four tributaries in its middle and lower reaches—Ash Creek (fig. 20.1), Hot Springs Canyon, Soza Canyon, and especially Aravaipa Creek—are noteworthy for their riparian vegetation.[1] The smaller tributaries arise from the Rincon Mountains to the west (Ash Creek, Soza Canyon) and the Galiuro Mountains to the east (Hot Springs Canyon). Aravaipa Creek flows through Aravaipa Canyon, which breaches the northern end of the Galiuro Mountains, and the headwaters of this watercourse are in the northern part of the Sulfur Springs Valley, the Santa Teresa Mountains, and the western side of the Pinaleño Mountains of south-central Arizona. None of these tributaries is regulated, although groundwater is used extensively in the headwaters of Aravaipa Canyon.

Ash Creek

Ash Creek is a western tributary of the San Pedro River that enters the middle reach of the river (fig. 19.1). Its headwaters are on the east side of the Rincon Mountains of Saguaro National Park. With a drainage area of about 52 square miles at its mouth, it has several reaches with perennial flow, and groundwater levels are high enough along most of its upper part to sustain a thriving riparian ecosystem.[2] There is no gaging record available for this stream. Ranching with low-yield wells for cattle tanks are the primary land uses in the middle part of this watershed, and the channel flows through agricultural fields at its confluence with the San Pedro River.

Four sets of repeat photographs document change in riparian vegetation in Ash Creek in the final third of the twentieth century. The upper part of the creek is lined with sycamore and Arizona ash trees, and mesquite occurs along the margins of the channel. Livestock were observed in this reach at the time of our photography. Downstream from the Ash Creek Ranch (fig. 20.2), the ephemeral stream becomes perennial within a canyon confined by schist rocks in the Little Rincon Mountains. In this reach, Arizona ash dominates a mixture of riparian species, including cottonwood, sycamore, netleaf hackberry, and seepwillow. Yewleaf willow, a largely tropical species that reaches its northerly distribution in southeastern Arizona, has increased in this canyon as well. No tamarisk was found in this reach, and livestock were present. As indicated in figure 20.1, riparian vegetation has increased along Ash Creek since the early 1960s, although large Arizona walnut trees are known to have died in the lower reaches of this creek.[3]

Tres Alamos Wash

Tres Alamos Wash enters the San Pedro River from the east, draining 134.8 square miles of the southern Galiuro Mountains (fig. 19.1). No gaging station records flow in this drainage basin, which has both perennial and ephemeral reaches that support varying amounts of woody riparian vegetation.[4] In approximately the geographic center of the watershed, the channel is confined within a shallow bedrock canyon through the Johnny Lyon Hills, and this reach has the greatest assortment of native riparian species.

Two sets of repeat photographs document changes in two ephemeral reaches of Tres Alamos Wash, where livestock grazing continues. In an upstream reach, native riparian species have increased along the channel (fig. 20.3). Although tamarisk may be present at discrete points along the channel, it is not continuously distributed along the channel within the watershed. The channel of Tres Alamos Wash near its confluence with the San Pedro River has narrowed as mesquite and more xerophytic species, such as catclaw and creosote bush, have become established along the channel margins.

Soza Canyon

Soza Canyon and Soza Wash meet the San Pedro River north of Cascabel, Arizona. Soza Canyon, which enters from the west, supports stands of native riparian vegetation, with no tamarisk present near the mouth. This canyon, with a drainage area of 46 square miles, occasionally produces large floods that affect riparian vegetation. In particular, the 1983 flood (chapter 19) was extremely large, stripping riparian vegetation from the floodplain

Figure 20.1 Photographs of Ash Creek.

A. (March 5, 1984.) This upstream view on Ash Creek shows seedlings of Arizona ash and Frémont cottonwood that became established after the floods of the late 1970s and early 1980s. The overhanging branches above the camera station are from an Arizona walnut established along the channel behind the camera station. Mesquite appears in relatively dense stands above the channel banks. (R. C. Zimmerman.)

B. (March 30, 2004.) Frémont cottonwood and Arizona walnut have increased within the view, obscuring most of the background. Many common riparian species—all native—are in the view or nearby, including Arizona ash, wild grape, netleaf hackberry, sycamore, live oak, seepwillow, mesquite, and catclaw. (D. Oldershaw, Stake 4476.)

A. (March 1965.) This upstream view shows an alluvial section of Ash Creek between the Rincon Mountains *(distance)* and a bedrock canyon just downstream. Significant base flow is in the channel at this time. Arizona ash is the dominant riparian tree, but there is also sycamore, cottonwood, netleaf hackberry, and black willow; dense mesquite occurs behind the floodplains. No tamarisk is present at this site. (R. C. Zimmerman.)

B. (February 28, 1984.) This match is forward and to the left of the original camera station. Careful location of common features in the 1965 and 1984 views indicates that the channel has aggraded slightly in the intervening nineteen years. Especially apparent in the center of the matched photograph, sycamore has greatly increased, blocking the view of the channel. (R. C. Zimmerman.)

C. (September 3, 2003.) This view, which matches photograph B, shows a large increase in riparian vegetation. As can be seen from blockage of the foothills of the Rincon Mountains, the height of the floodplain trees—mostly sycamore and Arizona ash—has greatly increased. The increase in size of mesquite on the slopes at lower left partially obscures the low-water channel. Tamarisk still is not present at this site. (D. Oldershaw, Stake 1375.)

Figure 20.2 Photographs of Ash Creek.

Figure 20.3 Photographs of Tres Alamos Wash.

A. (February 29, 1984.) This view of Tres Alamos Wash shows conditions in an ephemeral reach on the Three Links Ranch. The channel is wide and sandy, and the vegetation on the far bank includes Arizona ash *(left center)*, mesquite *(left)*, and netleaf hackberry *(center)*. (R. C. Zimmerman.)

B. (March 30, 2004.) Vegetation along the wash has increased in density, and the plants visible in 1984 have increased in size. Burrobrush is now common, including the plants in the right foreground. Seepwillow is now present in the reach, which suggests that water levels in the shallow alluvial aquifer may have risen higher than they were in 1984. (D. Oldershaw, Stake 4478.)

(fig. 20.4). Livestock grazing occurs in part of this reach, although recent fencing and property acquisition by The Nature Conservancy may decrease grazing intensity or eliminate it in the future.

Eight sets of repeat photographs show change in riparian vegetation at the mouth of Soza Canyon. Despite occasional setbacks, riparian vegetation near the mouth has increased significantly since the first photographs were taken in 1964. Frémont cottonwood, which was not present in 1964,[5] is now common at the mouth of the canyon and a short distance upcanyon to the head of permanent water. Black willow that became established in the 1980s and 1990s has recently died back, perhaps because of the ongoing drought beginning in 2003.

Hot Springs Canyon

Hot Springs Canyon enters the San Pedro River from its headwaters to the east in the southern Galiuro Mountains. With a drainage area of 114 square miles, Hot Springs Canyon is the largest of the three middle San Pedro River tributaries with photographic records, and four sets of repeat photographs document changes here. Grazing once was heavy in this watershed and focused on the riparian ecosystem. Current management of the lower parts of this watershed severely restrict grazing,[6] and The Nature Conservancy, which manages Hooker's Hot Springs in the headwaters, also restricts livestock grazing.

The 1983 flood in Hot Springs Canyon had a peak discharge of 20,000 ft^3/s,[7] and this flood apparently had substantial effects on riparian vegetation (fig. 20.5). Despite the occasional occurrence of large floods, which presumably would scour plants, woody riparian vegetation thrives in this canyon. The 1983 and 1993 flows in this canyon encouraged the germination and establishment of native species, in particular cottonwood and mesquite. Observations suggest that seepwillow expanded downstream about a half-mile between 1965 and 1984, and Arizona ash has remained about the same.[8] Tamarisk is not present in this reach, and the increases in riparian vegetation between 1964 and 2003 are wholly in native riparian species.

Aravaipa Canyon

Aravaipa Creek drains 537 square miles of agricultural lands and rangelands in Aravaipa Valley and the northern Galiuro Mountains east of the San Pedro River. This is the only tributary of the lower San Pedro that penetrates the mountain ranges that bound the east and west sides of the valley. The small town of Klondyke is the only significant settlement in the basin, but the history of this drainage basin is a long one with many observations of riparian vegetation.[9]

Aravaipa Canyon is considered one of the premier riparian habitats in Arizona, so much so that The Nature Conservancy purchased the property in 1971 and manages access to the canyon in coordination with the BLM.[10] In addition to a thriving woody riparian vegetation, seven species of native fish occur in the canyon, a diversity higher than in any other watercourse in the state. Flood records for Aravaipa Creek downstream from Aravaipa Canyon (fig. 20.6) show one large flood, in 1983, and an otherwise complacent annual flood series in comparison with other rivers in the region.

Early Observations and Land Uses

Despite the probable visits of Spaniards in the seventeenth and eighteenth centuries, the first reliable descriptions of Aravaipa Canyon come from the middle of the nineteenth century. At the western end of the canyon, flow has been intermittent downstream from the mouth of the canyon from the mid-1800s to the present day.[11] William Alexander Bell, an English physician who accompanied railroad surveys in the late 1860s, visited Aravaipa Canyon with his camera in 1867 and took some of the first photographs within the state of Arizona.[12] He described "luxuriant vegetation" within the canyon, an observation verified in his photography.

Several U.S. Army camps, culminating with Camp Grant, were established near the confluence of Aravaipa Creek and the San Pedro River. The purpose of Camp Grant was to oversee a reservation for the Aravaipa Apache between 1871 and 1873. In 1870, both watercourses were dry sand channels, although cottonwood and mesquite grew around the camp.[13] The water supply for this camp was a well, which at 90 feet was very deep for this time, and malaria was a severe problem (chapter 19). Following removal of the Apache to the San Carlos Apache Indian Reservation in 1873, the Aravaipa Creek watershed attracted American settlers, who came to it for both its water supply, which allowed irrigated agriculture, and its mineral resources.[14] Five short-lived mining camps were established in the tributaries of Aravaipa Creek.

Settlement took its toll on riparian vegetation. The combination of surface-water diversion and groundwater withdrawal resulted in the elimination of perennial waters, including ciénegas, upstream from Klondyke and at the upstream end of Aravaipa Canyon.[15] Grazing was ubiquitous through the canyon, and agriculture, with attendant clearing of riparian vegetation, was common in the wider reaches. During drought periods and particularly during the mid-twentieth-century drought, many large cottonwood were felled both to increase available surface water and to provide food for cattle. Woodcutters removed mesquite and walnut, in particular old trees, from the eastern end of Aravaipa Canon in the 1920s and 1930s.[16]

Aravaipa Canyon currently is managed for wilderness and environmental preservation. The central part of the canyon is the Aravaipa Wilderness Area, a 35,000-acre tract managed by the BLM. The Nature Conservancy owns a combined 7,000 acres on both the east and west ends of the canyon, and the BLM joins with The Nature Conservancy to limit hiker access to the canyon.[17] Before The Nature Conservancy obtained the property, goats heavily browsed some areas on the eastern end of Aravaipa Canyon, creating a "park-like regime of big trees and bermuda grass with very little underbrush."[18] After the goats were removed, shrub growth increased in the riparian zones. The alluvial reaches of Aravaipa Creek, upstream and downstream from Aravaipa Canyon, continue to sustain livestock.

Tributaries entering Aravaipa Creek within Aravaipa Canyon have significant amounts of riparian vegetation in their own right. Mesquite bosques are common in many of these tributar-

A. (March 1965.) This view across the San Pedro River looks into the mouth of Soza Canyon. A large sandbar deposited during a San Pedro River flood has filled the channel of this side canyon, and flow here is ephemeral at this time. Other than mesquite on the hillside, well above either channel, only scattered riparian trees appear in Soza Canyon. These trees likely are black willow. (R. C. Zimmerman.)

B. (April 19, 1984.) The channel of Soza Canyon has reestablished, and sediment appears to have moved the low-water channel of the San Pedro River behind the camera station. This sediment movement is undoubtedly the result of the 1983 flood, which had a substantial (but unknown) discharge from this tributary. R. C. Zimmerman found some startling changes in the riparian vegetation here during a 1984 visit. No Frémont cottonwood were present in 1965; cottonwood dominates in 1984. (R. M. Turner.)

C. (September 3, 2003.) The large increase in cottonwood and black willow is very apparent. Tamarisk is scarce at this site but is present upstream and downstream. Owing to drought conditions in 2003 and possibly increased water use by riparian vegetation, the channel of Soza Canyon is dry upstream from its mouth, in contrast to its usually perennial flow (Zimmerman 1969). (D. Oldershaw, Stake 1112.)

Figure 20.4 Photographs of Soza Canyon.

A. (February 9, 1965.) This upstream view of Hot Springs Canyon shows a perennial reach that is partially confined by bedrock. Although dormant in the winter season, Frémont cottonwood and mesquite are readily identifiable in this view. (R. C. Zimmerman.)

B. (March 6, 1984.) This view, taken after the devastating floods of October 1983, shows a relatively wide channel with a much coarser bed than in 1965. The cottonwood previously present at left center is dead, and the trunk remains; a damaged mesquite tree appears at right center. (R. C. Zimmerman.)

C. (October 17, 2003.) The increase in riparian vegetation is notable, especially because no nonnative woody species are present. Many cottonwood seedlings are apparent in this view. The persistent cottonwood trees have grown considerably, as have woody shrubs, and sycamore is now present in this reach. The increase in low vegetation along the channel is deceptive because this match was taken after the summer growing season, whereas the other two photographs were taken in winter. Cattle have been excluded from this reach for six years already. (D. Oldershaw, Stake 1371.)

Figure 20.5   Photographs of Hot Springs Canyon.

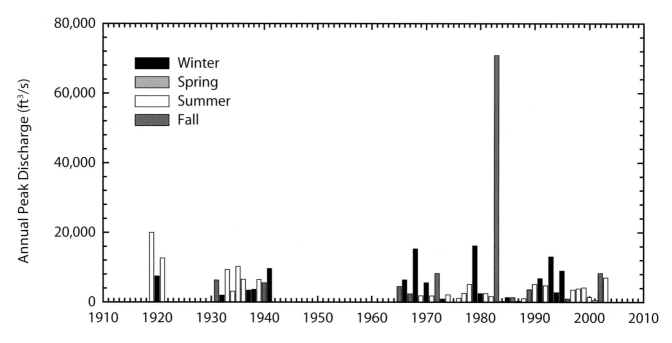

Figure 20.6 Annual flood series for Aravaipa Creek near Mammoth, Arizona (station 09473000; 1919–1921, 1931–1941, 1965–2003).

ies, as are many other riparian species, including Arizona walnut, sycamore, soapberry, netleaf hackberry, and Arizona ash. Although tamarisk has never been common in Aravaipa Canyon, eradication of it is continual on property owned by The Nature Conservancy, and their managers cut down any tamarisk they find and paint the stumps with an herbicide.

Floods

The gaging record for Aravaipa Creek, measured at the western end of the canyon, shows the discontinuous record of floods in this unregulated watershed (fig. 20.6). The average daily discharge is 37.1 ft³/s, and flow is perennial at the gaging station. Because of the canyon's long settlement history, the anecdotal history of flooding in it is extensive and much longer than the gaging record. The 1915 flood caused severe bank erosion in the agricultural lands at the west end of the canyon; another flood in 1923 had a similar effect.[19] Other damaging floods occurred in 1960 and 1977.

The 1983 flood, discussed more fully in chapter 21, had a large impact on Aravaipa Canyon and private property upstream and downstream. Although the 1983 flood appears to be far and away the largest flood that has

occurred in this watershed (fig. 20.6), that appearance is deceptive. One study, which used different high-water marks in a reach upstream from the gaging station, suggests that the peak discharge of the 1983 flood, estimated by the U.S. Geological Survey to be 70,800 ft³/s, actually was about 26,500 ft³/s.[20] In either case, the 1983 flood is the largest flood in the postsettlement history of Aravaipa Creek.

Changes in Riparian Vegetation

The photographic record of Aravaipa Creek begins at the upstream end of Aravaipa Canyon, and changes in ciénegas and other riparian habitat farther upstream in the alluvial reaches in Aravaipa Valley could not be evaluated. Upstream from the bedrock-confined channel, the once grassy channel and floodplain is now lined with mesquite and catclaw.[21] Within Aravaipa Canyon, woody riparian vegetation was dense and extensive in all thirty-one of the original photographs replicated here. Replication of the 1867 William Bell photographs shows that riparian vegetation has increased considerably in the past 135 years, notably Frémont cottonwood, Arizona ash, and sycamore (figs. 20.7 and 20.8). Despite intensive eradication efforts, tamarisk still occurs at four or five locations

evenly spaced through the canyon. These plants are likely to be removed when they are discovered by the managers of Aravaipa Canyon.

The middle of Aravaipa Canyon alternates between a narrow, bedrock-controlled channel with little room for riverine riparian vegetation and slightly wider reaches with significant growth of native riparian species. About 5 miles upstream from the western end of the canyon, the river valley widens, and agricultural fields and orchards are planted near the river. Riparian vegetation in this reach has increased significantly the damaging effects of floods, in particular the 1983 event (fig. 20.9). Near the former site of the gaging station, the amount of riparian vegetation has increased despite the death of large cottonwood trees once used for cable supports (fig. 20.10).

In the intermittent reach of Aravaipa Creek at the Arizona Highway 77 crossing, the presence of young trees well illustrates the positive effects of the 1983 and 1993 floods on germination and establishment of native species (fig. 20.11). However, the recent death of young black willow trees demonstrates that this reach is not able to sustain growth of obligate riparian trees unless newly established individuals have significant time for roots to reach reliable groundwater.

A

B

Figure 20.7 Photographs of Aravaipa Creek in the eastern end of Aravaipa Canyon.

A. (1867.) This downstream view, one of the earliest photographs taken in Arizona, shows a thriving riparian ecosystem along Aravaipa Creek at the upstream end of Aravaipa Canyon. The creek is just visible in the center of the view. The large trees are Frémont cottonwood; mesquite is present in the foreground and may comprise the low band of trees on the right side of the channel in the midground. (W. A. Bell F7313, courtesy of the Colorado Historical Society.)

B. (November 6, 2003.) The growth of cottonwood is enormous, completely obscuring the view of the channel. This reach also contains Arizona ash, netleaf hackberry, sycamore, and Arizona walnut. Tamarisk, once present in this reach, has been mostly eradicated. (D. Oldershaw, Stake 3722a.)

A

B

Figure 20.8 Photographs of Aravaipa Creek in the western end of Aravaipa Canyon.

A. (1867.) This downstream view shows a different angle on the presettlement riparian vegetation of Aravaipa Canyon. The creek is just visible at lower center, and a thriving forest of cottonwood and sycamore lines the canyon. (W. A. Bell F40589, courtesy of the Colorado Historical Society.)

B. (October 24, 2002.) All of the trees have increased in size along Aravaipa Creek, which is no longer visible from this camera station. Mesquite has grown up in the foreground, and Arizona ash, which was probably present in 1867, is now prominent in the middle foreground. One-seed junipers have increased on the far slope. (R. M. Turner, Stake 3223.)

Figure 20.9 Photographs of Aravaipa Creek in the western end of Aravaipa Canyon.

A. (1920.) This upstream view on Aravaipa Creek is just within the western boundary of The Nature Conservancy property in Aravaipa Canyon. The view shows scattered, extremely large cottonwood trees, with lower trees in continuous bands between them. These lower trees probably are black willow. (H. V. Brown, courtesy of D. E. Brown.)

B. (November 8, 2003.) The increase in riparian vegetation is extremely striking here. Besides the cottonwood *(right)* and black willow *(left)*, the riparian tree species include sycamore and netleaf hackberry. Although dwarfed by the trees, seepwillow and coyote willow are also common. (R. M. Turner, Stake 3236.)

Figure 20.10 Photographs of Aravaipa Creek in the western end of Aravaipa Canyon.

A. (May 22, 1931.) This upstream view shows Aravaipa Creek at the downstream end of Aravaipa Canyon. A worker is pouring cement to create a control for the gaging station. Frémont cottonwood and seepwillow are readily apparent on the low floodplain; a greythorn branch partially blocks the lower right. The cottonwood at left center is massive and was used to anchor the cableway for the gaging station. A large sycamore is vaguely apparent through the cottonwood branches at upper left. (W. S. Eisenlohr, courtesy of the U.S. Geological Survey.)

B. (October 23, 2002.) Riparian vegetation in the view has increased substantially despite the occurrence of extremely large floods, such as the 1983 event. The larger trees are cottonwood and sycamore; the large cottonwood tree in photograph A is dead. Black willow composes most of the band of trees nearest the creek, and netleaf hackberry is uncommon in the area. The channel has a substantial cover of coyote willow, seepwillow, and burrobrush. No tamarisk is present in this reach. The greythorn persists at lower right. (D. Oldershaw, Stake 4332.)

Figure 20.11 Photographs of Aravaipa Creek at the Arizona Route 77 crossing.

A. (January 8, 1941.) This upstream view on Aravaipa Creek at the Arizona Route 77 Bridge shows a low cement weir built as a control for low flows at the gaging station *(midground)*. The channel is wide and flat, with significant cobbles on the surface. Frémont cottonwood, mesquite, and catclaw acacia are apparent on the right bank *(left midground)*. (Unknown photographer 3227, courtesy of the U.S. Geological Survey.)

B. (October 23, 2002.) Woody riparian vegetation, mostly native species, has greatly increased in this reach. Young cottonwood trees, with a few sycamore, appear along the left side of the channel *(right side of the view),* and the mesquite trees that may have been in the original photograph appear in the left midground. Young black willow trees became established here but have died recently, possibly as a result of ongoing drought. Seepwillow is present in the immediate foreground, and a line of it extends off into the left center of the view; burrobrush is also common here. Tamarisk is relatively common in this reach of Aravaipa Creek, and tree-of-heaven is present as well. (D. Oldershaw, Stake 4338a.)

Summary. The Santa Cruz River is one of the few watercourses with headwaters in the United States that flows south into Mexico, only to reenter the United States to flow north to the Gila River. This diverse river supports lush riparian ecosystems in its upper reaches, then transitions to an ephemeral reach, and finally ceases any resemblance to a natural river as a result of groundwater overdraft, channelization, sewage effluent discharge, and irrigation-return flows. Riparian vegetation has increased in most of the reaches upstream from Tucson that have perennial flow, which originates from either base flow or sewage effluent. The Santa Cruz River at Tucson is one of the few well-documented cases of complete destruction of a riparian ecosystem in the Southwest owing to the combination of arroyo downcutting, water development, and channelization. Following the 1993 flood, the river deposited a low floodplain within the channelized reaches in Tucson, and sewage-effluent discharges north of the city sustain a newly established riparian ecosystem.

Historically and presently, the Santa Cruz River is an extremely important watershed in southern Arizona, in terms of both historical settlement and economic development (fig. 21.1). With a basin size of 8,581 square miles, this river drains the largest area south of the Gila River. Around 1700, Spaniards joined the resident Tohono O'odham, once called the Papago, along the river, making the Santa Cruz River basin the site of the first European colonization in Arizona. The descendents of Spaniards became Mexican citizens, who then were joined by Americans following the Gadsden Purchase in 1854.[1] As settlers continued to arrive in the late nineteenth century, they became increasingly dependent on water from the Santa Cruz River for

irrigation.[2] As a result, a long history of change in channel morphology and riparian vegetation is available for this river.[3]

The drainage basin of the Santa Cruz River at Tucson has similarities to that of the San Pedro River near Charleston (table 21.1), especially in terms of basin precipitation and elevation. In terms of runoff history and water-development history, the two rivers cannot be more different. The two drainage basins are adjacent and share headwaters in southern Arizona and northern Mexico (fig. 21.1). Like the San Pedro, the Santa Cruz was a discontinuous ephemeral stream in the 1800s with effluent-influent reaches that supported dense woody vegetation. With the exception of periods of flooding, there is no evidence that the Santa Cruz River had continuous flow from its headwaters to its terminus at the Gila River. Instead, local reaches of perennial flow punctuated an otherwise ephemeral stream.[4]

The story of change in riparian vegetation along the Santa Cruz and San Pedro Rivers reflects the differences in groundwater discharge as well as in water-development history. Although obligate riparian vegetation has generally increased along the San Pedro River, whether in the perennial or ephemeral reaches, riparian vegetation locally has been eliminated from many reaches along the Santa Cruz River. As such, the Santa Cruz River represents the type example of loss of obligate riparian vegetation in the Southwest. That loss is not evenly distributed along this river but is localized in the Tucson basin. Even so, this incredibly durable type of vegetation is making a comeback of sorts because of the sewage effluent discharged from the city of Tucson.

Early Observations of the Santa Cruz River

The history of the Santa Cruz River has been extensively analyzed for changes in hydrology related to past and present civilizations,[5] and only the salient points related to changes in riparian vegetation are given here. Spanish settlement in the basin began in 1700, and the presence of Spanish officials precluded American fur trappers, who sought to avoid authority even if they held valid trapping permits. Although beaver undoubtedly were in several headwater tributaries, the presence of long dry reaches likely precluded aquatic mammals such as beaver from much of the length of this river.[6] Statements such as "beaver, muskrat, and waterfowl were common"[7] are likely more fanciful than fact.

The Santa Cruz River once had a complicated series of perennial and ephemeral reaches that requires careful interpretation to understand changes.[8] Its reputation in the nineteenth century was that "at times huge volumes of water march majestically its whole extent, sweeping and destroying whatever is in its track."[9] Most observers agree that historically, as at present, flow in the Santa Cruz River was perennial from headwaters to north of Nogales, Arizona, except during extreme drought conditions.[10] Gaged flows at Lochiel and Nogales are intermittent, with annual daily discharges of 3.9 and 27.9 ft³/s, respectively, and periods with a dry channel are common.

North of Nogales, water levels were high in the nineteenth century, and those bestowed with land grants from the Spanish government brought in large numbers of livestock (e.g., 5,000 sheep were reported at Tubac in an 1804 census).[11] Father Eusebio Kino, who first visited Tumacácori in 1691,

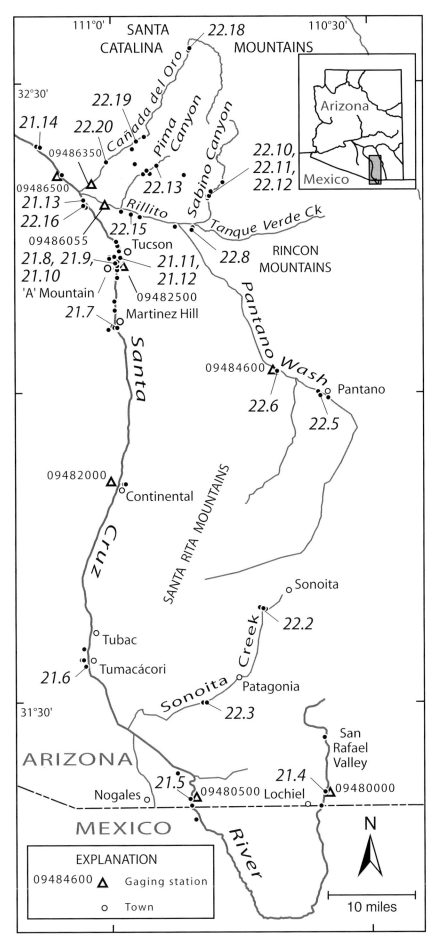

Figure 21.1 Map of the Santa Cruz River and tributaries, southern Arizona and northern Mexico.

reported that sheep and goats were present in 1695.[12] Reportedly, 3,500 cattle, 2,500 sheep, and 1,200 horses were at Tucson. In 1822, 4,000 head of cattle were transferred from the mission at Tumacácori to the San Pedro River watershed. These livestock used the riparian ecosystem well before arroyo downcutting commenced and before observers reported thriving riparian ecosystems along the river.

Observations consistently suggest that perennial flow did not occur in the reach from present-day Green Valley to Martinez Hill (fig. 21.1).[13] The gaging station at Continental was dry more than 80 percent of the time over a forty-six-year record.[14] South of Martinez Hill, a mesquite bosque of unknown size sustained more than seventy species of birds, some of which were rarely observed in Arizona.[15] At Martinez Hill, springs, including one named Punta de Agua, delivered effluent flow as a result of subsurface geologic structures that have been termed the "Santa Cruz Narrows." About 1,600 Tohono O'odham lived in this reach at the end of the seventeenth century, using this water for agriculture.[16] As many as five species of native fish occurred in the reach between Punta de Agua and Tucson.[17]

Influent conditions prevailed until the river reached the base of what is variously known as Sentinel Peak or "A" Mountain, where once again subsurface geologic structures created effluent conditions. In December 1864, "quite a respectable stream of water" ran in the Santa Cruz River at Tucson.[18] In the vicinity of Tucson, the river corridor "once was covered by sacaton grass with groves of mesquite and swampy areas of tule."[19] Downstream, particularly downstream from the north end of the Tucson Mountains, the Santa Cruz River was ephemeral to its confluence at the Gila River,[20] although in 1864 the channel was marked with a line of cottonwood between Tucson and Picacho Peak.[21]

Unlike the historical documentation of flow conditions, which were essential to the patterns of settlement in the nineteenth century, observations of the distribution of riparian vegetation are sparse. In the middle of the nineteenth century, explorers described dense mesquite forests

Table 21.1 Comparison of Drainage Characteristics Upstream from the Primary Gaging Stations on the Santa Cruz and San Pedro Rivers, Southern Arizona.

River	Gaging Station	Drainage Area (mi²)	Annual Runoff (acre-ft)	Mean Basin Elevation (ft.)	Mean Annual Precipitation (in.)
San Pedro River	Charleston	1,234	40,090	4,840	16.5
Santa Cruz River	Tucson	2,222	1,630	4,050	16.9

Sources: Pope, Rigas, and Smith 1998; U.S. Geological Survey data.

near Tubac and a narrow band of cottonwood near Calabazas.[22] A similar stand of mesquite was present in the vicinity of Tucson at the end of the nineteenth century.[23] In 1893, at present-day Lochiel, Edgar Mearns observed a "treeless plain," after which he followed the river into Mexico and through its 180-degree bend back into the United States. There, just upstream from the present-day gaging station of the Santa Cruz River near Nogales, he found "the river is heavily wooded with cottonwood, willow, walnut, very tall mesquites, and other trees."[24] The only information on the distribution of other species, in particular Frémont cottonwood, comes from historical photography.

Floods and Arroyo Downcutting

Like other watercourses in the Southwest, the Santa Cruz River periodically created and filled arroyos along its length during the Holocene. Near Martinez Hill, two springs known as Punta de Agua and Punta de Misión were present,[25] and three broad periods of channel change occurred during the Holocene.[26] From the end of the Pleistocene until 8,000 years ago, the channel and floodplain aggraded with sediment delivered by what has been characterized as a braided stream. From 8,000 to around 5,500 years ago, the channel underwent a general period of erosion, creating broad channels. After 5,500 years ago, the floodplain aggraded vertically, punctuated by short, prehistoric periods of arroyo cutting and filling. Periodic arroyo downcutting may have influenced Hohokam settlement of the floodplain in the vicinity of Martinez Hill.

Historical arroyo cutting on the Santa Cruz River has been well documented.[27] Headcuts were present in the vicinity of marshy reaches in 1871, but these channelized reaches were not connected. Beginning around 1878, the Santa Cruz River downcut into its floodplain in the vicinity of Tucson, and the headcut propagated upstream.[28] The downcutting and subsequent widening occurred during a series of floods, with most of the erosion taking place from 1878 through 1891. Nineteenth-century agricultural activities near Tucson were sustained by surface-water diversion, which ceased as a result of arroyo downcutting and the midcentury drought.[29] Floods in 1905 and 1916, which is the start of the gaging record at Tucson (fig. 21.2C), widened the channel and destroyed bridges by scouring piers and eroding abutments. Photographs taken during the period of arroyo entrenchment document recovery of riparian vegetation, with new cottonwood stands and mesquite bosques forming adjacent to the active channel.[30]

At some unknown time, but probably extending through the 1930s, the arroyo coalesced into a continuously incised channel from Tucson to the headwaters. The seasonality of flooding changed at about this time, with annual flood peaks changing from floods caused by fall and winter storms to summer floods, particularly at Tucson in the middle part of the twentieth century (fig. 21.2C).[31] This period coincides with the regional midcentury drought, which was more severe elsewhere than in Tucson. For example, in 1954 and 1955 during the driest part of the midcentury drought, the Santa Cruz River had more peaks above base discharge than at any other time in its gaging history.

Beginning with the 1977 event and extending through 1993 (fig. 21.2), a series of fall and winter floods widened the channel to its current extent.[32] The flood of October 9, 1977, on the Santa Cruz River, caused by Tropical Storm Heather, had severe effects in terms of inundation and channel change.[33] This flood scoured channel beds (by 4 to 6 feet at Tucson), eroded floodplains, and removed riparian trees, particularly in the reaches near Nogales and near Tumacácori. These erosional impacts led to the installation of soil cement to replace the natural banks in Tucson and to protect floodplain structures and development. "Soil cement" is created by adding less than 10 percent (and usually less than 5 percent) cement to channel materials to form a weakly cemented and steeply sloping bank. When used as part of an engineered, trapezoidal channel with grade-control structures, such channels transport flood waters without storing sediment. Secondary effects are that established riverine vegetation is destroyed and that the change in hydraulics retards reestablishment.

The largest flood of the gaging record of the Santa Cruz River at Tucson occurred in early October 1983, the result of the collision of moisture from dissipating Tropical Storm Octave and an unusually strong cutoff low-pressure system from the North Pacific.[34] This storm was related to the end of the strong 1982–1983 El Niño; the probability of tropical cyclones moving over the western United States is increased during El Niño periods. The peak discharge of 52,700 ft³/s at Tucson firmly entrenched the river-management policy of installing engineered channels. In contrast, the 1983 flood is only the third largest in the Nogales record (fig. 21.2B); the peak discharge of the 1977 flood dwarfs all other floods recorded near Nogales.

Another large flood occurred in January 1993, embedded in general flooding that affected the entire

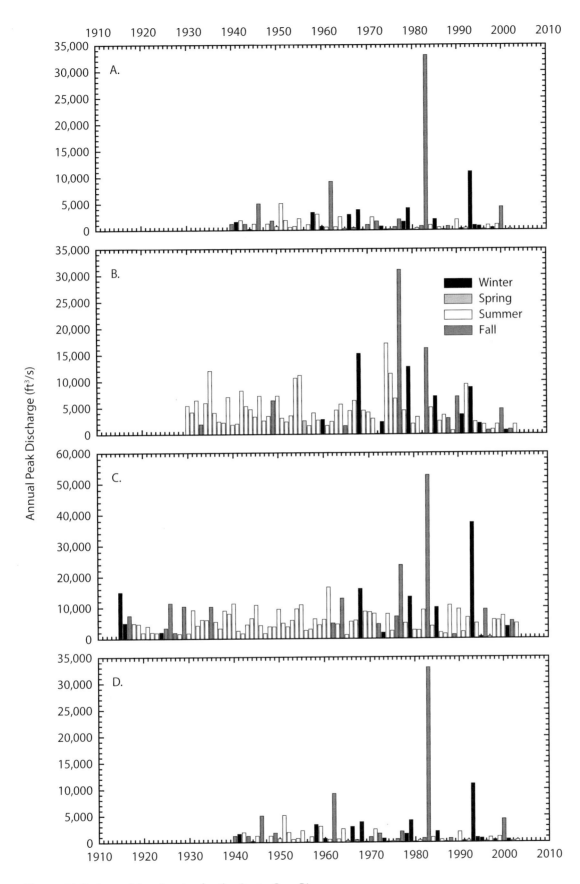

Figure 21.2 Annual flood series for the Santa Cruz River.

A. The Santa Cruz River near Lochiel, Arizona (station 09480000; 1940–2003).

B. The Santa Cruz River near Nogales, Arizona (station 09480500; 1930–2003).

C. The Santa Cruz River at Tucson, Arizona (station 09482500; 1915–2003).

D. The Santa Cruz River near Laveen, Arizona (station 09489000; 1940–2003).

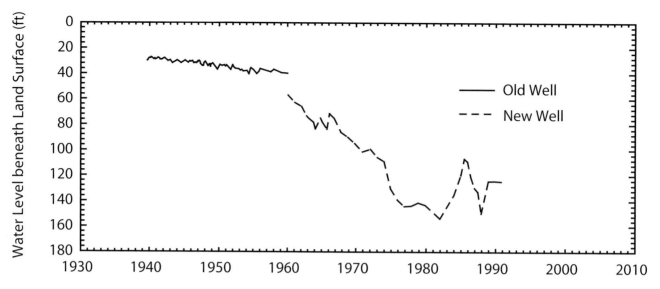

Figure 21.3 Groundwater levels for two wells, D-15-13 22DCC2 (old well) and D-15-13 23CCB2 (new well), along the Santa Cruz River north of Martinez Hill, Arizona.

state for the first time since 1891. El Niño conditions prevailed from 1991 through 1995, creating the longest sustained period of warm ENSO conditions in the twentieth century. Following a wet December 1992, three extremely large frontal systems, fueled with moisture from the tropical Pacific Ocean, passed through Arizona on around January 7, January 18, and February 20. The first and third of these storms caused severe flooding in central Arizona (chapters 23 and 24), and the first and second caused severe flooding on the upper Gila River (chapter 17). At Tucson, the flood peaked at 37,400 ft³/s on January 19, simultaneously being recorded as the second-largest flood in the gaging record (fig. 21.2C) and further justifying the installation of engineered channels and soil-cemented banks.

As shown in figure 21.2, the history of twentieth-century flooding reveals a change in seasonality of flooding and suggests that flood magnitude is nonstationary in time.[35] The San Pedro River also had seasonality change, but not the increased flood frequency (fig. 19.2), and the upper Gila River had the fluctuation in flood magnitude but not the seasonality change (fig. 2.2A). The broader flood record of the Santa Cruz River is similar to most other rivers in the region: floods in the late nineteenth century, starting around 1880 and persisting through 1891, followed

by a severe drought through 1904, then resurgence of flooding through about 1930, followed by the midcentury drought and summer-dominated small floods. Unusually large floods occurred again from the late-1970s through the mid-1990s, causing a new episode of channel downcutting and widening. That period may have ceased with the 1993 flood considering that the following decade represents a return to smaller annual floods, albeit retaining the seasonality change that began in 1960 (fig. 21.2C).

Groundwater Development

The early residents of Tucson fought a fruitless battle to divert surface water from the Santa Cruz River for the purpose of irrigation, power, and domestic supplies.[36] As surface-water supplies dwindled during the early-twentieth-century drought, they turned to development of infiltration galleries and wells that penetrated the alluvial aquifer along the river.[37] Creation of infiltration galleries required installation of more than a mile of horizontal-lying perforated pipes greater than 20 feet deep in the still-high alluvial aquifer near Martinez Hill. The pipes gathered groundwater, which then entered a distribution network that stretched northward to Tucson.

Use of surface water was abandoned by the 1920s, and water levels

had dropped beneath the infiltration galleries by this time as well. This technology was used, however, until the late 1940s on the Tohono O'odham Indian Reservation near Martinez Hill.[38] Well drilling accelerated as both the population of Tucson and the acreage of agriculture increased. As groundwater extraction increased, water levels dropped to the point where centrifugal pumps could no longer pump water to the surface.[39] Development of the deep turbine pump renewed use of groundwater, giving a shot in the arm to Tucson's agricultural economy and a long-term prognosis for growth.

Groundwater development accelerated into the 1950s, leading to lowering of the water table by several hundred feet in the center of the basin[40] and about 140 feet near the Santa Cruz River by the 1970s (fig. 21.3). Groundwater extraction in the Santa Cruz River basin peaked in 1975.[41] The 1983 flood temporarily reversed the downward trend in water levels as the aquifer immediately underneath the river was recharged, as depicted by a 40-foot rise in water level following that flood. This rise was merely a groundwater "mound" beneath the river that leveled out as water flowed into the broader alluvial aquifer. Water-level measurements ceased in the early 1990s, but the 1980s water-level increases were not sufficient to sustain growth of new riparian vegetation. Dewatering of the alluvial aquifer has caused extirpations of some obligate species, notably the checkered garter snake (*Thamnophis marcianus*), which once lived on the Santa Cruz River floodplain northwest of Tucson,[42] and native fish that occurred in the reach from Tucson south to Martinez Hill.[43]

The arrival of CAP water in 1992 meant that Tucson would increasingly rely on surface water from the Colorado River and decrease its use of groundwater from the Tucson basin.[44] After direct delivery of treated CAP water initially failed, nearly a decade passed before Tucson Water tried again to convince the public to accept Colorado River water. They accomplished the feat by artificially injecting CAP water into the alluvial aquifer of Avra Valley, the valley west of Tucson, and then delivering the blended CAP and groundwater to Tucson residents. Although the downward trend in the alluvial aquifer beneath Tucson was reversed as CAP water was delivered, there is no possibility that regional groundwater levels will rise along either the Santa Cruz River or Rillito Creek (chapter 22) to the extent that riparian vegetation can be sustained.

Changes in Riparian Vegetation

The long history of Tucson and a number of unique residents in town with an interest in the natural environment provide a long history of photography of the Santa Cruz River.[45] For the entire Santa Cruz River, 148 historical photographs document changes in riparian vegetation. The photographic record begins with an 1880 view from Sentinel Peak ("A" Mountain) by an unknown photographer, extends through the 1880s and 1890s with photographs of arroyo downcutting during major floods at Tucson, and finishes with a series of photographs at gaging stations that well document changes in riparian vegetation following the initial period of arroyo widening and a second period of widening that occurred during the floods of the 1970s through 1990s.

San Rafael Valley

Repeat photography at the monuments on the U.S.–Mexico border, where the Santa Cruz River both crosses going south from Lochiel and recrosses going north near Nogales, shows that woody riparian vegetation, primarily cottonwood, increased between 1893 and 1983.[46] Our rematches of these photographs (not shown) indicate that the increase continues into the twenty-first century. At the Lochiel gaging station, a short photographic record shows that the amount of woody riparian vegetation, with no nonnatives present, increased after 1968 (fig. 21.4). This increase occurred in the context of large, frequent floods that took place mostly during El Niño conditions (e.g., 1993 flood; fig. 21.2A).

The Santa Cruz River flows south across the U.S.–Mexico border, then loops to the west and back north to recross the border east of Nogales (fig. 21.1). The river is perennial in this reach, and diversions are made for agriculture in Sonora. Before arroyo downcutting in the 1820s, a ciénega was present in this reach.[47] An 1890 photograph taken near Mascarenas (not shown), just south of the border from the Nogales crossing, shows that black willow and cottonwood were common along the river in this reach amidst agricultural clearing.

Nogales to Tucson

The Santa Cruz at the U.S.–Mexico border east of Nogales sustains a gallery forest of cottonwood with a variety of other species, notably sycamore, Arizona walnut, and black willow. Indeed, the word *nogales* means "walnut" in Spanish. A photographic record of the river between the border and Arizona Route 82 shows several important changes.[48] Upstream from developments, cottonwood and willow, which were present in 1890, have increased along the river, but establishment of developments up to the river's banks have eliminated cottonwood trees locally. Repeat photography at the long-term gaging station near Nogales shows that obligate riparian vegetation has increased, although the channel has not narrowed concurrently (fig. 21.5). Part of the reason for the wide channel may be the legacy of the 1977 flood, which is the flood of record at this station (fig. 21.2B). Continued

development of Nogales and nearby Rio Rico are causing accelerated groundwater development, and late 1990s mortality of cottonwood in the vicinity of the confluence of Nogales Wash and the Santa Cruz River may be related to excessive pumping of the alluvial aquifer.

North of Nogales, the Santa Cruz River flows through a broad alluvial valley, and its channel is lined with a closed gallery forest of cottonwood and willow. These trees are aided by sewage effluent jointly discharged from the cross-border communities of Nogales, Arizona, and Nogales, Sonora.[49] During the first few years of the twenty-first century, a massive die-off of riparian trees occurred in the vicinity of Rio Rico. The cause for this die-off is presently unknown but may be related to the combination of groundwater use and the early-twenty-first-century drought, which reduced base flow in the river.

At Tumacácori, the amount of cottonwood along the river increased from 1930 through 1995 (fig. 21.6), possibly owing to retirement of agricultural lands along the river. Cottonwood were removed from the river in the 1930s as a result of clearing for agriculture.[50] Many of the new trees germinated following the 1977 and 1983 floods, which stripped vegetation from the floodplain (much of which was tamarisk) and saturated the alluvium, providing ideal conditions for growth of native species.

In the reach upstream from Continental to south of Martinez Hill, the river likely was ephemeral at the time of settlement and remains so. The river flows through pecan groves near Continental and sustains mostly xerophytic vegetation, such as desert broom, burrobrush, tamarisk, and tree tobacco. The channel morphology changes north of Continental from a wide, sandy channel to an arroyo deeply incised into Pleistocene alluvium.[51] This narrow, deeply incised channel lies several hundred feet west of its former course along the lowest part of the valley. The new channel follows the old Tucson-Nogales wagon road as shown on original land-survey maps. This road, progressively deepened by runoff from the nearby Sierritas, was finally captured by a flood on the Santa Cruz

A

B

Figure 21.4 Photographs of the Santa Cruz River near Lochiel, Arizona.

A. (June 1968.) The gaging station at Lochiel measures flow in the Santa Cruz River before it enters Mexico. The river winds through the broad San Rafael Valley, a grassland, and in 1968 the shallow channel has scattered cottonwood trees along it. This upstream view shows the gaging station and the low-water control. (R. L. T., courtesy of the U.S. Geological Survey.)

B. (October 8, 2000.) The channel has meandered to the left despite a grade-control structure just downstream from the bridge. Cottonwood have grown up throughout the view, and no woody nonnative riparian species are present. The bed sediment in the reach has coarsened, increasing channel roughness. (D. Oldershaw, Stake 1953.)

Figure 21.5 Photographs of the Santa Cruz River near Nogales, Arizona.

A. (November 24, 1930.) This downstream view, one of three photographs taken in the early 1930s showing this general perspective, shows the stilling well and cableway at the gaging station on the Santa Cruz River near Nogales. Frémont cottonwood and black willow appear along the river in the distance. The vegetation on the hillside behind the cableway tower consists of mesquite and catclaw. (W. E. Dickinson 875, courtesy of the U.S. Geological Survey.)

B. (May 21, 2001.) Riparian vegetation has increased in this reach despite the devastating floods of 1977 and 1983. Black willow blocks most of the downstream view, and cottonwood and black willow are more numerous along the right bank *(right side),* despite the appearance of one dead cottonwood on the extreme right. (D. Oldershaw, Stake 3781.)

Figure 21.6 Photographs of the Santa Cruz River near Tumacácori, Arizona.

A. (December 12, 1934.) This view, apparently taken to show the old mission at Tumacácori with the Santa Rita Mountains in the background, also documents channel conditions on the Santa Cruz River. Frémont cottonwood can be distinguished from the lower mesquite trees along the river, which has a wide, barren channel. (C. P. Russell B-166, courtesy of the National Park Service.)

B. (April 2, 1995.) The amount of cottonwood along the river has increased considerably, decreasing the view of the channel. The woody shrubs in the foreground are xerophytic mesquite and catclaw. (D. Oldershaw, Stake 3323.)

River, and the unnatural route has remained on the landscape.[52] This reach supports little riparian vegetation beyond mesquite and catclaw on its high banks.

The photographic record from Martinez Hill, upstream of Tucson (fig. 21.1), shows the demise of what once was known as the Grand Mesquite Forest.[53] In 1912, a shallowly incised channel wound through a corridor of cottonwood trees in a broad mesquite bosque (fig. 21.7A). Woodcutting reportedly had reduced the forest by 80 percent in 1917.[54] Despite water-level declines just downstream that accelerated in the 1920s,[55] the bosque was still present in 1942 and was highly valued as White-winged Dove habitat by the Arizona Game and Fish Department. Photographs taken in 1958[56] show healthy cottonwood in the channel, but also dead or dying cottonwood on the arroyo banks. By the 1970s, the riparian vegetation and mesquite bosques had been eliminated owing to the combination of woodcutting, channel downcutting and widening,[57] and groundwater development. The channel downcut and widened, primarily because of the 1977 and 1983 floods (figs. 21.7B, 21.7C), and now it is incapable of sustaining significant amounts of riparian vegetation because of the lowered water table (fig. 21.3).

Tucson to Marana

A total of 114 photographs documents changes in the Santa Cruz River and its floodplains in the vicinity of Tucson, especially through the twentieth century. Changes in riparian vegetation followed the related changes in channel dimension, flood frequency, and water development. Near Sentinel Peak, the first historical photographs document an open gallery forest of cottonwood trees with scattered shrubs and herbaceous vegetation.[58] Mesquite bosques were locally dense along the floodplain. At the start of the twentieth century, riparian vegetation along the river was not impressive (fig. 21.8), especially considering that the river was perennial at that time. However, the presence of Frémont cottonwood, black willow, an unknown willow, and netleaf hackberry was recorded in 1909.[59]

The story of the Santa Cruz River through Tucson is one of gradual conversion of a natural arroyo channel to a channelized river with ever-increasing development on its floodplains (fig. 21.9). In the 1910s, mesquite bosques were still common, although agricultural clearing had reduced much of their former area. By the late 1950s, most of the native trees were gone from along the river, and Athel tamarisk, touted as a xerophytic windbreak and effective erosion control, was planted instead (fig. 21.9B). Despite the river engineering of the 1970s and 1980s, the Santa Cruz River since the 1993 flood has begun to deposit new, low floodplains within its artificial banks (fig. 21.9D). Small amounts of riparian vegetation—including black willow, mesquite, desert broom, burrobrush, seepwillow, and tamarisk—are becoming established on these floodplains.

In the reach between Sentinel Peak and the Congress Street Bridge in Tucson, riparian vegetation has undergone severe fluctuations in density and species composition. Before groundwater development, effluent conditions increased around the foot of Sentinel Peak and extended downstream past the bridge. Despite arroyo downcutting, riparian vegetation was lush in 1904 and included dense mesquite bosques, cottonwood, and willows (figs. 21.10A and 21.12A). Because this reach was the focus of early-twentieth-century water development for Tucson, the alluvial aquifer quickly dropped, changing what appeared to be a thriving riparian ecosystem around the Congress Street Bridge to one that had only deep-rooted cottonwood with little understory (fig. 21.11A). In the early twenty-first century, the only cottonwood in this reach are planted and irrigated by the city of Tucson.

Despite the installation of soil cement, storm runoff from Tucson streets is sufficient to sustain growth of some obligate riparian trees within the arroyo downstream from Congress Street. Farther downstream, effluent from two wastewater-treatment facilities is sustaining a significant, new riparian area from just upstream of the confluence with Rillito Creek through Marana (figs. 21.13 and 21.14). This effluent discharge is approximately 50

to 75 ft³/s year round at the Cortaro Road gaging station. A great variety of riparian species is present in this newly perennial reach, including Frémont cottonwood, black willow, Bonpland willow, coyote willow, seepwillow, mesquite, tamarisk, and desert broom. These new riparian areas appear to be trapping sediments during summer thunderstorm runoff, leading to accelerated construction of low, in-channel floodplains. In the absence of large floods in the immediate future, this riparian assemblage seems poised to create a new gallery forest along the Santa Cruz River.

Marana to the Gila River

North of Marana, the Santa Cruz River flows across broad alluvial flats with a largely unincised channel except in two specific reaches, one near the confluence with the Gila River.[60] Groundwater use is heavy in this reach, leading to moderate amounts of ground subsidence southeast of the confluence with the Gila River.[61] A gaging station on the Santa Cruz River near Laveen (fig. 26.1), near its confluence with the Gila River, has recorded flow since 1940 (fig. 21.2D). The two largest floods in the record are the 1983 and 1993 events, respectively largest and second largest, and represent the only known times in the twentieth century when the river flowed continuously from its headwaters to the Gila River. Flow at the Laveen station usually consists of agricultural return flows, which combine groundwater with surface water from the Gila River and which typically are less than 20 ft³/s; the channel is dry in many years, especially during droughts.

The broad basins that the Santa Cruz River flows through north of Marana are not very scenic, nor are there interesting geological features that would attract photographers. Consequently, few photographs document long-term change in riparian vegetation in this reach. The photographic record here shows mostly increases in tamarisk (fig. 21.15).

A

B

Figure 21.7 Photographs of the Santa Cruz River at Martinez Hill, Arizona.

A. (1913.) This upstream view of the Santa Cruz River, from near the summit of Martinez Hill, shows the river flowing in a narrow, cottonwood-lined channel through a dense mesquite bosque. Open water is visible in the center of the view, showing that the alluvial aquifer here remained high. Punta de Agua, the water source for the Mission San Xavier del Bac, is out of the view to the right, and the mission would be visible if the camera were turned ninety degrees to the right. The land here is part of the San Xavier District of the Tohono O'odham Reservation. (Unknown photographer, Senate Document 973, 62-3 Plate I, courtesy of the National Archives.)

B. (December 15, 1981.) Groundwater overdraft, which became a significant problem just downstream from here in the 1920s, combined with floods, such as the 1977 event, to change riparian vegetation completely in this reach. The bosque had persisted into the 1940s but apparently declined steadily thereafter. The remains of a failed agricultural project appear in the center. The corridor for Interstate 19 crosses the upper right of the view. (R. M. Turner.)

C

D

C. (April 5, 1989.) The 1983 flood caused substantial widening in this reach, causing failure of one of the Interstate 19 bridges *(out of the view to the right)*. More agricultural clearing is apparent on the extreme right and to the left of the freeway corridor. (R. M. Turner.)

D. (November 25, 2002.) Other than the establishment of some xerophytic shrubs and the shifting of the channel thalweg to the left *(right in this upstream view)*, little has changed in this reach. Decline in groundwater levels and ephemeral flow in the river preclude establishment of riparian vegetation, even mesquite. (R. M. Turner, Stake 1057.)

A

B

Figure 21.8 Photographs of the Santa Cruz River at Tucson, Arizona.

A. (April 16, 1903.) This upstream view shows channel conditions in the Santa Cruz River in the vicinity of the present day Twenty-second Street Bridge in Tucson. Most of the trees present on the top of the eroding river banks appear to be mesquite. Rudimentary bank protection, in the form of poles driven into the channel bed, appears at lower right and in the center of the view. The Mannering artesian wells were near this camera station, suggesting that the water in the river here was perennial. (D. Griffiths 83-FB-2204, courtesy of the National Archives.)

B. (July 27, 2001.) The Santa Cruz River is now channelized in soil cement and is ephemeral in this reach. Riparian vegetation here is desert broom *(throughout the channel)* and Athel tamarisk *(top of bank at right and center)*. A solitary black willow appears in the distance and above the bank protection on the left side; this tree is immediately downstream from the Twenty-second Street Bridge. (D. Oldershaw, Stake 2483.)

Figure 21.9 Photographs of the Santa Cruz River at Tucson, Arizona.

A. (1919.) This upstream view, from Sentinel Peak ("A" Mountain), shows the Santa Cruz River in the same reach as figure 21.8A. The channel has widened significantly in sixteen years as a result of floods, especially in 1905 and 1916, damaging the agricultural lands in the midground on both sides of the channel. Mesquite is still present on the banks, and scattered groves of cottonwood are also present. A different view, taken by Mabel Cox from a slightly different camera position in 1917, shows similar conditions immediately following the floods of the 1910s. (G. G. Sykes, courtesy of the Arizona Historical Society, Tucson.)

B. (August 19, 1958.) This upstream view shows the same reach of the Santa Cruz River, but it is taken from a different camera position than photographs A, C, and D. The Twenty-second Street Bridge has not yet been built, but a bridge spans the river at Silverlake Road. The mesquite bosque present in 1919 is gone, replaced by agricultural fields that are now abandoned. The dark, conical trees in the river undoubtedly are Athel tamarisk. One of Tucson's landfills can be seen at the base of the hill in the right midground. (J. R. Hastings F IX 9, Stake 2496b.)

C. (January 6, 1988.) The Santa Cruz River was channelized in this reach a short time before this photograph was taken. Owing to this construction activity, the floods of 1977 and 1983, and substantial lowering of the alluvial aquifer, essentially no riparian vegetation appears in this view other than a grove of planted Athel tamarisk at lower left. The bridge in the midground is the Twenty-second Street crossing; a second bridge in the distance is on Silverlake Road. The tall building on the far right is the Pima County Prison, and the landfill formerly present downstream has been relocated. (R. M. Turner.)

D. (June 13, 2003.) Following the 1993 flood, the Santa Cruz River developed a low floodplain within its soil-cemented banks. Desert broom is the most common species in the channel, although a solitary black willow is present downstream from the Twenty-second Street Bridge. The Silverlake Road Bridge has been replaced with a new structure. (D. Oldershaw, Stake 1306.)

A. (1904.) In this downstream view of the confluence of the West Branch and the Santa Cruz River, looking northeast from the lower slope of Sentinel Peak, a thriving riparian ecosystem is present in 1904. A remnant of Warner's Dam, which impounded a water-supply reservoir and was destroyed by floods in the 1890s, is visible at left center, just upstream of the confluence. Small cottonwood trees are established at the center of the view and in the background, and a dense mesquite bosque is established on the margins away from the center of the channel. (W. Hadsell 24868, courtesy of the Arizona Historical Society, Tucson.)

B. (December 17, 1981.) The channel of the Santa Cruz River has shifted away from the camera station and is now difficult to see clearly owing to establishment of large Athel tamarisk on its banks. The number of cottonwood trees has declined markedly since 1904. The walls of the arroyo are visible in the middle distance beyond the Athel tamarisk on the left *(downstream)* side. Mission Road has been paved, but stoplights have not yet been installed at this busy intersection. (R. M. Turner.)

C. (September 17, 2000.) Groundwater pumping and channel stabilization completely eliminated natural riparian vegetation in this area in the middle of the twentieth century. By 2000, Mission Road has been widened and paved, and stoplights have been installed to regulate traffic. The banks of the river are soil cemented to eliminate lateral channel change; grade-control structures in the bed of the channel nearby prevent downcutting. The trees just beyond Mission Road are mesquite planted and irrigated as part of the Santa Cruz River Park, and the taller trees beyond are the remainder of the Athel tamarisk that were present in 1981. (D. Oldershaw, Stake 1026.)

Figure 21.10 Photographs of the Santa Cruz River at Tucson, Arizona.

A. (1914.) This downstream view shows the metal bridge spanning the Santa Cruz River at Congress Street. The flow in the channel is probably recessional from one of the numerous floods in this period; driftwood is wrapped around trees on both sides of the shallowly incised channel. The leafless trees appear to be black willow; Frémont cottonwood trees can be seen behind the bridge in the distance. (G. G. Sykes, courtesy of the Arizona Historical Society, Tucson.)

B. (November 29, 1930.) The camera station for this view of the concrete bridge at Congress Street is farther downstream than for photograph A. The cottonwood behind the bridge may be the same individuals that were present in 1914. Otherwise, the channel is much wider and appears to be deeper. (Unknown photographer 81, courtesy of the U.S. Geological Survey, Stake 3303.)

C. (September 1, 2003.) This view approximately matches photograph A. The banks are now soil cemented; the bridge has been replaced twice; and riparian vegetation is abundant on a low floodplain developed within the arroyo walls *(right foreground)*. The channel dropped 4 to 6 feet owing to the 1977 flood. Besides annual herbaceous species, the vegetation includes Mexican palo verde, tamarisk, desert broom, and carrizo grass. A black willow was present on the upstream pier in the distance before the 1993 flood killed it. (D. Oldershaw, Stake 4688.)

Figure 21.11 Photographs of the Santa Cruz River at Tucson, Arizona.

A

B

C

Figure 21.12 Photographs of the Santa Cruz River at Tucson, Arizona.

A. (November 1907.) This downstream view from the Congress Street Bridge (fig. 21.11A) shows the narrow channel of the Santa Cruz River that was deepened by the floods of 1905 and 1906 (Betancourt 1990). The relatively shallow arroyo began downcutting in 1878. The trees visible are Frémont cottonwood and black willow; many of the trees appear to be young. (W. T. Hornaday 11669, courtesy of the Arizona Historical Society, Tucson.)

B. (November 22, 1930.) This downstream view, taken from near the west abutment of the old concrete Congress Street Bridge (fig. 21.11B), shows a wider alluvial channel than in the photograph A view. At this time, the river flows through a rural area, and cottonwood, more mature than in 1907, line its banks. The low shrubs in the left foreground appear to be burrobrush, and mesquite grows on the banks below large cottonwood trees in the distance *(center)*. The 1914 flood of 15,000 ft^3/s and the 1917 flood of 7,500 ft^3/s probably caused most of the channel changes before 1930. (Unknown photographer 4299, courtesy of the U.S. Geological Survey, Stake 3300.)

C. (December 22, 2003.) The channel of the Santa Cruz River is now confined within soil-cemented banks. The open gallery forest of cottonwood trees in photograph B was destroyed by groundwater pumping and floodplain development, and the channel reached its present depth between 1977 and 1983. The riparian vegetation present now includes black willow, burrobrush *(on the low terrace at center),* tamarisk, desert broom, mesquite, and blue palo verde. The nitrogen bubbler line for the gaging station at Tucson appears on the right side on the soil-cemented banks. (D. Oldershaw, Stake 3744.)

A. (Ca. 1938.) This view looks downstream on the Santa Cruz River across its confluence with Rillito Creek *(right)*. Both channels are shallowly incised, and riparian vegetation remains dense on the adjacent channel banks. Besides the towering Frémont cottonwood trees, mesquite is dense on both sides of the Santa Cruz River, and what appears to be burrobrush is closest to the wide, sandy channel. Willis (1939) reported black willow at this site as well, although none is clearly visible in this view. (E. L. Willis 1939, Plate XV-A.)

B. (November 9, 1983.) At this time, the Santa Cruz River is wider and much deeper than in 1938, owing to the 1983 flood (recessional flow from that event is in the channel) and the 1977 flood. Garbage from a landfill is pushed up on the right side as bank protection. The only plants visible are scattered mesquite or catclaw or both. Owing to widening, the channel of Rillito Creek is no longer in the view. (R. M. Turner.)

C. (December 21, 2003.) The camera station cannot be accurately reoccupied owing to the increase in riparian vegetation. Sewage effluent flows perennially in this reach now, and riparian vegetation has benefited considerably from the stable water supply. The garbage from the old landfill has been removed, and bank protection now consists of soil cement *(right side)*. The vegetation present here includes black willow *(left)*, mesquite *(center)*, desert broom, burrobrush, blue palo verde, and tamarisk. (D. Oldershaw, Stake 1102.)

Figure 21.13 Photographs of the Santa Cruz River at its confluence with Rillito Creek.

A. (March 14, 1941.) This upstream view on the Santa Cruz River shows a broad, barren channel upstream from the former gaging station at Rillito. A solitary Athel tamarisk appears on the left bank *(right center)*. The small, regularly spaced trees along the far side of the river *(left bank)* appear to be Athel tamarisk planted for erosion control. (Unknown photographer A-358, courtesy of the U.S. Geological Survey.)

B. (February 9, 1995.) The channel has narrowed, and sewage effluent from Tucson creates perennial flow in this reach. Soil cement has not been installed; instead, levees are set back away from the channel to contain any potential for overbank flooding. Riparian vegetation includes coyote willow, black willow, and Frémont cottonwood trees *(in the distance)*. The Athel tamarisk persists at center. (D. Oldershaw.)

C. (April 9, 2004.) The persistent Athel tamarisk in photograph B is now surrounded by woody vegetation, including blue palo verde *(left foreground, center foreground)* and black willow *(left midground, right background)*. Some cottonwood remains visible in the right background, and small mesquite trees are present. This grove of riparian vegetation extends upstream about 1.5 miles and downstream about a half-mile and is sustained by wastewater effluent. (D. Oldershaw, Stake 3321B.)

Figure 21.14 Photographs of the Santa Cruz River at the Rillito gaging station.

Figure 21.15 Photographs of the Santa Cruz River near Laveen, Arizona.

A. (September 2, 1969.) This downstream view shows a bridge over the Santa Cruz River on the Gila River Indian Reservation south of Laveen. Downstream from the bridge, the channel has been bladed to create control for the gaging station. What appears to be a monospecific stand of tamarisk lies beyond the bladed area. (T. W. Anderson 9-4890, courtesy of the U.S. Geological Survey.)

B. (May 22, 2001.) Tamarisk has increased in both the foreground and the background, and growth of these trees is evident from the amount of sky that is now blocked beneath the bridge deck. Mesquite is present as well, notably beyond the right abutment of the bridge *(right background)*. (D. Oldershaw, Stake 2450.)

Summary. The tributaries of the Santa Cruz River vary considerably in geomorphic setting, flood history, channel-change history, and change in riparian vegetation. Other than Patagonia Lake, the only human structures that control these watercourses are small diversion dams or grade-control structures, and large floods have occurred in nearly all these tributaries. Sonoita Creek is one of the most important tributaries upstream from Tucson; Rillito Creek and Cañada del Oro Wash join the Santa Cruz just downstream from Tucson. Sonoita Creek has highly valued riparian areas and one of the few reservoirs in the watershed. Rillito Creek is a complex watershed draining the Santa Rita, Rincon, and Santa Catalina Mountains and includes several streams with significant amounts of riparian vegetation, including Pantano Wash, Tanque Verde Creek, Sabino Canyon, and Pima Canyon. Certain reaches confined by either bedrock or Pleistocene alluvial fill have perennial flow, and riparian vegetation has generally increased in these reaches. Where channels cross deep alluvial fill, particularly in the Tucson basin, regional groundwater declines combined with, in some cases, channel widening have caused decreases in riparian vegetation. In the wide reaches, facultative riparian species have established on low floodplains adjacent to the active channel.

The Santa Cruz River is fed by numerous tributaries from its headwaters through the Tucson basin (fig. 21.1). Two of the principal tributaries in the headwaters are Nogales Wash and Sonoita Creek, which enter the river north of the U.S.–Mexico border. Rillito Creek, which enters the Santa Cruz River north of Tucson, is the most complex tributary owing to additions from the Santa Rita, Rincon, and Santa Catalina Mountains (fig. 21.1). Tanque Verde Creek and Pantano Wash join to form Rillito Creek northeast of Tucson. Other tributaries of Rillito Creek include Sabino Creek as well as numerous smaller tributaries (e.g., Pima Canyon) that drain the Santa Catalina and Rincon Mountains. Finally, the Cañada del Oro Wash has its headwaters on the north side of the Santa Catalina Mountains and turns 180 degrees to flow south, then southwest to join the Santa Cruz River near its confluence with Rillito Creek. These tributaries support a variety of riparian vegetation along their lengths, ranging from closed gallery forests of Frémont cottonwood to mostly xerophytic assemblages taking advantage of the extra water supply offered by ephemeral streams. Riparian vegetation has undergone few changes along some reaches and large increases along others.

Sonoita Creek

Sonoita Creek drains about 209 square miles of the east and south sides of the Santa Rita Mountains and lower ranges in southern Arizona (fig. 21.1). This drainage is ephemeral upstream from Monkey Spring, where a large inflow from river left adds significant water while sustaining dense riparian vegetation.[1] This flow does not persist for more than several miles, and the creek is ephemeral up to near Patagonia, Arizona, where a fault causes groundwater to rise to the surface.[2] The perennial flow here sustains dense riparian vegetation that forms the basis of the Patagonia–Sonoita Creek Preserve managed by The Nature Conservancy.[3] However, influent conditions occur downstream of Patagonia Lake, and Sonoita Creek is ephemeral at its confluence with the Santa Cruz River.

Sonoita Creek is one of the major flood producers contributing to the upper Santa Cruz River. The flood record of Sonoita Creek, from 1930 through 1972, has a peak discharge of 14,000 ft³/s in September 1946 (fig. 22.1). This event is unusual in that no other significant floods occurred in the region on this date. The 1983 flood had a peak discharge of 16,000 ft³/s, which was estimated indirectly a decade after the gaging station ceased operations.[4] Patagonia Lake, which was built in 1969 for private development, does not provide significant flow regulation for Sonoita Creek. The lake is now owned by the Arizona State Parks and is managed as a recreational lake and wildlife area.

The first observers leaving written records described Sonoita Creek as having a discontinuous channel with swampy reaches from its headwaters downstream to present-day Patagonia.[5] The reaches downstream from Patagonia reportedly were lined with dense forests of cottonwood and willow.[6] At the Patagonia–Sonoita Creek Preserve, The Nature Conservancy claims to have some of the largest and oldest Frémont cottonwood trees in the Southwest, although at 130 years old[7] these trees are half the age of some in the Escalante River basin. The creek basin is also home to or visited by three hundred species of birds, and therefore it has one of the highest biodiversities of avifauna in the United States.

Repeat photography in the upper reaches of Sonoita Creek documents postarroyo changes in riparian vegetation in an ephemeral reach (fig. 22.2). Although some ephemeral reaches like this one appear to support only facultative riparian species, the decrease in channel width and increase in woody vegetation are in accord with what is generally happening throughout the region. Other reaches with more

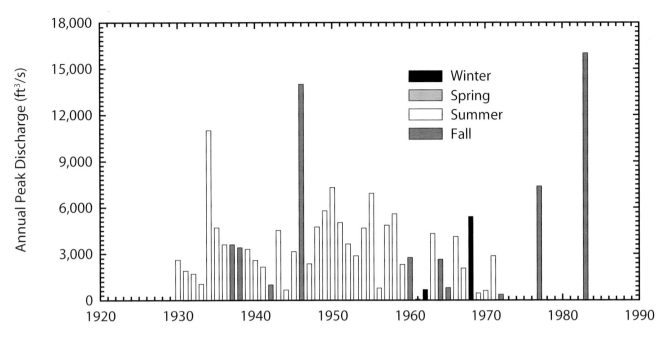

Figure 22.1 Annual flood series for Sonoita Creek near Patagonia, Arizona (station 09481500; 1930–1972, 1977, 1983).

water have increases in trees, including Frémont cottonwood and oaks.[8] Downstream from Patagonia and the Patagonia–Sonoita Creek Preserve, riparian vegetation change is documented in fourteen sets of repeat photographs (fig. 22.3). In this reach near the former gaging station, native riparian vegetation has greatly increased in density and biomass.

Pantano Wash

Pantano Wash has its headwaters in the high grasslands north of Sonoita, Arizona (fig. 21.1). Its principal tributary, Cienega Creek, joins with other drainages from the north side of the Santa Rita Mountains and the Whetstone Mountains and with Agua Verde Creek from the southern Rincon Mountains to form it. The combined watercourse alternates between perennial and ephemeral reaches until its confluence with Tanque Verde Creek (fig. 22.4). Pantano Wash was named for its marshy conditions in its middle reaches (*pantano* is the Spanish word for "swamp"),[9] and Cienega Creek once had its namesake riparian areas. However, like most other streams in southern Arizona, Pantano Wash and its principal tributaries were not perennial from their headwaters to Rillito Creek.

Cienega Creek is highly valued as a

riparian area by Pima County, which created the Cienega Creek Natural Preserve of 3,979 acres of primarily riparian habitat along its downstream reaches.[10] Where Cienega Creek emerges from canyons deeply incised into Pleistocene alluvium, the channel turns west at a point where the Southern Pacific Railroad crosses from the Santa Cruz to the San Pedro River basin. An early view by pioneer photographer Carleton Watkins documents pre-arroyo conditions on Cienega Creek (fig. 22.5). Where once the floodplain was shallowly incised and covered with herbaceous vegetation and grasses, it is now covered with dense mesquite and other riparian species.[11]

In the perennial reach in the vicinity of the gaging station near Vail, riparian vegetation has increased in Pantano Wash (fig. 22.6). In the 1960s, no tamarisk was present at this site, and no sycamore was observed in Pantano Wash.[12] However, nonnative tree tobacco and native Arizona walnut were observed here in 1984,[13] but probably were dead in 2005.[14] Downstream, Pantano Wash flows through a broad alluvial channel that is now part of suburban Tucson. The channel of Pantano Wash narrowed between 1941 and 1979, then increased in width during the wet period of the 1980s and 1990s.[15] Groundwater levels have declined in this general area, leading

to conditions that are not conducive to establishment and growth of obligate riparian species. Therefore, the increases observed near the gaging station are not representative of the channel downstream, where riparian vegetation likely has decreased.

Downstream to the confluence with Tanque Verde Creek, Pantano Wash is largely channelized with soil cement. A variety of channel disturbances occurred in this reach, including in-stream gravel mining, floodplain encroachment by development, and installation of bank protection. In the middle half of the twentieth century, an average downcutting of 8.5 feet has been estimated for Pantano Wash, with maximum degradation of 16 feet.[16] This downcutting likely would have occurred at the expense of riparian vegetation.

Tanque Verde Creek

Tanque Verde Creek drains 219 square miles of highlands between the Santa Catalina and Rincon Mountains, primarily the Redington Pass area and Agua Caliente Hill (fig. 21.1). Tanque Verde Creek obtained its name from a permanent, stagnant body of water in the alluvial floodplain.[17] This pool ultimately was breached as arroyo cutting on Rillito Creek propagated upstream into this tributary. The gaging record

A. (June 1, 1909.) This view across Sonoita Creek downstream from Sonoita, Arizona, and near old Fort Crittenden shows a wide, barren channel with perennial flow. Trees on the background slope include one-seed juniper, oaks, and mesquite. (J. M. Hill 6, courtesy of the U.S. Geological Survey Photographic Library.)

B. (July 9, 1968.) The channel has narrowed in the intervening fifty-nine years, and rubber rabbitbrush has colonized a low floodplain on river right *(left side).* (J. R. Hastings.)

C. (July 21, 1994.) The creek is ephemeral in this reach, and most of the vegetation in the foreground is rubber rabbitbrush. Tamarisk is not present in this reach. (D. Oldershaw, Stake 386.)

Figure 22.2 Photographs of Sonoita Creek near Patagonia, Arizona.

A. (May 1939.) This upstream view shows channel conditions in the vicinity of a gaging station on Sonoita Creek downstream from Patagonia. The station is built on the pier of a partially dismantled railroad bridge that once ran from Patagonia to Nogales. Cottonwood and sycamore trees are present along the wide channel, and a closed gallery forest appears in the background. (A. A. Fischback 2596, courtesy of the U.S. Geological Survey.)

B. (January 24, 1964.) Taken from a slightly different camera position, this view shows that the trees present in 1939 have grown but that few new individuals are present. (F. S. Anderson, courtesy of the U.S. Geological Survey.)

C. (October 8, 2000.) The creek is perennial in this reach, and The Nature Conservancy manages a dense riparian area several miles upstream. The gaging station has been removed, and riparian vegetation has increased considerably in this reach. The diverse array of species present includes Frémont cottonwood, sycamore, Arizona walnut, black willow, coyote willow, mesquite, and netleaf hackberry. In 2000, the tips of mesquite branches were being killed by girdle beetles. Tamarisk is not present in this reach. (D. Oldershaw, Stake 1951.)

Figure 22.3 Photographs of Sonoita Creek near Patagonia, Arizona.

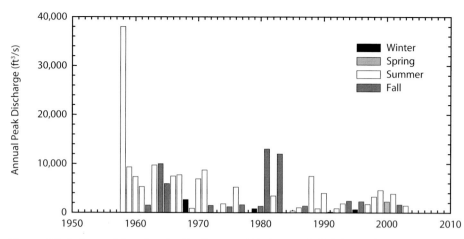

Figure 22.4 Annual flood series for Pantano Wash near Vail, Arizona (station 09484600; 1958–2003).

A. (1880.) This downstream view shows Cienega Creek, a major tributary of Pantano Wash east of Tucson, before arroyo downcutting affected the watershed. The Southern Pacific Railroad had constructed tracks here a short time before this photograph was taken. Cienega Creek has a narrow, treeless channel, and the floodplain has a grass cover with few woody plants. Mesquite appears on the slopes in the left midground. A primitive wood fence crosses the midground. This view supports the observations that Cienega Creek (the Spanish word *ciénega* means "marsh") had high groundwater conditions that could have excluded woody vegetation. (C. E. Watkins 1324, courtesy of the Huntington Library, San Marino, Calif.)

B. (January 24, 1998.) The increase in woody bottomland vegetation, primarily mesquite but also cottonwood and willow, is striking. The channel of Cienega Creek is deeper and wider owing to arroyo downcutting and is the riparian niche along which the cottonwood and willows grow. The channel changes prompted the Southern Pacific to move the tracks and bridge after the original bridge washed away. The current bridge was built upstream from this site in 1912. Increase in woody vegetation on the adjacent xeric hillslopes has been discussed elsewhere (Turner, Webb, and Bowers 2003). (R. M. Turner, Stake 3411.)

Figure 22.5 Photographs of Cienega Creek near Pantano Station, Arizona.

Figure 22.6 Photographs of Pantano Wash near Vail, Arizona.

A. (October 30, 1961.) Pantano Wash, a tributary of Rillito Creek southeast of Tucson, is perennial in this reach owing to a bedrock constriction with a concrete dam built within it. The stream is unregulated upstream from the gaging station; the concrete sill forms a weir that provides a low-water control for the gaging station. The flood of record, estimated to be 38,000 ft^3/s, occurred in 1958. The channel margins support mesquite. (Unknown photographer 4658, courtesy of the U.S. Geological Survey.)

B. (November 28, 2000.) In the thirty-nine years between the photographs, the largest flood was 13,000 ft^3/s in 1981. As much as 8 feet of channel scour is apparent downstream from the dam. Cottonwood trees are now established near the low-water channel, and mesquite trees have grown up on the banks. The density of saguaros has also increased, particularly at right center. (D. Oldershaw, Stake 2219.)

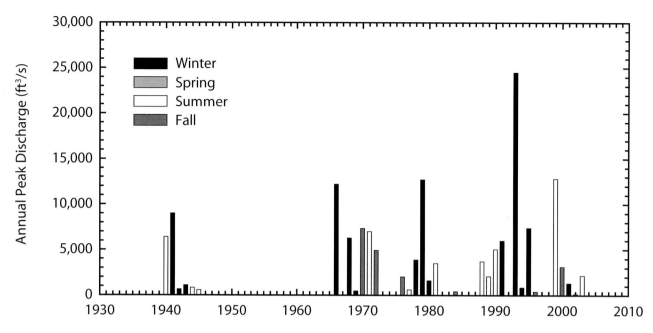

Figure 22.7 Annual flood series for Tanque Verde Creek at Tucson, Arizona (station 09484500; 1940–1945, 1965, 1968–2003).

for this watercourse is relatively short. The flood of record occurred in January 1993 and had a peak discharge of 24,500 ft³/s (fig. 22.7). Channel changes in Tanque Verde Creek mirror those in Rillito Creek downstream.[18]

At the confluence of Agua Caliente Wash and Tanque Verde Creek, upstream from where Sabino Creek and Pantano Wash join (fig. 21.1), groundwater pumping has caused near-lethal decline in a mesquite bosque[19] as well as dieback in old Frémont cottonwood. Riparian vegetation has decreased in this reach owing to the co-occurrence of groundwater withdrawals, large floods that caused channel widening, and bank protection (fig. 22.8). This channel widened following the 1965 and 1983 floods in the Rillito Creek system.[20] Although the combination of channel widening, bank protection, large floods, and groundwater declines is generally lethal to riverine riparian vegetation, cottonwood seedlings were present in 2002 in the channel of Tanque Verde Creek at the Sabino Canyon bridge.

Sabino Creek

Sabino Creek is a perennial stream in Sabino Canyon about 12 miles northeast of Tucson, Arizona. This creek drains 35.5 square miles of the Santa

Catalina Mountains, and Sabino Canyon is a major recreational area for residents of Pima County. The annual flood series for this canyon shows an apparent increase in the late twentieth century (fig. 22.9), culminating in the flood of record of 15,400 ft³/s in July 1999. No land-use changes in the headwaters are associated with this increase in flood frequency.[21] The channel of Sabino Creek is constrained by bridges and grade-control structures through Sabino Canyon, but otherwise flow in this stream is not regulated. A dam used to divert irrigation water remains across the creek at the mouth of Sabino Canyon.

Sabino Canyon is a recreation area valued for its riverine resources, in particular its lush riparian vegetation. A wide variety of native and non-native species occurs here, including sycamore, Frémont cottonwood, Arizona ash, mesquite, tamarisk, and giant reed. Despite the presence of nonnative species and occurrence of large floods, native riparian species are thriving along Sabino Creek, especially sycamore (fig. 22.10). Small forests of riparian trees, documented in photographs from the late nineteenth and early twentieth centuries,[22] remain in the channel despite shifts in its position (fig. 22.11). Tamarisk becomes more abundant where Sabino Canyon

widens and leaves its bedrock canyon (fig. 22.12), although native species have increased in this reach as well.

Pima Canyon

Pima Canyon has a drainage area of only 4.93 square miles, primarily the south side of the front range of the Santa Catalina Mountains. This canyon is an important recreational area in the Coronado National Forest north of Tucson. Pima Wash has a short gaging record, and the flood of record (1964–1983) is only 460 ft³/s during the October 1983 floods in the Tucson basin (see chapter 21). This creek has ephemeral flow in its short, rocky channel, and little change in riparian vegetation has occurred here (fig. 22.13).

Rillito Creek

Rillito Creek has a drainage area of 922 square miles, which includes the Pantano Wash system, Tanque Verde Creek, and the drainages from the south side of the Santa Catalina Mountains. Although many of these headwater tributaries are perennial for at least part of their length, Rillito Creek is fully ephemeral, in part because of heavy groundwater extraction along its length and in part because

A

B

Figure 22.8 Photographs of Tanque Verde Creek.

A. (December 13, 1940.) This upstream view on Tanque Verde Creek, from the Sabino Canyon Road Bridge, shows a small flood in progress in a channel braided among the riparian trees. The trees include Frémont cottonwood and black willow. (C. T. Pynchon 3217, courtesy of the U.S. Geological Survey.)

B. (October 25, 2002.) This reach is now channelized and much wider. A diverse riparian ecosystem is present on the extreme right and behind the soil cement on river right *(left side)*. Riparian vegetation growing in the soil cement includes Frémont cottonwood *(seedlings on extreme right)*, seepwillow, burrobrush, and desert willow; behind the soil cement are mesquite, palm trees, and cottonwood. No tamarisk is present in the vicinity of the bridge. (D. Oldershaw, Stake 4346.)

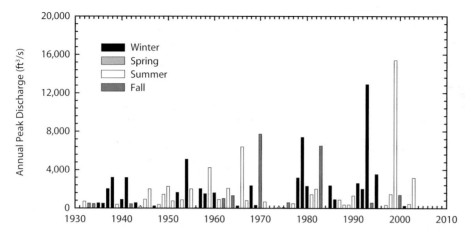

Figure 22.9 Annual flood series for Sabino Creek near Tucson, Arizona (station 09484000; 1932–2003).

A. (February 1900.) This upstream view on Sabino Creek shows a small grove of sycamore trees in the middle of the channel. These young trees are very close to water level and clearly are positioned to sustain major damage during floods. Desert broom is relatively dense on river right *(left side of the view)*. (S. J. Holsinger, courtesy of James Klein.)

B. (September 26, 2003.) The river channel has changed considerably, and deposition of boulders 1 to 3 feet in diameter has occurred in this reach. The sycamore tree in the center of the view appears, however, to have survived all the floods and channel change of the intervening 103 years. Most of the trees in the view are sycamores, and Bonpland willow appears on the extreme left and is common behind the camera station. Frémont cottonwood, Arizona ash, and coyote willow are also common in this reach. Although tamarisk is not present, nonnative giant reed and fountain grass are nearby. (R. M. Turner, Stake 4693.)

Figure 22.10 Photographs of Sabino Creek.

Figure 22.11 Photographs of Sabino Creek.

A. (February 1900.) This downstream view shows the road entering Sabino Canyon in the distant saddle and the channel of Sabino Creek in the foreground and midground. Low-water flow in Sabino Creek appears in the open area on the left. (S. J. Holsinger, courtesy of James Klein.)

B. (September 26, 2003.) The channel has shifted to the right side of the floodplain, and the open area on the left is now an overflow channel. The amount of riparian vegetation has not changed significantly, nor have nonnative species become established in significant numbers in this reach. Arizona ash dominates the riparian zone, but other species present include netleaf hackberry, desert hackberry, mesquite *(mostly to right center)*, sycamore, and Frémont cottonwood. (R. M. Turner, Stake 4692.)

A

B

Figure 22.12 Photographs of the gaging station on Sabino Creek.

A. (February 13, 1938.) This upstream view on Sabino Creek shows the gaging station at the downstream end of Sabino Canyon. Part of the attraction of Sabino Canyon is the robust riparian ecosystem, which lines a stream punctuated by dam-controlled pools and stands in stark contrast with the surrounding desert slopes, with saguaro cacti marking the ridgelines. (W. E. Dickinson 2357, courtesy of the U.S. Geological Survey.)

B. (April 2, 2001.) The gaging station was moved downstream about a thousand feet in 1974, and the stilling well was removed. A flood of 15,400 ft³/s occurred on July 15, 1999, causing considerable damage to roads and dams in the canyon upstream. The camera station is not exact, owing to both shifting of the channel and about 5 feet of deposition since 1938. The view is now largely obscured by thick stands of black willow and Arizona ash, both of which have increased nearby despite the recent floods. (D. Oldershaw, Stake 2439.)

Figure 22.13 Photographs of Pima Canyon.

A. (March 3, 1902.) This upstream view on Pima Canyon shows an ephemeral reach with mostly xerophytic vegetation, including ocotillo *(right foreground and left side),* sotol *(left and right foreground),* and several shrub species. A tall tree that has the appearance of an isolated sycamore appears in the left distance, and mesquite forms a dense stand at right center. (D. Griffiths 83-FB-938, courtesy of the National Archives.)

B. (November 14, 1998.) Despite the obvious growth of xerophytic vegetation, little has changed in this reach in ninety-six years. The tall tree in the distance is gone, and several new mesquites are present. (D. Oldershaw, Stake 3801.)

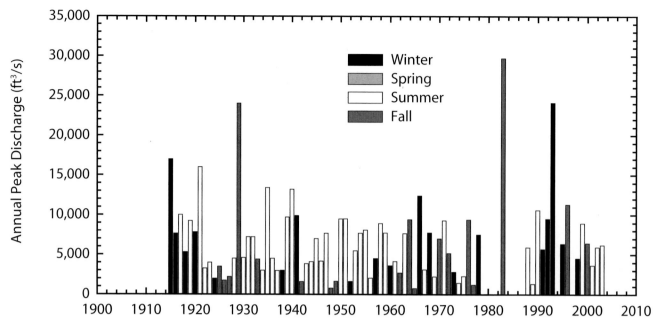

Figure 22.14 Combined annual flood series for Rillito Creek near Tucson, Arizona (station 09486000; 1915–1978, 1983), and for Rillito Creek at Dodge Boulevard at Tucson, Arizona (station 09485700; 1988–2003).

it flows through a deep alluvial basin. The word *rillito* means "little creek" in Spanish, so the name "Rillito Creek" is redundant, much like "Table Mesa" and "Picacho Peak."

Less is known about change in riparian vegetation along Rillito Creek than for the Santa Cruz River, in part because much of this drainage was unused before the twentieth century. The first attempt at settlement near Rillito Creek was made in 1858,[23] well after settlement of the Santa Cruz River south of present-day Tucson. Fort Lowell was built near perennial water that once was present near the confluence of Tanque Verde Creek and Pantano Wash. As many as three species of native fish occurred in Rillito Creek and its tributaries Sabino Creek and Tanque Verde Creek.[24]

Although anecdotal reports indicate that riparian vegetation at the time of settlement formed continuous groves of cottonwood, Arizona ash, black willow, and Arizona walnut, paintings from the time indicate scattered trees.[25] Mesquite bosques and grassland occupied the higher alluvial terraces.[26] It is doubtful that these conditions occurred the length of Rillito Creek, which likely was ephemeral from an unknown distance downstream of the Tanque Verde–Pantano confluence to the Santa Cruz River.[27] Cattle grazing

began in the vicinity of the floodplain in the 1870s, and agriculture followed establishment of Fort Lowell in 1872.

The present-day arroyo of Rillito Creek began to downcut in 1881, and by the early 1890s the arroyo had fully coalesced. A particularly severe flood occurred on September 11, 1887, when a flash flood killed livestock, destroyed property, and inundated the floodplain to a width of 2 miles in places.[28] Channel widening followed and then temporarily ceased by the early 1940s.[29] By the 1930s, riparian vegetation in the vicinity of Fort Lowell consisted of seepwillow near the active channel and Frémont cottonwood, Arizona ash, sycamore, Arizona walnut, and "willow" on low terraces.[30] A grove of cottonwood was present at the confluence of Rillito Creek and the Santa Cruz River, and groundwater levels were high along the length of the river. Some obligate riparian species have been extirpated, notably the Mexican garter snake (*Thamnophis eques*), which once lived on the floodplains of Rillito Creek.[31] Tree tobacco had already become established by 1939.

From 1941 through 1965, a period of small floods in the Santa Cruz River system (chapter 21), Rillito Creek decreased in width and changed from a braided channel to a single-channel thread.[32] Widening again occurred fol-

lowing 1965 and continued through the wet period of the 1980s and 1990s, which obliterated any riparian vegetation established within the channel. The 1983 flood on Rillito Creek, the largest event in the long gaging record (fig. 22.14), had a peak discharge of 29,700 ft³/s and caused extensive channel changes on Rillito Creek. The damages prompted installation of soil cement along the length of the river through the urbanized reaches of Tucson, and the 1993 flood, which had a peak discharge of 24,100 ft³/s, caused little damage. Reduction in peak discharges after 1993 has caused channel narrowing and development of a low floodplain within the soil-cemented banks.

As along the Santa Cruz River and throughout the Tucson basin, groundwater levels along the Rillito Creek steadily declined through the middle of the twentieth century, particularly in the late 1980s owing to high groundwater withdrawals.[33] Flow in Rillito Creek caused "wild fluctuations" during droughts and "rapid recovery" during runoff.[34] A water-level ridge developed beneath the channel, and the size of this mound was made larger in comparison to regional water-level lowering. Although isolated cottonwood were present in 1937 along Rillito Creek,[35] groundwater declines, floods, channel narrowing, and eventually installation of bank protection since then have limited the potential establishment of many obligate riparian species. Nonetheless, several isolated stands of young cottonwood, which probably germinated during the 1993 flood, and facultative riparian species have become established within the soil-cemented banks of Rillito Creek (fig. 22.15).

Near its confluence with the Santa Cruz River, Rillito Creek has fluctuated between a relatively wide channel (fig. 22.16A) and a relatively narrow one.[36] In the early 2000s, facultative riparian vegetation greatly increased on a low terrace developed adjacent to the active channel (fig. 22.16B). A solitary Frémont cottonwood tree has established and is thriving just upstream from the Interstate 10 bridges; this tree benefits from groundwater perched above thick clay layers beneath lower Rillito Creek[37] and from a groundwater mound formed beneath the Santa Cruz River created by sewage-effluent discharges upstream from the confluence.

Cañada del Oro Wash

Cañada del Oro Wash has its headwaters on the northern slopes of the Santa Catalina Mountains north of Tucson (fig. 21.1). At first, this wash flows north through a deep cleft in the mountains, then abruptly turns south, then southwest to join the Santa Cruz River just downstream from its confluence with Rillito Creek. The drainage area of Cañada del Oro Wash, 255 square miles, is much smaller than that of Rillito Creek, and floods along this watercourse are correspondingly smaller (fig. 22.17).[38] For example, the 1983 floods were devastating on the Santa Cruz River and Rillito Creek but were relatively minor on the Cañada del Oro Wash.

Downstream from where the Cañada del Oro Wash exits its mountainous canyon, the river flows along the range-bounding fault of the western Santa Catalina foothills (fig. 22.18). Flow in this reach is intermittent, and some permanent water bodies occur in side canyons. The channel was relatively wide and barren in 1930; now, woody riparian vegetation has increased as the channel has narrowed. Suburban development is booming in this part of the Cañada del Oro watershed (fig. 22.18B), and groundwater withdrawals associated with these developments may reverse this increase in riparian vegetation.

Groundwater is heavily used in the alluvial reaches of the Cañada del Oro, as in most of the Tucson basin. In the middle of the twentieth century, groundwater levels consistently declined at a rate of about 1 foot per year.[39] However, wells adjacent to the channel have sharp water-level rises following floods. Runoff in the winter of 1960 caused a 75-foot rise in the water level of one well. Despite such influxes, groundwater in the alluvial reaches is well beneath the depth range of most obligate riparian species.

As the channel of the Cañada del Oro Wash moves away from the front of the Santa Catalina Mountains, groundwater levels drop, flow is ephemeral, and riparian vegetation generally is facultative. At some sites, old obligate species have been replaced by shrubbier species (fig. 22.19). Increases in riparian vegetation where the channel is adjacent to Pusch Ridge are limited to shrubs such as burrobrush, desert broom, and mesquite.[40] Farther downstream, where the channel overlies the deep alluvial basin, riparian vegetation lining the channel is a combination of xerophytic trees characteristic of Sonoran Desert uplands and facultative shrubs, in particular palo verde and desert broom (fig. 22.20).

Figure 22.15 Photographs of Rillito Creek at First Avenue in Tucson, Arizona.

A. (May 1939.) This upstream view on Rillito Creek is taken from the First Avenue Bridge north of Tucson. The channel is wide, sandy, and barren, although a few scattered desert broom appear on the right side and in the center midground. Mesquite is present in a discontinuous band along the high banks. (C. T. Pynchon 2591, courtesy of the U.S. Geological Survey.)

B. (August 28, 2003.) Rillito Creek is now channelized with soil-cemented banks, and a low terrace has developed on the floodplain. Runoff in the channel appears black owing to the high ash content of runoff from areas burned by the Aspen Fire in the Santa Catalina Mountains this year. The channel has a variety of species, including a young cottonwood *(center)*, desert broom *(most abundant)*, and mesquite; Mexican palo verde is also present. Most of this vegetation became established after the 1993 flood. (D. Oldershaw, Stake 4684.)

A

B

Figure 22.16 Photographs of the confluence of Rillito Creek and the Santa Cruz River.

A. (Ca. 1938.) This upstream view is at the confluence of Rillito Creek and the Santa Cruz River northwest of Tucson. Rillito Creek has relatively low banks, and only a 3-foot-high bank separates the broad sandy channel from the surrounding desert. Isolated mesquites appear in the center of the view. (E. L. Willis 1939, Plate XV-B.)

B. (January 30, 2004.) This view is approximately from the same camera station, which has been altered owing to channel downcutting and channelization with soil cement *(left side)*. The channel thalweg is now to the right of the camera station. Vegetation in the channel includes burrobrush, desert broom, mesquite, desert willow, and blue palo verde. (D. Oldershaw, Stake 1270.)

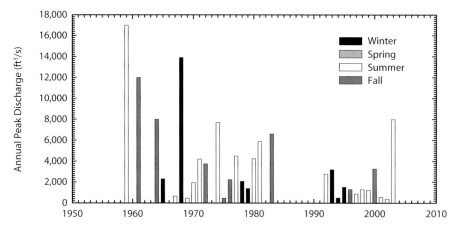

Figure 22.17 Combined annual flood series for Cañada del Oro Wash near Tucson, Arizona (station 09486300; 1959–1983) and for Cañada del Oro Wash below Ina Road near Tucson, Arizona (station 09486350; 1992–2003). The 2003 peak followed the devastating Aspen Fire in the Santa Catalina Mountains.

A. (May 1930.) This downstream view on the Cañada del Oro is part of a panoramic view of the western side of the Santa Catalina Mountains. Pusch Ridge, a major topographic feature north of Tucson, appears in the distance at center. The channel is lined with what appears from this distance to be a combination of mesquite and xerophytic shrubs, although a single tree—likely a small cottonwood—appears in the center of the view. (P. B. King 24, courtesy of the U.S. Geological Survey Photographic Library.)

B. (October 25, 2002.) The channel of the Cañada del Oro is now difficult to see owing to channel narrowing and the increase in riparian vegetation. Many species are present, including abundant cottonwood, mesquite, and burrobrush. A major new subdivision of the Saddlebrooke development in Catalina is under construction on the high terraces to the left of the channel. The saguaros that were in the foreground are now dead, and a new one appears in the left foreground. (D. Oldershaw, Stake 4343c.)

Figure 22.18 Photographs of the Cañada del Oro Wash near Rancho Solano, Arizona.

A

B

Figure 22.19 Photographs of the Cañada del Oro at Steam Pump Ranch, Arizona.

A. (Ca. 1910.) This view of a gnarled old cottonwood, with Pusch Ridge of the western Santa Catalina Mountains in the background, shows a wide, shallow Cañada del Oro downstream from Catalina, Arizona. The watercourse flows left to right across the view. Several mesquites appear in the background, lining the ephemeral channel. (G. G. Sykes, courtesy of the Arizona Historical Society, Tucson.)

B. (August 28, 2003.) The camera station is next to a ruined wood-frame structure that once was on the floodplain but is now protected by soil cement, which crosses the view in the foreground. Mesquite and desert broom have increased in the foreground, although the channel appears to have narrowed owing to bank protection and encroachment by riparian vegetation. (D. Oldershaw, Stake 4682.)

A

B

Figure 22.20 Photographs of the Cañada del Oro.

A. (February or March 1967.) This upstream view on the Cañada del Oro shows Overton Road crossing the center of the view with the gaging station on the upstream side. The trees on both sides of the channel are mostly blue palo verde *(left side),* and small individuals of desert broom are in the center of the channel. A mistletoe-infested mesquite tree appears on the extreme right side. (Photographer unknown, courtesy of the U.S. Geological Survey.)

B. (October 25, 2002.) The gaging station has been removed, and the road appears to be in the same place. Both the blue palo verde and the mesquite tree persist, and the desert broom are much larger and in full bloom at this time. Off-road vehicle use is common in the wash at this point. (D. Oldershaw, Stake 4340.)

Summary. The Verde River, which drains much of the highlands of central Arizona south of the Mogollon Rim, has the largest number of species of woody riparian vegetation that we observed in this region. Owing to flow regulation in its middle reaches, the Verde River is like two rivers, an unregulated one upstream and a regulated one downstream. Here, as along some other perennial rivers, riparian vegetation was thinned by extremely large floods between 1891 and around 1940. After this period, riparian vegetation generally increased along the full length of this river and its tributaries, despite heavy livestock grazing, flow regulation, the repeated occurrence of large floods, and floodplain agriculture. Although tamarisk, Russian olive, giant reed, and tree-of-heaven are present on this river, nonnative species—in particular tamarisk—are numerous only in the reaches downstream from Horseshoe Dam. Urban development and consequent increases in water use in the river's headwaters may threaten the springs that yield perennial flow to the upper Verde River.

The Verde River drains 6,615 square miles of central Arizona, including most of the area under the western section of the Mogollon Rim (fig. 23.1). This river is a highly valued source of domestic and irrigation water, both within the drainage basin and downstream in the Salt River Valley. It is known as well for its massive floods and perennial flow, so dams impound two lakes relatively close to its confluence with the Salt River. In addition, the towns in the upper part of the watershed are some of the fastest growing in Arizona, and demands for water are increasing in the basin. The net result is a drainage system with current surface-water demands that cause the lower third of the river

to be fully regulated except during large floods and with increasing water demands in the headwaters that may cause reduced base flow.

The Verde River sustains high-quality habitat for wildlife and riparian species, notably birds but also twelve nonavian species with special management status.[1] Five reaches of this river are closed or have restricted visitation from December 1 through June 15 owing to nesting of Bald Eagles.[2] Designated as a Wild and Scenic River,[3] the Verde is one of the only rivers in the region with such (albeit minimal) protection. Finally, the Verde River flows through the Mazatzal Wilderness Area (established in 1938), one of the oldest designated wilderness areas in the United States. The Verde River presents a serious challenge to future water management in Arizona because of the concurrent demands for its water and the need for protection of its resources.

Tributaries of the Verde River

Granite Creek near Prescott

Granite Creek drains only 36 square miles of central Arizona and is the main channel through Prescott. The average daily discharge of this watercourse is only 6 ft^3/s. Because Granite Creek flows through an urban area, it can produce floods that cause significant damage. One flood, on August 19, 1963, is the flood of record in a discontinuous gaging record (fig. 23.2) and caused substantial property damage to Prescott.[4] Photographs taken shortly after the gaging station was established in 1932 show a channel devoid of riparian vegetation (fig. 23.3). The presence of young cottonwood and other native species at this site now is one example of the general increase in riparian vegetation in this watershed.

Oak Creek

Oak Creek Canyon is considered to be one of the most beautiful riparian areas in Arizona. Visitation to the canyon, its perennial stream, and the dense riparian vegetation along its banks is high because Arizona Highway 89 passes close to Oak Creek and because the canyon is located between Sedona and Flagstaff. The gaging station near Cornville, which records runoff from 355 square miles, has an average daily discharge of only 90 ft^3/s, but this stream has produced several large floods, including five with peak discharges greater than 23,000 ft^3/s, in 1970, 1979, 1980, 1993, and 1995 (fig. 23.4). Despite the frequent floods, riparian vegetation downstream from the gaging station was dense when the station was established in 1940 and has increased through 2003 (fig. 23.5). This site is somewhat unusual in that the only nonnative species readily apparent in the vicinity of the gaging station, tree-of-heaven, is not widespread.

Sycamore Creek

The headwaters of Sycamore Creek, a tributary that joins the lower Verde River near Scottsdale (fig. 23.1), are in the rugged Mazatzal Mountains east of Phoenix.[5] This watershed, as its name suggests, supports abundant, although discontinuous, riparian vegetation along its channels. Near the Verde River, the riparian zone of Sycamore Creek sustains nesting sites for Bald Eagles, precipitating closures to visitation for half the year.[6]

Although its watershed is only 164 square miles and the average daily discharge is only 30 ft^3/s, this tributary can produce large floods, similar to the New River and the Agua Fria River to the west (chapter 25). The flood

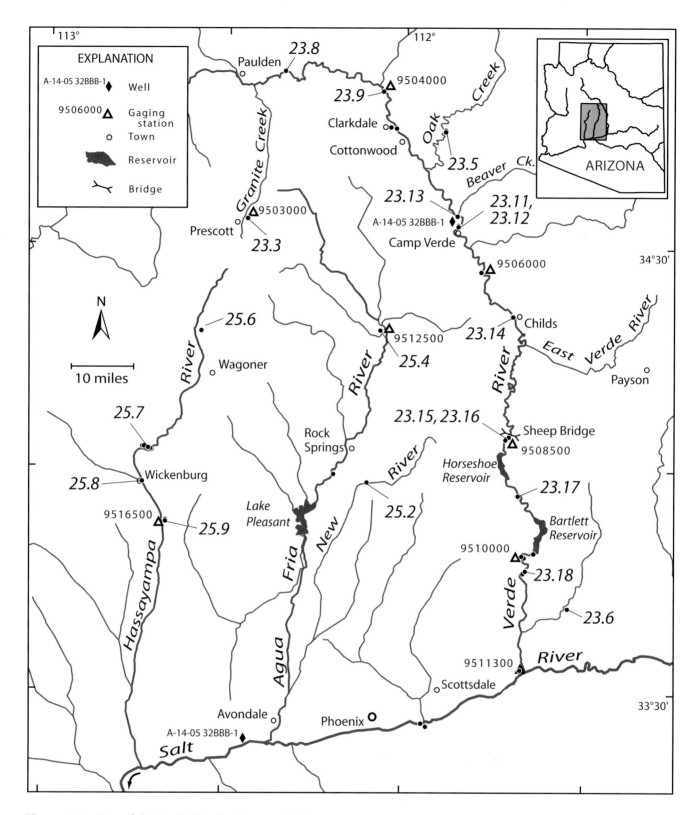

Figure 23.1 Map of the Verde River basin, central Arizona.

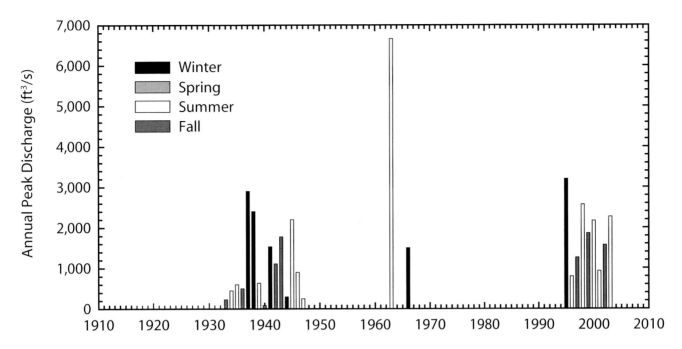

Figure 23.2 Annual flood series for Granite Creek near Prescott, Arizona (station 09503000; 1933–1977, 1963, 1965, 1995–2003).

spawned by Tropical Storm Norma in September 1970 peaked at 24,200 ft³/s on Sycamore Creek.[7] Because the drainage basin is mostly uninhabited, no property damage was sustained during this extremely large flood. Large floods also occurred in 1978, 1980, and 1993 during regional flooding that affected both the Salt and the Verde Rivers.

Sycamore Creek yields considerable base flow to the lower Verde River and therefore is important to water supply in the Salt River Valley. Approximately 4,000 acre-feet per year are discharged from its alluvial aquifer into the Verde River.[8] In the mid-1960s, evapotranspiration losses in the lower alluvial reach averaged 1,500 acre-feet per year. The ephemeral reach in the vicinity of the gaging station, which is in the lower part of the bedrock canyon, has had an increase in riparian vegetation despite the damaging floods that occur in this watershed (fig. 23.6). We obtained no photographic evidence of the riparian zone in the alluvial reach near the Verde River, although the trends on the Verde River and its other tributaries suggest that both native and nonnative vegetation probably increased there. The greater amount of vegetation may have increased the evapotranspiration losses in the lower alluvial reach.

Main Stem of the Verde River

The name "Verde River" elicits considerable discussion as to its origin. Some believe it originated from the Spanish explorer Padre Nentvig, who in the mid-1750s named the Río Verde after its abundant cottonwood trees.[9] Another Spaniard, don Antonio de Espejo, viewed it in 1583 and called it El Río de los Reyes.[10] U.S. explorers, such as Lieutenant Amiel Whipple, called it the San Francisco River because Oak Creek, with its origin near the San Francisco Peaks, was once mistakenly thought to be the headwaters reach of the Verde. Finally, Native Americans living along the river referred to it in Spanish as the "Verde" owing to the presence of malachite along the river (*verde* means "green"), which later was mined at Jerome. Whether the current name actually refers to copper ore or to trees, it is particularly appropriate in the latter sense today because riparian vegetation has generally increased along the Verde River.

Early Observations of the Verde River

Trappers found the Verde River in the 1820s, beginning with James Ohio Pattie and his companions.[11] They left no record of their takings in beaver pelts. Between 1884 and 1888, Edgar Mearns repeatedly visited the Verde River and made notes and collections of beaver.[12] Trappers still operated on the rivers of Arizona, and one individual obtained 120 skins in one winter. Beaver killed mature cottonwood along the Verde in this time period, many of these trees about 2 feet in diameter, and one was 89 inches in diameter. One visitor to the area recorded on April 3, 1887, that in the canyon reach downstream from present-day Camp Verde, "Beavers are numerous, and have cut much of the timber along the river bank."[13]

Several observers with science backgrounds followed a little later in the nineteenth century, providing limited but valuable information on the condition of the riparian ecosystem upstream from present-day Camp Verde.[14] The river meandered through a broad floodplain that supported highly variable assemblages of riverine riparian vegetation. Some reaches had mostly herbaceous vegetation, especially grasses, with locally heavy stands of woody riparian vegetation on higher terraces;[15] other reaches had willow thickets or were lined with cottonwood.[16] The various woody riparian species currently on these different reaches, including cottonwood, wil-

A

B

Figure 23.3 Photographs of Granite Creek near Prescott, Arizona.

A. (January 6, 1935.) This view across Granite Creek shows the bridge over U.S. Highway 89 about 2 miles downstream from Prescott. A low cobble-boulder dam, probably created to provide control for the gaging station, crosses the river in the center. The channel in this view is completely devoid of riparian vegetation. (W. E. Dickinson 1019, courtesy of the U.S. Geological Survey.)

B. (November 22, 2001.) Woody riparian vegetation is now dense in this reach. Young cottonwood trees appear at right; Arizona ash, black willow, and coyote willow are also in the view. The dam probably was eliminated during the first significant flow after photograph A was taken. (D. Oldershaw, Stake 4311.)

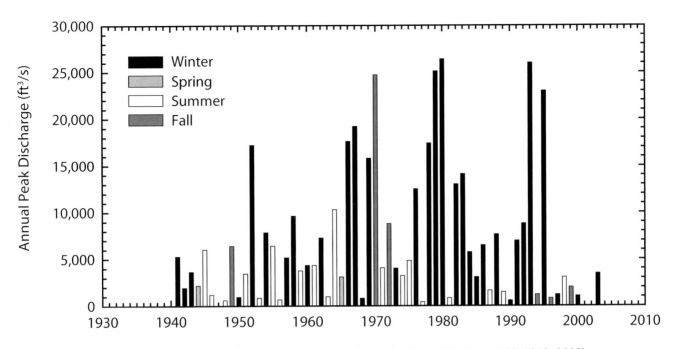

Figure 23.4 Annual flood series for Oak Creek near Cornville, Arizona (station 09504500; 1938, 1940–2003).

low, sycamore, alder, and wild grape, are mentioned in historical accounts; the only now common species not mentioned is tamarisk. The town of Cottonwood was built in 1878 near its namesake trees at the mouth of a tributary.

Oxbow lakes were and are common along the Verde River, including Pecks Lake and Tavasci Marsh. High water tables meant malaria, which plagued the early settlers along the Verde River.[17] The high water tables also trap pollen in sediments, which allows reconstruction of past vegetation changes.[18] Fluctuations in the amount of riparian-tree pollen appear roughly timed with water levels in Pecks Lake, suggesting that periods of high groundwater might discourage woody riparian vegetation but encourage open-water-type vegetation.

The floods that devastated other rivers in the region at the end of the nineteenth century also affected the Verde River. The 1862 flood caused significant channel change along the upper Verde River, flooding a swath estimated at a mile wide. Two years after that flood, Pratt Allyn described a driftwood logjam on the lower Verde River that was extensive enough to back water up, and this driftwood probably was produced during the 1862 event.[19] The February 1891 discharge, which peaked at 150,000 ft³/s

(fig. 23.7C), caused substantial damage to towns and floodplain structures. Other large and damaging floods occurred in 1906, 1916, and 1920.

Land use also affected the riverine environment. A copper smelter near Cottonwood emitted high-sulfur smoke, which denuded hills and stripped foliage from vegetation downwind.[20] Tailings ponds were built into the floodplain at the expense of the riverine environment. Agriculture dependent on surface-water diversion flourished adjacent to the channel in the middle reaches of the Verde River.

The result is all too familiar: early images of the Verde River show a wide, barren floodplain with remnants of riparian vegetation on higher terraces.[21] In one reach between the mouth of Oak Creek and Clarkdale (fig. 23.1), vegetation expanded as the channel narrowed during the low-flood period of the midcentury drought. The amount of cottonwood-willow vegetation increased until flooding increased in the late 1970s, then stabilized. Although mesquite's total acreage did not increase in one reach, its density increased considerably from 1940 through 1995.[22]

Little information is available on riparian vegetation in the middle reaches of the Verde River, and only sketchy information is available for the reaches in the vicinity of present-day

Horseshoe Dam and Bartlett Dam. Archaeological evidence suggests that the prehistoric river was slow moving and marshy and that over the past thousand years it sustained a wide variety of riparian vegetation, including cottonwood, willow, netleaf hackberry, arrowweed, and carrizo grass, as well as numerous aquatic mollusks and mammals.[23] The alluvial reach of the Verde River upstream from its confluence with the Salt River sustained cottonwood along the banks and thick mesquite on the higher terraces.[24]

Groundwater–Surface Water Interactions

Base flow in the Verde River increases downstream and is sustained primarily by groundwater discharge from the Big Chino Valley basin-fill aquifer and, to a lesser extent, by the Little Chino Valley basin-fill aquifer on the western side of the drainage basin.[25] Base flow increases from about 25 ft³/s near Paulden to 83 ft³/s at Clarkdale and about 200 ft³/s at Camp Verde (fig. 23.1). This increase results from discharge from the regional sedimentary aquifer underlying the Coconino Plateau to the north and delivered in tributaries, including Oak Creek, Wet Beaver Creek, and Dry Beaver Creek.

A. (July 13, 1940.) This downstream view, from a bridge on County Route 60 over Oak Creek, shows considerable riparian vegetation lining the channel. None of this riparian vegetation is nonnative. Instead, an extremely diverse riparian assemblage is present consisting of Frémont cottonwood, sycamore, Arizona ash, mesquite, black willow, seepwillow, and Arizona alder (which is locally abundant). (W. L. Heckler 3018, courtesy of the U.S. Geological Survey.)

B. (August 19, 1964.) The bridge has been replaced, eliminating the wood guardrail. Although woody riparian vegetation remains dense along the channel, dieback is apparent in one cottonwood (not visible in photograph A), and the channel appears to have been affected by flooding just before this photograph was taken. (Unknown photographer, courtesy of the U.S. Geological Survey.)

C. (October 2, 2002.) A new cable crosses but does not obscure the view, and a fish hatchery releases effluent into the stream at the downstream reach shown in this view. The existing riparian trees have increased in size in this reach, and young individuals, including tree-of-heaven *(not visible)*, have increased along the channel. Cattail are present in the center foreground and may be indicative of ponding associated with the hatchery effluent. (T. Brownold, Stake 4403.)

Figure 23.5 Photographs of Oak Creek from the Page Springs Bridge.

A. (January 1961.) This upstream view of the gaging station on Sycamore Creek, a tributary of the lower Verde River, shows the stilling well with catwalk on the left side and a low concrete weir across the low-flow part of the channel. This weir provides control for low-flow measurements of this creek. A line of mesquite appears behind the gaging station, extending upstream on the right bank. An unknown tree, possibly an Arizona ash, appears on the extreme right side. (O. M. Peterson, courtesy of the U.S. Geological Survey.)

B. (November 20, 1979.) A flood shortly before this photograph was taken destroyed the gaging station, and the stilling well and other twisted metal associated with the station appear in the left midground. The tree on the extreme right side of photograph A is gone, but other plants, including a mesquite that was upstream from the stilling well, have survived. A line of seepwillow appears in the left foreground. (R. M. Turner.)

C. (November 19, 2002.) By this time, the old gaging station has been replaced by a modern measurement system, and the pipes associated with this station are just visible in the left midground. The concrete control for the gage remains in place, and riparian trees have obviously grown up in the center and on the right side. Coyote willow appears in front of the new gaging station, and black willow *(upstream of the weir on the left)* and Arizona ash *(extreme right side)* are also in the view. The namesake tree for this tributary is not present near the gaging station. (D. Oldershaw, Stake 987.)

Figure 23.6 Photographs of Sycamore Creek near Fort McDowell, Arizona.

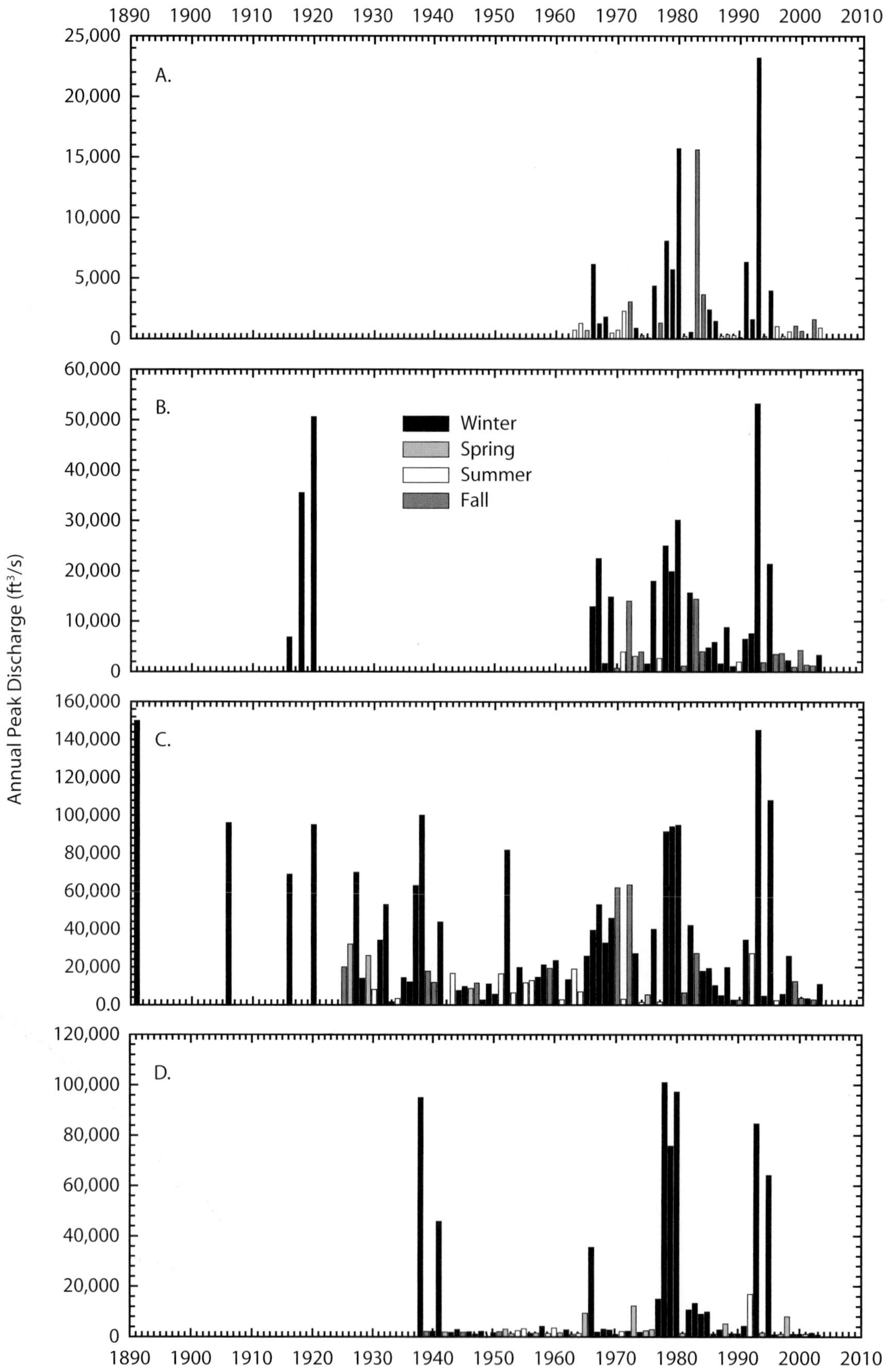

Figure 23.7 Annual flood series for the Verde River.

A. The Verde River near Paulden, Arizona (station 09503700; 1963–2003).

B. The Verde River near Clarkdale, Arizona (station 09504000; 1916, 1918, 1920, 1966–2003).

C. The Verde River downstream from Tangle Creek, Arizona (station 09508500; 1891, 1906, 1916, 1920, 1925–2003).

D. The Verde River below Bartlett Dam near Scottsdale, Arizona (station 09510000; 1938–2003).

In the early twenty-first century, the human population in the Verde River basin, particularly in Big Chino and Little Chino Valleys, is rapidly expanding, and groundwater is the main source of water to support development. Increased water-resources use eventually will reduce base flow in the Verde River and its tributaries and have deleterious effects on riparian vegetation. Because groundwater flow is dependent on a large variety of factors, and because the amount of groundwater pumping continues to change, it is difficult to determine how long it will be before base flow in the Verde River is affected.

Floods, Channel Change, and Flow Regulation

The Verde River, owing to its watershed position south of the Mogollon Rim in central Arizona, is highly affected by dissipating tropical cyclones and warm-winter storms. All of the large regional storms that entered the Southwest in the past 125 years have affected this watershed, in particular the nineteenth-century storms, but also storms from 1978 through 1980[26] and the 1993 storm (fig. 23.7). One storm that generated floods only in the Verde River and Bill Williams River systems (chapter 15) resulted from a dissipating tropical cyclone in August 1951. If not for this one event, the gaging record of the Verde River would mirror those of the Bill Williams River, the upper Gila (chapter 17), the San Francisco (chapter 18), and the Santa Cruz (chapter 21). All of these rivers had larger floods before 1941 and after 1975 and small floods during the mid-century drought.

Paleoflood records for the Verde River give limited information on floods in the past several thousand years. These data consist of dendrochronological records, which relate tree growth to annual flow volumes; flood deposits, which indicate peak discharges; and archaeological evidence, which can provide evidence of both peak discharge and flow volume. Tree-ring reconstructions of streamflow suggest that the period A.D. 665 through 950 had five periods of above-average runoff, that the most recent prehistoric period of high runoff occurred between A.D. 1354 and 1385, and that the period of 1905 through 1924 was the wettest period of the past fifteen hundred years.[27] Sediments interpreted to be flood deposits in archaeological sites provide some verification that large floods occurred in this above-average runoff period.[28] Paleoflood evidence verifies that the 1891 flood was the largest historic event, but that equivalent or slightly larger floods occurred prehistorically.[29]

Although most of the length of the Verde River is at least partially constrained within bedrock walls or against coarse-grained banks, some reaches have experienced considerable channel changes during floods.[30] Following the August 1951 flood, the channel narrowed within the Verde Valley until 1972, when another tropical cyclone–induced flood occurred. The floods of the late 1970s widened the channel considerably, probably at the expense of riparian vegetation and low floodplain deposits. Channel widths in 1892, when cadastral surveys were made in the Verde Valley, were wider than those measured from 1980; in the lower Verde Valley near Scottsdale, the channel narrowed between 1911 and 1980.

Because the Verde River joins the Salt River upstream from Phoenix, regulation for flood control and storage of water became a priority. The first dam constructed in the Verde River watershed was on Fossil Creek, a tributary in the middle reaches, in 1907.[31] Bartlett Dam was completed in 1939, and Horseshoe Dam, 20 miles upstream, was completed in 1946.[32] Horseshoe Dam, which is 194 feet high, was built by Phelps-Dodge Corporation for the Salt River Valley Water Users Association.[33] Bartlett Dam, which was raised to a height of 309 feet in 1995, impounds the larger Bartlett Lake. The primary purposes of these dams are to regulate flow for downstream water deliveries in the Salt River Valley and to provide flood control, but until recently this control affected only the small- and medium-size floods on the Verde River (fig. 23.7D). The two reservoirs inundate about 21 river miles of the Verde River, and their deltas currently sustain considerable populations of tamarisk as well as native vegetation.[34]

Changes in Riparian Vegetation

All of the forty-six repeat photographs of the Verde River show increases in riparian vegetation, irrespective of flow regulation, the occurrence of floods, or proximity to urban areas. At Paulden, a short photographic record illustrates that native species are rapidly becoming established along the upper Verde River (fig. 23.8). The average daily discharge at this gaging station is 45 ft³/s. The flood record at this site covers only the past forty years, and the largest flood in that record had a peak discharge of 23,200 ft³/s in 1993 (fig. 23.7A).

Near Clarkdale, the Verde River has an average daily discharge of 197 ft³/s, and the increase in flow between Paulden and Clarkdale reflects the increase in tributary inputs as well as in spring flow. Riparian vegetation along the river near Clarkdale increased considerably in the past thirty-three years (fig. 23.9). At least twenty native and nonnative riparian species are present in this reach, giving it one of the highest biodiversities of a riparian area in the Southwest. In the wider alluvial reaches downstream, analysis of historical aerial photography taken between 1940 and 1995 indicates that the amount of cottonwood-willow vegetation in the vicinity of Oak Creek peaked in 1977[35] before the start of a series of devastating floods between 1978 and 1995 (fig. 23.7). The photographic record suggests that the vegetation may have increased again after the 1995 flood.

Abundant historical photography in the vicinity of Camp Verde (fig. 23.1) shows large increases in riparian vegetation along a once-barren channel. The average daily flow in the river 9 miles downstream at the gaging station is 465 ft³/s. Groundwater levels remain high at Camp Verde and within the rooting depths of obligate riparian species in this reach (fig. 23.10), although 30-foot declines occurred during the drought years of 1989, 1991, and 1996. Photographs of or from bridges upstream and downstream from Camp Verde (figs. 23.11, 23.12, and 23.13) depict the same general theme of channel narrowing and large increases in riparian vegetation despite the floods from 1978 through

A. (April 21, 1969.) This downstream view of the Verde River shows a relatively narrow channel and canyon in the upper part of the watershed. A discontinuous line of shrubby riparian vegetation is along the channel. What appears to be a small cottonwood tree is present on the upstream side of the stilling well. (H. W. Hjalmarson 9-5037, courtesy of the U.S. Geological Survey.)

B. (November 23, 2001.) The channel has shifted, and the camera station is now on river right instead of river left. Heavy grazing continues in this reach, but, despite it, woody riparian vegetation continues to become established here. Many species have increased, including Frémont cottonwood, seepwillow, coyote willow, and black willow. Although no tamarisk is visible, groves are present downstream and out of the view. (D. Oldershaw, Stake 4313.)

Figure 23.8 Photographs of the Verde River near Paulden, Arizona.

Figure 23.9 Photographs of the Verde River near Clarkdale, Arizona.

A. (April 22, 1969.) Taken from a cliff overlooking the Verde River, this upstream view shows the gaging station on the outside of a right-hand bend in the river. Although the view is slightly out of focus, the channel appears to be mostly denuded of vegetation, and a barren, cobbly point bar appears in the left center. Upstream of the projecting cliff that protects the gaging station, a thin line of riparian trees is present on shallow alluvium and talus below the cliff. (H. W. Hjalmarson 9-5040, courtesy of the U.S. Geological Survey.)

B. (October 3, 2002.) A diverse riparian ecosystem is now present in this reach, and riparian vegetation has clearly increased. No one species is dominant, but mesquite, Arizona ash, sycamore, black willow, Frémont cottonwood, netleaf hackberry, Arizona alder, desert willow, Arizona walnut, wild grape, juniper, oak, and cattail are present. Tamarisk, tree-of-heaven, and Russian olive are also present but are not abundant. (T. Brownold, Stake 4405.)

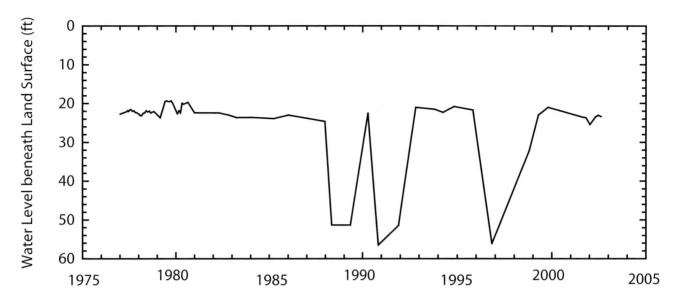

Figure 23.10 Groundwater levels for well A-14-05 32BBB-1 near Camp Verde, Arizona.

1995. Although several nonnative species are present in this reach—notably tamarisk and tree-of-heaven—most of the increase is in Frémont cottonwood, black willow, sycamore, and smaller-statured species such as seepwillow.

Downstream from Camp Verde, the river passes through a canyon partially constrained by basalt flows and other bedrock. At the gaging station near Camp Verde, repeat photos (not shown) reveal smaller increases in riparian vegetation than in the reaches upstream. This difference is probably related to the narrow canyon, higher velocities, and coarser substrate in this reach. Near Childs (fig. 23.14), native riparian vegetation has increased considerably in a lower-gradient reach.

The valley through which the Verde River passes widens as the river approaches the delta of Horseshoe Reservoir. At the long-term gaging station upstream from Tangle Creek, the average annual flow is 591 ft³/s. Repeat photography in the vicinity of this gaging station documents large increases in predominantly native vegetation in the once-wide river channel (figs. 23.15 and 23.16). In this reach, where the Verde River meets the Sonoran Desert, the longer growing season means faster growth rates

of riparian trees, and cottonwood and black willow have regrown on terraces damaged or denuded by the 1993 and 1995 floods.

The Verde River downstream from Horseshoe Dam and Bartlett Dam shows the twin effects of hydrologic stability: growth enhancement of riparian vegetation and large, clear-water releases from the dams. The releases coarsen floodplain substrate to the point where the habitat is unsuitable for most plants. The average daily discharge in these reaches is 663 ft³/s. At the former gaging station downstream from Horseshoe Dam, native and nonnative species have increased, but not to the same extent as at Camp Verde (fig. 23.17). Large dam releases, particularly in 1993, appear to have depleted the fines from the floodplain, leaving a cobble bar. The same phenomenon occurred below Bartlett Dam (fig. 23.18), where a thin band of native and nonnative species grows close to low-water levels and the larger cottonwood, black willow, sycamore, and mesquite are separated from the low-water channel by a largely barren floodplain.

The average daily flow of the Verde River near its confluence with the Salt River is 621 ft³/s. Much of the time, flow is relatively steady and less than

about 100 ft³/s, which allows riparian vegetation to become established close to the low-flow channel (fig. 23.19). As is the case with other reaches of the Verde River, both native and nonnative species have increased, although increases in Frémont cottonwood are somewhat restricted to discontinuous groves at discrete points along the channel.

The reaches of the Verde River covered by our repeat photography show increases in riparian vegetation. Increases are large where the river has a low gradient and smaller where it has a higher gradient. Both native species and nonnative species are increasing. Riparian vegetation has increased since the first third of the twentieth century despite the occurrence of extremely large floods during the last third of that century and despite the fact that dam releases during those floods created substrate that is less productive for riparian vegetation. Creation of functionally xeric floodplains by large, clear-water releases is a phenomenon common to both the Salt River (chapter 24) and the Verde River and is only locally important along other rivers in the region.

A

B

Figure 23.11 Photographs of the Verde River near Camp Verde, Arizona.

A. (March 3, 1938.) The White Bridge, on Arizona Highway 260 south of Camp Verde, is one of two bridges that cross the Verde River in this area. This view during a flood in the winter of 1938 shows the original bridge before an abutment was washed out. The river flows right to left in this view, which was photographed from the left bank. Frémont cottonwood in an open gallery forest lines the far side of the river; the flood inundates the bases of these trees as well as some unidentified low shrubs. A single tree in the river upstream from the bridge might be a leafless black willow, and the foreground plants are leafless mesquite. The background mountains on the left side of the view are obscured in low clouds. (Photographer unknown, courtesy of the U.S. Geological Survey.)

B. (October 28, 2001.) Because of channelization, replacement of the bridge, and the change in riparian vegetation, this match may be accurate only to within 30 feet of the original camera station. Dense riparian vegetation lines the Verde River in this reach, blocking the view of both the new White Bridge, the channel, and most of the lower hills visible in photograph A. Mesquite is the dominant plant across the foreground, and Frémont cottonwood, black willow, greythorn, catclaw, and Arizona ash are visible. (D. Oldershaw, Stake 3028.)

Figure 23.12 Photographs of the Verde River near Camp Verde, Arizona.

A. (November 1971.) This downstream view from the Black Bridge at Camp Verde shows the Verde River downstream from the gaging station. Cottonwood and willow trees line the left bank, and an open floodplain appears on the right with scattered trees and shrubs. A large flood of unknown size and associated with Tropical Storm Norma occurred in this reach thirteen months before this photograph was taken. (T. W. Anderson 9505550, courtesy of the U.S. Geological Survey.)

B. (October 28, 2001.) The channel has narrowed, and riparian vegetation has increased significantly in the intervening thirty years, despite a flood with a peak discharge of 119,000 ft^3/s in February 1993 and two other floods greater than 100,000 ft^3/s. Black willow is present in the left foreground and at various points in the view; the line of cottonwood present in 1971 is larger and denser now. Other species that are less obvious include Arizona ash, coyote willow, seepwillow, carrizo grass, and tamarisk. (D. Oldershaw, Stake 3029.)

Figure 23.13 Photographs of the Verde River near Camp Verde, Arizona.

A. (Late January 1917.) This upstream view shows the original Black Bridge at Camp Verde. The channel is wide and open, and what appears to be coyote willow is partially submerged in the foreground. (Unknown photographer, courtesy of the U.S. Geological Survey.)

B. (October 28, 2001.) The new bridge is farther upstream than the original one, and riparian vegetation has increased considerably. The channel is now lined with mesquite, tamarisk, Frémont cottonwood, Arizona ash, coyote willow, desert broom, black willow, box elder, seepwillow, and carrizo grass. Giant reed occurs along irrigation ditches in the vicinity of this camera station. (D. Oldershaw, Stake 3031.)

A

B

Figure 23.14 Photographs of the Verde River at the outlet of the Childs Power Plant.

A. (August 1939.) Fossil Creek, a tributary of the Verde River with a watershed extending northward to the Mogollon Rim, is diverted into a pipeline to generate power near Childs, Arizona. This view across the Verde River upstream from Childs shows a dense line of riparian vegetation that obscures the power plant. The species present include black willow, Frémont cottonwood, Arizona ash, seepwillow, sycamore, netleaf hackberry, and a low herbaceous species that might be horse-tail reed. (H. E. Dahman 2641, courtesy of the U.S. Geological Survey.)

B. (November 11, 2002.) The channel has narrowed, despite devastating floods in this reach in the 1990s. Riverine riparian vegetation has increased significantly in terms of both the density of plants and the heights of individuals that appear to be persistent. Tamarisk is not obvious in 1939 but is now present, and cattail appear in the foreground. (T. Brownold, Stake 4430.)

Figure 23.15 Photographs of the Verde River at Sheep Bridge.

A. (August 1945.) Sheep Bridge, built in 1943, is a suspension bridge over the Verde River upstream from Tangle Creek that allows pedestrians and livestock to cross. This view across the Verde River from river left shows the newly constructed bridge over a nearly barren channel. Several unidentified shrubs are present, probably seepwillow, downslope from a rich xerophytic assemblage characteristic of the upper Sonoran Desert. (Unknown photographer 3760, courtesy of the U.S. Geological Survey.)

B. (November 4, 2002.) Owing to flood damage, the bridge was rebuilt in 1989. Riparian vegetation has increased in this reach and includes black willow, tamarisk, seepwillow, and scattered cottonwood, with cattail and carrizo grass near the channel in pools. Mesquite and catclaw have increased on the foreground slopes. (D. Oldershaw, Stake 4353.)

Figure 23.16 Photographs of the Verde River downstream from Tangle Creek.

A. (August 1945.) This upstream view of the Verde River at the gaging station downstream from Tangle Creek shows the measurement cable crossing midground. Low woody riparian vegetation grows along the channel and on the floodplain. Coyote willow appears to form a dense patch on the far bank at center; a line of small trees might be Arizona alder. Mesquite and xerophytic shrubs occur in the foreground. (Unknown photographer 3759, courtesy of the U.S. Geological Survey.)

B. (November 4, 2002.) Riparian vegetation blocks the view of the channel but not the cable or the background mountains. The species present in the foreground include foothill palo verde, catclaw, and saguaro. The woody riparian vegetation includes mesquite, black willow, Arizona ash, sycamore, burrobrush, desert broom, and sweet bebbia. Grazing still occurs in this reach. (D. Oldershaw, Stake 4351.)

Figure 23.17 Photographs of the Verde River below Horseshoe Dam.

A. (September 8, 1938.) This view across the Verde River downstream from Horseshoe Dam shows the hydrographer's house and facilities as well as the stilling well *(extreme left)*. The river flows from right to left in this view. The floodplain is essentially barren, with only a couple of shrubs present, which appear to be young tamarisk. (Unknown photographer 2381, courtesy of the U.S. Geological Survey.)

B. (November 5, 2002.) The hydrographer's house has been removed, and the river has shifted closer to the left bank and the camera position. The riparian vegetation along the channel appears to have become established after the 1995 dam releases, which peaked at 84,700 ft³/s in 1993 and 61,400 ft³/s in 1995 in this reach, and it includes Frémont cottonwood, black willow, coyote willow, tamarisk, burrobrush, seepwillow, and desert broom. At this location, mesquite grows as a xerophytic shrub high on the channel margins. (D. Oldershaw, Stake 4354.)

A. (August 19, 1934.) This downstream view shows a reach of the Verde River just upstream from Camp Creek and before the river leaves its confining bedrock valley and flows across deep alluvium to its confluence with the Salt River. A low growth of shrubby riparian vegetation, which appears to be coyote willow, lines the channel, and what looks like seepwillow grows along the right side. (R. E. Cook 473, courtesy of the U.S. Geological Survey.)

B. (March 8, 1979.) This view, taken after the 1979 floods from a camera station in front of the original station, shows less on the right side. The floods have scoured channel and floodplain, reducing the amount of riparian vegetation. The plants survived the floods but appear to have sustained significant damage. Mesquite has increased on the floodplain on river left. (R. M. Turner.)

C. (November 20, 2002.) Riparian vegetation increased significantly in this reach following the floods of the 1970s through the mid-1990s. Coyote willow is clearly visible in the left foreground, and black willow and Frémont cottonwood form the open groves of trees in the middle distance and on the bend. Tamarisk is abundant in this reach but is confined to clumps instead of in dense thickets. (D. Oldershaw, Stake 965.)

Figure 23.18 Photographs of the Verde River below Bartlett Dam.

Figure 23.19 Photographs of the Verde River upstream from its confluence with the Salt River.

A. (March 27, 1945.) This upstream view of the Verde River was taken from a hillslope about a half-mile upstream from its confluence with the Salt River. The camera station is on the Fort McDowell Indian Reservation, and the channel is wide, with several flood-overflow channels in the view, particularly in the center and left midground. Frémont cottonwood is common in the view, and several groves are present as far up the channel as is visible. The lower trees along the river appear to be black willow, although tamarisk likely also is present. Mesquite is dense on the higher banks on the left side and in the right midground. A foothill palo verde appears in the left and right foreground. (H. D. Padgett Jr. A-562, courtesy of the U.S. Geological Survey.)

B. (April 8, 2004.) The channel has narrowed considerably, and riparian vegetation has increased in several parts of the view. The groves of cottonwood have increased in density, particularly in the bend in the distance. Black willow in particular has increased in the view, and burrobrush, cattail, seepwillow, and tamarisk are common. The foothill palo verde in the right foreground has died, and the stump is prominent. (D. Oldershaw, Stake 4766.)

Summary. The Salt River is arguably the most important drainage basin wholly within Arizona, providing water supply and hydroelectric power to the Salt River Valley. This river drains much of the highlands of eastern Arizona before flowing through deep bedrock canyons that end about 20 miles upstream from Phoenix. Most of the upper Salt River flows through narrow bedrock canyons, where both native riparian vegetation and nonnative riparian vegetation have increased modestly despite the occurrence of extremely large floods. Downstream from its confluence with the Verde River, the Salt River flows through the Salt River Valley in a meandering pattern over a broad floodplain. Despite full regulation by dams upstream from Phoenix, the combined flows of the Salt and the Verde Rivers produced damaging floods through the Salt River Valley as recently as 1993. Little water is normally released through the deep alluvial valleys of the Phoenix basin, which once supported scattered stands of cottonwood. It is likely that open groves of cottonwood once existed in the Salt River Valley, and they were eliminated by the combination of nearly complete surface-water diversion, except during extreme floods; agricultural clearing; urban development; and channelization of the Salt River.

Arising along the Mogollon Rim in east-central Arizona, the Black and White Rivers merge to form the Salt River, which then flows 200 miles to its confluence with the Gila River (fig. 24.1). The Salt's drainage area of 13,223 square miles includes some of the most rugged mountainous and desert terrain in Arizona. Downstream from the Black-White River confluence, the river is confined in Salt River Canyon before emerging into shallow alluvial basins upstream from Theodore Roosevelt Lake. A series of four

dams—three upstream from the confluence with the Verde River (chapter 23)—completely regulates flow from the mountain ranges on the east to the alluvial Phoenix basin. Crossing the Salt River Valley, the river flows ephemerally between soil-cemented banks during rare periods of storm runoff not captured by the upstream reservoirs.

Although this river has had many names, it is called the Salt River owing to the brackish nature of groundwater effluent during low-flow periods.[1] Because it provides most of the water supply for Phoenix, it is arguably Arizona's most important river, having headwaters wholly within the state. The Salt River was a benchmark in water development in the West because it was the first to be affected by the major flow-regulation projects. Granite Reef Dam, completed in 1908, was one of the first concrete structures built to divert surface water for irrigation. More than 1,000 miles of irrigation canals channel surface water through the Salt River Valley,[2] overshadowing Hohokam irrigation projects of the twelfth and thirteenth centuries.[3] In addition to the intensive use of surface water, groundwater resources were developed and used extensively in the twentieth century. By 1922, yearly groundwater withdrawals in the Salt River Valley exceeded combined groundwater use in the rest of Arizona.

Early Observations of the Salt River

The first visitors to the Salt River Valley guessed that the river had perennial flow that was about 200 feet wide with a depth of as much as 4 feet.[4] James Ohio Pattie and his companions arrived at the confluence of the Gila and Salt Rivers on February 1, 1826.

They saw the abundant beaver in the Salt River and traveled upstream more than 80 miles, trapping in both the Salt River Canyon and the Verde River Canyon. Pattie described the Salt River near present-day Phoenix as "a most beautiful stream, bounded on each side with high and rich bottoms,"[5] but he did not mention whether cottonwood trees were present through the Salt River Valley. Beaver sign were seen near the Salt-Gila confluence as late as 1885.[6]

Settlement of the Salt River Valley began with the arrival of Mormons to found Mesa, Arizona, in 1877.[7] They immediately began to regulate the Salt River, using an abandoned Hohokam canal to irrigate fields carved into the floodplain. Thus began a thirty-year struggle with this river that ended with construction of Granite Reef Dam and the establishment of the Salt River Project (SRP). Settlers observed abundant bonytail, Colorado River pikeminnow, and other native fishes in the river and irrigation canals before about 1915.[8] The last known Colorado River pikeminnow was caught in the Salt River upstream from Phoenix in 1937; this species, reportedly common in 1906, was the basis of a commercial fishery.[9]

Floods

Because the Salt River flowed perennially through an arid region, it attracted civilizations. The Hohokam established major cities fed by a network of canals from the Salt. This culture abandoned the river valley around A.D. 1350, leaving a desolate basin largely uninhabited by humans. The reason for the abandonment might have been large floods combined with sustained droughts. Twenty-seven floods between around A.D. 850 and

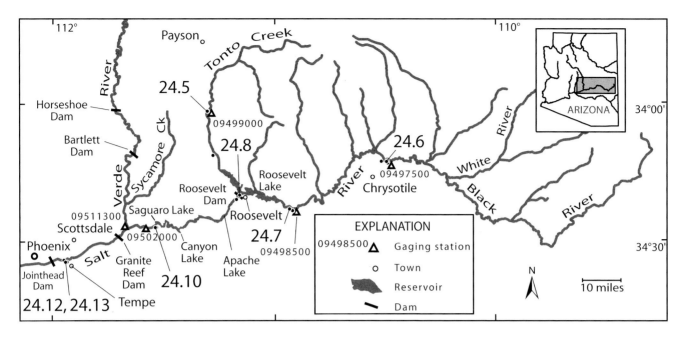

Figure 24.1 Map of the Salt River in central Arizona.

1976 exceeded a bankfull discharge threshold of 175,000 ft³/s on the Salt River upstream from Tempe.[10] In A.D. 899, the Salt River reportedly yielded 2.5 million acre-feet of runoff, based on tree-ring reconstructions,[11] and a flood with a peak discharge of 420,000 ft³/s occurred at about this time.[12] The long-term average flow volume of the Salt and Verde Rivers at gaging stations upstream from the Salt River Valley is 1.2 million acre-feet per year,[13] and the largest measured flood, in February 1891, had a peak discharge of about 300,000 ft³/s (see the next section).

The combined flow of the Salt and Verde Rivers has created some of the most devastating floods in Arizona history. For example, in February 1890, the Salt River rose 17 feet, washing out the bridge at Tempe and damaging floodplain structures all the way to Yuma.[14] In February 1891, the river rose one foot higher than this and flooded a swath of the valley up to 8 miles wide. The 1891 flood on the Salt River, with a peak discharge estimated to be 300,000 ft³/s,[15] had serious consequences for the fledgling town of Phoenix; the floodwaters reportedly destroyed one-third of the town.[16] Floods that caused inundation or other damage occurred in September 1897, January 1905, March 1905, January 1916, November 1919, and March 1938.[17]

Despite the presence of dams upstream on both the Salt and the Verde Rivers, floods in the last quarter of the twentieth century were severe through Phoenix. Generated mostly upstream in the Salt River (fig. 24.2), these floodwaters combined with waters pouring down the Verde River to create truly awesome floods through the urban area. In March 1978, floodwaters peaked at 125,000 ft³/s through Phoenix, and on January 8, 1993, the peak discharge was 129,000 ft³/s (see fig. 24.11). The need to protect floodplain structures, such as bridge abutments, and to increase the area for development next to the river drove channelization of the river and the installation of soil cements on its banks.

Flow Regulation and Groundwater Development

The first diversion of the Salt River for irrigation purposes occurred in 1885.[18] This structure paled in comparison to what came next: Granite Reef Dam (1908), Roosevelt Dam (1911), Mormon Flat Dam (1926), Horse Mesa Dam (1927), and Stewart Mountain Dam (1930). Other dams on the Verde River (see chapter 23) completed the full regulation of the Salt River by 1946. Roosevelt Dam, originally called Tonto Dam because it was built at the mouth of Tonto Creek, impounds

Theodore Roosevelt Lake and flooded the town of Roosevelt. In 1959, the dam was renamed for President Theodore Roosevelt. The original dam, 284 feet high, was raised to a height of 357 feet in a project completed in 1996.[19] Downstream from the former town of Roosevelt, the Salt River is completely regulated for the purposes of flood control, irrigation, and domestic water supply for the Salt River Valley.

The SRP, which began in 1903 as part of the Salt River Valley Water Users Association,[20] delivers a little less than one million acre-feet per year of water to central Arizona, primarily to Phoenix and its suburban satellite communities.[21] It also produces hydroelectric power from the dams it manages: potential power production of 36 megawatts (MW) from Roosevelt Dam; 129 MW from Horse Mesa Dam and its pumped storage unit; 60 MW from Mormon Flat Dam and its pumped storage unit; 13 MW from Stewart Mountain Dam;, and a combined output of less than 5 MW from several small structures on canals. The combination of flood control, water storage, and power generation makes this complex of dams on the Salt River extremely important to Arizona's development and economy.

Despite the abundance of surface-water diversion for irrigation, groundwater pumping in Arizona began to

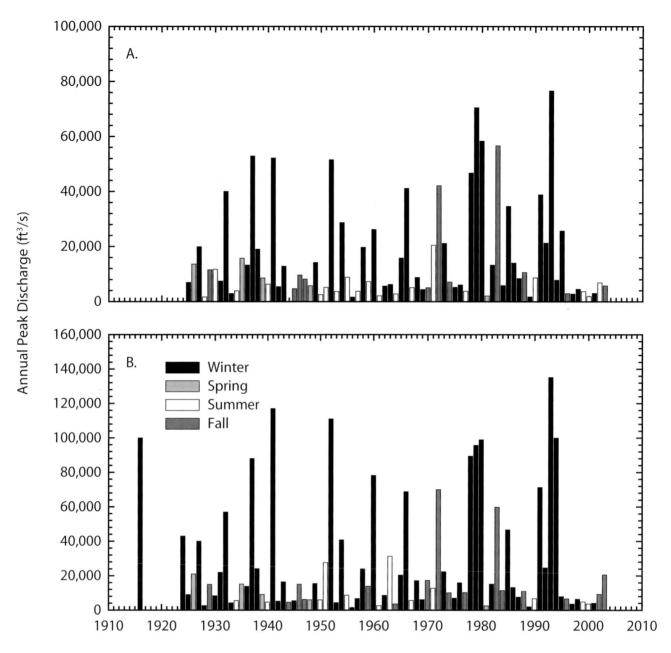

Figure 24.2 Annual flood series for the Salt River.

A. The Salt River near Chrysotile, Arizona (station 09497500; 1925–2003).

B. The Salt River near Roosevelt, Arizona (station 09498500; 1916, 1924–2003).

accelerate in the 1920s following the development of turbine pumps (fig. 24.3).[22] As water levels dropped, some fragile layers in the alluvial aquifer drained and compacted, causing subsidence of the ground surface. In the Salt River Valley, problems attributed to land subsidence, including earth fissures and collapsed well casings, were documented by the early 1960s.[23] Excessive groundwater development in the Salt River Valley led to the creation of the Phoenix Active Management Area in 1980.[24] Water levels in some areas had declined as much as 450

feet between 1923 and 1982. Nearer to the Salt River, water levels that had declined in the 1960s rebounded by the 1980s (fig. 24.3B), responding to the combination of flood releases and use of CAP water imported from the Colorado River.

Changes in Riparian Vegetation

Tonto Creek

Tonto Creek, named for the Tonto Apache who once inhabited this watershed,[25] drains an area of more than 675

square miles below the Mogollon Rim of central Arizona (fig. 24.1). As with many drainages in this region, floods can be extremely large during regional storms. The flood of record, which occurred during the January 1993 floods, had a peak discharge of 72,500 ft^3/s (fig. 24.4). The effect of the early-twenty-first-century drought is shown in the substantial decrease in peak discharges between 1996 and 2002 in this drainage basin. Despite the floods and perhaps because of a sustained drought period, photographs show that riparian vegetation has increased

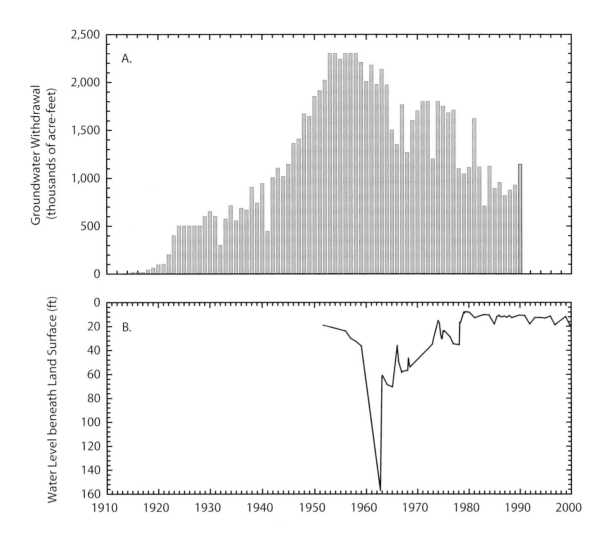

Figure 24.3 Groundwater records for the Salt River Valley near Phoenix.

A. Annual amount of groundwater used in Maricopa County (from Anning and Duet 1994).

B. Groundwater levels for well B-01-02 36BBC near the Salt River in Phoenix, Arizona.

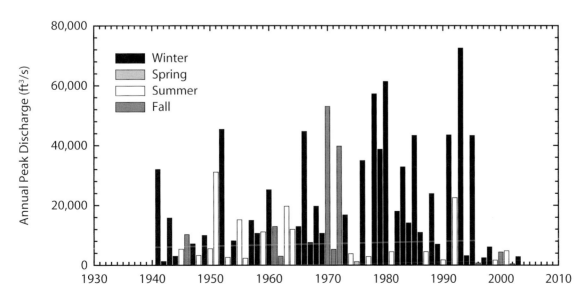

Figure 24.4 Annual flood series for Tonto Creek above Gun Creek, near Roosevelt, Arizona (station 09499000; 1941–2003).

along Tonto Creek, even in relatively narrow channel reaches (fig. 24.5). Other photographs near the delta with Theodore Roosevelt Lake show similar increases, although tamarisk is more abundant.

Salt River Upstream from Roosevelt Dam

About 30 river miles downstream from the confluence of the Black and White Rivers, the Salt River passes under a pair of bridges for the combined U.S. 60 and Arizona Highway 77. Photographs upstream from the older bridge (fig. 24.6) show conditions in the bedrock-controlled channel of the Salt River in the vicinity of the gaging station (near Chrysotile; fig. 24.2A). Riparian vegetation, mostly native species but also with significant amounts of tamarisk, has steadily increased along the channel. Interestingly, one of the species here is fan palm (fig. 24.6C). This tree likely was planted, although the species has been found in natural settings in central Arizona.[26]

Downstream from the bridges, the Salt River flows through a wilderness reach renowned for its whitewater characteristics. Several small tributaries add flow to the river, including Cibecue Creek, Canyon Creek, and Cherry Creek from the north and Pinal Creek from the south. A second gaging station is on the bridge for Highway 288 at the downstream end of this reach and just upstream from the delta of Theodore Roosevelt Lake. As shown in fig. 24.2, the increase in flood magnitude on the Salt River is dramatic owing to contributions by the intervening tributaries.

The valley widens downstream from the bridges, and the river flows through a shallow alluvial valley with considerable room for growth of riparian vegetation. Between 1937 and 2000, the channel width decreased, and riparian vegetation increased despite the spate of severe floods between 1978 and 1995 (fig. 24.7). Although most of the riparian vegetation in this reach is tamarisk, the presence of carrizo grass, seepwillow, and black willow suggests that native species are also increasing here.

Theodore Roosevelt Lake was the first large reservoir in Arizona, and it covers about 10 miles of the once free-flowing Salt River and 8 miles of Tonto Creek in a relatively wide reach. All woody riparian vegetation in these reaches was killed, but replacements occurred on the delta. As shown in figure 24.8, little riparian vegetation was present in 1910 at the confluence of the Salt River and Tonto Creek, just upstream from the present-day location of the dam. Fluctuations in lake level preclude establishment of significant riparian vegetation along the steep shoreline (fig. 24.8B), although tributary deltas can support dense stands of mostly nonnative vegetation.

Salt River Downstream from Stewart Mountain Dam

The reservoir system on the Salt River upstream from its confluence with the Verde River (fig. 24.1) controls small- and intermediate-size floods on the main stem. However, large floods have overwhelmed the combined storage capacity of the reservoirs, resulting in damaging releases downstream into the Salt River Valley. Ultimately, floods such as those shown in figure 24.9 prompted raising the height of Roosevelt Dam and the storage capacity of Theodore Roosevelt Lake in 1996.

Because the presence of dams regulated floods after 1911, and because releases were steady to supply irrigation water to the Salt River Valley, the channel along the lower Salt River was stable enough to grow significant amounts of riparian vegetation in the 1930s (fig. 24.10A). Tamarisk, which was introduced into the basin in the late nineteenth century, was well established here in 1938. The clear-water releases of the late 1970s, in particular the January 1979 release that peaked at 54,000 ft^3/s, scoured the channel in this reach and coarsened the substrate on the low floodplain, reducing its ability to sustain riparian vegetation (fig. 24.10B). By 1995, and following the 1980 and 1993 releases of 64,000 and 34,500 ft^3/s, respectively (fig. 24.9), the channel had widened, and floodplain substrate had coarsened even more (fig. 24.10C). Riparian vegetation here consists of tamarisk as well as native shrubs and Frémont cottonwood.

Granite Reef Dam, below the confluence of the Salt and Verde Rivers, diverts most of the surface flow from the channel into an intricate system of canals, and riparian vegetation quickly diminishes downstream in the main channel. The channel of the Salt River has migrated laterally up to a mile across a broad floodplain,[27] which would minimize the ability for riparian vegetation to remain established near the channel. This migration occurred during extremely large floods fed by the combination of the two rivers (fig. 24.11). Before modifications were completed in 1996, the dams upstream from the confluence of the Salt and Verde could not control the largest discharges, so that four floods with peak discharges greater than 100,000 ft^3/s passed through the Salt River Valley between 1978 and 1993.

Long-term changes in the Salt River corridor (1868–1969) have been described where it flows through Tempe.[28] Before channelization, the channel was braided and shifting. Cadastral survey notes from 1868 described the channel as being lined with "timber cottonwood along banks, and mesquite and willow brush." By 1934, the floodplain was devoid of vegetation, although riparian vegetation lined the low-flow channel.[29] By the late 1950s, urban and industrial encroachment had eliminated the original channel of the Salt River, constraining the ill-defined channel within discontinuous dikes.

Following heavy usage in the middle of the twentieth century, groundwater levels rebounded and remained high near the Salt River (fig. 24.3B). The declines were in part in response to the midcentury drought, and the rebound occurred at least in part because of importation of CAP water to the Phoenix basin in 1985.[30] In the latter half of the twentieth century, the Salt River alternated between extreme floods and drought (fig. 24.11), hardly conditions that would sustain growth of significant amounts of riparian vegetation. However, a time series of aerial photographs[31] shows the midcentury establishment and destruction of considerable riparian vegetation—possibly dominated by tamarisk—along the Salt River upstream from the Mill Avenue Bridge in Tempe. The destruction was largely complete at the time of a devastating flood in late December 1965.

Figure 24.5 Photographs of Tonto Creek.

A. (February 1, 1941.) This upstream view of Tonto Creek shows the approach to the gaging station within a bedrock canyon with a flow of 456 ft³/s. Riparian vegetation covers the few alluvial terraces, notably the one on river left *(photo right)*. The vegetation appears to be mostly mesquite with scattered willows. (J. A. Baumgartner 3403, courtesy of the U.S. Geological Survey.)

B. (November 10, 2002.) Flow at this time is low (less than 0.5 ft³/s) into a pool upstream of Gun Creek. The vegetation present is diverse and includes carrizo grass, tamarisk, seepwillow, and lesser amounts of mesquite, netleaf hackberry, and black willow. (T. Brownold, Stake 4427.)

A. (November 26, 1935.) This upstream view from the old, two-lane bridge that crosses the Salt River in Salt River Canyon shows a relatively small discharge of 277 ft^3/s. Scattered native shrubs, including willows and brickellbush, appear to occupy the floodplain at right center. The road leading to Show Low (the combined U.S. Highway 60 and Arizona Highway 77) appears as a one-lane cut through the hillslope at center. (R. E. Cook 2280, courtesy of the U.S. Geological Survey.)

B. (June 25, 1964.) The water is low (about 97 ft^3/s), exposing the bedrock that forms the channel bed and the low-water control downstream from the gaging station. In the intervening twenty-nine years, three floods with peaks of greater than 50,000 ft^3/s passed through this reach. At this time, tamarisk is interspersed with the native shrubs on the floodplain and lines river left, which was mostly devoid of woody vegetation in 1935. Fan palms *(lower right),* which are not native to this area, were planted as part of a roadside park well before this photograph was taken. The roadcut on the skyline has been widened. (R. M. Turner.)

C. (October 25, 2000.) The water level is only slightly higher in 2000 than it was in 1964. In the intervening thirty-six years, two floods have exceeded 70,000 ft^3/s, and four have exceeded 50,000 ft^3/s. Despite these floods, riparian vegetation along the banks has increased, in particular nonnative tamarisk. The palms have grown considerably. (D. Oldershaw, Stake 363.)

Figure 24.6 Photographs of the Salt River near Chrysotile, Arizona.

A. (April 22, 1937.) In this upstream view, the Salt River is flowing at 4,000 ft³/s. The long-term gaging station in this reach is associated with the bridge in the distance, and a diversion dam is present just downstream from the camera station. Two months before this photograph was taken, the brush-covered island at right center was submerged during a February flood; most floods on the Salt River occur during the winter months. This camera station is several miles upstream from the top of Roosevelt Lake, the first of the major flood-control and water-supply structures on the Salt River upstream from Phoenix. (W. E. Dickinson 2166, courtesy of the U.S. Geological Survey.)

B. (February 3, 1979.) The brush-covered island is now densely covered with mostly nonnative tamarisk, although many native species also occur in this reach, including cottonwood, coyote willow, black willow, and various species of brickellbush. The bar in the left foreground was scoured during large floods in both 1978 and 1979. (R. M. Turner.)

C. (November 25, 2000.) Flood frequency on the Salt River did not change significantly in the twentieth century, as it did on other rivers in the region, although four one-hundred-year floods did occur in a fifteen-year period. The 1993 flood, which had a peak discharge of 143,000 ft³/s at the gaging station on the bridge visible in the distance, did little to slow the advance of riparian vegetation—in particular tamarisk—at this site. Native species, notably carrizo grass, have also increased, although they are difficult to distinguish from the tamarisk in this view. (D. Oldershaw, Stake 955.)

Figure 24.7 Photographs of the Salt River near Roosevelt, Arizona.

A

B

Figure 24.8 Photographs of the Salt River at Roosevelt, Arizona.

A. (Ca. 1910.) This view and several others taken immediately before completion of Roosevelt Dam show the crossing of the Salt River at the town of Roosevelt. Tonto Creek enters the Salt River from the left. The channel is wide and mostly barren of riparian vegetation, although discontinuous patches of what likely is mesquite appear on the floodplain in the midground. (G. W. James P4539, courtesy of the Southwest Museum.)

B. (April 8, 2004.) Theodore Roosevelt Lake now covers the former town of Roosevelt, which was moved several miles toward Globe, Arizona, after construction of Roosevelt Dam was completed. As this view shows, lake level fluctuates, minimizing establishment of riparian vegetation on coarse talus slopes. (D. Oldershaw, Stake 4770.)

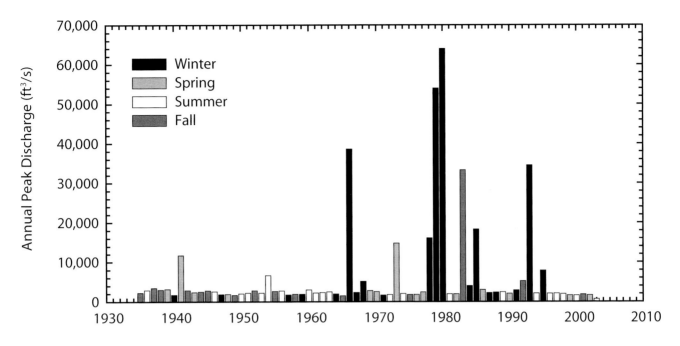

Figure 24.9 Annual flood series for the Salt River below Stewart Mountain Dam, Arizona (station 09502000; 1935–2003).

Nonetheless, it is difficult to assess the change in woody riparian vegetation in the roughly 35 miles of channel between Granite Reef Dam and the Salt's confluence with the Gila River. Most old photographs taken in this reach either cannot be replicated owing to blockage by structures (e.g., Sun Devil Stadium in Tempe), elimination of camera stations owing to extensive urbanization, or difficulty in determining the location of the original camera station. Moreover, although riparian vegetation quickly established, it persisted for only short periods before destruction from urban encroachment or floods.

Some photographs taken from prominent overlooks can be matched, yielding some information on the amount of change that has occurred in this reach (figs. 24.12 and 24.13). Many early photographs show an open gallery forest of cottonwood with low shrubs that may be coyote willow. Significantly, some of these stands of riparian vegetation appear to form a narrow band on a low floodplain beneath higher terraces, reminiscent of the condition present in arroyos. During the first third of the twentieth century, riparian vegetation was becoming reestablished within the eroded channel margin, as occurred along the Santa Cruz River at Tucson (chapter 21). Other photographs from the middle of the twentieth century (fig. 24.12A) show mostly herbaceous vegetation along the channel.

Although it is likely that at least some cottonwood groves were eliminated through the Salt River Valley, it is impossible to assess just how much change has occurred in this reach. The extensive channelization using concrete and rock, combined with the complete diversion of base flow, virtually guarantees that significant amounts of woody riparian vegetation are unlikely to become reestablished along this part of the Salt River despite high groundwater levels (fig. 24.3B). Downstream, wastewater effluent is discharged into the Salt River, and woody riparian vegetation has become established in that reach, but we have no photography to evaluate the amount of change.

A. (September 9, 1938.) This downstream view of the Salt River below Stewart Mountain Dam shows low, woody riparian vegetation encroaching to water level on both banks. Flow at the time of this photograph is 2,390 ft^3/s and is completely regulated by the four dams upstream. The vegetation consists of seepwillow and brickellbush, with scattered tamarisk trees on river right. (J. A. Baumgartner 2391, courtesy of the U.S. Geological Survey.)

B. (March 7, 1979.) This view, taken after the 1978 and 1979 floods (highest peak discharge was 54,000 ft^3/s, fig. 24.9), shows a wide, scoured floodplain with a low flow of 13 ft^3/s. Other photographs taken in this reach indicate that although the bars remain composed mainly of cobbles, the average particle size has coarsened, indicating further degradation of habitat within the low floodplains. Despite the floods, tamarisk remains on river right. (R. M. Turner.)

C. (January 31, 1995.) This view, taken after the 1993 peak discharge of 34,500 ft^3/s, shows an even coarser particle size on the floodplain. The flow is 470 ft^3/s. The channel position has shifted in response to rearrangement of the coarse sediment within the channel. One tree that survived the floods of the late 1970s and early 1980s is dead in the center of the view. Low shrubs, tamarisk, and cottonwood appear to be established. (D. Oldershaw, Stake 962.)

Figure 24.10 Photographs of the Salt River below Stewart Mountain Dam.

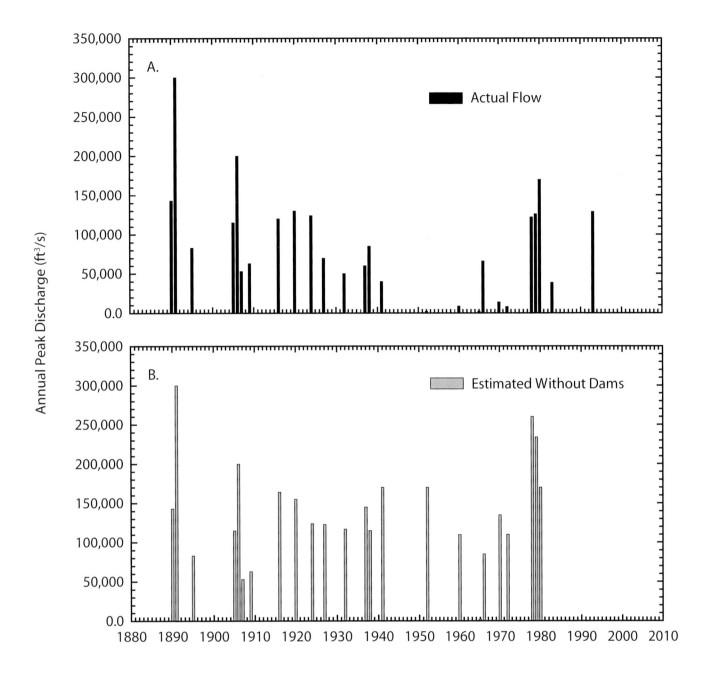

Figure 24.11

A. Annual flood series for the Salt River at Jointhead Dam near Phoenix, Arizona (station 09512060; various years). During many years, the Salt River does not flow here.

B. Annual flood series that would have occurred if the upstream dams were not in place (from Aldridge and Eychaner 1984). The data in series B represent only 1890–1980; clearly, floods would have occurred in other years if dams had not been completed upstream.

A

B

Figure 24.12 Upstream photographs of the Salt River in Tempe, Arizona.

A. (Just after December 11, 1959.) This upstream view from Tempe Butte shows the Salt River through Tempe after a relatively minor flood of less than 200 ft³/s released from dams on the Salt and Verde Rivers on December 11. The dike across the right midground likely provided minimal flood protection for part of Tempe. Older photographs of this reach suggest that the combination of high groundwater conditions and the wide channel allowed sparse, mainly nonwoody riparian vegetation to exist here. Scattered groves of cottonwood trees were once present. (J. A. Baumgartner 4624, courtesy of the U.S. Geological Survey.)

B. (December 4, 2002.) The city of Tempe has grown considerably, requiring bank protection along the Salt River to allow floodplain development. The Rio Salado Project, a river-restoration effort for the Salt River through Tempe, consists of a lake confined within the soil-cemented banks of the Salt River. The once wide and shallow channel is now lower into the floodplain and anchored in space, minimizing the potential for meander migration. (T. Brownold, Stake 4442.)

Approximately 70 S.F. at Tempe, noon, Mar. 21, 1926

A

B

Figure 24.13 Photographs of the Salt River from Tempe Butte.

A. (March 21, 1926.) This set of photographs makes a panorama of the Salt River Valley west of Tempe Butte. The closest bridge was destroyed by a flood many years before this photograph was taken. A discontinuous row of Frémont cottonwood trees appears along low flow in the channel, which the photographer estimated to be 70 ft^3/s. What appear to be more extensive stands of woody riparian vegetation appear in the background. From this distance, it is uncertain whether these stands are mesquite, cottonwood, or black willow. (Unknown photographer, courtesy of the Salt River Project, Phoenix, Arizona.)

B. (April 7, 2004.) Most of the channel of the Salt River in this view is part of Tempe Lake, impounded by an inflatable dam and part of the Rio Salado Project. Of the three bridges visible in 1926, only one remains the same, and three new highway bridges have been built. The river is channelized through the Salt River Valley, and development occurs up to the edge of the channel in many places. (D. Oldershaw, Stake 4764.)

Summary. Three tributaries of the Salt and Gila Rivers—the Agua Fria, New, and Hassayampa Rivers—support riverine riparian vegetation in a variety of habitats ranging from bedrock canyons to wide alluvial reaches. These channels produce sizeable floods owing to the topography of their headwaters, and despite such forces as development, large floods, flow regulation, off-road vehicle use, and grazing, woody riparian vegetation has increased. The effects of drought and future development of groundwater from the alluvial aquifer near Wickenburg, Arizona, may threaten the long-term stability of riparian vegetation along the lower Hassayampa River.

Several tributaries of the Salt and Gila Rivers arise from headwaters in the mountains of central Arizona and flow south across the Phoenix basin (fig. 23.1). These water courses are important for several reasons, mostly because they support (or once supported) significant amounts of riparian vegetation and because they occasionally cause flooding in urban areas. The Agua Fria River, with its major tributary the New River, drains landscapes on the western side of the lower Verde River. Farther west, the Hassayampa River drains a diverse landscape west of the Bradshaw Mountains (fig. 23.1). The Hassayampa River, in particular, is valued for its riparian resources, which include dense groves of native cottonwood-willow vegetation downstream from Wickenburg.

New River

The New River is the major tributary on the eastern side of the Agua Fria River basin that drains the New River Mountains, a low desert range north of Phoenix and Scottsdale (fig. 23.1). The gaging record for this river is short (fig. 25.1), but it shows that this watercourse can deliver large floods from its relatively small watershed (68 square miles). The mountains on the north side of the Phoenix basin have some of the highest rainfall intensities in Arizona owing to orographic lifting of tropical moisture tracking northward.[1] Tropical Storm Norma, in particular, caused severe flooding in the New River.[2]

Only two sets of photographs associated with the gaging operation of the New River have been matched (fig. 25.2). They show modest growth of riparian shrub species, notably seep-willow with occasional riparian trees. Upstream near Black Canyon City, the channel of the New River is lined with scattered young Frémont cottonwood, and date palms have become established on the floodplain as well.

Agua Fria River

The Agua Fria River is important for a variety of water-related uses in the Phoenix basin. Because of the interaction between topography and storms, floods on the Agua Fria River have been extremely large. The watershed particularly responds to winter rainfall, possibly because the orientation of mountains in the headwaters is perfect for orographic lifting of moisture trending in a northeasterly direction. For example, a flood in November 1919 destroyed bridges at Avondale and is the largest flood known to have occurred on this river.

Floods produced by the Agua Fria River and demand for water storage prompted flow regulation. Waddell Dam, on the Agua Fria River upstream from Phoenix, was completed in 1927 and is 35 miles upstream from the Salt River.[3] Its height was increased to 298 feet in 1992, and the structure was renamed New Waddell Dam. New Lake Pleasant, the reservoir impounded by this dam, has diverse uses that range from recreational boating to flood control and storage of Colorado River water from the CAP canal system.

Flow regulation did not stop flood damages. Flooding caused by Tropical Storm Norma in early September 1970 caused overbank flooding and movement of gravel and cobbles in the bed of the river near Rock Springs above the dam.[4] A discharge of 66,600 ft³/s was released into the Agua Fria River below Waddell Dam on February 20, 1980; this flood was partially responsible for the height of the dam being raised to increase its potential flood storage. The gaging station for the Agua Fria River near Mayer (fig. 25.3) records flow from a 585-square-mile watershed; the 1993 discharge, the second highest in the gaging record following the 1980 event, did not cause significant downstream flooding.

Nine repeat photographs document changes at the gaging station near Mayer (fig. 25.4). They show an increase in obligate riparian species—notably trees such as cottonwood—that likely germinated and became established following the 1980 and 1993 floods. Tamarisk is present in this drainage basin, but at this elevation (3,434 feet) it is not a large contributor to the amount of woody riparian vegetation. Downstream from New Waddell Dam, the amount of tamarisk in the channel increases significantly for a short distance and then is replaced mostly by shrubs and trees more characteristics of Sonoran Desert uplands. Unfortunately, we have not been able to obtain historical photography of this reach.

Hassayampa River

The Hassayampa River has headwaters in the Bradshaw Mountains of

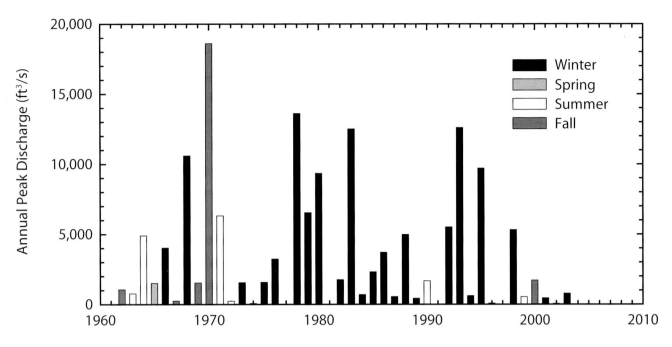

Figure 25.1 Annual flood series for New River near Rock Springs, Arizona (station 09513780, 1962–2003).

central Arizona and flows in a general southerly direction to the Gila River downstream of the Salt-Gila confluence (fig. 23.1). The name "Hassayampa" means "Place of the Big Rocks Water."[5] Its upper reaches are generally within canyons or deep valleys at least partially confined by bedrock. Downstream from Wickenburg, Arizona, groundwater rises to the surface over a 5-mile reach, creating ideal conditions for growth of woody riparian vegetation.[6] After the river passes its last bedrock control, the channel is broad and sandy to its confluence with the Gila River, a site largely obscured within a massive tamarisk thicket.

The riparian ecosystem along the Hassayampa River is one of the most valued in western Arizona. In its higher-elevation reaches, the channel is lined with a closed gallery forest of primarily Frémont cottonwood. The Hassayampa River Preserve, downstream from Wickenburg, was purchased by The Nature Conservancy in 1986 to manage the cottonwood-willow forest.[7] The Arizona skink (*Eumeces gilberti*) occurs only along the Hassayampa River.[8]

Floods

The Hassayampa River has a long-term annual flow of only 27 ft³/s

(1939–2002), but this watercourse has had several unusually large floods since settlement in the 1870s. For example, a thunderstorm upstream from Wickenburg on August 6, 1891, caused the Hassayampa River to rise from dry conditions to a stage of 15 feet and a width of about a mile in a short period of time.[9] A similar flood occurred in December 1883; the March 1916 flood was particularly damaging at Wickenburg. The flood generated by Tropical Storm Norma in September 1970 had a stage of 44 feet in Box Canyon and a peak discharge of 47,500 ft³/s downstream from Wickenburg from a drainage area of 796 square miles (fig. 25.5). Near Wickenburg, floodplains were overtopped with 4 to 6 feet of water, the channel was scoured, and numerous riparian trees were destroyed.[10]

Although flow in most reaches of this river is ephemeral, the upper perennial reaches enticed water development early in Arizona history. The Walnut Grove Dam, built by an irrigation supply company in 1887, was one of the first dams in Arizona[11] and was built to supply water for downstream mining operations.[12] The reservoir backed up water into upstream valleys, eliminating what one lithograph suggested was an open gallery forest of cottonwood. Poorly constructed, this 110-foot-high dam failed catastrophi-

cally on February 22, 1890, leading to disaster downstream at Wickenburg, where about fifty people perished.[13] The flood wave reportedly scoured the channel clear of riparian vegetation to the Gila River.

Change in Riparian Vegetation

Near Wagoner, the Hassayampa River flows through a series of wide and narrow valleys with limited bedrock control. At the gaging station on a still existing bridge, riparian vegetation has increased substantially as documented using five sets of repeat photographs (e.g., fig. 25.6). Cottonwood, black willow, and other important native riparian species have increased at this site, and tamarisk was not observed in the channel here. Bulldozing of the low floodplain to channelize flow for downstream irrigation projects continues, but the disturbance does not appear to be significantly affecting the riparian vegetation. Even if tamarisk were present, it likely would not be able to grow efficiently owing to the shading by taller trees.

Fourteen repeat photographs document change in a narrow bedrock reach at the Box Canyon dam site, about 10 miles upstream from Wickenburg. Flow is perennial here, and the stage of channel-scouring floods

Figure 25.2 Photographs of the New River near Rock Springs, Arizona.

A. (September 3, 1969.) The New River flows through shallow bedrock canyons for nearly its entire course in central Arizona. At this site, southwest of Rock Springs, the river takes a wide left turn against a bedrock slope. The riparian vegetation in this upstream view is sparse and consists mainly of mesquite, although a solitary sycamore is apparent at right center. (T. W. Anderson 9-5137.8, courtesy of the U.S. Geological Survey.)

B. (October 26, 2001.) The channel appears to have downcut by about 2 to 3 feet and narrowed. A riparian shrub assemblage, consisting mainly of seepwillow, has grown up on the floodplain. The sycamore has died back, but its crown is sprouting, and at least one mesquite has been cut down, probably for firewood. (D. Oldershaw, Stake 3019.)

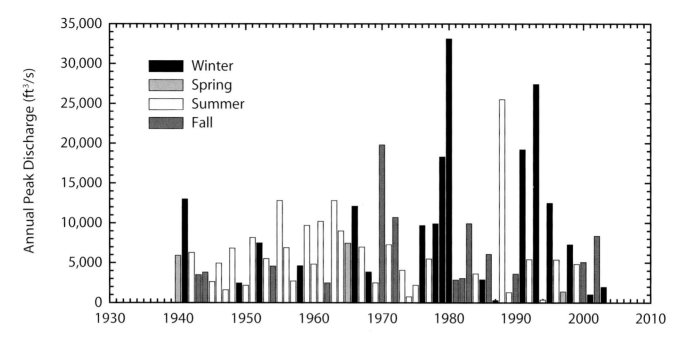

Figure 25.3 Annual flood series for the Agua Fria River near Mayer, Arizona (station 09512500; 1940–2003).

is high; therefore, riparian vegetation tends to be restricted to settings downstream from small projections in the bedrock walls. Despite significant floods in recent decades, including in 1970, 1983, and 1993, and what appears to be unregulated off-road vehicle traffic directly up the channel, woody riparian vegetation has increased in this reach (fig. 25.7).

The vicinity of Wickenburg is one of the fastest-growing areas in Arizona at the start of the twenty-first century. Established in 1864 and named for a miner-rancher named Henry Wickenburg,[14] this town is a major crossroads for travel in western Arizona. Photographs of the main bridge across the Hassayampa River (fig. 25.8) show that riparian vegetation has increased in the channel, despite numerous damaging floods in the twentieth century.

Future groundwater development to support Wickenburg's growing population might threaten this riparian vegetation if water levels are drawn down in the alluvial aquifer.

Downstream of Wickenburg, the river is perennial for a short distance before exiting its bedrock confinement and flowing across deep alluvium to the Gila River. Channel widening in this reach during the 1993 floods created conditions favorable for marshes, which increased five times in area.[15] Another flood in 1995 reversed the increase in marsh habitat, as deposition filled in the lower parts of the floodplain. The principal gaging station on the Hassayampa River (fig. 25.5) is on a railroad bridge at the downstream end of this canyon, and the reach here is intermittent. During wet years, such as the 1980s, flow past this gaging sta-

tion is perennial; in drought periods, such as the early 2000s, the channel is dry near the gaging station.

Nine repeat photographs document changes in this reach (e.g., fig. 25.9), including one set from the downstream end of the Hassayampa River Preserve. They show an increase in the major native species present in the region, including Frémont cottonwood, black willow, seepwillow, mesquite, and burrobrush, as well as an increase in the nonnative tamarisk. Although vegetation has increased in all the views matched in this reach, the presence of obligate riparian species along this influent reach portends that drought-induced change might be imminent. Upstream from this site, denuded or dead young trees were common in 2003 near the downstream end of the Hassayampa River Preserve.

A

B

Figure 25.4 Photographs of the Agua Fria River.

A. (February 1940.) This upstream view shows the Agua Fria River east of Cordes Junction and near Mayer, Arizona. Big Bug Creek, a major tributary, enters the Agua Fria River near the leafless Frémont cottonwood tree in the left midground. (A. A. Fischback 2796, courtesy of the U.S. Geological Survey.)

B. (October 27, 2001.) Although riparian vegetation has increased in all of the photographs taken in this reach, the increase is perhaps most dramatic in this view. The newly established vegetation in the foreground is mostly cottonwood, black willow, and seepwillow. Although the cottonwood near the mouth of Big Bug Creek is not visible, it is persistent. Coyote willow also is present in this reach, and tamarisk appears on the left side of the view. (D. Oldershaw, Stake 3025.)

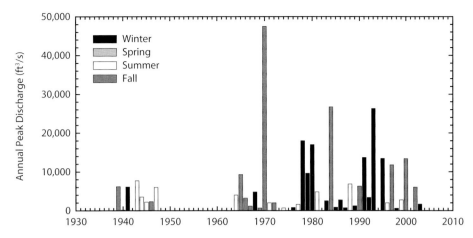

Figure 25.5 Annual flood series for the Hassayampa River near Morristown, Arizona (station 09516500; 1939–1947, 1964–2003). Conversion to hydroclimatic year eliminated the annual peak for 1973.

A. (March 1940.) This downstream view on the Hassayampa River was taken from a bridge (now gone) over the channel. Although floods during this time were mostly small (excluding the 1938 flood of unknown but potentially large size), the channel appears scoured and mostly barren of woody riparian vegetation. Scattered Frémont cottonwood and mesquite are apparent in the view, and some of the smaller-looking shrubs or trees may be black willow. (A. A. Fischback 2773, courtesy of the U.S. Geological Survey.)

B. (May 25, 2001.) The bridge has not been replaced; as a consequence, this photograph is not an exact match of the original. The channel has been recently bulldozed, probably to direct flow into downstream irrigation ditches. Despite this manipulation, the increase in riparian vegetation is startling, including cottonwood, black willow, netleaf hackberry, mesquite, and burrobrush. (D. Oldershaw, Stake 2459.)

Figure 25.6 Photographs of the Hassayampa River near Wagoner, Arizona.

Figure 25.7 Photographs of the Hassayampa River in Box Canyon.

A. (January 6, 1938.) This upstream view of the Hassayampa River at the Box Canyon gaging station shows a denuded channel with margins mostly devoid of riparian vegetation. A small grove of mesquite appears at left center, and Frémont cottonwood trees are visible in the distance. (R. E. Marsh 2224, courtesy of the U.S. Geological Survey.)

B. (May 25, 2001.) Because of the confined nature of the channel within this bedrock canyon, there is little room for woody riparian vegetation. Nonetheless, some new cottonwood trees are obvious, particularly on the left side of the channel. Black willows also are common downstream of projections on the canyon walls. Burrobrush and tamarisk are dense in the wide reaches downstream from this site. The channel is now a regularly used off-road vehicle route, although this use does not appear to be affecting woody riparian vegetation in this reach. (D. Oldershaw, Stake 2472.)

Cable bridge across the Hassayampa River 1918

Figure 25.8 Photographs of the Hassayampa River at Wickenburg, Arizona.

A. (1918.) This view across the Hassayampa River at Wickenburg documents failure of the road bridge leading to town *(behind the camera station)* during a March 1916 flood, which affected much of southern and central Arizona. A photograph taken at that time showed essentially no woody riparian vegetation along the river channel. Taken two years after this flood, this photograph shows a suspension bridge temporarily replacing the destroyed bridge and young riparian trees dense on the floodplain *(river left)*. These plants are leafless, but their branch architecture suggests a combination of Frémont cottonwood and black willow. (Unknown photographer, 96.24.41.11, courtesy of the Cline Library.)

B. (April 23, 2004.) The bridge for U.S. Highway 93, the main route between Phoenix and Las Vegas, was eventually moved downstream to its current location *(out of view to right)*. Riparian vegetation has increased in this reach, as documented by several other photographs. The species present include Frémont cottonwood *(left background)*, black willow, seepwillow *(foreground)*, mesquite, catclaw, and tamarisk *(foreground)*. Eucalyptus trees, possibly river red gum *(Eucalyptus camaldulensis)*, have been planted out of view to the right. (D. Oldershaw, Stake 4772.)

A

B

Figure 25.9 Photographs of the Hassayampa River downstream from Wickenburg, Arizona.

A. (January 1933.) This upstream view is from beneath the railroad bridge over the Hassayampa River downstream from Wickenburg. The river is exiting a right-hand bend before leaving the bedrock canyon that confines it and that creates effluent groundwater conditions that sustain a riparian forest of mostly Frémont cottonwood. Here, proximity to the unconfined alluvial basin downstream creates influent conditions, and the riparian vegetation is mostly mesquite, with scattered small cottonwood in the distance. (W. E. Dickinson 2304, courtesy of the U.S. Geological Survey.)

B. (May 25, 2001.) Recent floods have created a low, sandy floodplain along the Hassayampa River, narrowing the channel from its 1933 width. At the time of this photograph, the river was dry, which is not unexpected given that this is the downstream end of the influent reach and a severe drought was occurring. Nonetheless, the presence of burrobrush and the cottonwood at left center indicates an increase in riparian vegetation here. Mesquite at left has also increased. (D. Oldershaw, Stake 2478.)

Summary. The Gila River between Florence and its confluence with the Salt River is one of the few reaches where riparian vegetation has markedly declined. The evidence for this decline consists of written descriptions of channel conditions in the sixteenth through nineteenth centuries compared with written descriptions from the early twentieth century and current conditions. The Gila River in this reach is fully regulated and managed primarily for delivery of irrigation water and for minimization of damage from infrequent, large floods. Complete diversion of surface water and heavy groundwater use have prevented reestablishment of the once extensive riparian vegetation near Maricopa, which was largely eliminated by large floods at the end of the nineteenth century. Wastewater effluent from the city of Phoenix and irrigation returns sustain incipient to fully developed stands of woody riparian vegetation downstream from the confluence of the Salt and Gila Rivers. Dense tamarisk thickets are the basis for several wildlife areas west of the Salt-Gila confluence; historic descriptions indicate widely divergent vegetation in this reach, ranging from cottonwood-willow forests to dense carrizo grass. Riparian vegetation has increased downstream from the Salt-Gila confluence, although tamarisk accounts for most of the increase. Sparse photographic evidence and vague written accounts diminish the reliability of any conclusions concerning change in riparian vegetation downstream from the confluence beyond the increase in tamarisk. Near the Colorado River, irrigation returns to the Gila River again sustain woody riparian vegetation.

The lower Gila River, especially downstream of its confluence with the Salt River, is arguably the most important river in Arizona (fig. 26.1). With the exception of the Little Colorado and Bill Williams Rivers, all of the major river systems in Arizona converge here, making the Gila the primary drainage system for Arizona. The watershed area upstream of Dome, Arizona, is 57,850 square miles and includes parts of western New Mexico. The Gila River once provided sustenance for the Pima, then was the lifeline for southern travelers from the East to California, and now is little more than a drainage ditch across the south-central Sonoran Desert of Arizona. As a result, some have called it a "dead" river.[1] Despite large losses in some reaches, however, significant and valued stands of riparian vegetation are present along this archetypal desert river.

Early Observations of the Lower Gila River

Spaniards first arrived at the Gila River in 1694 and found a thriving village of Pima near present-day Maricopa (fig. 26.1).[2] These native people had occupied the Gila River floodplain since around A.D. 1540, or about 150 years before the arrival of the Spaniards. They practiced irrigated agriculture and, following the Hohokam, partially regulated the low-water flow of the Gila River. They also included fish from the Gila as a major part of their diet.

Spanish missionaries and travelers noted the riparian vegetation, which was a beacon across an otherwise waterless desert. Father Eusebio Kino, during his second visit, traveled from the mouth of the San Pedro to the Piman villages following the Gila River and "its very large cottonwood groves." In 1744, the confluence of the Salt and Gila Rivers had "an abundant growth of [willows] and cottonwood."[3] These observations are mostly limited to about a 30-mile reach of the river centered on the Piman villages.

In February 1826, at the Salt-Gila confluence, James Ohio Pattie wrote that the Gila River was "about 200 yards wide, with heavily timbered bottoms."[4] Descriptions of river width varied from this figure to as little as 40 feet in the nineteenth century, which reinforces the idea that the amount of flowing water depended on where the river was viewed in the alternating effluent-influent reaches. There is also little doubt that the channel consisted of braided, weaving strands through sandy islands in the center. Several marshy areas, possibly sustaining alkali sacaton, were present, including near present-day Sacaton, at the Santa Cruz–Gila confluence and near the mouth of the Salt River.[5]

Lieutenant Colonel W. H. Emory, of the Advanced Guard of the Army of the West, traveled down the Gila River in 1846 en route to California, all the way commenting on flora and fauna. Near present-day Florence, he mentioned dense growth of willows but not cottonwood. He estimated the population of Pima and Maricopa to be up to ten thousand people near the confluence of the Gila and Salt Rivers and stated that "a great deal of the land is cultivated." Near present-day Gila Bend, the course of the river could readily be discerned from the line of green cottonwood lining its banks, but to the west, near present-day Painted Rock Dam, "the bottoms of the river are wide, rich, and thickly overgrown with willow" and "the river spread over a greater surface, about 100 yards wide, and flowing gently over a sandy bottom, the banks fringed with cane, willow, and myrtle."[6]

In December 1846, the Mormon Battalion traveled from the Piman villages north of present-day Maricopa, Arizona, to the Yuma crossing of the Colorado River.[7] Their descriptions,

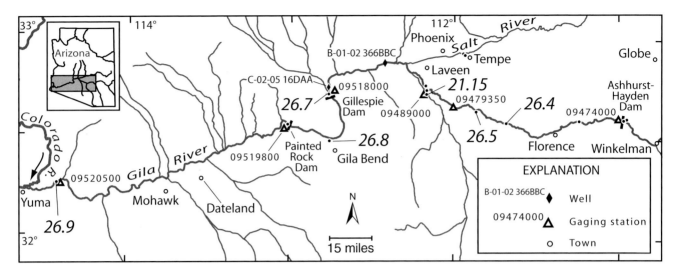

Figure 26.1 Map of the Gila River from Florence to Yuma, Arizona.

combined with later visits by Forty-niners, suggest that the Gila River was lined with a thin band of cottonwood and black willows; seepwillow was common nearest the permanent water sources. Dense mesquite bosques were common near the floodplain;[8] in 1938, the confluence of the Gila and Santa Cruz Rivers was described as a dense mesquite thicket (both screwbean and velvet mesquite). In 1854, John Bartlett found the Gila River to be dry at this point owing to complete diversion of low flow to Piman crops.[9] Upstream from the Piman villages, and presumably upstream from the flow diversions, the river had 15-foot-high banks and a closed-gallery cottonwood-willow forest.

In the late nineteenth century, the lower Gila River was perennial—or at least intermittent—in this reach, sustaining open cottonwood galleries punctuated with bottomlands vegetated with grasslands and willows.[10] In 1864, Pratt Allyn described the Gila River near its confluence with the Salt River as having a margin of willows and cottonwood.[11] Near Powers Butte, one observer in 1889 found the river lined with "cottonwoods and bushes."[12] Downstream from the Salt-Gila confluence, the Mormon Battalion found grass to be scarce and instead fed cottonwood bark to their stock. Lieutenant Cave Couts, en route to California, wrote a cryptic note in 1848, stating that the river had "salt grass. . . . The remainder is nothing but cottonwood

(thinly scattered along the margin), coarse chaparral bushes and weeds, and the water willow."[13] Others noted the presence of Arizona ash in small numbers.

In 1894, Edgar Mearns visited the Gila River and described riparian vegetation along its lower 20 miles above the confluence with the Colorado River. He stated later: "The stream, as usual, is bordered by cottonwood and willow trees. Mesquite and screw bean are the common trees of the river bottom . . . [and there are] numerous sloughs, bordered with cat-tail, tule, cane, sedge and rush."[14] By the 1920s, arrowweed dominated floodplain vegetation,[15] to be supplemented with (or replaced by) nearly monospecific stands of tamarisk by the early 1950s.

Some observations are baffling in their seeming contradictions. One report from 1879 said the Gila River near Florence was nothing but a channel of dry sand, yet a ferry operated at this town, linking it to other towns in the Salt River Valley.[16] Presumably, this ferry operated during the predictable spring runoff and during unpredictable floods. At the end of the nineteenth century, all rivers in the region were experiencing alternating periods of extreme floods and extreme droughts. Although it is convenient to blame the apparent initial decline in riparian vegetation on water diversions, floods likely had at least as much of an impact.[17]

At the start of the twentieth century, a dense stand of carrizo grass was present near the Salt-Gila confluence.[18] Maps made around 1900 show extensive stands of "mesquite timber" lining both sides of the Gila River from its confluence with the Salt River upstream and past its confluence with the Santa Cruz River.[19] Groundwater levels were less than 50 feet below land surface in a broad swath centered on the river from the Salt-Gila confluence upstream to Florence, and effluent conditions were reported in the river at two points, near Sacaton and in the vicinity of the current Interstate 10 crossing. However, few cottonwood and willow trees remained, despite the fact that large-scale water development would not begin for another decade. Descriptions such as "cottonwood occurs in a thin fringe . . . here and there a grove along the Gila and Salt Rivers" are consistent with other nineteenth-century observations. By 1923, the river was described as intermittent, and settlers described the cottonwood as having disappeared forty years earlier (1880s), but fish were reportedly still present in the river.[20]

Before extensive urbanization in the Salt River Valley, mesquite grew in extensive stands in the vicinity of and downstream from the confluence of the Salt and Gila Rivers.[21] Many of these bosques, in particular those downstream from the present-day site of Painted Rock Dam (fig. 26.1), have been cleared for agricultural lands.[22]

Others died owing to water-level declines in the alluvial aquifer by the 1970s.[23]

The native fisheries of the Gila River have been largely eliminated. In the middle of the nineteenth century, many native species, in particular Colorado River pikeminnow, were common in much of the river basin.[24] Many species of birds occur along the lower Gila River, particularly where riparian habitat remains. On the Gila River Indian Reservation, where a perennial stream supporting a cottonwood-willow forest is now a xerophytic riverbed, twenty-nine bird species are thought to have been locally extirpated.[25] Although bird life along the lower Gila River was abundant, aquatic mammals were relatively sparse, as in most other riparian areas in the region. Beaver became much less numerous along the lower Gila River following the 1891 flood (see the next section), although a colony was reported to be at Mohawk in 1894.[26] Few trappers reported beaver in abundance on the lower Gila River, although beaver and muskrat, now uncommon in most Arizona rivers, remain.[27]

Floods and Channel Change

The first evidence of large floods along the lower Gila River are preserved in paleoflood records in damaged irrigation canals once used by the Pima and show variation in flood occurrence in the late Holocene.[28] In particular, flooding was relatively low from around four thousand to one thousand years ago, followed by a period of high flood frequency, channel instability, and damage to irrigation canals. Lack of evidence suggests a period of low flood frequency and high channel stability leading to the nineteenth century.

As the first observations suggest, the Gila River was a shallow, braided stream, particularly between Florence and the confluence with the Salt River.[29] Some researchers erroneously believe that the Gila River once flowed with a volume sufficient to sustain steamboat traffic.[30] The origin of this myth may be W. H. Emory's report, which stated that "[t]he Gila, at certain stages, might be navigated up to the Pimos [Pima] village."[31] Although this claim is clearly an exaggeration, two early expeditions built boats to float from near the present site of Gila Bend to the Colorado River. The first attempt was made by the Mormon Battalion in 1845.[32] In a failed attempt to avoid slow travel through deep sands, they converted a wagon into a boat, but flow was shallow, and the wagon was repeatedly stranded on sandbars. The second attempt, this time successful, was made by a Forty-niner group who turned their wagon into a scow.[33] After the *Explorer* (see chapters 12 and 27) was salvaged in the late 1850s, it reportedly was used to haul firewood down the Gila River near its mouth.[34] Clearly, seasonal flow in the river was sufficient to tempt river traffic and to sustain native fish populations, but insufficient to allow regular boat traffic.

Floods were once common on the lower Gila River and are known mostly because of damage at Yuma (see chapter 27). Some floods, such as an 1833 event, are thought to be larger than twentieth-century floods but are poorly known because written records were not made and discharge measurements are not available.[35] Two floods, in September 1868 and again in 1869,[36] also caused significant inundation along the lower Gila River; the 1868 flood reportedly was 4 miles wide. The 1891 flood, one of the most significant floods in Arizona history, had peak discharges of 250,000 ft^3/s at Gillespie Dam and 280,000 ft^3/s at Dome (fig. 26.2). With the construction of bridges and other floodplain structures, flood damage became a serious problem. For example, in October 1895, a flood on the Gila River destroyed a railroad bridge near Maricopa;[37] another flood in July 1898 destroyed a stagecoach. The 1905 flood, a relatively small event at Dome, widened the Gila River and converted it into a braided form.[38] More damage occurred during the January 1916 floods. Channel width increased by a factor of four to five between 1868 and 1923.[39]

Since construction of Coolidge Dam in 1928, only two significant floods have occurred in the reach upstream from the Salt-Gila confluence. The 1983 flood came from the combination of the San Pedro and Santa Cruz Rivers (chapters 19 and 21), and the 1993 flood also came from these tributaries, combined with emergency releases from Coolidge Dam. Downstream from the Salt-Gila confluence, floods have been more frequent owing to the effects of large regional storms on the Verde and Salt Rivers (chapters 23 and 24). Extremely large events that rivaled the size of predam floods occurred in 1978 (March and December), 1980, and 1993 (fig. 26.2b).

The channel of the Gila River upstream from the Salt-Gila confluence responded differently to the 1983 and 1993 floods. The 1983 flood, with a peak discharge of 35,000 ft^3/s, was not sustained and caused little channel change.[40] In contrast, the 1993 flood peaked at 41,600 ft^3/s, and the sustained flows in the late winter and spring of 1993 caused significant channel widening. The post-1993 channel reverted to its appearance following the 1905 floods (see fig. 26.4).

Flow Regulation

For several hundred years, the Pima diverted water from the Gila River for irrigation purposes. As others moved into the region in the 1870s, diversions increased, but the dams frequently washed out. The 1891 flood reportedly destroyed all the irrigation dams along the Gila River; a small flood in 1900 cut a 20-foot-wide breach in the irrigation dam at Florence.[41] The joint needs for flood control and irrigation diversion prompted a basinwide development of water resources in the early twentieth century.

Roosevelt Dam, completed in 1911, regulates much of the flow on the lower Gila River (chapter 24). As noted in chapter 17, Ashhurst-Hayden Dam (1922) was the first concrete-and-steel dam to regulate at least partially the Gila River upstream from its confluence with the Salt River. The lower Gila River has been fully regulated by dams on its main stem and principal tributaries since completion of Coolidge Dam in 1928. Releases from Coolidge Dam are typically diverted into a canal network at Ashhurst-Hayden Dam upstream from Florence, leaving the channel downstream dry most of the time. This dam is designed to pass flood releases from Coolidge Dam—as occurred in 1993—and occasional large floods from the San Pedro River.

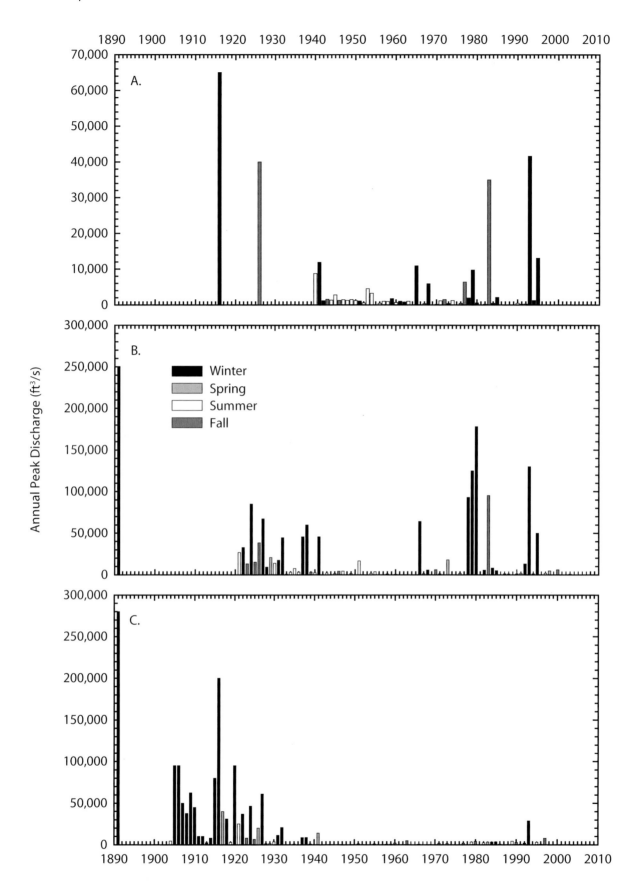

Figure 26.2 Annual flood series for the Gila River. Many years have no flow.

A. The Gila River near Laveen, Arizona (station 09479500; 1916, 1926, 1940–1995) combined with Gila River near Maricopa (station 09479350; 1995–2000).

B. The Gila River below Gillespie Dam, Arizona (station 09519500; 1891, 1921–2003).

C. The Gila River near Dome, Arizona (station 09520500 1891, 1904–2003).

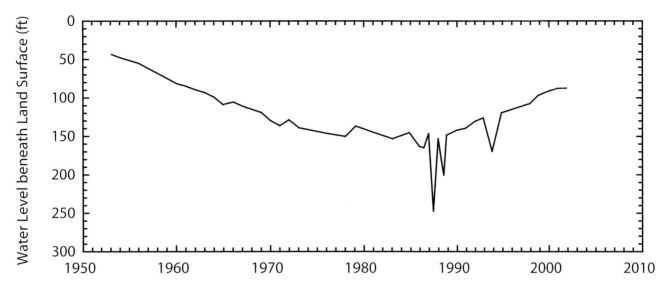

Figure 26.3 Groundwater levels for well D-04-03 16CBB along the Gila River near Maricopa, Arizona.

Widespread development of the Gila River below present-day Painted Rock Dam began in 1857 with small-scale diversions. Floods destroyed these early dams, but the fertile valleys remained magnets to farmers. The first solution was drilling of irrigation wells in the 1910s, but ultimately surface water from the Gila and Colorado Rivers was needed. The Wellton-Mohawk Irrigation District was formed in the 1940s to irrigate 75,000 acres along the Gila River.[42] Painted Rock Dam, 181 feet high, was completed in 1960 to provide flood control for the lower Gila River.[43] This dam potentially impounds one of the largest reservoirs wholly within Arizona (by volume, 2.5 million acre-feet), but the 53,200-acre reservoir usually is dry. The largest discharge entering the reservoir was 220,000 ft³/s during the February 1980 floods in central Arizona; the dam and spillways released 26,500 ft³/s in February 1993 into a channel designed to pass 10,000 ft³/s.[44] The sustained releases from Painted Rock Dam in 1993 damaged levees and inundated farmland, but the releases were an order of magnitude smaller than historical flood peaks in this reach.

Changes in Riparian Vegetation

Florence to the Salt-Gila Confluence

Although once lush with woody riparian vegetation, at least within certain specific reaches of perennial flow, the Gila River between Florence and its confluence with the Salt River is now usually dry. Flood flows from the Santa Cruz River basin have occasionally reached the Gila River near Laveen, notably in 1983 and 1993 (chapter 21). Otherwise, irrigation returns are the only sources of water in this reach. Groundwater levels near the river steadily declined as aquifers were pumped to supplement surface water (fig. 26.3). However, beginning in the early 1990s, water levels have rebounded, but not to heights sufficient to sustain riparian vegetation. The increase likely is the result of importation of CAP water extracted from the Colorado River combined with recharge of the alluvial aquifer during the 1993 flood.

Fourteen photographs at two locations document twentieth-century changes in riparian vegetation; no historical photographs show the riparian conditions orally described at the Piman villages. One of the few historical photographs in this reach shows an extremely wide and barren channel in 1915 (fig. 26.4A). Changes in the twentieth century, reminiscent of those in the San Juan River (chapter 8), are steady revegetation of the channel during the low-flow period of the midcentury drought (fig. 26.2A), but owing to groundwater declines and decreased surface flow the vegetation is mostly mesquite and tamarisk in

xerophytic positions or plants typical of the Sonoran Desert. The 1993 flood release, which was sustained over a two-month period in February and March, simultaneously destroyed some of the xerophytic vegetation (fig. 26.4D) and caused some germination of cottonwood; several isolated individuals survive in this reach.

Where Interstate 10 currently crosses the Gila River, the channel already had become nearly devoid of riparian vegetation by 1903 (fig. 26.5A). The channel of the Gila River at the former gaging station at Laveen (fig. 26.1) sustains a nearly monospecific stand of tamarisk on flow that is either irrigation returns or infrequent floods on the Santa Cruz River (chapter 21). Downstream, the Gila River flows through a narrow valley between the Sierra Estrella and South Mountain, which may force groundwater toward the surface. Near the Gila's confluence with the Salt River, Frémont cottonwood has returned to the channel banks as a result of a change in both groundwater levels and Salt River flow regime.

The Gila-Salt Confluence to Gillespie Dam

The Gila River has perennial flow from its confluence with the Salt River to near Gila Bend owing to the combination of irrigation returns and sewage effluent from Phoenix and its suburbs

Figure 26.4 Photographs of the Gila River at Olberg, Arizona.

A. (March 17, 1915.) This downstream view on the Gila River, from a low butte adjacent to the right bank, shows an extremely wide, denuded channel with a thread of flow occurring in the right midground. The line of horse-drawn wagons is in place to repair a diversion dam that serves the canal leading to the right. An abandoned canal lies at the base of the butte below the camera station. This photograph was taken a few months after the devastating floods of February 1915 in the Gila River basin (see fig. 26.2C.) A few mesquite trees appear along the right bank, and an open grove of Frémont cottonwood appears on both banks in the background. (H. L. Shantz I-7-1915, courtesy of the Homer Shantz Collection, University of Arizona Herbarium, Tucson.)

B. (March 6, 1974.) The Gila River seldom flows in this reach owing to total diversion of surface water upstream at the Ashhurst-Hayden Dam, about 40 miles upstream. Xerophytic shrubs, including mesquite, catclaw, and creosote bush characteristic of the Sonoran Desert lowlands, appear where the river once flowed. A hint of obligate riparian trees remains in the background, although they may be trees planted around houses. (R. M. Turner.)

C

D

C. (March 3, 1984.) The channel is no longer visible, and xerophytic shrubs completely dominate the scene. The town of Olberg now is apparent across the vegetated watercourse. (R. M. Turner.)

D. (November 11, 2002.) The top of the butte, and the water tower that once was behind the camera station, have been removed. Owing to effects of the 1993 flood, the channel has reappeared in a narrower form but has also been denuded. (D. Oldershaw, Stake 736a.)

A

B

Figure 26.5 Photographs of the Gila River at Gila Butte, Arizona.

A. (1903.) This northerly view across the Gila River and its floodplain shows Gila Butte, a low hill north of Bapchule on the Gila River Indian Reservation. The channel is relatively wide and devoid of woody riparian vegetation, although two palo verde trees appear in the left foreground. A grove of Frémont cottonwood appears in the distance on the right. (W. T. Lee, courtesy of the U.S. Geological Survey Photographic Library.)

B. (April 7, 2004.) The camera station is now under the southbound lanes of Interstate 10 at the river left abutment of the bridges over the Gila River. The river in this reach supports mostly tamarisk, desert broom, and mesquite; a grove of cottonwood is upstream, and isolated seedlings that germinated following the 1993 floods remain in the now ill-defined channel. Channel-stabilization efforts to protect the highway bridges here have locally modified the floodplain. (D. Oldershaw, Stake 4761.)

into the Salt River. Groundwater levels have been high at this confluence, both historically and currently, and a cottonwood-willow forest persists at this point within a sea of tamarisk in the wide floodplain. One analysis shows that riparian vegetation in the channel was nearly nonexistent around 1900 but peaked in the 1930s immediately following construction of Coolidge Dam.[45] In 1971, the area of tamarisk with greater than 50 percent cover was 5,900 acres in this reach, and a total of 11,540 acres had tamarisk.[46] This tamarisk grove is highly valued for its White-winged and Mourning Dove populations, as indicated by the establishment of the Robbins Butte and Arlington State Wildlife Areas in this reach. Thick groves of mesquite and catclaw are behind the tamarisk-dominated floodplain.

Gillespie Dam, built as an irrigation diversion structure in 1921,[47] almost immediately filled with sediment (fig. 26.6A). Nine photographs document changes in riparian vegetation at this site. Riparian vegetation began growing almost immediately in the wide delta area, and this environment became prime habitat for tamarisk. Gillespie Dam breached during the 1993 flood (fig. 26.6C), which caused the channel to downcut through the former delta, lowering its groundwater level. Fires here have burned some of the dense stands of tamarisk, which may not regain their former density owing to the lowered water table. The filled reservoir supports mostly dense groves of tamarisk, but some areas,

closer to the active channel level, support cattail.

Gillespie Dam to Yuma

Downstream from Gillespie Dam, the Gila River flows south to Gila Bend and the delta of Painted Rock Reservoir. This reach—including the usually dry reservoir—is lined with agricultural fields that grow primarily alfalfa and cotton. These fields are irrigated with a combination of groundwater and surface water, and as in many other alluvial aquifers in the desert region of Arizona, groundwater levels have had precipitous declines (fig. 26.7). Water levels rebounded with the frequent floods between 1977 and 1983 (fig. 26.2B). Native and nonnative riparian vegetation is common along the river upstream from the reservoir.

At Gila Bend, the river's course swings westerly for its last reach of more than 100 miles to its confluence with the Colorado River. Its first 30 miles are within the boundaries of Painted Rock Reservoir. Four historical photographs document postdam changes in riparian vegetation here (fig. 26.8) and downstream from Painted Rock Dam. Owing to periodic inundation—the most recent occurred in 1993—the vegetation here is either tamarisk, riparian shrubs, or xerophytic species associated with disturbance.

Below Painted Rock Dam, the Gila River is mostly dry until irrigation returns associated with the Wellton-Mohawk Irrigation District add some flow to the channel. The combination

of flow regulation upstream and agricultural clearing changed much of the riparian vegetation in this reach. Tamarisk encroachment was significant enough to prompt large-scale clearing of 2,700 acres along 142 miles of channel between 1958 and 1959.[48] By 1970, most of the vegetation in this reach was tamarisk within the floodplain and mesquite on the nearby uplands.[49] Riparian vegetation occupying 16,400 acres along the Gila River upstream from Dome was mapped on 1970 aerial photography.[50] At that time, tamarisk occupied half of the mapped area. Other nonnative species of note were giant reed, found in association with cattail assemblages adjacent to standing or slow-moving water, and tree tobacco, associated with tamarisk-arrowweed assemblages.

Twenty-two photographs associated with the gaging station on the McPhaul Bridge near Dome, Arizona, document changes in riparian vegetation (fig. 26.9). Flow regulation by Roosevelt and Coolidge Dams and the midcentury drought have drastically reduced the size of floods at this site (fig. 26.2C). The channel has narrowed considerably through the combination of reduced flow and channel stabilization (fig. 26.9B), and agricultural development of floodplains has reduced the amount of riparian vegetation, including both cottonwood and mesquite. Riparian vegetation now consists of the combination of a narrow line of cottonwood, with dense tamarisk behind and mesquite on the far margins of the channel.

Figure 26.6 Photographs of the Gila River at Gillespie Dam, Arizona.

A. (1924.) Gillespie Dam was built in 1921 to divert water into canals for irrigation near Gila Bend, which is about 25 miles downstream. Three years after its construction, the reservoir is full of sediment, but its intent never was to store water, only to divert it. Riparian vegetation, which appears to be a combination of arrowweed and cattail, appears in the reservoir, and the far slopes have few shrubs or trees. (Photographer unknown, 1382, courtesy of the U.S. Geological Survey.)

B. (December 18, 1930.) This view is from a slightly higher elevation and a different angle across the dam. The passage of six years has allowed woody riparian vegetation to encroach on the now-dry delta that has formed upstream from the dam. The vegetation appears to be a mixture of tamarisk, black willow, and arrowweed. (W. E. Dickinson 1372, courtesy of the U.S. Geological Survey.)

C. (November 29, 2001.) This view matches photograph B. The 1993 flood caused failure of the center of Gillespie Dam, and diversions now occur downstream of the ruined structure. A small dam backs water up in front of and behind the former impoundment, and the deepest part of the channel has shifted closer to the camera station. Most of the vegetation is tamarisk, and a large area in the midground has recently burned. Black willow also is abundant in this reach, as is arrowweed; mesquite typically occurs behind the wall of tamarisk. (D. Oldershaw, Stake 3042d.)

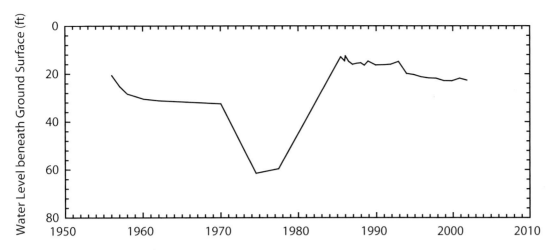

Figure 26.7 Groundwater levels for well C-02-05 16DAA near Gila Bend, Arizona.

A. (May 7, 1937.) This downstream view along the Gila River is from a low basalt butte on its north side *(right bank)*. The river is in flood, and water is flowing through an ill-defined channel lined with young tamarisk trees and what appears to be scattered mesquite. (R. R. Humphrey 2634, courtesy of the photographer.)

B. (November 29, 2001.) This reach of channel is periodically flooded by Painted Rock Reservoir, which fills only during periods of flooding that cannot be contained by the upstream dams. This reach was last inundated in 1993. The channel has moved closer to the butte and has become more incised. Tamarisk is dense in the floodplain, and arrowweed is also present. The saguaros that once were on the midground slopes have disappeared, and many rocks that once were in the foreground are now gone. (D. Oldershaw, Stake 3044.)

Figure 26.8 Photographs of the Gila River near Gila Bend, Arizona.

A

B

Figure 26.9 Photographs of the Gila River at Dome, Arizona.

A. (December 23, 1930.) This downstream view on the Gila River shows the McPhaul Bridge in the left foreground and the confluence with the Colorado River in the background. A considerable number of young cottonwood trees line a relatively narrow channel in this reach. Although it is difficult to determine from this distance, it is possible that the low shrubs on the low floodplain above the bridge abutment are tamarisk; at least some arrowweed is most likely present as well. Clumps of mesquite appear on the back of the floodplains on both the right and left sides of the view. A match of an upstream view from the bridge shows that tamarisk formed a nearly monospecific stand here in 1970 (Office of Arid Lands Studies 1970). (W. E. Dickinson 1215, courtesy of the U.S. Geological Survey.)

B. (November 30, 2001.) The channel is much wider than in 1930, owing to the combination of channelization and the 1993 flood. The channelization protects agricultural fields, which have replaced the open stands of mesquite and cottonwood that once grew here. The Wellton-Mohawk Canal appears in the left distance. The channel is marshy and supports cattail, and isolated cottonwood trees appear on both banks. Both Athel tamarisk *(larger trees)* and tamarisk are present here now, the latter in dense stands where mesquite once was present. This is an unusual situation where Athel tamarisk appears to have become established from seed. A few low mesquite clumps appear in the right midground just above the bedrock slope. (D. Oldershaw, Stake 3045g.)

Summary. The lower Colorado River, downstream from Hoover Dam, is primarily a conduit for water transfers to Arizona, California, and Mexico. The amount and timing of dam releases bear little resemblance to flows in the predam river, which is now intensely regulated and lined with highly altered stands of woody vegetation. Nonetheless, the combination of flood control, incidental dam releases, and leakage from unlined canals provides water for riparian ecosystems along most of this reach upstream from Imperial Dam. Because floods no longer scour banks, woody vegetation has changed, and channels have been narrowed by levees and encroachment of riparian vegetation consisting of both native species and tamarisk. Change in riparian vegetation along a river reach as diverse as the lower Colorado River is extremely difficult to determine. Narrow canyon reaches had little riparian vegetation before flood-control operations began at Hoover Dam; now these reaches support significantly more riparian vegetation, but much of the new vegetation is tamarisk. General descriptions of nineteenth-century riparian vegetation are vague and somewhat misleading because most of them appear to focus on wide sections of the river valley or in the delta and merge consideration of large areas of herbaceous species with observation of areas of woody species. Use of wood for steamboat fuel suggests that woody riparian vegetation once was extensive, and reestablishment of this vegetation, whether in response to human disturbance or channel change during floods, was rapid. Tamarisk apparently did not arrive in the lower Colorado River reach until at least the 1920s, or much later than some previous reports have suggested and much later than in upstream parts of the watershed. Construction of dams completely changed the flow regime into the delta, directly eliminated native riparian vegetation, and encouraged shoreline stands of tamarisk, mesquite, and arrowweed. Changes to the extent of riparian vegetation in the delta is largely unknown, but the available scientific literature suggests that a large decrease in riparian vegetation occurred in Mexico.

Downstream from Hoover Dam, the Colorado River meanders south in a reach that has been called "the Nile of America."[1] Here, we consider the lower Colorado River to stretch from Hoover Dam to the Gulf of California (fig. 27.1). This reach, fully regulated by Hoover Dam since 1935, is one of the most fascinating parts of the Southwest, in part because of the juxtaposition of lush riparian habitat against hyperarid surroundings and in part because of the claims of losses in riparian habitat. The Colorado River delta, which is one of the most regulated river reaches in the United States and Mexico, has been proclaimed to be "dead,"[2] and the "legendary forests"[3] claimed to have once been in the delta have supposedly been extirpated by water diversions. Although there is little question that the species composition of riparian vegetation has changed owing to the introduction of tamarisk and that the amount of area occupied by this vegetation has decreased, the magnitude of the changes are open to question.

Significant losses in native fishes have occurred throughout the lower Colorado River.[4] Native fishes, in particular Colorado River pikeminnow and bonytail, flowed with the water into irrigation canals; they were extracted and used for fertilizer.[5] The initial decline in the native fisheries has been attributed to the 1934 drought and construction of Hoover Dam, completed in 1935; few records of Colorado River pikeminnow in this reach are available after 1949.[6] Although reduction in the native fisheries has been well documented, changes in riparian vegetation are less known.

Few scientists who made early visits to the lower Colorado River collected botanical specimens or took precise notes on the ecology of this reach of river, and none did in the nineteenth century, creating a void in knowledge that invites speculation about the amount of change that has occurred along this heavily regulated river. However, explorers traversed the river, looking for everything from beaver to railroad routes to a cheap transportation route. Although the Colorado River provided a sometimes formidable barrier to east-west travel, it also offered a viable north-south route, and steamships began plying its waters in the 1850s.[7] With the advent of the railroad and especially with construction of dams, steamship travel ended, and flow regulation began. The once powerful river, with its raging floods during three seasons, became nothing more than an irrigation conveyance channel. At present, demands for irrigation water, calls for ecological restoration of the delta, and severe drought have created a situation that does not bode well for the future of riparian vegetation. What we know of the extent of woody riparian vegetation along the lower Colorado River—and hence whether restoration is possible and appropriate—comes from three sources: written accounts, historical images, and the amount of wood used for commerce.[8]

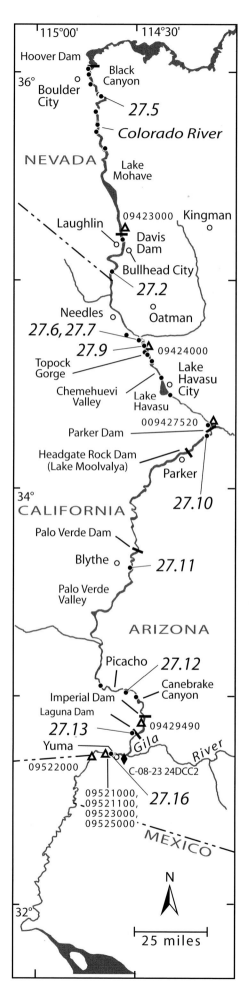

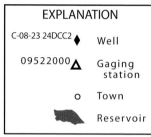

Figure 27.1 Map of the lower Colorado River from Lake Mead to its delta.

Historical Observations and Images of Riparian Vegetation

The Cucapá

Many indigenous tribes lived along the lower Colorado River for perhaps a thousand years before the Spaniards first arrived, and they made use of riparian resources for food, building materials, and clothing.[9] Arrowweed became thatch for the roofs of their lodges, mesquite seeds were food, and willows provided the structural material for baskets. The Cucapá used mesquite beans for food and willow bark to create skirts and vests. They burned the riparian vegetation for a variety of reasons, including to thin it for access, and built their houses on cottonwood poles to be above flood stage. Ultimately, as discussed later, they cut wood to sell as fuel for steamboats. Before 1850, the Mojave Indians used the floodplain in the vicinity of present-day Needles, California, for agriculture.[10] Therefore, humans have utilized woody vegetation along the lower Colorado River for perhaps a millennium.

The Trappers

In 1826, Jedediah Smith traveled down the Virgin River (chapter 13) to the Colorado River, crossed to the south side, and then traveled to the Mojave villages near present-day Bullhead City, Arizona (fig. 27.1).[11] He observed riparian vegetation in Cottonwood Canyon, now submerged beneath Lake Mohave; most of the trees he saw along the river were cottonwood and mesquite, and "some willow [probably coyote willow] extends entirely along the river varying in width from ½ to 2½ miles in width[,] the river winding through woodland from one side to the other." He found no beaver to trap on the Colorado River and headed west across the Mojave Desert.

James Ohio Pattie visited the lower Colorado River twice, first in 1826 and again in 1829.[12] On his first visit, near the confluence with the Gila River, he saw "very heavy" woody riparian trees and found abundant beaver. Moving upstream, his group trapped thirty beaver, possibly in the present-day Palo Verde Valley. They returned in

December 1829, this time traveling downstream from the Yuma, Arizona, crossing into the delta.[13] During their adventures, they had both hostile and friendly exchanges with Native Americans and a near-fatal experience with the tidal bore in the delta. In a single night just downstream from Yuma, they trapped thirty-six beaver, and "sometimes [they] brought in 60 [beaver] in a morning." Pattie complained about the number of raccoons who got into their traps and reduced their take of beaver. The floodplain had "chiefly cotton-wood, which in the bottoms is lofty and thick set. . . . [T]he bottoms are from six to ten miles wide."[14] In an area where the cottonwood were larger but more sparse, they killed a jaguar.

Paintings and Photographs

In the steamship *Explorer*, Lieutenant Joseph Ives explored the lower Colorado River from the delta to Black Canyon in 1857.[15] Ives and his photographer, Frederick von Egglofstein, failed to obtain photographs of the lower Colorado River in 1857 (chapter 5), although Ives's artist, Heinrich Balduin Möllhausen, painted stunning watercolors.[16] They show a variety of riparian vegetation from Yuma to present-day Hoover Dam. In some cases, such as at Yuma, the watercolors document cottonwood-willow forests; one watercolor documents carrizo grass lining the river near Picacho.

In nearly all the watercolors showing a distant vista, the river is lined with a thin band of riparian trees, not an immensely wide forest as some, including Pattie and Smith, reported in their diaries. Many of the watercolors show dead stumps rising above smaller trees, suggesting either a fanciful artistic interpretation of the riverine environment or an indication that something killed mature riparian trees. The river had low water at the time of this expedition, and nearly all the watercolors showing the water suggest a heavy load of driftwood. In most of the watercolors, however, repeating patterns of dead stumps and tree shapes that suggest conifers reduce the interpretive usefulness of these images.

Ives described riparian vegetation that differs somewhat from the images depicted by Möllhausen. In 1858, he generally observed "luxuriant growth of willows, cottonwoods, [mesquite] . . . and dense thickets of arrow-wood" along the lower Colorado River.[17] At Picacho Peak, Ives saw "occasional growth of mesquite, cottonwood, or willow." Near present-day Needles, he found "plenty of timber growing in the valley," but he also saw considerable agricultural lands on the floodplain and tended by the Mojave. The Wheeler survey of 1869 also observed considerable mesquite and willows, with scattered areas cleared for agriculture.[18]

Rudolf D'Heureuse took the first photograph of the Colorado River and the first known landscape photograph of the state of Arizona in March 1863 (fig. 27.2). Fort Mohave, now abandoned, once stood on the high terraces above the Colorado River south of present-day Bullhead City, Arizona (fig. 27.1). D'Heureuse's photographs, some close-ups of Native Americans at Fort Mohave, clearly show isolated cottonwood trees, with dense forests in the background, across a wide channel that is now lined with bank protection.

Numerous photographs of the lower Colorado River were taken in the late nineteenth and early twentieth centuries, but the views for most cannot be relocated owing to the lack of a suitable background.[19] Perhaps the most famous photographer to record the lower Colorado River was Timothy O'Sullivan of the Wheeler Expedition. In 1871, O'Sullivan was part of a crew who launched from the mouth of El Dorado Canyon and traveled upstream, exiting the Colorado River at Diamond Creek. His images record the stark nature of canyons now mostly submerged beneath reservoirs. Other photographers include D. T. MacDougal, an ecologist from the Desert Laboratory in Tucson; Godfrey Sykes, MacDougal's occasional assistant and Colorado River explorer in his own right; Eugene C. La Rue, a U.S. Geological Survey hydrographer interested in damming the Colorado River; and other travelers to the region who were impressed enough to photograph the mighty Colorado River. In total, 115 replicated photographs and perhaps another 20 views that cannot be replicated document change in riparian vegetation along the lower Colorado River between Hoover Dam and Yuma (fig. 27.1).

Steamships, Mines, and Riparian Vegetation

Riparian vegetation along the Colorado River beckoned to travelers, foretelling the presence of permanent water eagerly sought as desolate landscapes both east and west were crossed. The same vegetation offered economic prosperity to miners and steamboat operators as well. The lower Colorado River passes through mountains of the Basin and Range that contained economically viable ores, notably gold and silver. Strikes in the Picacho Mountains of California, in the Kofa Mountains of Arizona, at Oatman, Arizona, and in El Dorado Canyon in Nevada required transportation of supplies to the mines and transport of ores—smelted or raw—to market.

Transportation of raw ore was extremely expensive, particularly before construction of the transcontinental railroads. Therefore, miners sought to smelt ores on site to lower the weight and bulk of their product. Whether they used stamp mills to crush ore or added furnaces for smelting the crushed rock, their mining operations needed large quantities of wood for fuel. For example, smelters and mills associated with gold mining at Picacho Peak used significant quantities of wood—most likely mesquite from nearby bosques—between the late 1870s and 1910.[20] The effect of this woodcutting undoubtedly was severe but local in extent.

Transport of supplies to towns and mines along the Colorado River was extremely expensive by freight wagon. Beginning in 1852, steamships were introduced to the lower Colorado River.[21] Steamships are most efficiently operated using coal, but because no significant coal deposits were present along the lower Colorado River, the ship captains used wood to run their boilers. Between 1852 and 1916, nineteen paddlewheel or screw-driven steamboats were used on the river. Before the arrival of the Southern Pacific Railroad in 1877, the ships were built and maintained at the shipyard of Port Isabel, near the mouth of the river in Mexico, where arriving ocean-

A

B

Figure 27.2 Photographs showing the Colorado River at Fort Mohave, Arizona.

A. (March 1863.) This view, the first landscape photograph taken in Arizona, shows the Colorado River behind some of the buildings at Fort Mohave, near present-day Bullhead City. The buildings were constructed on an active floodplain and undoubtedly were destroyed by flooding, no later than 1884. A nearly closed gallery forest of cottonwood and willow appears across the river in the background. (R. D'Heureuse, 1905.16894-A[4:37], courtesy of the Bancroft Library.)

B. (September 26, 2000.) Fort Mohave, relocated to the terrace behind the camera station, is now only foundations and rubble. This reach is now channelized, and the cottonwood-willow forest is gone. A thin band of mesquite—some with parasitic mistletoe in their crowns—lines the terrace base in the foreground, and coyote willow, carrizo grass, and desert broom appear in patches near the channel. Tamarisk is abundant in this reach, and fountain grass is locally heavy on the artificial levees. (D. Oldershaw, Stake 2122.)

going ships off-loaded supplies to be transported upriver. After 1877, those supplies were off-loaded from railcars in Yuma or Needles.

The trees and driftwood found in abundance along the lower Colorado River had several important uses that some believe caused reductions in woody riparian vegetation. The first ferryboat was built of planks sawn from cottonwood trees at Yuma.[22] The steamship companies that plied the river from 1857 through the early twentieth century had designated wood yards at regular intervals to supply the boats with fuelwood, typically harvested by the Cucapá.[23] In the 1850s, cottonwood was sawn into lumber for buildings in Yuma, and kilns rendered an unspecified wood (probably mesquite) into charcoal.[24] Some houses were built of adobe, with willow (presumably black willow) poles used for lintels and rafters.[25] Cucapá woodcutters would cut, peel, and dry willow logs, then would build a raft of the poles for transport downstream. The source of the willow logs apparently was in the woodlands upstream from Canebrake Canyon.

For twenty-five years, steamships required wood for fuel to traverse from the Gulf of California to Yuma. Seven wood yards were built and supplied by Native American labor to fuel those ships, and two to four operated at any one time. It is unknown how much wood was cut for these ships, but the local effects must have been devastating to the riparian vegetation. Upstream from Yuma, steamships regularly went to El Dorado Canyon, off-loading supplies at a wharf now submerged beneath Lake Mohave. On several occasions, steamships traveled to Callville, Nevada, near the mouth of the Virgin River. The locations of wood yards for resupply of steamships in those reaches is not well known, although one was located at the mouth of the Bill Williams River and another was on Cottonwood Island, now submerged beneath Lake Mohave.[26]

Few riverboat travelers described riparian vegetation in detail. T. S. Van Dyke's observation from about 1895 is general: he saw "mesquite groves" teeming with quail, "cottonwoods you could almost touch" as the boat passed, and "dense willows that nearly brush the boat."[27] Even in the predam era, the distribution of vegetation was unstable; the extent of cottonwood and willow increased about 10 miles downstream from Yuma at the end of the nineteenth century.[28]

Other Observations of Riparian Vegetation

Many late-nineteenth-century observers mentioned the riparian vegetation of the lower Colorado River,[29] but none of the descriptions gave significant detail. For example, in 1846 Lieutenant Colonel W. H. Emory crossed the river at Yuma with the Army of the West and observed cottonwood, willow, and mesquite.[30] In 1865, Pratt Allyn observed "dense growth of willows, cottonwood and mesquite" in the vicinity of Fort Mohave near present-day Bullhead City.[31] The presence of willow (presumably black willow), cottonwood, and mesquite was commonly noted, and the width of the riparian belt ranged from hundreds of feet to many miles wide, depending on where the observations were made. The channel shifted continuously, eating into its floodplain periodically to destroy riparian vegetation.[32] Riparian vegetation quickly became reestablished following abandonment of channels, either where backwaters formed in meander cutoffs[33] or when the main channel was abandoned.[34]

In 1894, Edgar Mearns visited the Colorado River from its confluence with the Gila River to the delta and made observations on the species of riparian vegetation present.[35] At Yuma, he observed that "[t]he channels of the Gila and Colorado Rivers are marked by lines of tall cottonwoods and a lesser fringe of willows," "the adjacent bottom lands . . . are more or less covered with mistletoe-matted mesquites and screwbeans," and "the commonest shrubs of the low ground are the arrowwood and *Baccharis*." Traveling downstream, he observed, "The river channel is marked by a line of unusually tall cottonwoods and a lesser fringe of willows [coyote willows]. . . . The common shrubbery is a dense and monotonous growth of arrowwood and, in places, of *Baccharis*." Mesquite occurred in a "park-landscape" on "less well watered surfaces."[36]

Nearer to the Gulf of California, Mearns noted, "As the river broadens toward its mouth, vast savannas, canebrakes, and tule and carrizo marshes are encountered . . . its most peculiar feature consisting in the dense growth of willows . . . making jungles almost as difficult to penetrate as the canebrakes." At Colonia Diaz, 30 miles south of the Southerly International Boundary, he observed, "The Colorado bottom is many miles in width at this point and covered by carrizo, cane, tule, and other semiaquatic vegetation, with mesquites on the drier places and willow and cottonwood beside the marshes and lagunas." "Extensive grassy plains" were present; large areas "were largely vegetationless"; and "cottonwood and willow were restricted to the main channel while arrowweed and tule were found only in the area of the extreme upper delta."[37] Shifting of the channel westward in 1922 caused reduction in the amount of grassy area and increases in woody vegetation. In 1928, another observer noted, "the cottonwood or Alamo is largely restricted to a narrow margin along both banks of the old channel of the Colorado."[38]

These descriptions indicate that forests in the delta were not as extensive as previously claimed;[39] instead, the river was bordered with a gallery of woody vegetation, mesquite occupied higher terraces, and the wetlands created by high groundwater levels supported mostly herbaceous or aquatic vegetation. Grasslands are claimed to have been extensive in the nineteenth century, maintained by fires first set by the Cucapá and later by ranchers.[40]

In 1910, Joseph Grinnell described the flora and fauna of the lower Colorado River during a river trip from Needles to Pilot Knob (opposite Yuma).[41] He observed that meander migration in the open valleys of the lower Colorado River created a pattern of even-aged stands of cottonwood, black willow, and various shrub species; he did not report tamarisk along the river. Carrizo grass formed dense stands lining Canebrake Canyon downstream from Picacho, as it does today. Disturbance had killed native riparian plants and initiated a cycle of replacement; for example, he noted that channel avulsions in the delta had caused the river channel to deepen 7

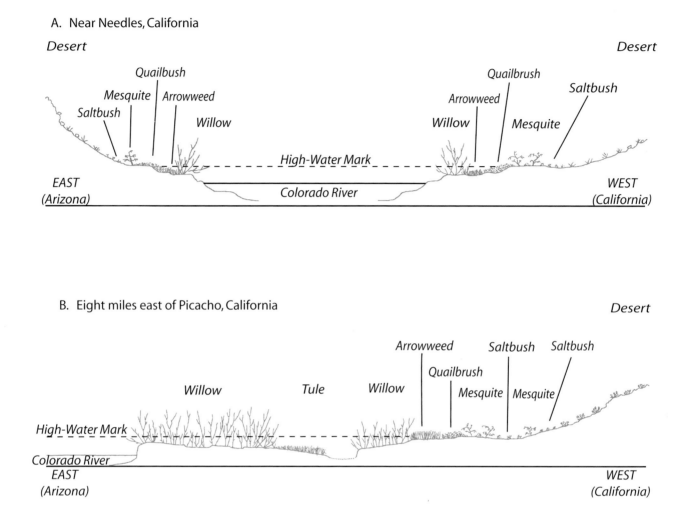

A. Near Needles, California

B. Eight miles east of Picacho, California

Figure 27.3 Cross sections of the Colorado River showing generalized relations between river stage and riverine riparian vegetation on the predam river (from Grinnell 1914, pp. 88–89).

A. Near Needles, California.

B. Eight miles east of Picacho, California.

feet, resulting in a conversion of the lower floodplain from cottonwood and willow to mesquite. Grinnell's representative cross sections (fig. 27.3) show only aquatic vegetation and riparian shrubs lining the channel and backwaters, suggesting that gallery forests were not common in the reach he represented. In the 1950s, cottonwood and black willow were abundant between Fort Mohave and the delta.[42]

Willis Jepson, noted California botanist, followed Grinnell down the Colorado River in October 1912.[43] Jepson's purpose was to collect plants and describe the botany of the lower Colorado River and its desert environment. Arrowweed was the most common species Jepson observed along the river, and they saw "thousands of small cottonwoods" in the Palo Verde Valley. All along the river, he noted the presence of Frémont cottonwood, both black and coyote willow, the ubiquitous arrowweed, and screwbean mesquite but did not mention the presence of tamarisk.

The most famous account of the ecological resources of the delta comes from Aldo Leopold, who canoed through its labyrinth of channels and sloughs in 1922.[44] Leopold described the delta as a veritable garden of Eden, but the purpose of his description was to praise wildlands and to champion their preservation in his concept of a new land ethic. He mentioned extensive wildlife—he hunted waterfowl—and described predators such as the bobcat, coyote, and, most important, the jaguar, now believed to be extirpated. He described mesquite but not cottonwood; he saw dense arrowweed thickets but did not mention dense forests. Mesquite undoubtedly was the most common tree in the delta; one observer called it an "orchard forest" covering many square miles.[45]

The arrival of tamarisk appears to have been delayed here compared to other river reaches in the Southwest. It does not appear on riparian species lists compiled in 1910 and 1923.[46] No tamarisk was observed on the banks of the Salton Sea in 1915.[47] Ten years after Hoover Dam began to fill, it still was not common along the lower Colorado River. At the earliest, it came to the lower Colorado River in the mid-1920s,[48] contradicting claims of earlier arrivals.[49] Another account states that tamarisk became established "several decades" before the early 1950s.[50]

Area of Riparian Vegetation

Historical change in the area of riparian vegetation along the lower Colorado River is difficult to determine, in part because maps created before 1962 are not geospatially reliable, map units are inconsistent, and different reaches have been mapped at different times. The first maps showing the distribution of riparian vegetation in the delta are crude and difficult to interpret quantitatively for change. One map dated 1900 shows undefined areas of "heavy mesquite and willow timber," "mesquite timber," and upstream of the "head of tide waters" (near Colonia Lerdo, Mexico) a band about 4 to 6 miles wide of "cottonwood, willow, and mesquite timber." Downstream of Colonia Lerdo, the vegetation was "tule lands."[51]

In 1902, the amount of alluvial bottomland along the Colorado River between Fort Mohave and Yuma was estimated from these unreliable maps to be 400,000 to 500,000 acres.[52] From a mapped area of 218,000 acres of the delta in Mexico, J. Garnett Holmes found that most of the area (65 percent) was "untimbered," that mesquite stands of varying density occupied 24

percent, and that cottonwood-willow stands accounted for only 1.6 percent.[53] In addition to mesquite, stands of arrowweed[54] were probably more common than cottonwood.

Beginning in 1962, riparian vegetation along the lower Colorado River was mapped at various times and at different scales with different units of riparian vegetation. In 1962, out of 294,000 acres of bottomlands, 167,000 acres of riparian vegetation were mapped from Davis Dam (roughly at Fort Mohave) and the Southerly International Boundary.[55] Of these 167,000 acres, only 50 acres were mapped as willow; cottonwood was not specifically mentioned; 53,180 acres were mapped as tamarisk; and 140,000 acres were proposed for eradication to conserve water.

In 1981, from Davis Dam to the Southerly International Boundary, riparian vegetation occupied a total of 112,000 acres of floodplain along the Colorado River.[56] Monospecific stands of tamarisk covered 35 percent of the area, followed by screwbean mesquite with tamarisk (22 percent) and monospecific honey mesquite (13 percent). Cottonwood-willow stands occupied 8,500 acres or 8 percent of the area. The total area of both species of mesquite—with or without tamarisk—was 43 percent of the area, showing the dominance of mesquite along the Colorado River despite agricultural clearing, which most likely occurred in mesquite bosques.

Another map, made using Landsat imagery taken in 1984, shows 115,000 total acres of riparian vegetation and 275,000 acres of agricultural land in the same reach as the 1981 map.[57] Data from 1986[58] indicate that 107,700 acres of riparian vegetation were present, and monospecific stands of tamarisk and mixtures of tamarisk and other native species, primarily mesquite, dominated with 68,400 acres (63.5 percent). Arrowweed, which reportedly did not contribute a significant area in 1981, occupied 7,500 acres (6.9 percent) in 1986. The 1986 map delineated only 5,750 acres of cottonwood-willow assemblages.

Finally, data on the area of riparian vegetation in 1997, measured from Davis Dam to the Southerly International Boundary, provide information

on 106,200 acres of riverine habitat in forty categories.[59] Approximately one-fifth of those acres were river, marsh, agricultural fields, or urban development, leaving 83,400 acres of total riparian vegetation. A total of 68,100 acres (65 percent) contained tamarisk, either in monospecific stands (43,390 acres) or in combination with mesquite. Cottonwood-willow assemblages covered a total of 3,786 acres (3.6 percent).

Taken literally, these data indicate that the cottonwood-willow assemblage continues to decline on the lower Colorado River, whereas the amount of tamarisk has remained about the same. The decline in cottonwood is in accord with our previous observations of the Green River (chapter 6) in Stillwater and Labyrinth Canyons, the upper Colorado River (chapter 7), and the Gila River (chapters 17 and 26), while serving as counterpoint to the Bill Williams River (chapter 15) and other tributaries of the Colorado River. Alternatively, uncertainties in the mapped areas of riparian vegetation may prevent meaningful comparison of change. For example, no cottonwood units were mapped in 1962. Does this mean that cottonwood-willow assemblages were not present in that year? Instead, the map scale used in 1962 probably made delineation of cottonwood-willow assemblages difficult.

According to one estimate, the potential area of riparian vegetation in the delta before completion of Hoover Dam was 768,000 acres.[60] Cottonwood-willow assemblages occupied only the immediate banks of channels and backwaters; most of the area was occupied by cattail, carrizo grass, and other plants variously called "wild rice" and "wild flax." Vegetation encroachment on any bare surface was so rapid that one observer claimed that "successive floods fought for mastery with the impending vegetation" for control of the delta.[61]

Flood control and irrigation diversions were expected to decrease the potential area of riparian vegetation by 93 percent and to cause severe reductions in or extirpation of aquatic mammals, freshwater fish, and waterbirds. By the early 1940s, some observers described forests of dead trees—"willows, cottonwoods, thorny Osage oranges [mesquite?]"—in the delta, apparently because of salt-water intrusion allowed by reduced river flow.[62] In the 1990s, native riparian trees, mostly black willows, germinated along the channels through the delta in response to modest flood releases from dams in the United States.[63] The dynamic germination-establishment-mortality regime described for the predam delta continues, although over a smaller area.

Floods and Flow Regulation

Floods

Historically, perennial flow adjacent to fertile floodplains in the desert Southwest attracted irrigation-based agriculture. From 1904 through 1934, the annual flow volume at Yuma was 21,300 ft³/s. Even though floods were to be expected, they ravaged newly created towns along the lower Colorado River. The 1862 flood, which devastated settlements in the Virgin River basin (chapter 13) and reportedly crested at 400,000 ft³/s at Topock (see fig. 27.8), severely damaged Yuma on January 2.[64] Would-be irrigators were warned about how difficult it would be to divert water from the Colorado River to nearby fields, but they persisted in their attempts to utilize river water.

In January 1874, the combined flows of the Colorado and Gila Rivers inundated about three-quarters of Yuma. The year 1884 had two notable floods: one on March 10–11, when levees broke at Yuma, flooding the town, and one in June and July that damaged the railroad bridge across the Colorado River. The latter event reportedly had a peak discharge of 300,000 ft³/s at Topock. In 1891, floods generated largely within the Gila River basin (chapters 23, 24, and 26) combined with the Colorado to flood Yuma after protective levees broke twice.

Devastating floods during the first decade of the twentieth century, combined with failed attempts to divert the river for irrigation in California and Arizona, led to a call for flow regulation of the Colorado River. Beginning in January 1905, the Colorado River had a series of floods downstream from its confluence with the Gila that damaged

Yuma, destroyed bridges, destroyed a canal system to the Imperial Valley, and ultimately led to the filling of the Salton Sea (see the next section). Notable flood peaks occurred in March 1905, April 1905, November 1905, and December 1906.[65] Other significant floods prior to completion of Hoover Dam occurred in January 1916 and in 1922.

Channel Changes

Before construction of dams, in particular Hoover, the Colorado River shifted alluvial channels across its floodplain and delta.[66] Historical backwaters along the Colorado River in the vicinity of Needles had formed, had repeatedly filled, and had an average lifespan of fifty to seventy years.[67] Because of low topography in the delta, periodic breaches diverted most or all of the river's water and sediment into the Salton Sink.[68] The Salton Sink periodically filled during the Holocene, forming Lake Cahuilla at least four times between A.D. 700 and 1580.[69] This occurrence suggests that the Colorado River may have abandoned its delta as much as one-third of the time in the past two thousand years. Spanish expeditions in the seventeenth century reported a lake in the sink with a water elevation at about sea level; this lake reportedly began to recede by 1640.[70]

The floods of 1905 flooded the Colorado and Gila Rivers, causing failure of an unregulated canal gateway and leading to the filling of the Salton Sea.[71] This event was hardly unprecedented; in recorded history, the Colorado River downstream of Yuma was fickle as to the channel in which it flowed. In most years, river flow rushed southward to the Gulf of California. However, in 1828, 1840, 1849, 1862, 1867, 1884, and 1891, most of which were El Niño years, the river broke through its natural levees and flowed west and north into the California desert.[72] The 1891 breach, caused by one of the largest Gila River floods in Arizona history, caused sufficient flow to the Salton Sink to allow several boats to pass successfully.[73]

Before the 1905 flood, humans inadvertently aided the Colorado in its long-standing attempt to maintain an inland sea in southern California.

Potential farmland surrounded the Salton Sink, which was a small, salty lake; irrigation water was all that was required to create an agrarian bonanza. By renaming the more accurate "Colorado Desert" and "Salton Sink" the Imperial Valley, investors attained funding to divert Colorado River water into the natural depression. Agriculture began to thrive because of the long irrigation ditch. But the years of drought had lulled the irrigators into complacency, and the flood of 1905 cut through the unregulated headgates of the canal with a vengeance; within a year, the Colorado River abandoned its delta.[74] The entire flow entered the Imperial Valley and filled the sink, creating the Salton Sea.

The Colorado River was finally forced back into its channel in 1908, but the damage to the fledgling agricultural industry was in the millions of dollars. Deeply incised channels were cut into alluvium leading northward into the Imperial Valley, and residents feared that the Colorado River would downcut into its bed upstream from Yuma, stranding irrigation headgates. The threat of headcut migration eventually led to construction of Laguna Dam, a concrete flow-diversion structure that once filled the All American Canal. Laguna Dam remains across the Colorado River, essentially filling the role of a grade-control structure.

Subsequent dams then caused their own problems with channel stability. Beginning with the completion of Hoover Dam in 1935, clear-water releases scoured the channel bed and floodplain margins, causing several types of sedimentation problems. Scouring of the river bed affected the stability of bridge piers and flood-protection dikes, and the scoured sediment accumulated in unwanted reaches, negatively affecting river navigation.[75] To minimize channel change and dredging of the main channel and canals, bank protection was installed to keep the channel in place, to minimize erosion of banks, and to decrease the probability of overbank flooding of croplands and municipalities.

Dams and Flow Regulation

To settlers along the lower Colorado River, the presence of abundant water adjacent to arid but fertile land was both a blessing and a curse. Diverting water for irrigation was difficult and dangerous, and construction of poorly designed canals ultimately caused millions of dollars in destruction in the Imperial Valley of California. The first attempt at diverting the Colorado River for irrigation occurred in 1857. The canal, which featured viaducts over ephemeral washes, was an engineering failure.[76] Brush dams were quickly destroyed during the annual flood, and the 1905 disaster and the filling of the Salton Sea raised flood control on the Colorado River to a national issue.

The call for flood control on the Colorado River began in earnest in 1905, although the rationale for controlling the nation's fifth-largest river had begun earlier. Laguna Dam, 13 miles upstream from Yuma, was the first permanent dam built across the Colorado River (completed in 1909). Only 10 feet high, the original purpose of this dam was to divert irrigation water to Arizona and California.[77] The backwater created by Laguna Dam, despite its small size, inundated the low floodplain about 10 river miles upstream, killing cottonwood, willow, and mesquite trees.[78] Imperial Dam, completed in 1938 and only 23 feet high, replaced Laguna Dam as the diversion point for the All American Canal to the Imperial Valley of California and for the Gila Main Gravity Canal for Arizona.[79]

One response to the 1905 disaster was development of an overall plan for flood control and flow regulation in the Colorado River drainage.[80] Instead of merely plugging holes in levees, the U.S. Reclamation Service, which later became the Bureau of Reclamation, decided that a comprehensive plan of flood control and water supply was required.[81] Such an expensive plan would necessarily involve the federal government as well as states that would benefit from water development. Because the dams would produce hydroelectric power, private power companies, in particular those in southern California, would eagerly become involved. Congress agreed and authorized the necessary studies.

The decision about where to locate the first dam was politically driven to be near the then-fledgling town of Las Vegas, Nevada, which had railroad access and was the closest site to southern California, which had the largest economic interests and political power.[82] In 1922, a damsite in Boulder Canyon was recommended in a report to Congress. The bill authorizing the Boulder Canyon Project was signed by President Calvin Coolidge on December 14, 1928. Construction of the 726-foot-high dam near Las Vegas is one of the greatest engineering feats of the twentieth century.[83] Boulder Dam, later renamed Hoover Dam, was built in Black Canyon because the bedrock that underlies its foundation was more favorable than the foundation of Boulder Canyon. When the dam's gates were closed on February 1, 1935, the dam impounded what at the time was the world's largest reservoir, extending eastward into Grand Canyon and northward up the Virgin River.

Completion of the high dam in Black Canyon led to other structures, in particular those designed to transfer water to California and Arizona. Parker Dam, completed in 1938, impounds Lake Havasu with a height of 320 feet above the former river level (fig. 27.1).[84] This lake has the intakes for both the California Aqueduct and the CAP. It has a storage capacity of 648,000 acre-feet, and its deltas create the environment for two NWRs: the Bill Williams River NWR extends up the Bill Williams River (chapter 15), and the Lake Havasu NWR extends upstream from the delta along the Colorado River.

Davis Dam, which impounds Lake Mohave, was completed in 1950. This 200-foot-high dam is 67 river miles downstream from Hoover Dam[85] and is part of Lake Mead National Recreation Area. Headgate Rock Dam, which impounds the small reservoir of Lake Moolvalya, was built downstream from Parker Dam in 1944 to divert flows into irrigation canals for agriculture on the Arizona side of the river. Finally, Palo Verde Diversion Dam, located 9 miles northeast of Blythe, California (fig. 27.1), was built in 1957 to divert irrigation water into a canal system that feeds the California side of the river.

Flow Diversions

The "plumbing system" of canals and aqueducts that remove and return flow to the Colorado River is extremely complicated and governed by multistate and international agreements.[86] By treaty, the Lower Basin (defined as downstream from Lee's Ferry) is apportioned 7.5 million acre-feet per year of water, and Mexico is allocated an additional 1.5 million acre-feet per year. Nevada extracts 0.3 million acre-feet per year from Lake Mead, meaning theoretically that 8.7 million acre-feet per year should flow into the lower Colorado River. In fact, on average 10.1 million acre-feet per year passed through Hoover Dam from 1935 through 2003 (fig. 27.4), leading to a surplus-flow condition in many years.

Diversions to California, Arizona, and ultimately Mexico sequentially reduce the annual flow volume downstream. Evaporation from the open water of reservoirs and evapotranspiration of native riparian vegetation and agricultural crops use still more water. With the period of 1950 to 2003 used as a baseline, it is estimated that the average annual flow volume decreases from 10.1 million acre-feet per year passing through Hoover Dam to averages of 8.86, 7.74, and 2.84 million acre-feet per year passing through Parker Dam, through Imperial Dam, and past Yuma, respectively (fig. 27.4). Given that the long-term average flow into Arizona, without diversions, has been estimated at 15.1 million acre-feet per year, the reduction in flow volume to the lower Colorado River is substantial; at Yuma, the decrease is more than 80 percent.

With this magnitude of decrease in available river water, the negative effects on riparian vegetation might be expected to be large. However, several factors mitigate the effects of loss of flow volume. Reduction in annual peak flow, discussed in the next section, minimized channel change and other destructive effects of floods. Reduction in floods and flow volume led to an increase in the area of channel available for colonization by riparian vegetation. Tamarisk, because of its life-history strategy, arrived first and became established in monospecific stands, and much of the remaining riparian vegetation continued to have access to groundwater in the alluvial aquifer. Much of that native vegetation—in particular mesquite—was

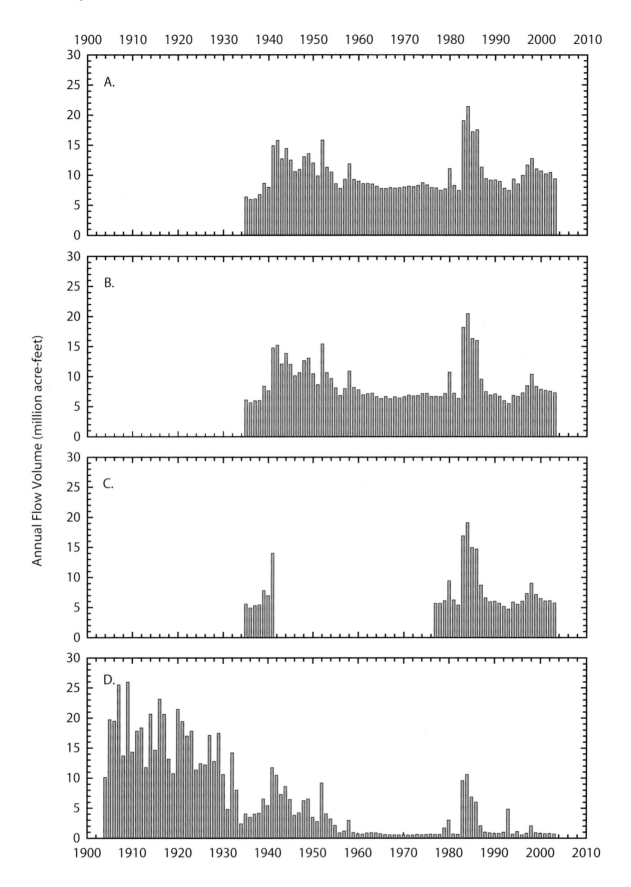

Figure 27.4 Annual flow volumes of the Colorado River.

A. Colorado River below Hoover Dam, Arizona-Nevada (station 09421500; 1935–2003.)

B. Colorado River below Parker Dam, Arizona-California (station 09427520; 1935–2003).

C. Colorado River above Imperial Dam, Arizona-California (station 09429490; 1935–1941, 1977–2003).

D. Colorado River below Yuma Main Canal Wasteway at Yuma, Arizona (station 09521100; 1904–2003).

A   B

Figure 27.5 Photographs showing the Colorado River in Black Canyon.

A. (1871.) In October 1871, Lieutenant George Wheeler used boats to go upstream on the Colorado River from El Dorado Canyon to Diamond Creek in Grand Canyon. This view shows what Wheeler apparently named "Camp Big Horn" in the heart of Black Canyon. The sparse riparian vegetation and widespread bare sand is typical of conditions in canyon reaches along the lower Colorado River before completion of Hoover Dam. (T. O'Sullivan 114, courtesy of the National Archives.)

B. (February 18, 2003.) This reach is now near the head of Lake Mohave about 5 miles downstream from Hoover Dam. Capture of sediment in Lake Mead, combined with few sizeable tributaries in Black Canyon, make this a sediment-starved reach with little remaining sand, except in tributary mouths. The bare rock slopes allow minimal establishment of riparian vegetation beyond the occasional tamarisk, mesquite, or grove of coyote willows. (R. H. Webb, Stake 4460.)

destroyed to create agricultural lands, especially in the wide valleys fed by irrigation canals. These changes inevitably led to the paradox of an increase in riparian vegetation irrespective of species composition along the river and an overall decrease in native species, notably mesquite.

Changes in Flow and Riparian Vegetation

Hoover Dam to Lake Havasu

Hoover Dam was built at the upstream end of Black Canyon to take advantage of extremely steep cliffs next to the river. Most of Black Canyon is similar, consisting of steep cliffs with little room for storage of alluvium on which riparian vegetation can grow (fig.

27.5).[87] Because clear-water releases from Hoover Dam scour this reach,[88] less alluvium was available after flow regulation, and completion of Davis Dam in 1950 inundated most of the remaining parts of Black Canyon. In the 1930s, the U.S. Geological Survey established a gaging station just upstream of one of the few ingress points, Willow Beach. Photographs of this gaging site (fig. 27.6) show the top of Lake Mohave and the growth of riparian vegetation—both tamarisk and mesquite—along the high-water line of the reservoir.

The valleys widen toward Laughlin, Nevada, and Bullhead City, Arizona, but despite names such as Cottonwood Canyon and Cottonwood Cove, little riparian vegetation grew in these reaches historically because there was

little room for deposition of a low floodplain (fig. 14.9). Downstream from Bullhead City, the river corridor widens into a broad alluvial valley. The Colorado River in this reach is considered to be influent and a major recharge source for the local alluvial aquifers.[89] Releases from Davis Dam scour the upper part of this reach, but tributary additions, notably from Sacramento Wash downstream from Needles, add considerable sand downstream. Date palms and fountain grass are common nonnative species along the levees that line the river.

Lake Havasu NWR begins in Topock Marsh and extends through the delta of Lake Havasu to near Lake Havasu City, Arizona (fig. 27.1). The area now called Topock Marsh was, at low water, a braided reach where numerous channels meandered among numerous barren sandbars (fig. 27.6).[90] The river is channelized through this reach by passing the marsh to the west, but it remains a sandy-bottom channel with vegetation-stabilized islands. In 2002, about 10,400 acres of open water and riparian vegetation were present in this part of Lake Havasu NWR, mostly (85 percent) tamarisk-arrowweed assemblages, with less than 2 percent cottonwood-willow assemblages.[91] Tributary additions from the Arizona side indicate that Topock Marsh is filling with sediment at an unknown rate.[92] At the downstream end of this marsh, bridges and pipelines span the Colorado River, and numerous photographs document change in riparian vegetation here (fig. 27.7).

Two gaging stations with nineteenth-century flood records are available for the lower Colorado River. The longest annual flood series for the river, or indeed for anywhere in the region, is the Colorado River at Topock, Arizona (fig. 27.8). Two noteworthy floods were recorded indirectly at this gaging station: the 1862 event had a discharge estimated to be 400,000 ft³/s, and the 1884 event was 300,000 ft³/s. Completion of Hoover Dam reduced flood peaks at this site, and filling of Lake Havasu hampered accurate operation of this gaging station by 1982.[93]

In the narrow Topock Gorge, riparian vegetation—mostly tamarisk—has increased significantly since bank-scouring floods were mostly eliminated (fig. 27.9). Downstream, the valley widens into the Chemehuevi Valley, and the amount of riparian vegetation is considerable in Lake Havasu NWR. Repeat photography in this reach shows that the formerly barren channel has narrowed appreciably; that tamarisk, cottonwood, and willow have all increased; and that marshlands have formed. The wider sections of the river corridor, such as the Pittsburg Flat area, now under Lake Havasu near Lake Havasu City, once supported patchy stands of cottonwood and black willow and nearly continuous bands of coyote willow and arrowweed; mesquite bosques formed in isolated, wide bottomlands (fig. 14.10). Construction of Parker Dam and filling of Lake Havasu eliminated these patches of riparian vegetation, and shoreline veg-

etation is now mostly tamarisk, with a thin band of mesquite and arrowweed upslope.

Parker Dam to Imperial Dam

Downstream of Parker Dam, the Colorado River flows through a narrow, shallow canyon and emerges into the Parker and Palo Verde Valleys (fig. 27.10). As in the Topock Gorge upstream from Lake Havasu, the width of the floodplain immediately downstream from Parker Dam is constrained by bedrock, but with more room for deposition of alluvium. Non-native vegetation—notably date palm, giant reed, and fan palm—is abundant, but mesquite and arrowweed are also common on the distal margins of the floodplain.

Downstream from Headgate Rock Dam, the river corridor widens into the Parker and Palo Verde Valleys. Riparian vegetation in this reach was mapped in 1962 and covered an estimated 108,000 acres, of which 73,000 acres was estimated to be in mesquite bosque and most of the remaining 35,000 acres was arrowweed.[94] This reach is now heavily channelized, eliminating once-thriving stands of woody riparian vegetation (fig. 27.11); clearing for agriculture tripled after about 1940. Despite these changes, some reaches still support significant amounts of native riparian vegetation among the ubiquitous presence of monospecific stands of tamarisk. Cibola NWR was established in 1964 to protect habitat for migrating waterfowl over a 12-mile reach of the Colorado River in the lower Palo Verde Valley.[95] This reach includes significant amounts of riparian vegetation, including dense stands of carrizo grass.

At the downstream end of the Palo Verde Valley, the river enters Canebrake Canyon, a short bedrock canyon. The once-thriving gold-mining town of Picacho, California, was built along the right bank of the Colorado River at the upstream end of Canebrake Canyon.[96] A former gaging station, the Colorado River near Picacho, was also present in this reach, and photographs show that the channel has narrowed significantly (fig. 27.12). Although tamarisk is common in this reach, carrizo grass, the dominant species before

flow regulation, remains dominant. At the downstream end of Canebrake Canyon, the valley widens one final time into the delta reach. The river, impounded by Imperial Dam, flows through Imperial NWR, established in 1941 for the same reason as the Cibola and Lake Havasu NWRs. Imperial NWR encompasses 30 river miles of the Colorado River upstream from Imperial Dam through Canebrake Canyon.

Imperial Dam to Yuma

At Imperial Dam, the All American Canal extracts most of the remaining flow from the Colorado River—5.2 million acre-feet per year—leaving only a meager 2.84 million acre-feet to pass through Yuma. Immediately after construction of Laguna Dam, the channel here was wide and barren, but it is one of the few places where a wide gallery forest of cottonwood and willow was present on the floodplain (fig. 27.13A). Now, the once-barren channel has greatly decreased in width, with the abandoned area now providing habitat for black willow, cottonwood, and tamarisk (fig. 27.13B). The closed gallery forests are gone, at least temporarily.

As shown in figure 27.14,[97] the annual flood series for the Colorado River at Yuma dramatically demonstrates the combined effects of flood-control operations, primarily at Hoover and Glen Canyon Dams, and water diversion from Lake Havasu and Imperial Dam. The average annual flood from 1903 to 1934 was 99,000 ft³/s; after 1935, the average annual flood decreased to 11,000 ft³/s. The river continues to recharge the alluvial aquifer, and groundwater levels remain high (fig. 27.15) despite regional water development, which includes water withdrawals. The highest water levels are associated with the extremely wet El Niño years of 1983 and 1993, which included large flow releases from dams on the Colorado and Gila Rivers.

As a result of flood control and bank protection, the channel through Yuma has narrowed considerably, with significant changes to the riparian vegetation (fig. 27.16). Several photographs taken upstream and downstream from the former gaging station,

Figure 27.6 Photographs showing the Colorado River upstream from the Needles Bridge.

A. (November 11, 1917.) This view, upstream and across the Colorado River, shows the approach of the Colorado River to a series of bridges downstream from Needles, California. The river channel is at least a third of a mile wide, barren, and braided in this low-water scene. A thin band of tall woody riparian vegetation—probably Frémont cottonwood and black willow—is apparent on either side of the river, and mesquite and arrowweed are the likely shorter species behind the trees. The mountains to the north include the Black Mountains of Arizona *(right background)* and the Newberry Mountains of California *(left background)*. (D. G. Thompson 128, courtesy of the U.S. Geological Survey Photographic Library.)

B. (April 22, 2004.) The Colorado River was channelized in this reach in 1966, separating the downstream-flowing channel *(nearest camera station)* from a 4,000-acre network of sloughs and oxbow channel remnants *(behind the river, in the distance)*. This reach is now the lower end of the Topock Marsh, a major feature of the Lake Havasu NWR. Cottonwood, black willow, coyote willow, seepwillow, mesquite, carrizo grass, and tamarisk are common here along with other aquatic plants, including cattail and reeds. (D. Oldershaw, Stake 4771B.)

A

B

Figure 27.7 Photographs showing the Colorado River downstream from the Needles Bridge.

A. (March 1905.) This view across the Colorado River was taken from river left about a quarter-mile downstream from the former bridge for U.S. Highway 66. An unusual winter flood occurred earlier in March and is the likely reason for the driftwood strewn across the wide, barren floodplain. The riparian vegetation on the downstream side of the far *(right)* bank is probably a small grove of black willow. (D. T. MacDougal A4-38, courtesy of the Desert Laboratory Collection.)

B. (December 18, 2002.) The channel, which is not channelized in this reach, has narrowed, and a fire occurred right before we took our photograph. The view is blocked mostly by a combination of arrowweed *(foreground)* and burned and dead tamarisk *(midground);* several young mesquite trees are also present in the vicinity. A natural-gas pipeline spans the river at upper right. (D. Oldershaw, Stake 4489.)

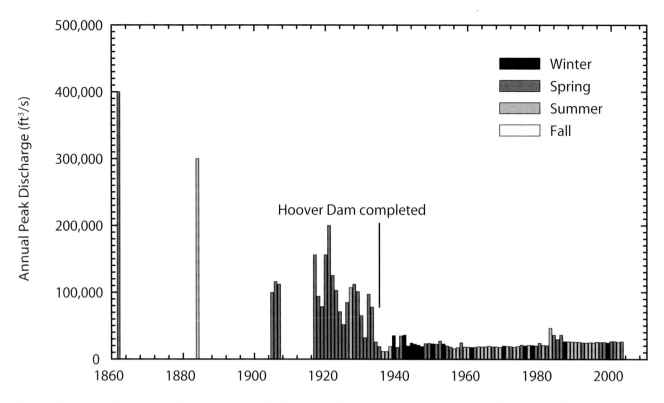

Figure 27.8 Combined annual flood series for the Colorado River near Topock, Arizona (station 09424000; 1862, 1884, 1905–1907, 1917–1982), and for the Colorado River at Davis Dam, Arizona (station 09423000; 1982–2003).

Colorado River at Yuma, show a thin band of riparian vegetation lining the channel, with noticeable quantities of black willow. Now, the combination of flow regulation that increases the area for vegetation growth and more stable hydrologic conditions has led to an increase in riparian vegetation, notably arrowweed. Breaks in this riparian vegetation upstream are associated with city parks.

The Delta and the Salton Sea

At a hard-to-define point around the mouth of the Gila River, or as far downstream as Pilot Knob west of Yuma, the Colorado River enters its delta.[98] For the next 150 river miles, the river meanders through alluvial channels to the Gulf of California. Before flow regulation and construction of irrigation canals, the channel in the delta was in some places braided and in others anastomosing around fixed islands. The generally slight slopes meant that channel avulsions were common, and the river shifted position numerous times.[99] Riparian vegetation quickly became established in abandoned channels.[100] The channel in this reach continues to shift, although it is constrained within high levees.[101]

As shown in figures 27.4 and 27.14, the Colorado River pushed tremendous quantities of water and accompanying sediment through its delta. The size of the annual flood and the annual flow volume decreased with closure of Hoover Dam and then decreased further as additional dams and flow diversions became operational. Only in periods of large storms or when maintenance of dams called for large releases has significant amounts of water passed through Yuma or past Morales Dam and into the river's former channel.

Leakage of irrigation canals—most notably the All American Canal—has raised groundwater levels downgradient, creating wetlands that can be extensive. In 2002, a total of 15,500 acres of wetlands in the United States and Mexico resulted from sixty-two years of leakage of the unlined All America Canal west of Yuma.[102] This formerly arid area, consisting of dune and interdune areas, was likely barren before local groundwater levels rose sufficiently to sustain standing water. These wetlands support extensive stands of native and nonnative riparian vegetation and are significant at-

tractants for resident and migratory birds.[103] Lining of the All American Canal has been proposed as a means of conserving water for irrigation in the Imperial Valley, and this removal of the "recharge" source for the Andrade Wetlands would eventually spell their elimination.

Some observers have predicted elimination or large reduction of cottonwood-willow forests in the delta.[104] At present, the delta has woody riparian vegetation covering approximately 150,000 acres, a small amount of what some researchers estimate as a former extent of 1.9 million acres.[105] Because riparian vegetation responds quickly to winter flood events, researchers have noticed that relatively small flows passing Yuma and the international border caused germination and establishment of trees within the channel system of the delta.[106] Some are now calling for only small annual flow releases, 32,000 acre-feet, into the delta, with a larger discharge of 260,000 acre-feet released about every four years.[107] Others have explicit plans for purchase of water rights for transfer to the riparian ecosystem of the delta.[108]

A

B

Figure 27.9 Photographs showing the Colorado River at the former Topock gaging station.

A. (September 9, 1923.) This across-channel view shows the former gaging station Colorado River at Topock, including the stream hydrographer's residence, a tent residence for visitors, a woodpile, the masonry gaging station, and the cableway for discharge measurements. In the foreground, a small tributary enters the Colorado River. A solitary tamarisk, flanked by small black willows and with arrowweed behind it, appears in the mouth of this tributary. Similarly, a small tributary mouth across the river appears to contain arrowweed, with mesquite behind it. (G. E. Smith 8, courtesy of the U.S. Geological Survey.)

B. (October 29, 2000.) This reach is now in the delta of Lake Havasu, and the gaging station has been abandoned but not completely dismantled. The tributary mouth is now choked with riparian vegetation, notably tamarisk (a single Athel tamarisk is out of the view to the right) mixed with carrizo grass and arrowweed. Upslope of the tamarisk, a combination of mesquite and catclaw forms a nearly impenetrable thicket. (S. Young, Stake 1669.)

Figure 27.10 Photographs showing the Colorado River below Parker Dam.

A. (July 26, 1924.) This upstream view from the site of the former gaging station, Colorado River near Parker, is part of a series of four views that document blasting associated with channel maintenance to increase the reliability of the gaging record. A small emergent outcrop at left center was removed to allow low flow to come closer to the gage house, thereby reducing sedimentation in the vicinity. Black willow undoubtedly dominates the taller vegetation on both banks, with arrowweed and mesquite also contributing significant biomass. (R. E. Cook 368, courtesy of the U.S. Geological Survey.)

B. (October 27, 2000.) This reach is now a recreational hotspot for water sports. Flood-control operations of the dams on the lower Colorado River have allowed extensive residential development on both sides of the river, and native and nonnative palms—both planted and escaped from cultivation—are now common along the river corridor. Mesquite, arrowweed, and carrizo grass remain common, and tamarisk and Athel tamarisk are present in large numbers. (D. Oldershaw, Stake 2126.)

Figure 27.11 Photographs showing the Colorado River at Ehrenberg, Arizona.

A. (September 17, 1910.) This view was taken with the intent of documenting riparian vegetation along the lower Colorado River. The taller trees that form the backdrop are black willows, with arrowweed in a band in front; a solitary, small cottonwood appears in the left distance. The lower herbaceous species undoubtedly include horsetails. No tamarisk is apparent in the view. (D. T. MacDougal A4-73, courtesy of the Desert Laboratory Collection.)

B. (February 20, 2003.) Channelization is discontinuous in this reach, and this view shows nonstabilized banks approaching the inside of a bend upstream from the Interstate 10 bridge connecting Blythe, California, with Arizona. Tamarisk is the most common species in a cosmopolitan mixture of riparian species, including mesquite, screwbean mesquite, seepwillow, arrowweed, carrizo grass, fan palms, and scattered coyote willow. No willows or cottonwood are present in this reach. (R. H. Webb, Stake 4467a.)

Figure 27.12 Photographs of the Colorado River near Picacho, California.

A. (July 20, 1935.) This view, from a peak in the Picacho Mountains in California, shows the gaging-station complex of the Colorado River near Picacho, with the Trigo Mountains of Arizona in the distance. This view, taken in the year that Hoover Dam began operations, shows a low band of riparian vegetation dominated by carrizo grass near the river and mesquite behind it. The ubiquitous presence of carrizo grass in this reach reinforces the name "Canebreak Canyon" that was given by steamboat captains in the mid-1800s. Scattered trees—probably black willows and older mesquites—are also present, and arrowweed forms thickets in little tributary mouths. About fifteen other photographs were taken near the river at this time, but none shows tamarisk in 1935. (J. A. Baumgartner, courtesy of the U.S. Geological Survey.)

B. (December 19, 2002.) The channel, now much narrower, is still lined with carrizo grass, separating "lakes"—actually backwaters formed at the expense of the former floodplain—from the main channel. Mesquite, catclaw, and other species typical of the Sonoran Desert create locally dense stands in the former tributary mouths. (D. Oldershaw, Stake 4486b.)

Figure 27.13 Photographs of Laguna Dam.

A. (November 18, 1922.) Laguna Dam, constructed in 1909, was the first permanent control structure built on the Colorado River. This wide, low structure diverted river flow into the All American Canal, shown in the immediate foreground. The wide channel is mostly barren, with a large number of driftwood piles. The dense grove of trees in the left midground consists mostly of black willow and a few Frémont cottonwood. (E. C. La Rue, courtesy of the U.S. Geological Survey.)

B. (November 22, 2002.) Most of the river flow has been diverted into the All American Canal at Imperial Dam, about 5 miles upstream; this canal now flows behind the camera station. Vegetation chokes the former river channel and includes tamarisk, arrowweed, mesquite, and cottonwood. Palms have been planted and have escaped along this reach. (D. Oldershaw, Stake 4384.)

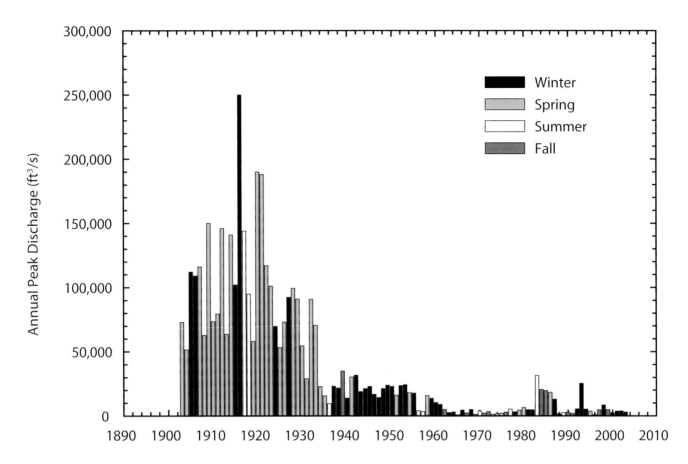

Figure 27.14 Annual flood series for the Colorado River at Yuma, Arizona (station 09521000; 1903–1965), combined with the annual flood series for the Colorado River below Yuma Main Canal Wasteway at Yuma, Arizona (station 09521100; 1966–2002) and corrected for canal additions (see text).

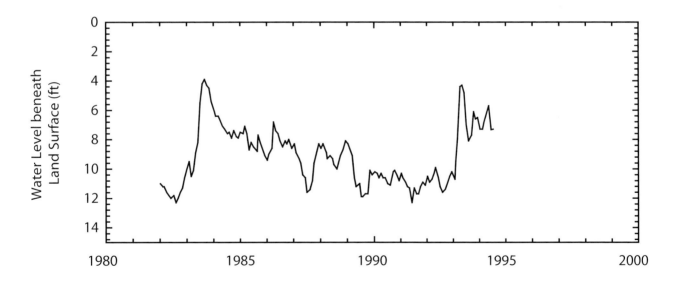

Figure 27.15 Groundwater levels for well C-08-23 24DCC2 near Yuma, Arizona.

A

B

Figure 27.16 Photographs of the Colorado River at Yuma.

A. (December 23, 1930.) This upstream view shows the road *(front)* and railroad *(back)* bridges across the Colorado River at Yuma. The wide channel has open stands of cottonwood and willow. (W. E. Dickinson 1572, courtesy of the U.S. Geological Survey.)

B. (November 22, 2002.) Owing to disturbance and new, dense vegetation, the camera station is in front of the original. The bridges for Interstate 8 now block the view of the two older bridges, both of which remain in place. Owing to channelization and decreased flow, the Colorado River has narrowed and is very shallow here most of the time. Carrizo grass and tamarisk dominate the riparian vegetation; the tall trees in the distance are growing in parks or other landscaping. (D. Oldershaw, Stake 4387a.)

Summary. The Mojave River is unusual among southwestern rivers for several reasons, including the fact that, instead of draining to the Gulf of California, it terminates in Soda Lake and Silver Lake playas. This river arises in the Transverse Ranges of southern California and historically had several perennial reaches separated by ephemeral reaches. Water development in the Mojave River basin—including operations of flood-control structures in the headwaters, extensive groundwater usage, and artificial recharge of water imported into the basin—has greatly changed the amount and distribution of woody riparian vegetation. In the perennial reach within and downstream from Victorville, California, woody vegetation has increased, with cottonwood assuming dominance over an assemblage once dominated by black willow. In the reach from downstream of Helendale to Camp Cady Ranch, most obligate riparian vegetation has been eliminated, and tamarisk (both *Tamarix aphylla* and *T. ramosissima*) and mesquite are the only remaining species with widespread distribution. Increases also occurred in the vicinity of Camp Cady, although lowering of the alluvial aquifer continues, threatening in particular mesquite and cottonwood on the upstream and downstream sides and the cottonwood-willow assemblage in the middle of the reach. Afton Canyon, once a perennial reach with scattered trees, now supports mainly tamarisk stands that are the target of an active eradication program.

The Mojave River, along with Animas Creek and tributaries of the Río Yaqui (chapter 16), are the only watercourses discussed in this book that do not drain to the Colorado River. The Mojave River has its headwaters in the San Bernardino and San Gabriel Mountains of southern California, and its terminus is a playa system composed of Soda and Silver Lakes (fig. 28.1). Downstream from its headwaters, this river once had as many as three perennial reaches separated by ephemeral reaches that crossed deep alluvial basins. Those perennial reaches, which were lined with obligate riparian vegetation, attracted settlers interested in agriculture in an otherwise arid region, and as a result the surface-water and groundwater systems of the Mojave River were highly developed. The alluvial aquifer of the Mojave River is a primary source of water for domestic supplies and irrigation in the western Mojave Desert.

The river supports about 10,000 acres of riparian communities unique within the larger Mojave Desert ecosystem.[1] Extensive stands of cottonwood, black willow, mesquite, and tamarisk occur along this river, as well as smaller-stature species that include coyote willow and seepwillow. According to maps of the riverine environment,[2] other woody riparian species exist in the basin, mostly in small amounts, including sycamore, velvet ash, cattail, and bulrush. In addition to tamarisk, at least one other nonnative species (giant reed) occurs upstream from and at the Lower Narrows, which is downstream from Victorville, California, and in other perennial reaches.

Three aquatic animals of note also once lived in this river but are now rare. The western pond turtle (*Clemmys marmorata*), a species of special concern in the state of California, occurs in the perennial-flow reaches of Afton Canyon, and a smaller population is held in artificial ponds at Camp Cady.[3] The Mohave tui chub (*Gila bicolor*), once widely distributed in the Great Basin and in the Mojave River north of Victorville,[4] occurs only in natural ponds adjacent to Soda Lake and in artificial ponds and relocation sites. Finally, beaver once were present as well.[5] The combined effect of high water use, the presence of unique plant assemblages, and threats to rare animals makes water-resources management of this river a high priority for state and federal agencies.

Surface Water and Flow Regulation

Most of the Mojave River's drainage basin—in contrast to the climate of its headwaters, which receive more than 20 inches of annual precipitation—is arid, with an annual rainfall of less than 6 inches.[6] The drainage areas at the long-term gaging stations are 513 square miles at the Lower Narrows near Victorville, 1,290 square miles at Barstow, and 2,121 square miles at Afton Canyon. At the Lower Narrows gaging station north of Victorville, the river has a highly variable annual flood series, with some years having only base flow and other years having floods measuring as high as 70,600 ft^3/s (fig. 28.2). It has been argued that owing to climatic fluctuations, twentieth-century floods were much larger and more frequent than ones that occurred under presettlement conditions in the late Holocene.[7]

Although the largest flood in the gaging record occurred in 1938, which was not an El Niño year, most floods on the Mojave River occur during El Niño conditions. El Niño years with large floods include 1891, 1905, and 1916. In recent decades, the relation between flooding and El Niño has strengthened, with floods in 1978, 1983, 1993, and 1998 (fig. 28.2). The Mojave River flows continuously from its source to its terminous in the Silver Lake playa during these years.

The Mojave River and its tributaries have three dams in the headwaters that

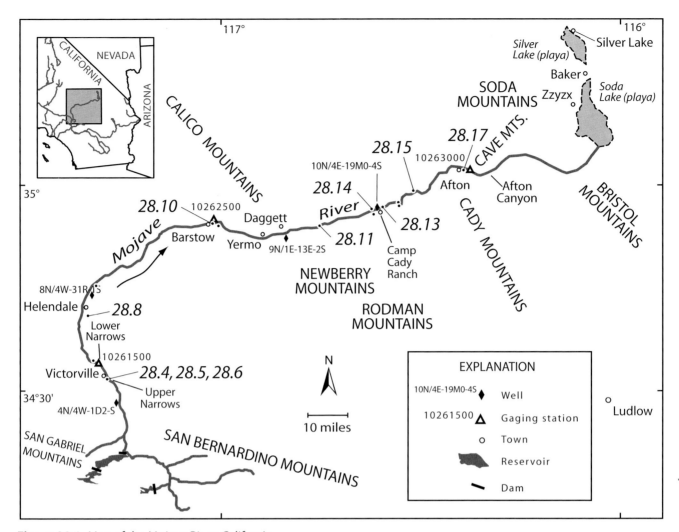

Figure 28.1 Map of the Mojave River, California.

store water and provide some flood control for the reaches in the Mojave Desert. Both the Mojave River Forks Reservoir and Silverwood Lake (not shown in fig. 28.1) were completed in 1971.[8] The Mojave River Dam, which impounds the Forks Reservoir, allows a maximum release of 23,500 ft[3]/s (excluding spillways).[9] The presence of these reservoirs and the capping release through the Mojave River Dam may be the reason why the size of floods appears to have declined in the latter part of the twentieth century, although this decline also might be the result of climatic fluctuations. Lake Arrowhead Reservoir, built in 1922, provides only minimal flow regulation. Most of the flow decreases are likely the result of local groundwater pumping; the headwater dams have not had a significant impact on flow in the Mojave River other than to attenuate flood peaks.[10]

Downstream from the Mojave River Forks Reservoir, the Mojave River crosses its first reach of deep alluvium, resulting in a channel that ordinarily is dry. Floods periodically sweep through this reach (fig. 28.2), despite flood-control operations at the Mojave River Dam, and these floods recharge floodplain sediments and the alluvial aquifer.[11] A fish hatchery in this reach extracts groundwater and discharges it downstream as irrigation return or directly into the river above Victorville. Limited amounts of water imported from the California Aqueduct and stored in Silverwood Lake are released into the Mojave River, and water is also released from the Mojave Water Agency's Morongo Basin Pipeline 5 miles below the Mojave River Forks Reservoir.[12] The combination of climatic fluctuations, artificial recharge, flood recharge, and high groundwater use has led to large fluctuations in groundwater levels (fig. 28.3).

North of the front of the San Bernardino and San Gabriel Mountains, perennial flow begins south of Victorville and extends reliably downstream to the end of the Lower Narrows (fig. 28.1). The average daily discharge at the Lower Narrows is 79.8 ft[3]/s. Flow can be intermittent downstream from Helendale during high runoff years, notably during El Niño episodes. Flow also is perennial in Afton Canyon, where the river passes through a short bedrock canyon en route to Soda Lake (fig. 28.1). Before groundwater development, flow was perennial at Point of Rocks, north of Helendale; between Barstow and Daggett; and at Camp Cady Ranch. A large grove of cottonwood once served as a picnic ground west of Daggett around the turn of the twentieth century.[13] Surface water was sufficient in this reach to support diversion for irrigation as well as habitat for western pond turtle. Camp

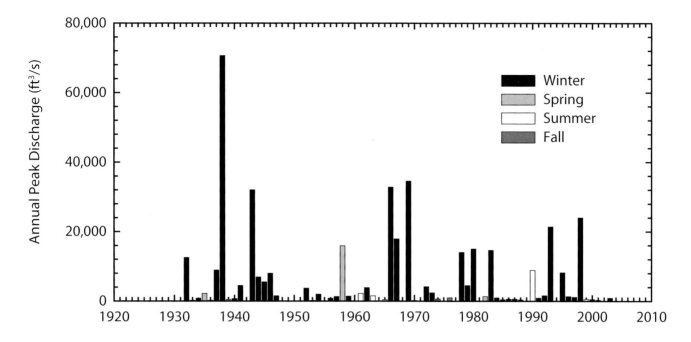

Figure 28.2 Annual flood series for the Mojave River at Victorville, California (station 10261500; 1931–2003).

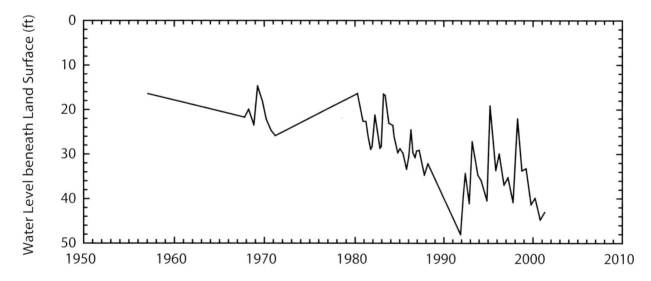

Figure 28.3 Groundwater levels for well 4N/4W-1D2-S south of Victorville, California.

Cady, established at a short perennial reach about halfway between Daggett and Afton Canyon, was a nineteenth-century army outpost that protected travelers on the Mojave Road.

Geology and Groundwater

The hydrogeologic framework of the Mojave River is well documented.[14] The parts of the Mojave Desert through which the Mojave River flows are relatively young and geologically active. The San Andreas fault largely defines the western edge of the river's headwaters, and the channel passes through bedrock reaches near Victorville and in Afton Canyon. Faults, which may or may not have surface expressions, force groundwater toward the surface near Daggett and Camp Cady (fig. 28.1). Development of the drainage basin is thought to have been driven by the filling of lakes between the fault-block mountains, followed by spilling through passes that created canyons, such as Afton Canyon.[15] As a result, the basin fill through which the river flows has numerous faults and clay layers that force groundwater to the surface.

No stronger contrast exists between headwaters and terminus in the Southwest than for the Mojave River. Nearly all of the discharge above base flow comes from its headwaters, a 215-square-mile area that is a mere 4.7 percent of its total drainage. Even in the desert region near Victorville, recharge occurs only along the river or in ephemeral washes, and infiltrating water may require hundreds of years before it reaches the saturated water table.[16] Water in most of the regional aquifer west of Victorville may have been recharged twenty thousand years before the present under a different climate regime, whereas the age of groundwater beneath the river is recent.[17]

Groundwater supplies both domestic and irrigation to water users in the basin, particularly upstream from Upper Narrows and to a lesser extent between Barstow and Camp Cady. Groundwater extraction is heavy in the basin, particularly upstream from Upper Narrows and to a lesser extent between Barstow and Camp Cady. Extensive groundwater develop-ment has led to declining water levels through much of the basin.[18] Artificial recharge in the basin, using water imported from outside the basin, is a possible way to restore water levels in the alluvial aquifer,[19] with potential side benefits to riparian vegetation. A groundwater model confirms that water levels in the regional aquifer declined steadily in the latter part of the twentieth century, as determined from previous water-level measurements.[20] Groundwater changes are of concern to land-management agencies because they have led to decreases in or elimination of riparian vegetation in some reaches.[21] Some water is imported into the basin to recharge the alluvial aquifer artificially; this water also benefits riparian ecosystems.

Well-documented losses and changes along the Mojave River, combined with a need for development of a sustainable groundwater resource, have led to intensive management of water in this basin.[22] Six hydrologic subsections of the 1,400-square-mile basin have been defined: Oeste, Este, Alto, Transition Zone, Centro, and Baja.[23] Of these areas, Este and Oeste are groundwater basins east and west of the Mojave River, respectively; Alto represents the Mojave River from the Mojave River Dam to about Helendale; the Transition Zone is the northern part of the Alto subsection from the Narrows to about Helendale; Centro covers the reach from about Helendale to near Yermo; and Baja represents the reach from near Yermo to Afton Canyon.[24]

Historical Photography of the Mojave River

The photographic history of the Mojave River is the longest for any riverine riparian ecosystem in the Southwest. In March 1863, Rudolf D'Heureuse, in collaboration with the California Geological Survey,[25] crossed the Mojave Desert from Cajon Pass through present-day Victorville and Barstow and then followed the Mojave Road—with notable side trips—to the Colorado River at Fort Mohave (see chapter 27). He followed the Mojave River from north of Victorville to Afton Canyon and photographed scenes using a large-format, glass-plate cam-era. His photographs include the river and its attributes at Point of Rocks near present-day Helendale, at two sites near Barstow, at the old Camp Cady, at Afton Canyon, and at the site of present day Zzyzx on the west edge of Soda Lake playa.[26]

Early geologists and hydrologists took numerous landscape photographs of the Mojave Desert. David G. Thompson, a hydrologist with the U.S. Geological Survey, studied surface-water and groundwater hydrology of the Mojave River in the 1910s and 1920s.[27] He photographed key reaches of the Mojave River, including the Upper Narrows and the Lower Narrows near Victorville; the reach upstream of Helendale, Barstow, and the Yermo area; and the reach between Camp Cady and Afton Canyon. Also, D. T. MacDougal, a pioneering desert ecologist, photographed the Mojave River downstream from Camp Cady Ranch in 1914.

Changes in Riparian Vegetation

Victorville to Helendale

Water rises to the surface in Victorville, contributing to perennial flow also fed by discharge from the fish hatchery. This flow extends through the Upper Narrows downstream through the Lower Narrows (fig. 28.1), which compose the Alto management reach. Water is forced up by granitic bedrock at the Upper Narrows, and the reach upstream from this geologic structure once supported dense stands of what appears to be primarily black willow (figs. 28.4, 28.5). Because of the perennial water, cattle grazing was heavy, but cottonwood have increased significantly here despite the grazing pressure. The closed gallery forest that grows here is now part of Mojave Narrows Regional Park, showing the high value placed on woody riparian vegetation in the Mojave Desert.

The largest increases occurred in the area that now is Mojave Narrows Regional Park, upstream of the Upper Narrows between Victorville and Apple Valley. Riparian trees in this reach were mostly black willows in 1917, although Frémont cottonwood was common here in 1893;[28] now cottonwood greatly exceed the willows

A

B

Figure 28.4 Photographs of the Upper Narrows at Victorville, California.

A. (December 15, 1919.) D. G. Thompson photographed this upstream view of the Mojave River at the Upper Narrows as part of a three-part 360-degree panorama. This photograph is the rightmost view. The trees around the ranch buildings and near the river are leafless, which indicates that the season is winter. The band of trees nearest the photograph appears to be mostly black willows, but scattered cottonwood are also in the view. (D. G. Thompson 445, courtesy of the U.S. Geological Survey Photographic Library.)

B. (October 24, 2000.) Until recently, the ranch that appears in this view remained in the same family that owned the property in 1917. Most of the land in the view has been grazed continuously, but likely at varying intensities, since 1917. Irrigation canals and water sources remain in approximately the same places, indicating that groundwater levels likely have not fluctuated significantly. Cottonwood are now the dominant trees in the view, and scattered black willows are still present. (D. Oldershaw, Stake 2110a.)

A

B

Figure 28.5 Photographs of the Upper Narrows at Victorville, California.

A. (December 15, 1919.) In this view, which pans left of the pair of photographs in figure 28.4, the trees appear to be mostly black willow. A few cottonwood appear in this view, in particular one at lower center, and some low shrubs along the channel may be coyote willows. The Mojave River is flowing toward the camera and has a wide, mostly denuded floodplain. Vegetation on this floodplain may be recovering from damages sustained in the 1916 flood, which was one of the largest on the Mojave River. (D. G. Thompson 447, courtesy of the U.S. Geological Survey Photographic Library.)

B. (October 24, 2000.) Most of the riparian forest shown in this view is in the Mojave Narrows Regional Park. The channel of the Mojave River has decreased to a mere fraction of its former width, in part because of regulation by dams upstream and the influence of the riparian vegetation. The trees present are mostly cottonwood, but very large black willows and other species are also present. The white haze in the background is smog over Cajon Pass north of San Bernardino. (D. Oldershaw, Stake 2110e.)

Figure 28.6 Photographs of the Upper Narrows at Victorville, California.

A. (1901.) This downstream view toward the Lower Narrows shows the former highway bridge across the Mojave River below the Narrows as well as the main railroad line leading toward Barstow. The trees in the middle distance and on river left are a mixture of cottonwood and black willows. The river channel is very wide, probably in response to nineteenth-century floods including the 1891 event. (M. R. Campbell 173, courtesy of the U.S. Geological Survey Photographic Library.)

B. (October 24, 2000.) Cottonwood and developments now dominate the view, and the channel of the Mojave River cannot be seen. The old bridge was replaced twice, once by a similar two-lane bridge and then more recently by the four-lane bridge shown at lower right. (D. Oldershaw, Stake 2111.)

in abundance and stature. Few woody plants were present in the Lower Narrows in 1917, with the exception of coyote willow, but many new cottonwood and tamarisk trees are now in this reach. South of Helendale, the historical photographs show the riparian corridor in the distance, but it is clear from a comparison of the photographs that cottonwood trees increased in height between 1917 and 2000. Downstream from the Upper Narrows, the Mojave River is perennial as it flows across shallow granitic bedrock. Few woody plants were present in this reach at the turn of the century (fig. 28.6). Now, woody riparian trees, in particular cottonwood, have increased

along with urban development in Victorville. Similarly, few woody plants were present in the Lower Narrows in 1919, with the exception of coyote willow, but many new cottonwood and sycamore, as well as tamarisk, are now in this reach.[29]

North of the Lower Narrows, in the Transition Zone, the river crosses deep alluvium, leading to influent flow conditions. Decreases in the alluvial-aquifer level create unstable conditions for woody riparian vegetation. Owing to flood-related downcutting in the channel of Oro Grande in 1998, the depth to water on cottonwood-dominated floodplains increased, leading to widespread mortality of this species.[30] Most

black willows in this reach, in contrast, were damaged but survived, again suggesting that this species has a greater environmental tolerance than does Frémont cottonwood. Large fluctuations in water level (more than 10 feet; fig. 28.7) underscore the potential for combinations of groundwater withdrawal, drought, and channel changes to affect established cottonwood. Despite these changes, which would be particularly deleterious to riparian vegetation, the historical photographs taken south of Helendale show the riparian corridor in the distance (fig. 28.8), but it is clear from a comparison of the photographs that cottonwood trees increased in height between 1917 and 2000.

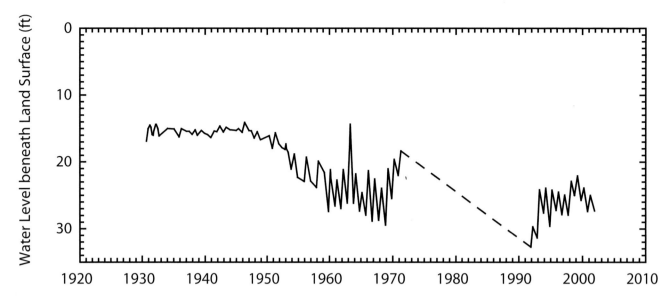

Figure 28.7 Groundwater levels in well 8N/4W-31R-1S near Helendale, California. The dashed line indicates a gap in the record.

Although base flow and floods recharge water to the alluvial aquifer along the Mojave River, larger floods play a greater role in maintenance of the riparian ecosystem. Germination of riparian species occurs in the wet sediments deposited during floods; the sequence of floods and the amount of time between them determine whether the plants that germinate become established. Large floods also supply a greater quantity of water to trees on the distal margins of floodplains, aiding their survival. These floods also destroy smaller plants, and lateral channel change may undercut larger trees. Significant floods have occurred about every five years in recent decades (fig. 28.2) and may have contributed to the increases in riparian vegetation in some reaches.

Barstow to Camp Cady

Groundwater declines along the Mojave River are most severe in the reach between Helendale and Camp Cady Ranch, mostly as a result of extensive agriculture in the vicinity of Barstow.[31] Woody riparian species decrease markedly downstream from Helendale, with dense tamarisk thickets replacing cottonwood-willow stands, which then recede to scattered xerophytic tamarisk stands.[32] Downstream from Barstow, water levels in one long-term well de-

clined from around 60 feet below land surface in the first half of the twentieth century to 100 feet below land surface by 1970 (fig. 28.9). These declines have been verified for the regional aquifer system using a groundwater model.[33]

At Barstow, the Mojave River is dry most of the time. Native riparian vegetation, which once was mostly mesquite, has been replaced by tamarisk (fig. 28.10). Athel tamarisk, commonly planted in and around Barstow, appears to have become locally established along the river. Between Barstow and Camp Cady Ranch, which is in the Centro subsection, several camera stations could not be relocated because of unstable sand dunes and channel margins, but most of the mesquite and other riparian trees that grew in this reach in 1919 are dead. At the "Meeting of the Roads" near present-day Yermo, a D'Heureuse photograph from 1863 (fig. 28.11) shows that what once was a dense grove of black willows has been replaced with unstable sand dunes with scattered mesquite groves. The stumps of numerous dead trees line the river channel in this reach, underscoring the magnitude of riparian tree mortality.

Camp Cady to Afton Canyon

At Camp Cady Ranch, in the Baja subsection, riparian vegetation increased

between 1919 and 2000, although declines upstream and downstream from the camera stations suggest that the vegetation in this reach is in peril. Woody perennial vegetation—notably cottonwood, black willow, and mesquite—became established in response to a historical water source at Camp Cady (fig. 28.1). The water level of the regional aquifer system rises here in response to a fault and clay-rich lake beds that rise in the downstream direction.[34] Groundwater levels are generally less than 20 feet deep in this reach (fig. 28.12), but they have declined recently, eliminating perennial flow in the river. Camp Cady Ranch, a former livestock operation and now a California Fish and Game facility devoted to endangered species such as the Mohave tui chub and western pond turtle, is upstream from the former army outpost. A closed gallery forest of cottonwood and black willow occurs at the former ranch.

At the old military outpost (fig. 28.13), woody vegetation was sparse in 1863, although small trees and mesquite are apparent in the view. Now, mesquite has increased, mostly blocking the historical view of the floodplain. As groundwater levels rise toward Camp Cady Ranch, riparian vegetation increases from primarily xerophytic tamarisk to a mixture of tamarisk and mesquite and finally to the closed

Figure 28.8 Photographs of the Mojave River at Helendale, California.

A. (December 1919.) This view, from the east side of the Mojave River valley looking west, shows the Mojave River in the distance with low bluffs of fine-grained river and lake deposits on the skyline. The river is lined with trees that are likely a mixture of cottonwood *(taller, darker trees)* and black willows *(lower, lighter trees)*. Both old Route 66 and the railroad bisect this view in the midground. The vegetation in the foreground is mostly creosote bush. (D. G. Thompson 441, courtesy of the U.S. Geological Survey Photographic Library.)

B. (October 24, 2000.) The trees along the river have increased in size and number despite groundwater pumpage in the vicinity. Cottonwood is now the most obvious species from this distance. The number of creosote bushes and other small shrubs in the foreground has greatly increased. (D. Oldershaw, Stake 2114a.)

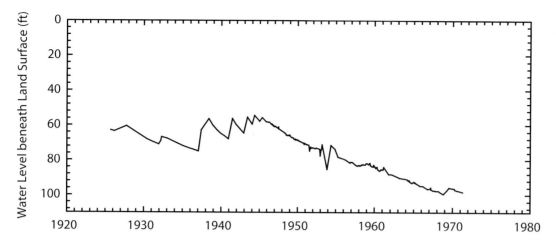

Figure 28.9 Groundwater levels for well 9N/1E-13E-2S near the Mojave River downstream from Barstow, California.

A. (October 1919.) D. G. Thompson took this upstream view of the Mojave River from a small hill on its north bank at the main road crossing at Barstow. Current deflectors appear at right center, designed to minimize erosion of the north bank and its roads. The low vegetation at left appears to be mesquite on sandy hummocks, and scattered shrubs of unknown species appear throughout the open, sandy channel. What look like cottonwood trees are visible in the distance. (D. G. Thompson 234, courtesy of the U.S. Geological Survey Photographic Library.)

B. (October 22, 2000.) Nearly all the native riparian plants in the previous view are dead, victims of groundwater overdraft in the reach upstream from Barstow. The low, hemispheric shrubs in the foreground are tamarisk, which accumulate eolian sand entrained from the open riverbed and deposited around the trees. The vegetation at left is planted and irrigated, and most of the trees are Athel tamarisk. (D. Oldershaw, Stake 2106b.)

Figure 28.10 Photographs of the Mojave River at Barstow, California.

A

B

Figure 28.11 Photographs of the "Meeting of the Roads" near Yermo, California.

A. (March 1863.) In 1863, the Mojave Road and the Spanish Trail met at the Mojave River east of Barstow, near the current site of Yermo. This photograph shows that the Mojave River was lined with short trees, probably black willows. These trees suggest that reliable surface water once was present here. (R. D'Heureuse 1905.16894-A[1:10], courtesy of the Bancroft Library.)

B. (April 25, 2001.) During the past 138 years, hummocks with tamarisk, mesquite, desert willow, and California scale broom have replaced the willows. Stumps of now-dead trees are present behind the encroaching dunes shown in the foreground. Although this photo is of mesquite-covered dunes with a few thorn bushes, tamarisk is the dominant woody shrub along this stretch of the Mojave riverbed. (D. Oldershaw, Stake 4085.)

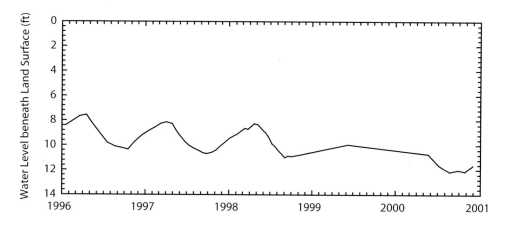

Figure 28.12 Groundwater levels for well 10N/4E-19M0-4S near Camp Cady, California.

A. (March 1863.) In the early 1860s, the U.S. Cavalry established several outposts at perennial water sources in the Mojave Desert for the protection of travelers along the Mojave Road, then known as the Government Road. The structure shown in this view is Camp Cady, on the north bank of the Mojave River northwest of the Cady Mountains. Behind and to the right of the fortification are riparian trees, which appear to be mesquite *(plants just behind the fort)*, black willows *(both sides of fort)*, and coyote willow *(right)*. (R. D'Heureuse 1905-16894-A[1:11], courtesy of the Bancroft Library.)

B. (October 23, 2000.) Mesquite now blocks the view of the fort, which survives only in some broken foundations and a small commemorative sign. Mistletoe appears in some of these mesquite trees, which were very healthy at the time of this photograph. The foreground shrubs are cattle spinach. Cottonwood trees rise above some of the mesquite trees. (D. Oldershaw, Stake 2107.)

Figure 28.13 Photographs of the Mojave River at Old Camp Cady.

gallery forest, and this sequence reverses in the downstream direction past the former military outpost.

Several panning photographs document long-term change at this site. Two views from 1917 show that the cottonwood-willow assemblage here dates to the early twentieth century (fig. 28.14). A 2000 match of the more upstream of the 1917 views (fig. 28.14B) shows that the once-native vegetation assemblage now contains significant tamarisk and that dieback and mortality are common in what once was an extensive mesquite bosque. Comparison of the 1863 and 1917 views is not valid because the older view is of what likely was an effluent reach with lower potential for growth of woody riparian species other than mesquite. Both views indicate that channel narrowing has occurred in the reach in response to establishment of native vegetation. They also show that riparian vegetation increases have occurred despite livestock operations.

Afton Canyon

Downstream from Camp Cady Ranch, depth to groundwater increases in the deep alluvial fill, creating an ephemeral stream environment. Mesquite grew in xerophytic settings here and likely received significant additional water only during episodic floods, when overbank flooding would temporarily recharge the upper parts of the alluvial aquifer. The once-dense mesquite thickets, which never were as extensive here as in southern Arizona, now appear to be declining, and tamarisk has colonized the formerly barren channel (fig. 28.15). Large stumps and fallen tree trunks suggest that obligate riparian vegetation historically occurred in this reach.

Surface water rises at the head of Afton Canyon and flows for a little more than 3 river miles. This reach has an average daily discharge of 6.88 ft³/s and a typical base flow of between 0.2 and 5 ft³/s. During the 2003 drought, the channel was dry at the gaging station.[35] Floods as large as 18,000 ft³/s (in 1969) and probably much larger have occurred here (fig. 28.16). This reach is possibly one of the "flashiest" in the western United States in terms of the contrast between the low-flow and high-flow periods, creating an environment that encourages growth of woody species during low flow but destroys vegetation during high flow. Most of the floods occur in fall and winter, which is consistent with the winter-dominated rainfall regime of the Mojave Desert. Approximately 700 acres of riparian habitat occur in Afton Canyon, and this unusual habitat in a desert canyon prompted the BLM to designate Afton Canyon as an Area of Critical Environmental Concern.[36] Upstream water use and tamarisk establishment have decreased base flow in this reach.[37]

Little riparian vegetation was present in Afton Canyon in 1863, when D'Heureuse photographed the head of the canyon (fig. 28.17). The few trees present in 1863 were either black willow or Frémont cottonwood, and these trees are now gone. The channel appeared to be scoured in 1863, probably in response to the 1862 flood. This event was not recorded for the Mojave River but likely was extremely large, given the widespread nature of the storms over the western United States.[38] Perennial water attracted ranchers, who grazed cattle extensively in this canyon, and tamarisk became established at an unknown time after 1929.[39] The site was recently fenced to eliminate livestock.

Because of the high value of riparian habitat in a desert setting, eradication of tamarisk in this reach is a management priority. Eradication efforts began with herbicide applications in the 1950s, which were ineffectual for long-term control. They were renewed in the 1990s to increase perennial flow and to provide space for native species.[40] Removal methods included burning, which killed 10 to 25 percent of the individuals in a 40-acre area, and cutting followed by application of herbicide to the stump surface. The match of D'Heureuse's view (fig. 28.17B), taken in 2000 after tamarisk eradication had occurred in the foreground, shows an assemblage dominated by arrowweed with a single black willow in the foreground, cottonwood and mesquite in the distance, and reeds and cattail ringing open pools of water.

A

B

Figure 28.14 Photographs of the Mojave River at Camp Cady Ranch, California.

A. (September 5, 1917.) Downstream of Barstow, the Mojave River crosses a wide, sandy valley that in 1917 had sand dunes and mesquite hummocks. Where the river approaches the northwest side of the Cady Mountains, its course apparently crosses a subsurface geological structure (either fault or geologic strata) that forces groundwater toward the surface. A zone of riparian vegetation consisting of mesquite, cottonwood, and black willow was well established here in 1917. The channel of the Mojave River, which flows right to left across the center of this view, is relatively wide. (D. G. Thompson 15, courtesy of the U.S. Geological Survey Photographic Library.)

B. (October 25, 2000.) This view across the upstream part of Camp Cady Ranch shows a reach of the Mojave River that has become much narrower. Mortality of mesquite has been high here owing to upstream groundwater pumping, much of it immediately upstream from the reach in this view. Tamarisk has increased considerably in the midground. At the time of this photograph, tamarisk removal was occurring in this reach. Despite the mortality, riparian vegetation increased during the middle of the twentieth century before the dieback. (D. Oldershaw, Stake 2117.)

A

B

Figure 28.15 Photographs of the Mojave River downstream from Camp Cady, California.

A. (April 22, 1915.) In this upstream view of the Mojave River, recessional flow from a winter flood appears in the channel. Historically, this reach is not known to have had perennial flow despite its location between Camp Cady and Afton Canyon. The riparian area at Camp Cady appears in the distance at right. The larger plants lining the otherwise barren channel are western honey mesquite; some of the midground plants appear to be in old coppice mounds that are 30 to 50 feet across. (D. T. MacDougal A3-67, courtesy of the Desert Laboratory Collection.)

B. (November 25, 2003.) The channel is now indistinct, with small, shrublike tamarisk in the channel. Tamarisk decreases downstream of this point until the Mojave River reaches Afton Canyon. The mesquite mounds appear to be deteriorating, which may signal a lowering of the groundwater table here. (D. Oldershaw, Stake 4756b.)

Figure 28.16 Annual flood series for the Mojave River at Afton Canyon in California (station 10263000; 1930–1932, 1953–2003). Many years have little or no flow.

A. (March 1863.) This downstream view of the Mojave River was taken at the head of Afton Canyon about a year after the 1862 flood. A wide, scoured channel with little riparian vegetation is present in the foreground, and larger trees of unknown species are in the middle distance. From this distance, it is impossible to determine whether these trees are cottonwood or black willows. (R. D'Heureuse 1905.16894-A[2:12], courtesy of the Bancroft Library.)

B. (October 25, 2000.) Many changes have occurred in the intervening 137 years in Afton Canyon, spurred by large floods, cattle grazing, groundwater depletion upstream, and manipulations of the riparian ecosystem. Changes to the riparian plants encompass the cumulative results of establishment of nonnative tamarisk in the middle part of the twentieth century, herbicide treatments to remove tamarisk in the 1950s, and mechanical removal of tamarisk that began in the 1990s before this photograph was taken. Despite all these changes, the amount of woody vegetation has obviously increased. The plants in the foreground are arrowweed, with a small black willow in the center. Mesquites are also present in the middle distance, and the large-statured trees in the background are cottonwood. At this site, the channel of the Mojave River contains perennial water, with reeds and cattail but no open channel. (D. Oldershaw, Stake 2116.)

Figure 28.17 Photographs of the Mojave River at Afton Canyon.

Summary. Repeat photography shows that woody riparian vegetation has changed along the rivers we have documented in the southwestern United States. Change was evaluated in 2,724 sets of repeat photographs using a qualitative assessment. Woody riparian vegetation had increases in density and biomass in 73 percent of the views and no change in 15 percent of the views. Camera stations with net decreases in vegetation generally are in reaches inundated by reservoirs or affected by excessive groundwater pumping or affected by total surface-water diversion, and they were concentrated in several reaches: notably the Santa Cruz River near Tucson, Arizona; the middle and lower Gila River; the lower Colorado River and the Mojave River at and downstream from Barstow, California. Frémont cottonwood increased in 69 percent of the views but had notable decreases on larger river reaches at lower elevations, in particular the Gila and Colorado Rivers. Black willow increased in 80 percent of the photographic matches. Tamarisk appeared in 1,577 matches and increased in 88 percent of the views because it generally was not present or was present only in small amounts at the start of the photographic record. Tamarisk seldom was viewed alone but typically occurred with one or more natives species. The time series of repeat photography suggests that the increases in woody riparian vegetation began after 1940, and the greatest amount of change occurred in the last third of the twentieth century.

Repeat photography in the southwestern United States has shown a variety of changes in woody riparian vegetation. Here, we examine changes over the region as a whole instead of concentrating on specific rivers or reaches. As discussed in chapter 5, previous studies using repeat photography[1] have extracted quantitative, qualitative, and categorical data from the matches, yielding information on long-term change or frequency of geomorphic processes. However, the views from many of our relocated camera stations were blocked by new vegetation, and the information content of repeat photography decreases with distance from the camera station; therefore, we cannot make estimates of change on the basis of area. We use a compilation of these interpretations to look at the overall patterns of change in woody riparian vegetation as a group and as selected species for which we have widespread photographic documentation.

Interpretation of Repeat Photography

Repeat photography is a powerful tool for documenting change in riparian vegetation, but this method has some important limitations (chapter 5). It provides long records of change in the riverine environment at specific camera stations, and species-specific changes can be interpreted from the matches. In addition, it documents geomorphic or land-use changes that may affect the stability of riparian vegetation.

Unlike analysis of aerial photography or data from satellite-borne spectrometers, for which change can be established per unit area, oblique ground photography is difficult to repeat accurately unless the original photograph was taken from a vantage point overlooking relatively flat terrain.[2] Moreover, most repeat photography of riparian areas shows such large changes that it would be difficult, if not impossible, to determine exactly the area within the view. Qualitatively, however, the trends of change discussed throughout this book compare favorably with analysis of aerial photography of the Bill Williams River (chapter 15) and the upper Gila River (chapter 17) as well as of satellite imagery of the San Pedro River (chapter 19).

Assessments of percentage area occupied by plants, calculated by digital image analysis, would be fruitless for most ground photography. The relative size of plants is a function of their distance into the view, and quantification of that distance would be required to compare different plants. Particularly in situations where riparian vegetation has increased, the base—or even the entire plant—is obscured by growth, which eliminates quantitative comparison. However, as shown throughout this book, changes in riparian vegetation as documented in repeat photography are seldom subtle, and objective qualitative interpretation is an effective means of portraying regional change.

In July 2004, the Desert Laboratory Collection of Repeat Photography contained 3,067 sets of repeat photographs from the southwestern United States that show riparian vegetation in the reaches we discuss in this book. We first eliminated duplicate matches—ones that show the same river reach over about the same time period—in order to achieve a set of independent observations of change and to minimize bias. Because of the large number of available repeat photographs from Grand Canyon (chapter 11), we reduced the number of photographs interpreted from this reach by concentrating only on original photographs taken by E. C. La Rue in 1923 and ignoring the remainder, thus lowering the number of matches for this reach from 1,251 to 396. Finally, we interpreted multiple photographs spaced through time at a camera station as independent observations, thereby increasing the number of matches evaluated for change. Incorporation of these three elements reduced the total number of interpreted repeat photographs to 2,724.

Following previous work,[3] the change we evaluate is a combination of density and biomass within the views, both for all riparian vegetation and for selected species. Particularly because of the large, positive increases in vegetation in most views, and because few individual plants persisted between views, we could not keep separate statistics on individuals, as would be needed to assess density changes accurately.[4] We chose five categories into which we placed each view: *large decrease, decrease, no interpretable change, increase,* and *large increase.* Where one category changes to another is necessarily fuzzy even though the categories of change are broad.

Our interpretations have several important limitations. We assume that our camera stations provide an unbiased sample of riparian vegetation, and this assumption cannot be tested regionally. In the cases of Cataract Canyon and Grand Canyon, photographs were taken systematically through the canyon, minimizing the bias introduced for other rivers by reliance on time series of photography at current or former gaging stations. In many cases, these gaging stations are sited where flow is most easily measured, and those reaches may in some cases bias the representation of reach-scale change in riparian vegetation. In particular, many photographs are taken from bridges, where channel form and flow have been altered.

We could not control the dates of the original photography, except in the case of the Green and Colorado Rivers, where abundant photography from 1889 through 1890 and 1921 through 1923 provided reach-scale coverage in a brief period. For the region, we had too few photographs from a specific time period to make meaningful comparison of changes. For example, we could not collect photography throughout the region to document the effects of the midcentury drought of the mid-1940s through early 1960s, a period when groundwater development rapidly increased.

The maps in the individual chapters (e.g., fig. 6.1) show the locations of individual camera stations and our interpretation of changes in woody riparian vegetation. In this chapter, our interpretative results are mapped at a scale of 1:6.6 million to provide a regional view of changes. The species for which we portray spatial changes, with the exception of Russian olive and carrizo grass, were selected because a large number of sets of repeat photographs containing those species (greater than 200) were interpreted.

Overall Change in Riparian Vegetation

The total change in riparian vegetation interpreted from all repeat photography, irrespective of species composition or dates of the original photographs, is given in figure 29.1. In general, riparian vegetation has had either an increase (49 percent) or a large increase (24 percent) in comparisons involving all years (fig. 29.2). Camera stations showing decreased vegetation are interspersed with ones showing increases except in areas along the Colorado River where reservoirs are now present (chapter 14), the Santa Cruz River at Tucson, Arizona (chapter 21), the Salt and Gila Rivers above their confluence (chapters 24 and 26), and the Mojave River downstream from Barstow, California (chapter 28).

We found no relation between change in riparian vegetation and elevation, latitude, or longitude. In other words, neither increases nor decreases in riparian vegetation were related to a specific elevation range or geographic part of the region. The time series of change, either from specific camera stations (e.g., the San Pedro River at Palominas, chapter 19) or from non-related photographs taken at different times in the same reach (e.g., the San Juan River in the vicinity of Bluff, chapter 8), suggests that, at least in the specific reaches for which we have documentation, riparian vegetation began to increase after about 1940. Increases in density of plants appear to have accelerated after the 1970s, followed by increases in the size of plants.

Many deviations from this general pattern have been noted in the chapters on specific reaches. Fire and the changes that occur afterward can alter the course of change in riparian vegetation. In some cases, native species may recover quickly following a fire (Bill Williams River, chapter 15); in others, either the vegetation does not recover, or nonnative species become established (Spanish Bottom, chapter 7). Flood damage and in particular the channel change caused by floods can temporarily offset gains in vegetation, but the photography of the Gila River (chapter 17), the Verde River (chapter 23), and the Salt River (chapter 24), all of which had extremely large floods as recently as 1993, still show overall gains.

The general impression from most of our repeat photographs is that the amount of space available for woody, obligate riparian vegetation is occupied mostly by either native or nonnative species or by a mixture of both types. Particularly within arroyos, the amount of space available with access to shallow groundwater is limited for riparian vegetation, and most of these floodplains appear to be fully occupied in the absence of flood damage. With the exception of floodplains coarsened by clear-water releases from dams (e.g., the Salt River downstream from Stewart Mountain Dam, chapter 24, and the Verde River downstream from Bartlett Dam, chapter 23), floodplains adjacent to perennial or intermittent channels are almost fully vegetated in our matched photographs. Further increases in the amount of woody riparian vegetation, given the unlikely occurrence of stable hydrologic conditions into the future, would have to result from channel narrowing.

Changes in Selected Woody Riparian Species

Native Obligate Riparian Trees

Change in Frémont cottonwood was interpreted from 1,128 sets of repeat photographs. Widely distributed in the region (fig. 29.3), cottonwood increased in 59 percent of the views (fig. 29.4). Most of the increase is along the smaller tributaries of the Green, Colorado, and Gila Rivers, although increases occurred in Desolation Canyon (chapter 6). The decrease in cottonwood at 24 percent of sites is the largest decrease regionally for any native species that we observed.

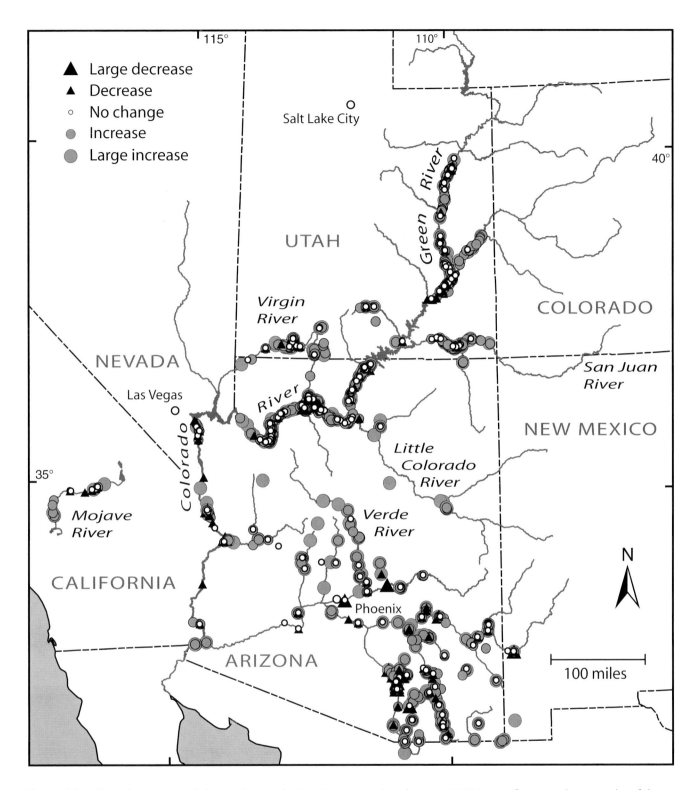

Figure 29.1 Map showing total change in woody riparian vegetation shown in 2,724 sets of repeat photographs of the southwestern United States. Scale is 1:6.6 million.

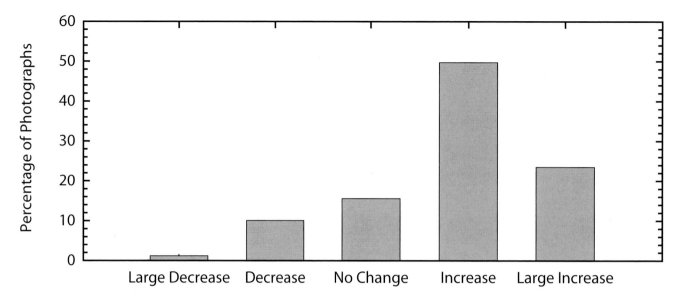

Figure 29.2 Histogram of the percentage of photographs showing change in the total amount of riparian vegetation as evaluated in five categories for all years.

The increase in cottonwood along the Mojave River is particularly striking in reaches where use of groundwater is either not excessive or has not yet affected the vegetation. Along the San Pedro and Escalante Rivers, increase in cottonwood began while livestock still grazed in the riparian zone. For many rivers, notably the Verde, the increase occurred despite recent extremely large floods combined with livestock grazing. For the Green River in Stillwater and Labyrinth Canyons and for the upper Colorado River, we have no satisfactory explanation for the decreases we observed in Frémont cottonwood, but these decreases do not appear to be associated with land use other than the possibility that flow regulation reduced the size of the largest floods, thereby limiting the potential for germination and establishment.

Black willow was observed in 489 sets of repeat photographs, mostly from lower-elevation reaches in the region (fig. 29.5). We interpreted increases in 80 percent and decreases in only 10 percent of the views, with the remaining 10 percent remaining about the same (fig. 29.6). The increases were particularly striking on the larger rivers in the region, notably the Gila River, the Gila's lower-elevation tributaries, and the Virgin River. Significant decreases were observed along the Mojave River, although in the case of

the reach upstream from the Narrows in Victorville, California, black willow appears to have been replaced with cottonwood.

Although netleaf hackberry is widely distributed in the region, it has a conspicuous presence only in the riparian zone of the Colorado River in Cataract Canyon (fig. 29.7). In this reach, hackberry generally is upslope from tamarisk and occupies substrate with more rocks, such as debris fans and talus cones. In other reaches where it is present, it ranges from subdominant locally to uncommon. We observed hackberry in 214 sets of repeat photographs (fig. 29.7); it increased in 59 percent of those matches and remained about the same in another 31 percent. Views with decreases in netleaf hackberry were mostly in the same reach where other photographs recorded no change or increases. An exception was at the beginning of Lake Powell, where hackberry has been replaced with tamarisk in the fluctuating zone of high lake level.

We also noted increases in Arizona ash and box elder, which appear in 196 and 96 sets of repeat photographs, respectively. We observed peachleaf willow in only 4 photographs from the upper Colorado River, and the photography contained too few observations of sycamore or Arizona walnut to interpret change in those species.

Desert olive is restricted to Stillwater Canyon on the Green River and to Meander and Cataract Canyons on the Colorado River. Although this species is one of the few obligate riparian species to persist for more than 130 years, too few observations were made of it to assess changes.

Native Obligate Riparian Shrubs and Carrizo Grass

Change in coyote willow was interpreted from 689 repeat photographs (fig. 29.8). Although this species extends into the Colorado River delta (chapter 27), most of our observations were in the intermediate-elevation to higher-elevation reaches in the region. Seventy-six percent of our repeat photographs show an increase in coyote willow (fig. 29.9), despite its close association with tamarisk; coyote willow commonly occurs on the streamward side of floodplains occupied by tamarisk or in a mixture with tamarisk. We observed no systematic decreases in coyote willow on any river reach in the region, although we found it to be less common along the lower Colorado River than suggested by historical observations.

Change in the presence of seepwillow[5] was interpreted from 479 sets of repeat photographs (fig. 29.10). Seepwillow, which in our record generally occurred at lower elevations than coyote willow,

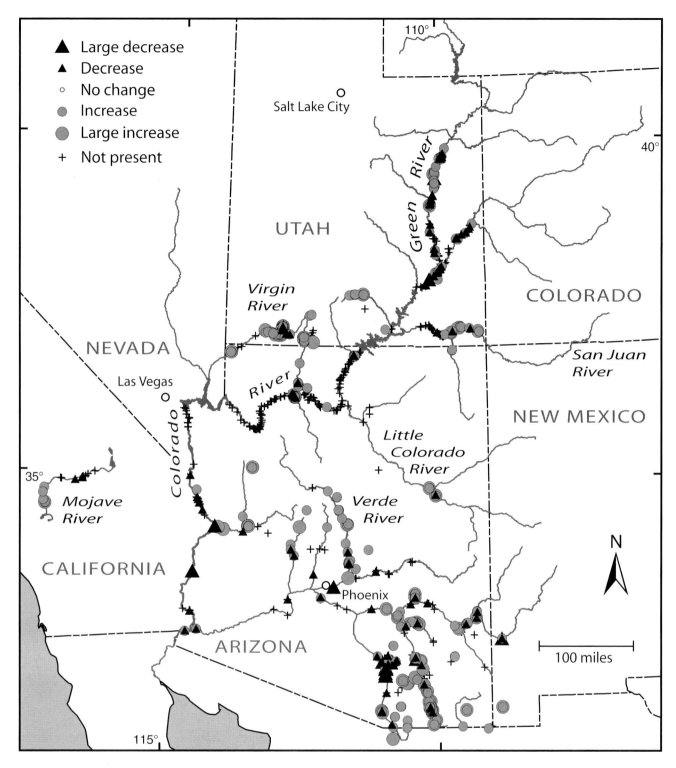

Figure 29.3 Map showing total change in Frémont cottonwood in 1,128 sets of repeat photographs in the Southwest. Scale is 1:6.6 million.

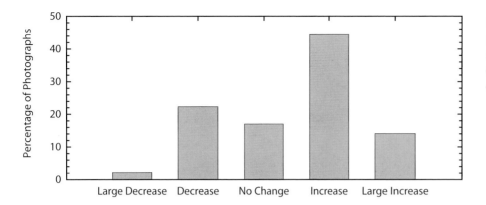

Figure 29.4 Histogram of the percentage of photographs showing change in Frémont cottonwood as evaluated in five categories for all years.

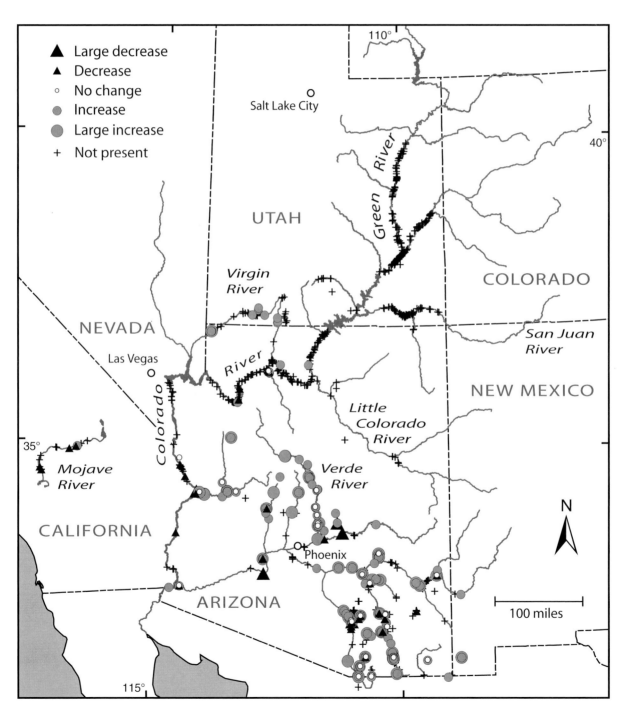

Figure 29.5 Map showing total change in black willow in 489 sets of repeat photographs in the Southwest. Scale is 1:6.6 million.

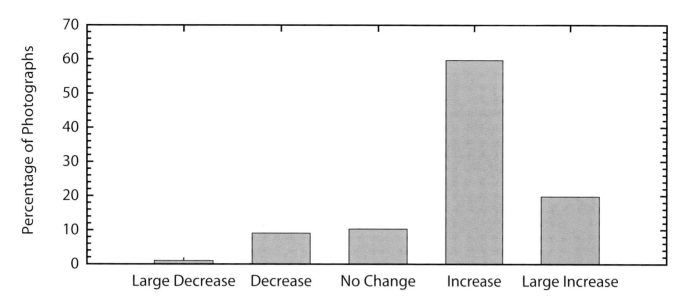

Figure 29.6 Histogram of the percentage of photographs showing change in black willow as evaluated in five categories for all years.

increased in 84 percent of our repeat photographs. We observed large increases in the more xerophytic burrobrush and arrowweed, although these species appeared in fewer repeat photographs (126 and 179, respectively). Burrobrush, in particular, increased along ephemeral channels in southern, central, and western Arizona and may have responded positively to the occurrence of floods in the last third of the twentieth century.

Carrizo grass, the only native herbaceous species that we included in our regional survey of woody riparian vegetation, was observed in 119 sets of repeat photographs (fig. 29.11). Carrizo grass increased in 84 percent of the repeat photographs, making it the native species with the largest overall increase that we observed. Much of the change in carrizo occurred along the Colorado River in Grand Canyon or downstream from Hoover Dam. Its largest increases were along the lower Colorado River upstream from Imperial Dam (chapter 27), where it appeared to be contributing to channel narrowing.

Native Facultative Riparian Trees and Shrubs

As discussed in chapter 3, we did not separate the three species of mesquite in the region (*Prosopis velutina, P.*

glandulosa, P. pubescens) because they cannot be reliably distinguished in historical photographs. Mesquite trees, which were observed in 779 sets of repeat photographs from the region, primarily at lower elevations in Arizona and along the Mojave River in California (fig. 29.12), increased in 61 percent of the views. This increase reflects the available photography, which is centered on river channels; we did not observe many cases of new mesquite bosques, although one was documented on Cienega Creek (chapter 22). Floodplain clearing for agriculture and urban development had the largest effect on mesquite and other facultative species that grew densely on the distal margins of floodplains. This part of the riverine setting was not routinely photographed and represents one bias in our assessments of regional change in riparian vegetation (fig. 29.13).

Catclaw was observed in 346 sets of repeat photographs from the region (fig. 29.14). Like mesquite, catclaw is restricted mostly to the southern part of the region, but, unlike mesquite, it is not present along the Mojave River. Catclaw increased in 54 percent of our matches, but 39 percent of the matches showed no changes, indicating that this species is one of the few in the region that has had a large, stable component of the population. Because catclaw is a common xerophytic species

in the Sonoran Desert, we restricted our interpretation of it to its occurrence either along the distal margin of alluvial floodplains or along the high-water line of the Colorado River in Grand Canyon.

Nonnative Species

Tamarisk is the most widespread riparian species in the Southwest, appearing in 1,577 sets of repeat photographs (fig. 29.15). The earliest photograph that shows tamarisk established in a nonhorticultural setting is a 1914 view taken at the Confluence of the Green and Colorado Rivers.[6] Not surprisingly, tamarisk increased in 88 percent of the repeat photographs (fig. 29.16). Only a few sites had dense stands consisting only of tamarisk, notably in the deltas of reservoirs (e.g., Gila River at Calva, chapter 17). Instead, tamarisk generally occurred in mixtures, either as a dominant (e.g., the San Juan River near Bluff, chapter 8) or in combination with more numerous native species (e.g., the Verde River, chapter 23).

Athel tamarisk was observed only along the Santa Cruz River near Tucson, along the Mojave River at Barstow, along the Gila River at Dome, and at two sites along the lower Colorado River. Of the 40 sets of repeat photographs that show Athel tamarisk, 75 percent show an increase in this

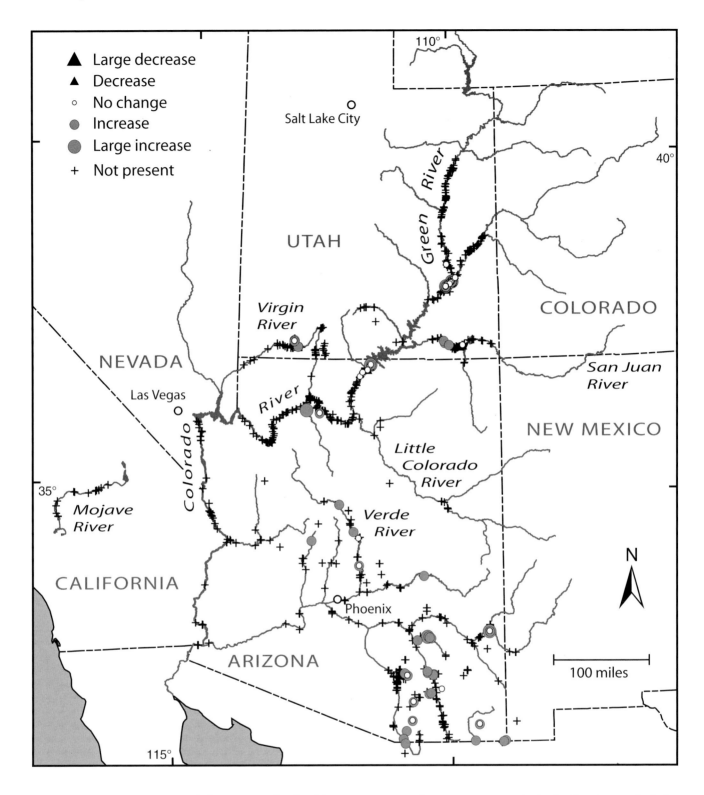

Figure 29.7 Map showing total change in netleaf hackberry in 214 sets of repeat photographs in the Southwest. Scale is 1:6.6 million.

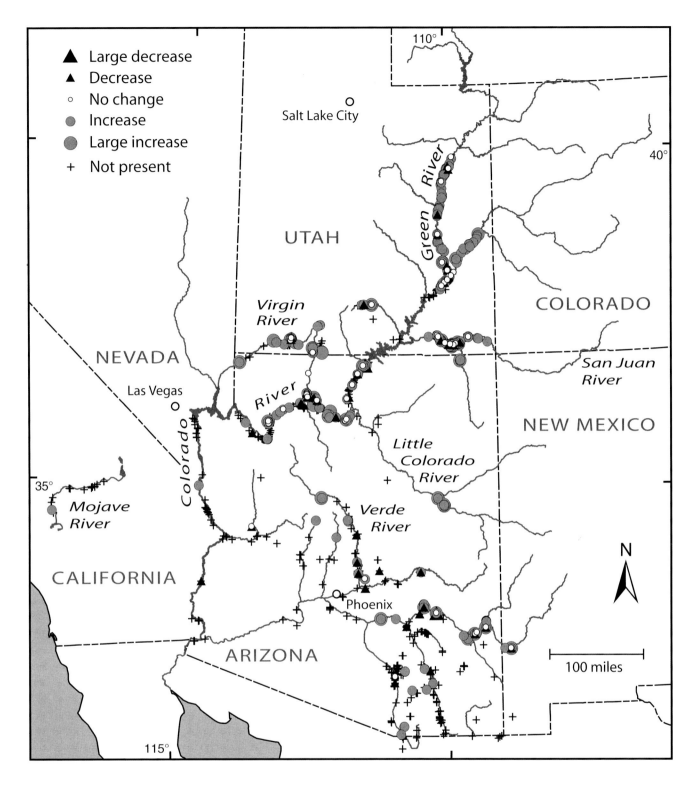

Figure 29.8 Map showing total change in coyote willow in 689 sets of repeat photographs in the Southwest. Scale is 1:6.6 million.

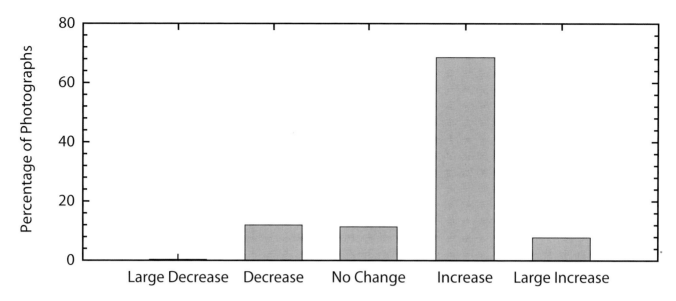

Figure 29.9 Histogram of the percentage of photographs showing change in coyote willow as evaluated in five categories for all years.

species. We could not distinguish Athel tamarisk planted along channels from individuals established vegetatively. Few sites had increases where Athel tamarisk had not been planted first; a notable exception was the Topock Gorge at the head of Lake Havasu (chapters 14 and 27). Our limited observations of this species suggest that it is not highly invasive because it mostly propagates vegetatively (chapter 4).

Russian olive is visible in 108 sets of repeat photographs, primarily in southern Utah, the Little Colorado River, and the upper Verde River (fig. 29.17). Although it has never fully approached a density excluding other species, it has achieved high densities in reaches with photographic coverage along the Escalante and San Juan

Rivers. This species occurs in lower densities in the higher-velocity reaches of these rivers and the Colorado River but appears to prefer wider reaches in alluvial valleys. Its distribution as observed in repeat photography fits its overall distribution pattern in the western United States.[7]

Giant reed was observed in 24 sets of repeat photographs widely separated in southern and western Arizona, notably along the slower-moving reaches of the lower Colorado River. This species increased in 94 percent of the views and was first visible in our matches from the 1990s and 2000s.

Although we found other nonnative species at our camera stations, none appeared in more than about 10 photographic matches. Tree-of-heaven was

observed primarily in the Verde River watershed, and tree tobacco occurred primarily along ephemeral washes in southeastern Arizona and within the Bill Williams River drainage.

Palms—both the nonnative date palm and the native California palm—occurred along several reaches, notably the lower Colorado River. Date palms are now common in reaches where houses are built close to the channel (e.g., near Needles, California, and Parker, Arizona). California palm was at one time restricted in its natural distribution.[8] This species has increased along a number of low-elevation reaches owing to its extensive horticultural use.

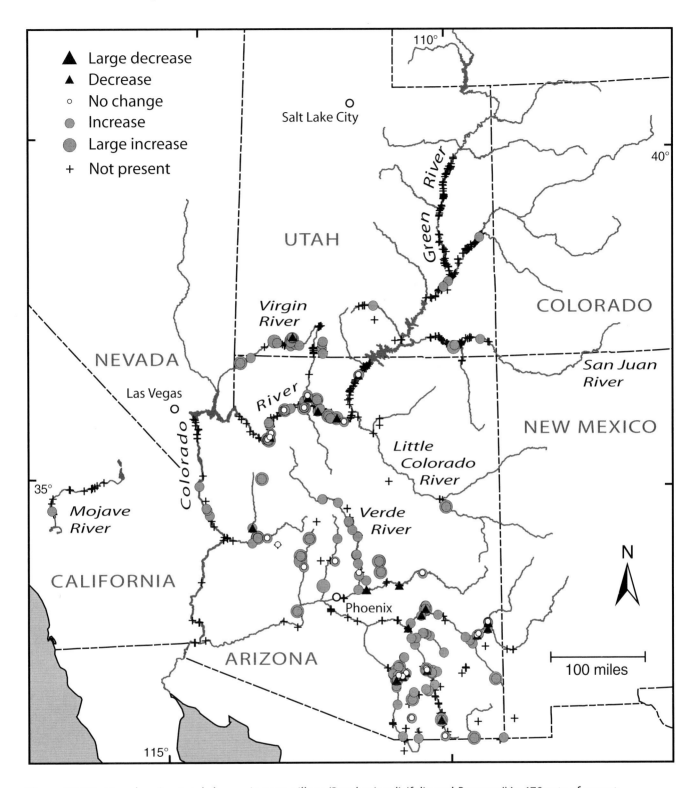

Figure 29.10 Map showing total change in seepwillow *(Baccharis salicifolia* and *B. emoryi)* in 479 sets of repeat photographs in the Southwest. Scale is 1:6.6 million.

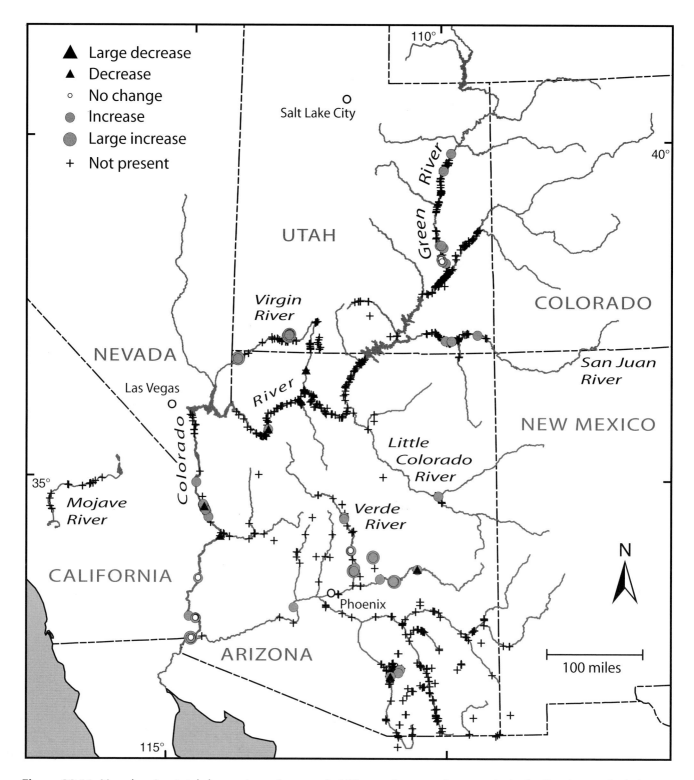

Figure 29.11 Map showing total change in carrizo grass in 119 sets of repeat photographs in the Southwest. Scale is 1:6.6 million.

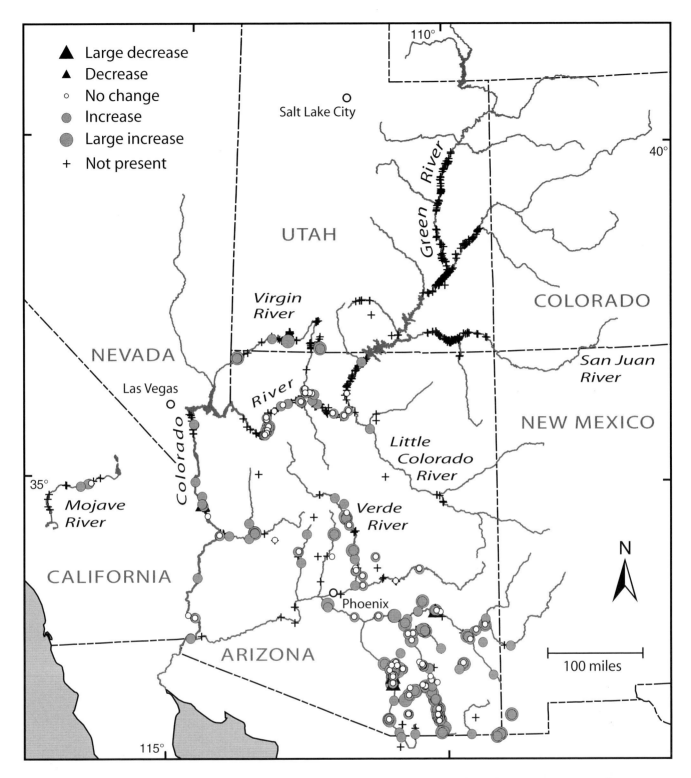

Figure 29.12 Map showing total change in mesquite *(P. glandulosa, P. velutina,* and *P. pubescens)* in 779 sets of repeat photographs in the Southwest. Scale is 1:6.6 million.

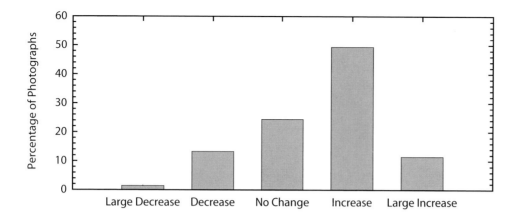

Figure 29.13 Histogram of the percentage of photographs showing change in mesquite *(P. glandulosa, P. velutina,* and *P. pubescens)* as evaluated in five categories for all years.

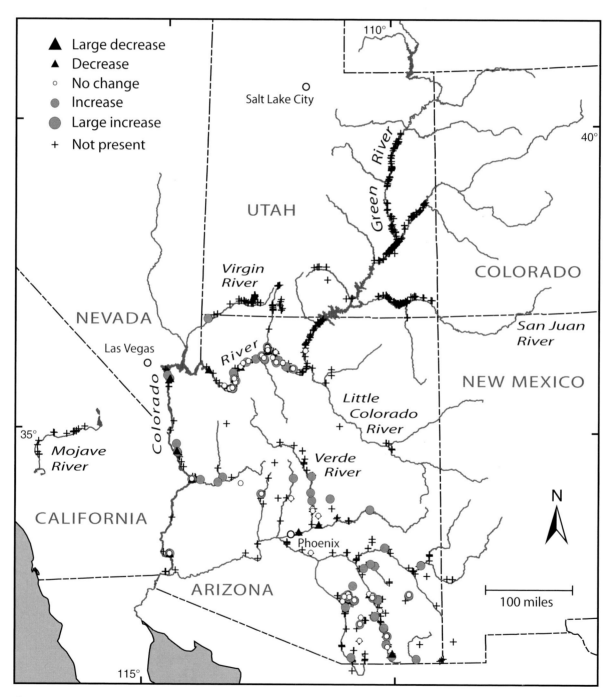

Figure 29.14 Map showing total change in catclaw in 346 sets of repeat photographs in the Southwest. Scale is 1:6.6 million.

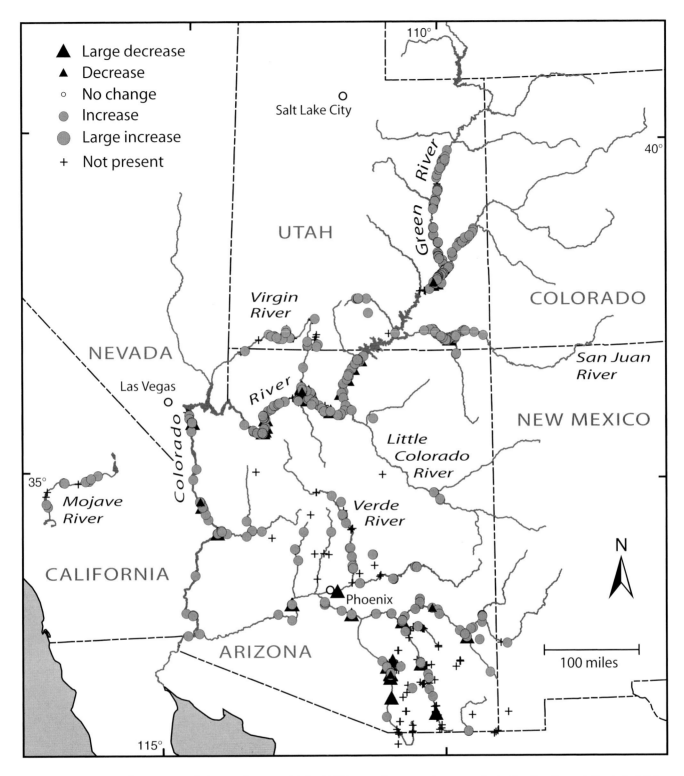

Figure 29.15 Map showing total change in tamarisk in 1,577 sets of repeat photographs in the Southwest. Scale is 1:6.6 million.

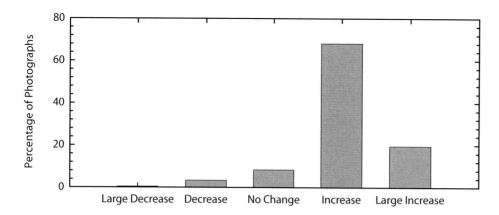

Figure 29.16 Histogram of the percentage of photographs showing change in tamarisk as evaluated in five categories for all years.

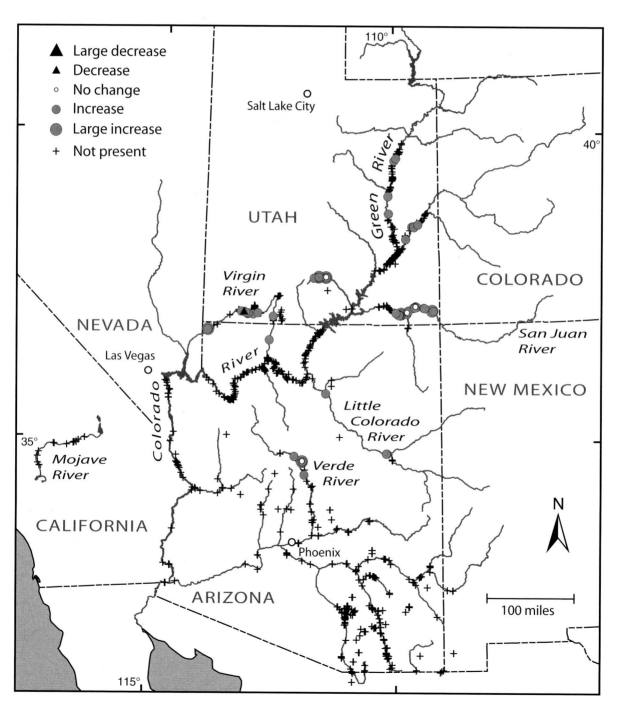

Figure 29.17 Map showing total change in Russian olive in 108 sets of repeat photographs in the Southwest. Scale is 1:6.6 million.

Summary. Woody riparian vegetation has changed throughout the region discussed in this book, with net gains in the total amount of riparian vegetation along many regulated and unregulated reaches and net losses along other reaches. The gains are associated with a large number of factors, beginning with partial dewatering of saturated floodplains caused by arroyo downcutting and continuing with climatic conditions conducive to large late-winter floods during the late twentieth century. In many cases, water levels in alluvial aquifers have remained high, sustaining established woody plants. Losses of riparian vegetation have occurred along the middle reaches of the Santa Cruz, Gila, and Mojave Rivers where water-resources development has lowered groundwater levels and depleted surface water; where reservoirs are now present, particularly along the Colorado River; and where intensive agriculture and channelization have occurred along the lower Gila and Colorado Rivers. Near-total elimination of woody riparian vegetation is uncommon and results from the combination of complete base-flow diversion and heavy groundwater usage, as exemplified by the middle reaches of the Gila River. Flood control and low-variability flow releases initially benefit woody riparian vegetation, although flood control ultimately minimizes the potential for germination and establishment of native species; artificial flood releases timed for seed production of native species can help reestablish senescent stands. Substantial decreases in flow volume may encourage replacement of existing species by making barren channel banks and low floodplains available for colonization. Tamarisk expanded rapidly in the middle of the twentieth century, occupying barren channel banks, and is expected to continue its domination of some floodplains with saline soils. In some reaches, recent changes suggest the potential for natural replacement of nonnative woody species with native species.

In this book, we used thousands of repeat photographs to address the claim that 90 percent of riparian vegetation has been lost in the southwestern United States. Our analyses were restricted to elevations below 5,000 feet, and we focused on the Colorado Plateau in Utah, in most of Arizona, and along the Mojave River in California. For pragmatic reasons related to our interpretative tool, we further restricted our treatment to change in woody riparian vegetation that can be interpreted from historical photographs, leaving the interpretation of changes in most herbaceous riparian vegetation to other researchers. We cited studies using repeat aerial photography, where it is available, that show change in woody riparian vegetation for reaches where we had repeat photography.

As discussed in chapter 29, we interpreted changes in fourteen woody species, including three common nonnative ones, and these results show that change is complex but has relatively well-defined patterns along watercourses in the region. Because repeat photography cannot easily be interpreted on a per-unit-area basis, and because we do not have full regional coverage, we cannot fully address the 90 percent loss statistic. Our interpretations of the available imagery, however, suggest more gains than losses within our photographic views and provide the more relevant observation that nearly all of the woody riparian vegetation documented in our photographic record has changed.

Few individual plants occupying space in the riparian zone have persisted through the photographic record; instead, nearly complete turnovers in riparian vegetation have occurred regionally, with or without directional changes. Riparian vegetation as viewed through the lenses of our photographers has increased at more camera stations than it has decreased, with some notable exceptions. The repeat photography shown in this book suggests that the claim that "since 1870 riparian wetlands have lost their once-luxuriant aspect"[1] is overly simplistic. The patterns of change and the reasons behind it are much more complex and subtle.

The nearly complete change in woody riparian vegetation in the region is commonly thought to have been caused in one way or another by human land use. As noted in the preface and chapter 3, native woody riparian vegetation is commonly described as threatened, degraded, or destroyed, regardless of any information on long-term trajectories of change. For example, floods described as damaging are usually attributed to human land-use practices[2] with little regard for their climatic context. The claim that poor land-use practices have caused excessive floods is made despite evidence suggesting that floods within the known historical period are not necessarily even the largest to affect these rivers in the past four thousand years.[3] Here, we emphasize a combination of human influences, climatic fluctuations, hydrologic changes, and the life-history strategies of dominant species in the riparian ecosystem—all of which vary across the region.

Repeat photography establishes the visual condition of woody riparian vegetation at the end of the nineteenth century and the beginning of the twentieth century. For many watercourses, this observation is flawed because the initial photography postdates a strong human influence on the watershed. Other photographs, such as for the Colorado River in Grand Canyon and Utah, predate extensive human land uses, either on floodplains or within the watershed. All river reaches discussed in this book have had at least some human modification, ranging from minor to extensive; similarly, all watersheds have experienced climatic fluctuations and attendant hydrologic changes. As implied from the repeat photography in a variety of settings, change should be expected in these ecosystems, and the consequences of change can be a consideration in strategies for management of woody riparian vegetation.

Causes of Changes in Woody Riparian Vegetation

Nineteenth-Century Status

Our most extensive understanding of presettlement vegetation comes from the bedrock canyons of the Colorado Plateau as well as the photography of Aravaipa Canyon (chapter 20). The extensive photography accrued during the first expeditions by land or water—whose leaders were interested in scientific documentation of the landscape, technical documentation of potential railroad grades, or documentation of mining claims—documents what was on at least part of the landscape in this region. Photographers arrived on the scene in these canyons before water developments and livestock grazing, and in many places domestic grazing animals had never been present.

Photographs from 1863 through 1875 show river channels with sparse amounts of riparian vegetation compared with conditions today, and some of the native trees in these canyons are among the few persistent riparian individuals that remain on the landscape after more than a century. Among the species with persistent individuals are Frémont cottonwood, desert olive, Gambel oak, and netleaf hackberry. As a group, even these species, with the possible exception of desert olive, show large changes over the period of photographic history.

Typical bedrock canyon reaches had mostly barren channel banks, particularly where perennial discharge was low or stage changes during floods were large, or both. Kanab Canyon, the most heavily photographed reach in the early 1870s, generally had sparse riparian vegetation, and coyote willow and seepwillow are the most common species observed in the photographs. It bears noting that historical photography mostly postdates several significant floods, especially the 1884 event that remains the flood of record on the Colorado River and several of its tributaries.

Coyote willow is the most common species observed in historical photographs of the Colorado Plateau. Photographs taken just before or just after settlement show what appear to be extensive, monospecific stands of coyote willow that obscure or closely hug the channels of the Escalante River, Kanab Creek, and Havasu Creek as well as the banks of the Green, Colorado, and San Juan Rivers. With the exception of the Green River in Desolation Canyon, trees were relatively rare in the reaches of the plateau, and some species, such as netleaf hackberry, had distributions restricted to certain reaches, such as Cataract Canyon. Cottonwood and black willow were uncommon in many of these canyons, with only one historical photograph showing cottonwood in Grand Canyon and only one showing cottonwood along the bedrock canyons of the San Juan River. Cottonwood was present at specific bottomlands along the canyons of the Colorado and lower Green Rivers in Utah. The 1873 views of the North Fork and the East Fork of the Virgin River show dense growth of cottonwood and other species as well, although those canyons had been settled for at least twelve years before the first photographs were taken.

In southern Arizona, both the historical photography and the anecdotal records show that some reaches supported growth of riparian trees in the nineteenth century. Aravaipa Canyon in southern Arizona had dense groves of cottonwood and a variety of other riparian trees in 1867. In 1885, Havasu Canyon had dense groves of cottonwood, Arizona ash, and other riparian trees despite the limited agricultural clearing by the Havasupai. Trapper diaries indicate that beaver were common but localized in their distributions to certain reaches; for example, they were not observed on the San Carlos River but were on the main stem of the Gila River, the San Francisco River, and the San Pedro River.

Descriptions of the Little Colorado River, the Gila River above and at its confluence with the Salt River, and the Santa Cruz River downstream from Tucson indicate that groves of cottonwood were present at least locally. As described by Spanish explorers, the middle reaches of the Little Colorado River supported cottonwood, although of course no historical photography documents the distribution of these trees, and the groves reportedly were separated by expanses of barren channel. The San Simon River, described as the "Valley of Willows," had ciénegas and high groundwater levels.

Photographs taken of bedrock canyons up until the start of the twentieth century do not show significant differences in riparian vegetation from those taken before or just after settlement. The Green and Colorado Rivers, in particular, appear to have changed little during the first thirty to fifty years of their photographic history. Riparian vegetation remained relatively dense on the North Fork of the Virgin River and on Havasu Creek, despite the occurrence of floods in 1909 and 1910, respectively, that caused property damage and loss of agricultural fields. Repeated images of Aravaipa Canyon show little significant overall change or may even show increases in riparian vegetation despite the effects of livestock grazing, mining operations, water diversions, and agricultural clearing.

Most of the photography documenting changes in alluvial reaches postdates settlement. Many alluvial reaches had sparse tree cover, with scattered, open groves of cottonwood and black willow. Along the San Pedro River in southern Arizona, water levels in the alluvial aquifer were high,[4] which is consistent with reports of infestations of malaria-bearing mosquitoes. Cottonwood and other obligate riparian species occurred on higher terraces, but with a few exceptions, notably near the confluence with the Gila River, they did not form gallery forests. Mesquite bosques occurred above the saturated floodplain, but there is no evidence to suggest that these bosques were continuous. Reaches near mining operations, such as the former mills in the vicinity of Charleston, had no more or less vegetation than reaches far removed from mining operations.[5] An exception appears to be the upper Verde River, where acidic exhaust from smelter operations killed trees.

Whereas most alluvial reaches had variable amounts of woody riparian vegetation, several notable exceptions occurred where this vegetation was established (for example, the Mojave River at Victorville and downstream near Yermo, California). Some of these reaches reportedly had large beaver populations, such as the Gila River through present-day Safford Valley and near Maricopa, and the lower San Pedro River. Other reaches, such as the Santa Cruz River through Tucson and the Mojave River in the vicinity of Barstow, had open gallery forests of cottonwood and willow as well as mesquite bosques, but beaver have not been documented in these reaches. Some reaches with substantial woody vegetation were separated by relatively barren reaches, including the Gila River above its confluence with the Salt River. Written descriptions indicate that the channel was barren immediately downstream from present-day Florence, that dense riparian vegetation was present upstream from the Salt-Gila confluence, and that relatively sparse woody vegetation, characterized by scattered cottonwood, occurred downstream from present-day Gila Bend to just upstream from the Gila's confluence with the Colorado River.

The available information suggests that reaches with dense woody riparian vegetation were discontinuous arroyos, separated from similar reaches by either ciénegas or dry channels. The mere presence of dense woody vegetation on a river did not guarantee that tributaries joining at that point would have similar characteristics; for example, the Gila River beneath present-day San Carlos Reservoir had open cottonwood galleries and substantial beaver populations, whereas the San Carlos River apparently had neither. The amount of riparian vegetation clearly was patchy and reach specific, and generalizations cannot readily be extended to entire watersheds.

What we know of the distribution of woody riparian vegetation at the time of settlement and just afterward suggests that local hydrologic conditions were the most important determinants of the amount of vegetation in any given reach. Bedrock canyons with high base flow and (apparently) relatively small floods (for example, Havasu Canyon, Aravaipa Canyon) supported dense woody riparian vegetation. Bedrock canyons with low base flow, large annual stage changes, and coarse substrate had sparse woody vegetation (for example, Kanab Canyon). Alluvial valleys with high groundwater tables and shallow channels generally had ciénegas, and woody vegetation was locally abundant on unsaturated floodplains.

Changes around the Start of the Twentieth Century

There is a high degree of association between the periods of arroyo cutting and filling (fig. 30.1), the floods that drove this geomorphic process, and changes in woody riparian vegetation. Most photographic evidence from the first third of the twentieth century shows barren channel banks in most bedrock canyons (the exceptions are noted earlier) and scoured and deepened alluvial channels associated with the period of arroyo downcutting and widening, which ended in around the early 1940s. For many reaches, the first photographs that show channel conditions were taken at the end of the period of large floods associated with arroyo downcutting or with channel widening or with both.

Bedrock canyons with dense riparian vegetation at the beginning of the photographic record generally had increases in riparian vegetation through the twentieth century despite floods that were the largest in human experience. The prime example is Havasu Canyon, where a flood in September 1910 altered waterfalls and swept away riparian trees.[6] Other bedrock reaches—such as the Gila, Verde, and Salt Rivers upstream from dams and the Escalante River—have relatively vegetation-free channel margins in the first photographs, which generally were taken from 1930 through 1950. These reaches, largely unregulated by dams or flow diversions, had large increases in riparian vegetation to around the

start of the twenty-first century, as shown in our repeat photography. Both Havasu and Aravaipa Creeks had large floods in the late twentieth century (September 1990 and October 1983, respectively) that did little to diminish the amount of woody vegetation. Even in reaches such as the Colorado River through Cataract Canyon, where the annual flood continues to cause large stage changes, native species, primarily coyote willow and netleaf hackberry, have increased.[7]

Although we observed local decreases in riparian vegetation in many places, reach-scale decreases in the twentieth century were present only in a few locations. The most striking decreases occurred along the Santa Cruz River in the vicinity of Tucson and along the Mojave River in a long reach centered on Barstow. Inundation of reaches beneath reservoirs, which were filled mostly in the middle of the twentieth century, also caused decreases in riparian vegetation, although the amount of vegetation once present generally was not substantial except in wide alluvial valleys. Other decreases are associated with changes driven by climate, although the changes likely were amplified by land-use practices. The decline in woody vegetation along the middle Gila River, for example, occurred before 1900 and was associated with large floods and the initial period of arroyo downcutting; the extensive water developments of the middle twentieth century prevented reestablishment of this woody vegetation. Cottonwood groves were no longer prominent along the Little Colorado River even before the Mormons settled there in the 1870s.

As discussed in chapter 29, reaches with overall increases in riparian vegetation had important changes in species composition as well as in the relative contributions of native and nonnative plants. In general, Frémont cottonwood declined along the largest rivers despite overall increases in native and nonnative riparian vegetation. This decline occurred in the lower Green River in Labyrinth and Stillwater Canyons, the Colorado River in Cataract Canyon, the Gila River in the Safford Valley, and the lower Colorado River. In the same period, cottonwood has greatly increased in many of these

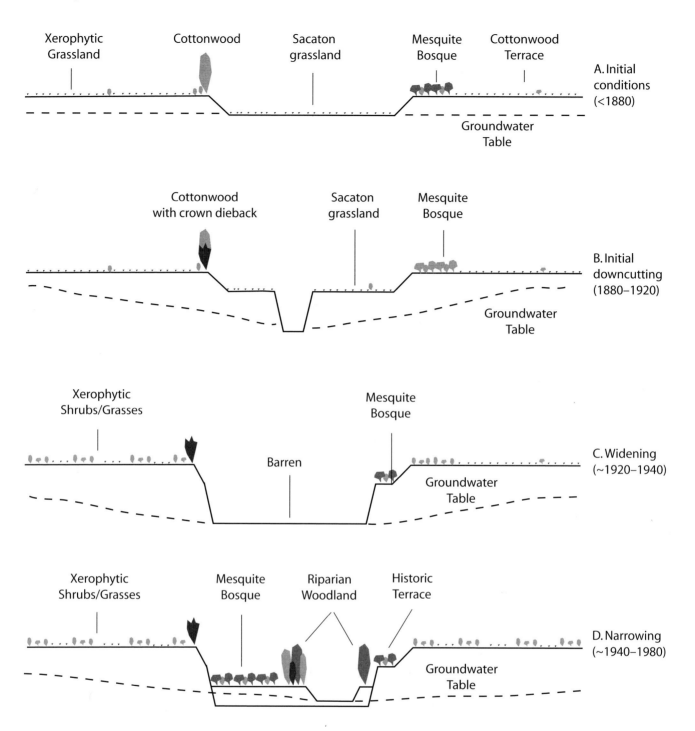

Figure 30.1 Generalized diagram showing the interrelations between arroyo changes and riparian vegetation in southern Arizona. Similar changes are expected in alluvial systems in other parts of the Southwest, although with different species of woody riparian vegetation. Terrace names, designated for the Colorado Plateau, follow Hereford (1984, 2002).

rivers' tributaries, especially along the Escalante River, the Virgin River, tributaries in Grand Canyon, the Bill Williams River, and Aravaipa Creek.

Several important contradictions appear in the interpretation of change in woody riparian vegetation in this region. The common assertion that tamarisk replaced stands of cottonwood-willow along many rivers[8] is generally unsubstantiated except for parts of the lower Colorado River. Historical photography shows that much of the area colonized by tamarisk was barren, and we found no evidence of a conversion of native to nonnative species without an intervening period of flood-induced bare substrate. There is evidence along many rivers, notably the Colorado and the Gila, that channel widening and avulsions swept away tracts of native vegetation and that tamarisk became established afterward.

It is difficult to find evidence that cottonwood-willow stands were once extensive in the region with the exception of the Colorado River delta in Mexico. Within the United States, the only locations where such forests may have decreased would be at the Gila-Colorado confluence and upstream from the Salt-Gila confluence. The cottonwood-willow assemblage along the Santa Cruz River at Tucson appears to have been narrow and extended perhaps 10 river miles. Narrow bands of cottonwood-willow assemblages were submerged beneath reservoirs along the lower Colorado River. Abundant evidence suggests that this type of assemblage, or cottonwood along with other species such as Arizona ash, has increased, for example, along several reaches of the San Pedro River and its tributaries, Havasu Canyon, the Virgin River, the Gila River between Coolidge and Ashhurst-Hayden Dams, tributaries of the Santa Cruz River, the Bill Williams River, and the Mojave River.

Establishment and maintenance of stands consisting only of tamarisk are limited to very specific places, most notably the deltas of reservoirs and reaches where salinity is high in either surface water or groundwater. Reaches with such stands include the Gila River upstream from Gillespie Dam; the Salt River at the head of Roosevelt Lake; and the Colorado River at the heads

of Lakes Powell, Mead, and Havasu. High salinity can result from natural discharge of groundwater high in salts, from agricultural effluent, or from surface water with salts concentrated as the result of reservoir evaporation. An example of a reach with high soil or water salinity is the Gila River upstream from San Carlos Reservoir. Other reaches for which we have no photographic documentation but that are known to support dense tamarisk include the Virgin River at the head of Lake Mead[9] and the Verde River upstream from Horseshoe Reservoir.

A more commonly observed condition is tamarisk that has become established on previously barren banks, in some cases furthering channel narrowing that was initiated by climatically induced decreases in flood frequency and enhanced by flood control (for example, the Green River[10]). Native riparian species establish more slowly, in some cases (for example, the Escalante River) on the streamward side of aggrading floodplains, thereby contributing to more channel narrowing. We have observed and support the contention that native species may eventually dominate or even replace tamarisk under certain conditions.

Nonnative vegetation cannot be considered as a group in terms of change in the Southwest. Russian olive, which is pervasive in slower-velocity reaches of the Green, the upper Colorado, the San Juan, and the Little Colorado Rivers, appears to be increasing in density, and the increases seem to be occurring on floodplain settings with minimal disturbance. Tamarisk appears to increase only following disturbances or in places adjacent to channels where flow has been reduced by drought or flow regulation. In the case of the alluvial reach of the San Juan River, where we found the most Russian olive, native riparian vegetation has also increased. We have no evidence that Russian olive has recently increased along the lower-elevation watercourses where it previously had been documented, such as the Salt and Verde Rivers, the lower Gila River, the Colorado River in Grand Canyon, and the lower Colorado River, and this observation agrees with the proposed climatic limitation on this species.[11]

Factors Associated with Change in Riparian Vegetation

Previous research has documented and hypothesized about a variety of human, physical, biological, and climatic factors that might be expected to cause reach-scale stability or change in woody riparian vegetation. In many cases, forces that affect riparian vegetation occurred simultaneously, confounding assessments of causation. In others, the effects of some of these forces may be isolated or at least minimized in terms of their effects on the riverine environment.

Livestock Grazing

Overgrazing is frequently cited as one of the largest threats to riparian ecosystems, either directly or indirectly.[12] The effects of grazing include overall watershed degradation, leading to increased runoff and sediment yields to riverine systems; direct consumption of seedlings of native woody species; and trampling and damage of established individuals on floodplains. Although there is little question that overgrazing can degrade riparian ecosystems, the question is whether grazing has had long-term negative effects on woody riparian vegetation in our region.

Livestock were introduced to southern Arizona before 1700, but the animals were restricted to the San Pedro and Santa Cruz River watersheds. After around 1880, livestock were heavily stocked to take advantage of favorable forage conditions, but lack of extensive water development constrained the animals' range to near permanent water.[13] High stocking levels at the end of the nineteenth century coincided with extreme drought, and range degradation and livestock deaths followed. Following the Taylor Grazing Act of 1932, which was passed near the end of the period of arroyo widening, domestic animals typically were stocked at lower levels in most of the Southwest. Reductions continued through the twentieth century, especially after the mid-1980s and particularly in riparian areas.[14]

Riparian vegetation has increased significantly in many reaches irrespective of the presence of grazing animals. In Cataract Canyon and

Grand Canyon, where the distribution of livestock is known and well constrained, riparian vegetation has increased, and the increase is in both nonnative and native species. Our photography captured an increase in riparian vegetation where ranches have been continuously present in riparian zones, such as the Mojave River at Victorville. Cottonwood has decreased in places where livestock have never grazed and has increased where grazing was heavy; likewise, decreases, such as in Cataract Canyon, occurred where livestock never have grazed. The large increases in woody vegetation along the Escalante River began well before livestock numbers were reduced.

Woody riparian vegetation began increasing along many reaches of the San Pedro River during the period when livestock numbers were at their peak stocking densities and well before grazing was intensively managed. Establishment of the San Pedro Riparian National Conservation Area, with its restrictions on grazing in riparian areas, followed the increase in the highly valued cottonwood and other native trees. It would be expected but not documented that the increase in woody riparian vegetation accelerated after livestock were partially or totally removed, but this additional increase may have been influenced by other factors as well.

Livestock grazing continues in areas that have had significant increases in woody native species as well as in nonnative species such as tamarisk. Examples include the Escalante River; the upper Gila River in most of the reaches for which we have repeat photography; the San Francisco and San Carlos Rivers; the middle reaches of the San Pedro River; the upper reaches of the Santa Cruz River and parts of Sonoita Creek; the San Carlos and San Francisco Rivers; and the Little Colorado River. We stress not only that current grazing management of these areas is intensive and use along some reaches is seasonally restricted, but also that the mere presence of grazing animals in a riverine setting does not portend declines or collapse of woody riparian vegetation. We also stress that our results do not justify either allowing livestock grazing or excluding it from riparian areas. How much the late-twentieth-century increases in riparian vegetation can be attributed either to grazing reductions or to other climatic and hydrologic factors remains an open question.

Agricultural Clearing

Clearing of floodplain vegetation for agriculture had demonstrable local effects on riparian vegetation, although the extent of clearing—with some notable exceptions—was relatively small by the start of the twentieth century. For example, agricultural clearing along the Santa Cruz River at Tucson, which dates to presettlement use by the Tohono O'odham, is apparent in early-twentieth-century photographs, but the fields were small in comparison with the total riparian area in the views. Agricultural clearing by the Hopi and Havasupai is apparent in old photographs and lithographs made from old photographs of Moenkopi Wash and Havasu Creek, respectively. Agricultural clearing occurred near every town along a watercourse, but these fields supplied mostly local produce, not commercial crops grown for long-distance transport. Flood damage to agricultural fields in many cases prompted farmers to move away from the floodplain to higher terraces and to irrigate either with canals or wells (for example, Kanab Creek at Kanab and the San Simon River near its confluence with the Gila River).

In southern Utah, much of the bottomland vegetation cleared along the major rivers was formerly dominated by saltbush, greasewood, and big sagebrush. Along the alluvial reaches of the San Juan River, cottonwood were cut not to clear land for agriculture, because the floodplains that sustained cottonwood were subject to inundation and scour, but locally to build bank-stabilization structures. Otherwise, cottonwood trees were treasured for the shade they provided in an otherwise treeless desert, and their loss during floods was frequently noted.

Agriculture on or adjacent to floodplains increased in the second half of the twentieth century in response to increased construction of water-diversion structures on the lower Colorado River or of well fields on floodplains removed from channels. For example, extensive agricultural development of the middle and lower reaches of the Gila River and of the lower Colorado River followed construction of irrigation canals; much of the land that was cleared formerly supported mesquite bosques or arrowweed. However, in one of our matches near Dome, Arizona, agricultural clearing removed mature cottonwood. The amount of land cleared for agriculture increased along the middle reaches of the San Pedro River, again mostly at the expense of mesquite.[15] The broad valleys of southern Arizona and the lower Colorado River had extensive agricultural development, but most of that development came at the expense of xerophytic vegetation. Where agriculture encroached on floodplains, mesquite suffered the highest losses.

Fire

Fires started either by lightning or by aboriginals is one of the frequently cited reasons for the extensive areas of prehistoric grasslands in southeastern Arizona,[16] and fire suppression is considered to be a primary reason why woody vegetation has increased in areas that formerly sustained grasslands.[17] Although no evidence exists to support this hypothesis, grassland fires probably swept into riparian zones and may possibly have helped to maintain the woody-free vegetation that once dominated ciénegas and floodplains in our region. In bedrock canyons or in riverine settings through more xeric landscapes, fire probably had minimal influence on the pattern of presettlement woody vegetation.

In the twentieth century, especially following the introduction of tamarisk and the increase in recreation, fire has had major, although spatially limited, effects on woody riparian vegetation in the reaches we have studied.[18] We cannot name a river where fires have not occurred during the period covered by our repeat photography. In some areas, such as Spanish Bottom in Cataract Canyon, a fire may cause partial or directional change in species composition from native to nonnative species (in this example, cottonwood to tamarisk). As shown by the Bill Williams River downstream from Planet Ranch, what might appear to be a devastating fire ultimately may result in a strong response from native species. Experience with historical fires in reaches dominated by tamarisk suggests that this nonnative species benefits from this type of disturbance,[19] although the occurrence of a late-winter flood or dam release following a fire may enable native species to become established as well.

The current management emphasis of riparian areas as recreation areas indicates that fire may play an increasing future role in shaping riparian areas. Although lightning-sparked fires are common, human-caused fires in riparian areas are increasing throughout the Southwest. Fire is still used as a vegetation-manipulation tool in riparian zones, primarily as a way to thwart senescence in native plant assemblages, but also to reduce tamarisk. In both prescribed burns and human-caused fires, the potential for regeneration of native species, particularly at the expense of nonnatives, is a consideration for managers.

Woodcutting

There is no evidence from photographs, land-use histories, or anecdotal observations to support the often-stated contention that humans caused widespread elimination of woody riparian vegetation at the time of settlement.[20] Small amounts of wood were used in construction; for example, the lintels above doors and windows may have been cottonwood. The only reach with explicit documentation of systematic woodcutting within the riparian zone is along the lower Gila and Colorado Rivers, where wood was consumed to power steamboats and smelters and cut to provide building materials for the expanding town of Yuma. Even in these reaches, the earliest photographs clearly show extensive stands of woody riparian vegetation, which indicates that this resource was not completely exploited. When an alternate source of higher-quality wood was available, such as oak and pine in southern Arizona mountain ranges, it was used. In former Glen Canyon, one steamboat and the gold dredge it supplied used coal mined from a tributary canyon, not wood harvested from the floodplain of the Colorado River.[21]

Woodcutting did occur locally on floodplains, and the extent of this cutting—as well as its impact on the riparian zone—cannot be known. Mesquite was the firewood of choice, not cottonwood or black willow. In some cases, woodcutting followed destruction of riparian vegetation by other means. South of Martinez Hill along the Santa Cruz River, an extensive mesquite bosque was affected by local woodcutters but ultimately was eliminated by excessive groundwater use, and the dead snags subsequently were collected for firewood. Similarly, wood was harvested along reaches of the Mojave River following the death of trees caused by declining groundwater levels. There is no evidence from the region that riparian trees were clear-cut or even significantly thinned from floodplains on a reach scale.

Floods

The destructive and regenerative effects of floods on woody riparian vegetation are usually emphasized in evaluations of long-term change. As discussed in the next section, the occurrence of floods is one potential explanation for why channel banks and floodplains were mostly devoid of woody vegetation in the first photographs of the region. The occurrence of mainly small summer floods is associated with the expansion of tamarisk in the middle part of the twentieth century, and fall- and winter-dominated flooding is one potential reason for the widespread establishment of native woody species in the late twentieth century. Examples of these changes include the upper Gila River (fig. 2.2) and the Santa Cruz River (fig. 21.2).

Particularly on larger rivers, floods are known to cause channel avulsions, resulting in both erosion of existing floodplains and deposition of new ones. This type of disturbance in the riverine environment resets woody riparian assemblages and results in the commonly observed even-aged stands of riparian trees on floodplains, particularly along the lower Colorado River and its delta. Historical photographs taken after the floods in late winter 1916 (for example, of the Hassayampa River) show that even in that period woody riparian vegetation became

well established within two years of the event.

The absence of significant floods after the early 1940s on some rivers of the Colorado Plateau may help to explain the large amount of growth in woody riparian vegetation in that part of the region. Plants that germinated before around 1940 grew to large size in the absence of continuing flood disturbance. However, resurgence of large fall and winter floods between 1977 and 1995 affected southern and central Arizona as well as the Mojave River. From our repeat photography, these floods—in particular the winter events—resulted in long-term increases in woody riparian vegetation that greatly offset any declines that occurred during the floods. Examples of this response include the upper Gila River, Aravaipa Creek, the upper Santa Cruz River and its tributaries, the Verde River, and the Salt River upstream from Theodore Roosevelt Lake. The positive effects of winter floods and the potential for summer floods to germinate and establish tamarisk are reasons cited for controlled flood releases from dams to be timed seasonally and scheduled regularly.

Arroyo Downcutting and Filling

The repeated downcutting and filling of arroyos in the late Holocene are well known in the geological record of the southwestern United States. The prehistoric episodes of downcutting occurred long before domestic livestock were introduced, and the latest episode occurred many decades after their introduction. As a result, the introduction of livestock likely neither initiated the changes nor significantly changed the outcome.[22] In addition, decimation or extirpation of beaver, depending on the river reach and intensity of trapping, occurred as much as seventy years before initiation of arroyos. Instead, arroyo downcutting occurred during a period of increased frequency of large regional storms. Floodplain aggradation, which presently is occurring throughout our region, is associated both with climatic fluctuation and the general increase in riparian vegetation, which traps sediments on floodplains during periodic overbank flow.

Arroyo downcutting is a major reason why woody riparian vegetation has increased in many alluvial reaches. Especially in southeastern Arizona, but also in southern Utah, high groundwater levels in alluvial aquifers, which may or may not have been affected by beaver populations, discouraged woody vegetation except on higher terraces. Arroyo downcutting drained the alluvial aquifers (fig. 30.1), and the former floodplains were either eliminated during the period of channel widening or reverted to facultative riparian or xerophytic vegetation. In southern Arizona, mesquite and cat-claw dominated the new assemblages on the formerly saturated floodplain; in northern Arizona and southern Utah, greasewood, saltbush, rubber rabbitbrush, and sagebrush occupied similar sites.

As new floodplains were deposited after around 1940 (fig. 30.1), woody riparian vegetation became established on the unsaturated sediments. There are no examples from the region where woody riparian vegetation—either native or nonnative—did not become established on low floodplains deposited after the period of channel widening. The important determinants of the species composition of vegetation occupying the new floodplains are the depth to groundwater and the amount of surface flow, and these factors are affected by water development.

The interrelation between arroyo processes and riparian vegetation may help explain the course of channel aggradation and arroyo filling, which appears to have occurred rapidly in the geologic record. Coyote willow and seepwillow are the most likely species to have become established prehistorically on unsaturated floodplains. These species would increase surface roughness on floodplains, thereby increasing deposition and channel filling. The presence of these species may have attracted beaver, assuming these animals could get to the reach through the connectivity of perennial flow. Beaver dams would have encouraged channel aggradation and accelerated the filling of arroyos. Rising groundwater levels in the alluvial aquifer created long-term saturated floodplains, which excluded woody vegetation, thus completing the cycle.

We emphasize that this interrelation between arroyo processes and woody riparian vegetation is only partially known and speculative. Paleoecological techniques for reconstructing riverine ecological conditions, in particular analysis of fluvial pollen, have significant problems in distinguishing regional and watershed changes from local changes on floodplains.[23] Although the association between late-twentieth-century floodplain development and establishment of woody vegetation is strong, the effects of land-use management cannot readily be unraveled from climatically driven changes.

Flow Regulation and Diversion

The effect of surface-water developments on woody riparian vegetation varies considerably and depends on how base flow is changed seasonally and annually and on the extent of flood control. In the case of the Colorado River through Grand Canyon and the Gila River near Kelvin, flood control, combined with reduced variation in seasonal flow and minimal diversion, has resulted in at least short-term increases in woody vegetation. In the case of the lower Verde River and the Salt River below Stewart Mountain Dam, the occurrence of exceptionally large floods coarsened channel and floodplain sediments. Woody riparian vegetation has become established in this coarse substrate because base flow has remained high.

Complete or near-complete diversion of base flow, irrespective of flood control, has large effects on riparian vegetation. In the case of the Gila River from Florence to its confluence with the Salt River, a once-thriving riparian ecosystem inhabited by the Pima was eliminated initially by large floods in the late nineteenth century, and reestablishment of woody obligate vegetation here is extremely unlikely owing to complete base-flow diversion at Ashhurst-Hayden Dam. Diversion of Colorado River water into the All American Canal at Imperial Dam has greatly reduced the amount of surface flow through Yuma, although water deliveries to Mexico, which include irrigation returns, sustain some woody riparian vegetation.

One of the largest decreases in woody vegetation occurred when reservoirs were filled, although the extent of net change in riparian vegetation remains uncertain. Most of the reaches inundated by Colorado River reservoirs had little woody vegetation, and what was present was mainly coyote willow or arrowweed, in accord with observations of still free-flowing reaches such as Grand Canyon. Tributary mouths may have had substantial riparian vegetation, but, overall, the earliest photography shows these reaches to have been largely barren, with dense mesquite in wider reaches and locally dense stands of cottonwood and black willow along the lower Colorado River. Construction of reservoirs increased the shoreline area, but fluctuations in shorelines create a hydrologically unstable environment that favors facultative riparian species, in particular tamarisk, that can germinate in summer when shorelines generally recede and can withstand the effect that lake-level fluctuations have on available water.

Channelization

Channelization of watercourses takes many forms. Installation of large-scale bank protection began with the accidental diversion of the Colorado River into the Salton Sink in 1905. By 1908, extensive levees were built along the lower Colorado River near Yuma to protect irrigation canals and to force the river to remain within its banks. The period of arroyo widening and attendant regional losses of agricultural lands and water-diversion structures prompted local installation of bank protection and grade-control structures. Projects initiated by the Soil Conservation Service and the Civilian Conservation Corps in the 1930s were regional in scale and followed the period of arroyo downcutting. This work, designed to encourage channel filling, promoted establishment of riparian vegetation by allowing floodplain deposition. Notable examples include the San Simon River and Upper Valley Creek, the principal tributary of the Escalante River.

Increasing agricultural and urban development along the lower Colorado River prompted installation of extensive levee systems, both to minimize the potential for lateral channel change and to increase the efficiency of water transfers. Former broad, braided channels with limited backwaters were artificially narrowed, and former channel and floodplain areas segregated from the main channel became backwaters or marshes owing to continuing high water levels in the alluvial aquifer (for example, the Topock Marsh near Needles). Levees are constructed with cobble and boulder fill, which is much too coarse to provide substrate for native species, and levees are designed to exclude floodwaters, but the presence of open water impounded behind the levees and high groundwater levels promote tamarisk. Through the Parker and Palo Verde Valleys, tamarisk is established in dense stands behind the protective levees, and native species can become established only in reaches without bank protection.

Because of the spate of large floods in southern Arizona between 1977 and 1995, bank protection was installed along most reaches of watercourses that passed through the Phoenix and Tucson metropolitan areas. These channels were designed to transfer floodwaters efficiently past floodplain infrastructure and development, and part of that efficiency was elimination of riparian vegetation that increased channel roughness. Woody vegetation has reestablished where wastewater is discharged, notably in the lower Salt River and the Santa Cruz River, and river-restoration plans that call for the planting of riparian trees may once again increase both channel roughness and the potential for flood damages.

The effects of channelization are complex and dependent on the hydrologic setting. For some reaches, such as the Salt River through Phoenix or the Santa Cruz River through Tucson, channelization greatly reduces the potential for obligate woody riparian vegetation. This potential is further reduced in the case of the Santa Cruz River, where lowering of the regional aquifer deters establishment of obligate riparian vegetation except where groundwater is locally perched or where effluent from treatment plants sustains a groundwater mound beneath the channel. In other reaches, highly valued marshes containing native species can develop with the protection afforded by bank stabilization. Along the lower Colorado River, for example, the Topock Marsh owes its existence to a protective levee, flat topography, and perennial flow.

Groundwater Depletion

Lowering of groundwater levels by excessive groundwater use is a well-known cause of decline in woody riparian vegetation. Examples include the Santa Cruz River through Tucson, where historically perennial flow sustained both cottonwood-willow assemblages and mesquite bosques for short distances; the middle Gila River, where declining groundwater levels caused heavy mortality of mesquite in the 1950s; and the Mojave River, where riparian vegetation has been eliminated over relatively long reaches both upstream and downstream from Barstow. These examples indicate that a large reduction in groundwater level alone is sufficient to eliminate riparian vegetation, regardless of the amount of surface-water development or use.

Changes in groundwater levels commonly result from pumping of the alluvial aquifer, but channel changes —in particular downcutting—affect water levels as well.[24] Although there is little question that water-level declines exceeding 30 feet below land surface are sufficient to kill many obligate riparian species, fast rates of water-level decline, as illustrated by one reach of the Mojave River downstream from Victorville, may have extreme effects on established trees. Root growth may not be rapid enough to follow the declining water table, even if the water levels are within 30 feet of land surface. Excessive groundwater use and diminished water levels minimize or prevent reestablishment of seedlings that germinate during rare winter floods or dam releases.

For the reaches we have studied, we present long-term water-level records that show trends, if any, in groundwater near watercourses (for example, fig. 2.3). In areas of high groundwater utilization, these water-level records show declines that can be linked to declines in or elimination of woody riparian vegetation (for example, fig. 21.3). Most water-level records generally mirror the climatic periods of the twentieth century, with lower levels during periods of regional drought, such as the midcentury drought, and higher levels during wet periods, such as the period from 1977 through 1995 (for example, fig. 17.4). This interpretation is complicated by other factors, such as more reliance on groundwater during periods of drought, less reliance during wet periods, and reduction in pumping rates. It does indicate, however, that not all the groundwater levels along watercourses in the region are in decline and may be showing some effect of the overall reduction in water usage in the region (see, for example, fig. 2.4).

The Future of Woody Riparian Vegetation in the Southwest

Although we have shown regional patterns of change in woody riparian vegetation using considerable repeat photography, none of these patterns is sufficient to anticipate or to predict future changes. Too many potential factors can affect woody riparian vegetation to know exactly what the future may hold in any given reach. This conclusion is perhaps best illustrated by the history of the Colorado River delta, where significant germination of cottonwood and willow trees in recent decades has followed inadvertent winter dam releases.[25] In the case of the Bill Williams River, for example, flow is completely regulated and capped by the capabilities of Alamo Dam, but recent inadvertent and deliberate flow releases have created and sustained native woody vegetation. The surface-water experience in the late twentieth century indicates that large winter floods coincidentally timed with seed availability might have effects that shatter expectations.

Past and future water developments threaten woody riparian vegetation, even in some cases where groundwater use is decreased. For example, the effect of groundwater use at a distance from river channels may not influence floodplain aquifers for many years, irrespective of decreases in the amount of pumpage. The lowering of the water table upgradient from rivers may take decades to propagate down-gradient to riparian ecosystems; for this reason, currently thriving reaches, such as the upper San Pedro River, may be imperiled regardless of future decreases in groundwater use in the vicinity of Sierra Vista. Similarly, riparian ecosystems developed in reaches with significant wastewater effluent, such as the Santa Cruz River downstream from Tucson, may be threatened if water conservation or reclamation of wastewater increases, so that less water is released into the channel.

Several examples from the region show that surface-water regulation involving a combination of flood control and decreased annual variability in flows may greatly enhance the establishment and growth of native and nonnative riparian vegetation.[26] As the example in Grand Canyon shows, increases in vegetation may result from flow regulation with flood control, particularly if low flows are increased and large dam releases are timed to discourage tamarisk germination. In the absence of flooding, senescence in woody vegetation might be expected in the future because of lack of opportunity for establishment.[27] We have no experience from which to gage how long a disturbance-free assemblage of woody vegetation might last or what the long-term consequences of no flooding might be. One possibility is conversion of mesic floodplain assemblages to xeric ones as mortality without replacement occurs.[28] Another possibility is that facultative species

that are less dependent on disturbance, such as mesquite or catclaw, may dominate. Given how quickly assemblages of woody riparian vegetation change in the Southwest, the timescale of senescence may be expected to be on the order of decades, not centuries.

Accompanying the potential for senescence is the potential for increased fire frequency, particularly given the high recreational use of riparian zones. Increasing recreational use already affects most river reaches, notably the Green and Colorado Rivers upstream from their confluence, and will undoubtedly result in more fires. The potential for lightning-sparked fires, either within the riparian zone or sweeping in from the surrounding landscape, remains high. Should future fires be allowed to burn at presettlement frequencies in the predominantly native vegetation, native and non-native species that can tolerate burning, such as desert olive and tamarisk, may be favored. If fires occur where flow regulation minimizes the potential for winter floods, or if fires occur during climatic periods with minimal winter storms, the potential for at least short-term directional changes in woody riparian vegetation is large.

The current emphasis on eradication of tamarisk, using mechanical means or herbicide application, presents several cautionary notes. Previous large-scale tamarisk eradication

efforts failed (for example, the Gila River Phreatophyte Project on the upper Gila River), largely because the replacement native species did not survive and tamarisk recolonized the open spaces. Eradication efforts must be accompanied by long-term and intensive maintenance that emphasizes establishment of native species or long-term conversion to agricultural production. Because the amount of water used by the replacement species may approach, if not exceed, usage by the vegetation that was removed, water-conservation efforts based on tamarisk eradication may not have the expected benefit over the long term.

Our experience with evaluating long-term regional change in woody riparian vegetation suggests several caveats for both management practices and prediction of future changes. How this highly valued resource will change in the future cannot be generalized using simplified assessments of threats or knowledge of past losses. The number of surprising changes that have occurred in the region suggests that this type of ecosystem is extremely resilient and responsive to favorable climatic and hydrologic events, except extreme hydrologic events. Considerable uncertainty as to the course of future changes should be expected irrespective of management practices that currently are considered to benefit or diminish woody riparian vegetation.

Notes

Authors' Preface

1. Works that describe extreme losses of riparian vegetation in Arizona and elsewhere include Arizona State Parks (1989), Johnson and Simpson (1988); R. Johnson (1991); National Research Council (1992); Braatne, Rood, and Heilman (1996); and Tellman, Yarde, and Wallace (1997). For example, Kusler (1985) simply claims "huge acreages continue to be destroyed" without citing a source or data. An internal memorandum from the Arizona Department of Game and Fish, dated February 14, 1992, notes the lack of evidence explicitly supporting the 90 percent loss claim and attempts to restrict the statement to "major desert watercourses" and "includes habitat that is altered or degraded." Childs claims, "Many 150-year-old cottonwoods at desert washes stand barren and dead. In Arizona, nearly all the once perennial desert streams have been robbed" (2000, p. 134). An article entitled "Study: Native Fish Being Allowed to Die Off" states, "In Arizona, more than 90 percent of riparian habitat is gone" (2003, p. A8). Interestingly, Mitsch and Gosselink (1993) report a loss of 36 percent of wetland acreage in Arizona from the 1780s through the 1980s. D. Brown (1995) decries the exaggerated claim of riparian loss and states that gains occurred as a result of favorable climate and grazing management.

2. Executive Order No. 91-6, issued in Phoenix, Arizona, on February 14, 1991, by Governor Rose Mofford.

3. The authors particularly thank S. A. Bengson for use of his unpublished manuscript submitted to the Arizona Chamber of Commerce entitled "In Quest of the Mystical '90 Percent'" for some of the ideas presented here. Bengson traces the 90 percent loss of wetlands in Arizona to Ohmart, Deason, and Burke (1977), who reported on changes along the lower Colorado River (chapter 27).

4. Hastings and Turner (1965); Webb (1996); Turner, Webb, and Bowers (2003); Webb, Belnap, and Weisheit (2004).

5. Kepner and others (2000); Kepner, Edmonds, and Watts (2002).

6. Dobyns (1981).

7. Bahre (1991).

8. Turner, Webb, and Bowers (2003).

9. Lingenfelter (1978).

10. Blake and Steinhart (1994), p. 94; McNamee (1994).

11. Both surface-water and groundwater data are available online via http://water.usgs.gov, accessed November 4, 2005.

12. Webb and Betancourt (1992).

13. From http://www.azwater.gov/dwr/, accessed November 4, 2005.

14. Garrett and Gellenbeck (1991); Pope, Rigas, and Smith (1998). Also see http://water.usgs.gov (accessed November 8, 2005) for online water and basin-characteristic data.

15. Kearney and Peebles (1960); Hickman (1993); Welsh and others (1993); Turner, Bowers, and Burgess (1995); Felger, Johnson, and Wilson (2001).

16. Webb 1996; Webb, Melis, and others (1999).

Chapter 1

1. Gregory and others (1991).
2. Carothers, Johnson, and Aitchison (1974).
3. Gregory and others (1991).
4. Cowardin and others (1979); D. Brown 1985; Mitsch and Gosselink (1993); Minckley and Brown (1994). For a map of Arizona showing these areas, see Brown, Carmony, and Turner (1981).
5. Hendrickson and Minckley (1985). Clearly, one person's ciénega can be another person's arroyo system, and we address that question later in this book.
6. Minckley (1973); Holden and Stalnaker (1975); Minckley and Deacon (1991); Rinne and Minckley (1991); Quartarone (1995).
7. Blinn and Cole (1991); Carothers and Brown (1991); Haden (1997); Shannon and Benenati (2002).
8. Meinzer (1927).
9. Malanson (1995); Ohmart (1996).
10. Sands and Howe (1977).
11. Strong and Bock (1990). Ohmart and Anderson (1982) suggest the amount of riparian vegetation in Arizona is closer to 1 percent.

12. McLaughlin (2004).
13. Trimble (1997).
14. Hendrickson and Minckley (1985).
15. Stromberg, Fry, and Patten (1997).
16. Stevens and others (1996).
17. Ohmart, Deason, and Freeland (1975); Guay (2001).
18. Cooper and others (1987); Naiman and Décamps (1997).
19. Haden (1997).
20. Meehan, Swanson, and Sedell (1977).
21. Lowe (1980), p. 153.
22. Lowe (1985).
23. Jakle and Gatz (1985).
24. Warren and Schwalbe (1985).
25. Naiman and Décamps (1997).
26. Suttkus, Clemmer, and Jones (1978).
27. Hanson (2001), p. 102.
28. Szaro and Belfit (1987).
29. Armstrong (1982).
30. Hanson (2001), p. 74.
31. McGrath and van Riper (2005).
32. K. Blair, U.S. Fish and Wildlife Service, personal communication (2004).
33. Hardy and others (2004).
34. Skagen and others (1998).
35. Carothers, Johnson, and Aitchison (1974); Johnson and Haight (1985).
36. Stevens and others (1977); Ohmart and Anderson (1982).
37. Hanson (2001), p. 14.
38. McGrath and van Riper (2005).
39. Brown, Carothers, and Johnson (1987).
40. Anderson, Ohmart, and Rice (1983).
41. Whitmore (1975).
42. D. Brown (1985).
43. Rice, Anderson, and Ohmart (1980).
44. Carothers, Johnson, and Aitchison (1974).
45. Naiman, Décamps, and Pollock (1993).
46. Carothers, Johnson, and Aitchison (1974); Ohmart and Anderson (1982); Ohmart (1996).
47. Lowe (1980).
48. Belsky, Matzke, and Uselman (1999).
49. Webb, Smith, and McCord (1991).
50. Hendrickson and Minckley (1984).
51. J. Stevens (1988).
52. Culler and others (1970); Gesink, Tomanek, and Hulett (1970); Burkham (1976a, 1976b).

53. Barnes (1967); Phillips and Ingersoll (1998); Phillips and others (1998).

54. Phillips and others (1998).

55. Meyboom (1964).

56. Bureau of Reclamation (1963); Babcock (1968); Culler and others (1970).

57. Blaney (1961), p. 37.

58. Ritzi, Bouwer, and Sorooshian (1985).

59. Brown, Carmony, and Turner (1977); Dick-Peddie and Hubbard (1977); Cowardin and others (1979); U.S. Fish and Wildlife Service (1998).

60. Brown, Carmony, and Turner (1977, 1981). The first-edition map was published in 1972.

61. Brown and Lowe (1980); D. Brown (1994).

62. Brown and Lowe (1974).

63. Arizona Department of Game and Fish (1993); see also http://www.land.state .az.us/alris/index.html, accessed November 4, 2005.

64. G. Williams (1978); Eschner, Hadley, and Crowley (1983).

65. W. Johnson (1994).

66. R. Hadley and others (1987).

67. W. Johnson (1992).

68. Schumm and Lichty (1963).

69. Osterkamp (1998).

70. Friedman, Osterkamp, and Lewis (1996a, 1996b).

71. Kay (1990).

72. Despain and others (1986).

73. Chadde and Kay (1991).

74. W. Johnson and others (1995).

75. Collier, Webb, and Schmidt (1996).

76. Hill and others (2002).

77. Brothers (1981).

78. Schultz (2001).

Chapter 2

1. Sellers, Hill, and Sanderson-Rae (1985).

2. Turner, Bowers, and Burgess (1995); Turner, Webb, and Bowers (2003).

3. W. Smith (1986).

4. Webb, Belnap, and Weisheit (2004).

5. Andrade and Sellers (1988); Webb and Betancourt (1992); Cayan and Webb (1992); Turner, Webb, and Bowers (2003); Hereford, Webb, and Graham (2002); Hereford, Webb, and Longpré (2004).

6. N. Roberts (1989), pp. 158–62.

7. Sheridan (1986).

8. Green and Sellers (1964); Sellers and Hill (1974); Sellers, Hill, and Sanderson-Rae (1985).

9. Webb and Betancourt (1992); Hereford and Webb (1992); Hereford, Webb, and Graham (2002); Hereford, Webb, and Longpré (2004); Turner, Webb, and Bowers (2003).

10. This peak discharge was approxi-

mately 10,500 ft³/s and is partially affected by flow regulation and diversions within the watershed.

11. Turner, Webb, and Bowers (2003).

12. Webb (1996).

13. G. M. Garfin, Institute for the Study of Planet Earth, University of Arizona, written communication (2006).

14. Hanks and Webb (in press).

15. Melis (1997).

16. Hefley (1937a, 1937b); Ohmart and Anderson (1982).

17. Brown, Carmony, and Turner (1981).

18. Winter and others (2002).

19. Webb and Betancourt (1992).

20. Webb and Betancourt (1992).

21. Rantz and others (1982).

22. For information on paleoflood hydrology, see the chapters in House and others (2002).

23. Cayan and Webb (1992); Webb and Betancourt (1992).

24. The partial-duration series represents peaks above a base discharge that is fixed for a given gaging station. More than one peak can be recorded per year in the partial-duration series.

25. Webb and Betancourt (1992).

26. Cayan and Webb (1992).

27. The data for Gila River stage depicted in figure 2.2D were calculated by applying the most recent stage-discharge record, effective in this case from 1993 to the present, to the entire daily discharge record. The most accurate method would be to apply the stage-discharge relation effective at the time to the daily discharges. Owing to channel changes, this liberty results in an unknown error, which may be largely masked in a duration analysis, where discharges are grouped and are not continuous.

28. Caldwell, Dawson, and Richards (1998).

29. See Kupel (2003).

30. Renard and others (1964).

31. Aldridge and Hales (1984), p. 36.

32. Aldridge and Eychaner (1984).

33. Aldridge and Hales (1984).

34. Konieczki and Heilman (2004).

35. Konieczki and Heilman (2004).

36. Konieczki and Heilman (2004).

37. Scott, Shafroth, and Auble (1999).

38. Stromberg and others (1992).

39. Judd and others (1971).

40. Shafroth, Stromberg, and Patten (2000).

41. Scott, Shafroth, and Auble (1999).

42. Scott, Lines, and Auble (2000).

43. Shafroth, Stromberg, and Patten (2000).

44. Albertson and Weaver (1945).

45. Theis (1940); Bredehoeft, Papadopulus, and Cooper (1982).

46. DeBano and others (2004).

Chapter 3

1. D. Williams and others (1998).

2. Spigler (2000).

3. Felger, Johnson, and Wilson (2001).

4. Kearney and Peebles (1960).

5. Benson and Darrow (1981); Albee, Shultz, and Goodrich (1988).

6. Patten (1998).

7. Benson and Darrow (1981), pp. 363–66.

8. Clark (1987).

9. Hanson (2001), p. 59.

10. Kearney and Peebles (1960); Welsh and others (1993).

11. Felger, Johnson, and Wilson (2001), p. 298.

12. Formerly known as *Salix taxifolia*; Felger, Johnson, and Wilson (2001).

13. Stromberg (1993a).

14. Albee, Shultz, and Goodrich (1988).

15. Felger, Johnson, and Wilson (2001).

16. Stromberg (1993a).

17. Rea (1997), pp. 194–95.

18. Albee, Shultz, and Goodrich (1988).

19. Müller-Schwarze and Sun (2003), pp. 107–9.

20. Benson and Darrow (1981), p. 361.

21. Felger, Johnson, and Wilson (2001).

22. Kearney and Peebles (1960).

23. Felger, Johnson, and Wilson (2001).

24. Albee, Shultz, and Goodrich (1988).

25. Benson and Darrow (1981), p. 186.

26. Turner, Bowers, and Burgess (1995), pp. 326–33.

27. Benson and Darrow (1981), pp. 243–44.

28. Peacock and McMillan (1965).

29. Glinski and Brown (1982).

30. Webb (1996). Repeat photography is the only reliable means for dating mesquite because mesquite does not reliably produce annual rings suitable for dendrochronology.

31. Turner, Bowers, and Burgess (1995), pp. 330–32.

32. Judd and others (1971).

33. Felger, Johnson, and Wilson (2001), p. 207.

34. Turner, Bowers, and Burgess (1995), pp. 329–30.

35. Felger, Johnson, and Wilson (2001), pp. 207–8.

36. Judd and others (1971); Minckley and Clark (1984); Stromberg (1993b); Felger, Johnson, and Wilson (2001).

37. Benson and Darrow (1981), p. 223.

38. Webb (1996).

39. Benson and Darrow (1981); Albee, Shultz, and Goodrich (1988).

40. DeBolt and McCune (1995).

41. Webb, Belnap, and Weisheit (2004).

42. Salzer and others (1996).

43. Albee, Shultz, and Goodrich (1988).

44. Rea (1997), pp. 145–49.

45. Busch and Smith (1995).

46. Leopold (1966).

47. According to www.usda.gov, *Baccharis glutinosa* is considered synonymous with *B. salicifolia.* But *B. glutinosa* is the most commonly used species name in the literature on this group.

48. *Baccharis emoryi* and *B. salicifolia.*

49. Benson and Darrow (1981); Albee, Schultz, and Goodrich (1988).

50. Meinzer (1927), p. 75.

51. Turner, Bowers, and Burgess (1995), p. 236.

52. Benson and Darrow (1981).

53. Gould (1951), pp. 93–94.

54. Felger (2000).

55. Albee, Schultz, and Goodrich (1988).

56. Benson and Darrow (1981).

57. Huseman (1995), p. 138.

58. Felger (2000).

59. Both *Typha latifolia* and *T. domingensis* occur in the southwestern United States.

60. Felger (2000).

61. Albee, Schultz, and Goodrich (1988), p. 1.

62. Benson and Darrow (1981), p. 144.

63. Benson and Darrow (1981), p. 371.

64. Albee, Schultz, and Goodrich (1988); Webb, Belnap, and Weisheit (2004).

65. Braatne, Rood, and Heilman (1996); Shafroth and others (1998). Stromberg (1993a, 1993b) reviews the ecology and life-history traits of Frémont cottonwood, black willow, and mesquite.

66. Bessey (1904). Reichenbacher (1984) summarizes seed production for cottonwood, willow, and mesquite.

67. Stromberg (1993b).

68. Warren and Turner (1975).

69. Braatne, Rood, and Heilman (1996).

70. Braatne, Rood, and Heilman (1996), p. 63.

71. Stromberg, Patten, and Richter (1991).

72. Reichenbacher (1984); Everitt (1995). See the generalized diagram figure 4 in Braatne, Rood, and Heilman (1996), p. 64.

73. Moss (1938). Fenner, Brady, and Patton (1984) provide data on seed viability of Frémont cottonwood in the Salt River Valley of Arizona. Reichenbacher (1984) summarizes the state of knowledge on seed longevity of *Populus, Salix,* and *Prosopis.*

74. Braatne, Rood, and Heilman (1996).

75. Siegel and Brock (1990).

76. Gladwin and Roelle (1998).

77. Stromberg (1993b); Turner, Bowers, and Burgess (1995).

78. Scott, Friedman, and Auble (1996); Shafroth and others (1998).

79. Everitt (1968).

80. Zimmerman (1969), p. 29.

81. Shafroth and others (1998).

82. Braatne, Rood, and Heilman (1996), p. 63.

83. For information on establishment or regeneration of Frémont cottonwood, see Fenner, Brady, and Patton (1985); Everitt (1995); and D. Cooper and others (1999). For plains cottonwood, see Shafroth, Auble, and Scott (1995); Scott, Friedman, and Auble (1996); Auble and Scott (1998). For a variety of cottonwood species, see Mahoney and Rood (1998); Rood, Kalischuk, and Mahoney (1998).

84. Rood, Kalischuk, and Mahoney (1998).

85. D. Cooper and others (1999).

86. Turner, Bowers, and Burgess (1995); Wilson, Webb, and Thompson (2001).

87. R. Turner (1963).

88. Turner, Webb, and Bowers (2003).

89. Siegel and Brock (1990).

90. Gary (1963).

91. Dawson and Ehleringer (1991).

92. Caldwell, Dawson, and Richards (1998).

93. McQueen and Miller (1972).

94. T. Robinson (1958); Tromble (1972).

95. Gary (1963).

96. Friedman, Vincent, and Shafroth (2005).

97. J. Phillips and others (1998), p. 10.

98. Shafroth, Stromberg, and Patten (2000).

99. Stromberg (1993a); Stromberg, Tiller, and Richter (1996).

100. Lite and Stromberg (2005).

101. Lines (1999).

102. Leffler and Evans (1999).

103. Potts (2002).

104. One example of this is in the delta of Lake Mead in western Grand Canyon, where high lake levels in 1997 and 1998 killed many black willow trees established on emergent deltaic sediments.

105. Clark (1987).

106. A review of mesquite rooting depths appears in Turner, Bowers, and Burgess (1995).

107. Stromberg, Tiller, and Richter (1996).

108. Drexler and others (2004).

109. Fenstermaker (2003); Nagler, Glenn, and Thompson (2003); Nagler, Cleverly, and others (2005); Nagler, Glenn, and others (2005). For a review of the most commonly used methods, see Drexler and others (2004).

110. Snyder and Williams (2000).

111. R. Scott and others (2004).

112. Lines (1999).

113. Dahm and others (2002).

114. Schaeffer and Williams (1998).

115. Schaeffer, Williams, and Goodrich (2000).

116. Schaeffer, Williams, and Goodrich (2000).

117. Nagler, Glenn, and Thompson (2003).

118. Spigler (2000).

119. Cleverly and others (1997).

120. Stromberg, Wilkins, and Tress (1993).

121. Jarrell and Virginia (1990); Scott and others (2004).

122. Tromble (1972).

123. Glenn and Nagler (2005).

124. Gay and Sammis (1977).

125. Johnson and Simpson (1988); Stromberg (1993a, 1993b, 2001); Braatne, Rood, and Heilman (1996), p. 75.

126. Fleischner (1994); Ohmart (1996); Feller (1998); Belsky, Matzke, and Uselman (1999); Saar (2002); Clary and Kruse (2004).

127. Reichenbacher (1984).

128. Lusby, Reid, and Knipe (1971); Ohmart (1996).

129. Skartvedt (2000).

130. Platts and Nelson (1985).

131. Brookshire and others (2002).

132. Auble and Scott (1998).

133. Glinski (1977).

134. Schulz and Leininger (1990).

135. Glinski (1977).

136. Skartvedt (2000); Brookshire and others (2002).

137. Reichenbacher (1984).

138. Krueper (1993); Saar (2002).

139. Note the extensive discussion of the history and impacts of livestock grazing in southern Arizona reported in Turner, Webb, and Bowers (2003).

140. Ohmart (1996).

141. McAuliffe (1997).

142. Stromberg, Patten, and Richter (1991).

143. Stromberg and others (1993).

144. One vivid description of flood impacts and resultant reorganization of riparian ecosystems is given in Hefley (1937b) for the Canadian River in Oklahoma. J. Phillips and others (1998) show response of riparian and xerophytic vegetation to flooding in central Arizona.

145. Minckley and Clark (1984).

146. Griffin and Smith (2004); J. D. Smith (2004).

147. Scott, Auble, and Friedman (1997).

148. Sparks and others (1990).

149. Stromberg and others (1993); Scott, Auble, and Friedman (1997); Rood, Kalischuk, and Mahoney (1998). Saar (2002) presents a conceptual model of riparian vegetation response to the combination of grazing history and flooding.

150. Friedman and Lee (2002).

151. Junk, Bayley, and Sparks (1989); Sparks and others (1990); Malanson (1995); Ohmart (1996).

152. Junk, Bayley, and Sparks (1989); Bornette and Amoros (1996).

153. Grimm (1993).

154. R. Turner (1974).

155. Irvine and West (1979).

156. Irvine and West (1979).

157. Webb, Belnap, and Weisheit (2004).

158. Campbell and Green (1968).

159. Webb (1985); Webb and Baker (1987); Betancourt (1990).

160. Several authors have summarized the possible causes of arroyo downcutting: see Hastings and Turner (1965); Cooke and Reeves (1976); W. Graf (1983); Webb (1985); Turner, Webb, and Bowers (2003).

161. A summary of grazing history in the Sonoran Desert appears in Turner, Webb, and Bowers (2003).

162. Hereford and Webb (1992); Hereford, Webb, and Graham (2002); Hereford, Webb, and Longpré (in press).

163. Waters (1985, 1988).

164. According to the classic study by Albertson and Weaver, whereas 90 to 100 percent of trees died in upland environments during the Dust Bowl in the upper Midwest, "death of trees varied greatly but the average was about 10 percent" along perennial streams (1945, 420). Along ephemeral streams, the amount of death was higher (about 50 percent).

165. Webb (1985); Webb and Baker (1987); Betancourt (1990); Ely and others (1993); Turner, Webb, and Bowers (2003).

166. Stahle and others (1998); Hereford (2002).

167. Cooke and Reeves (1976).

168. Hereford and Webb (1992); Hereford, Webb, and Graham (2002); Turner, Webb, and Bowers (2003).

169. Webb and Betancourt (1992).

170. Bull and Scott (1974); Graf (1979).

171. Dobyns (1981); Parker and others (1985).

172. Hereford (1984, 1993).

173. Bryan (1928).

174. Davis (1986).

175. Ohmart (1996) provides an extensive review of the effects of beaver on riverine ecosystems, but none of the studies is from the Southwest.

176. Müller-Schwarze and Sun (2003), 125.

177. Ohmart (1996), p. 267.

178. Apple (1985).

179. Hanson (2001), p. 123.

180. Bailey and others (2004).

181. Müller-Schwarze and Sun (2003).

182. Müller-Schwarze and Sun (2003), p. 109.

183. Mearns (1907), p. 395.

184. Pattie (2001).

185. Hafen (1997).

186. Cleland (1950).

187. Turner, Webb, and Bowers (2003).

188. Busch (1995).

189. Webb, Belnap, and Weisheit (2004).

190. Busch (1995).

191. Turner, Webb, and Bowers (2003).

192. Salzer (2000).

193. Turner, Webb, and Bowers (2003).

194. DeBano, Brejda, and Brock (1984).

195. See Lite and Stromberg (2005), table 8, p. 163, for a review of groundwater depth and lowering-rate thresholds for survival of Frémont cottonwood and black willow.

196. Scott, Shafroth, and Auble (1999).

197. Shafroth, Stromberg, and Patten (2000).

198. Scott, Shafroth, and Auble (1999); Scott, Lines, and Auble (2000).

199. Shafroth, Stromberg, and Patten (2000).

200. Stromberg and others (1992).

201. Judd and others (1971).

202. Collier, Webb, and Schmidt (1996).

203. Magilligan, Nislow, and Graber (2003).

204. W. Johnson (1998).

205. Friedman and others (1998).

206. Allred and Schmidt (1999).

207. Magilligan, Nislow, and Graber (2003).

208. Fenner, Brady, and Patton (1985).

209. Molles and others (1998).

210. Merritt and Cooper (2000). Stromberg, Tiller, and Richter (1996) predicted "desertification" of floodplains in the event of sustained groundwater withdrawal instead of flow regulation.

211. Salzer and others (1996).

212. Fenner, Brady, and Patton (1985).

213. S. D. Smith and others (1991).

214. Molles and others (1998); Stromberg (2001); Scheurer and Molinari (2003).

215. Sher, Marshall, and Taylor (2002) found that if cottonwood and willow germinate at the same time as tamarisk, the native species can outcompete the nonnative.

216. Friedman, Osterkamp, and Lewis (1996a, 1996b).

217. Brady and others (1985).

218. Campbell and Green (1968).

Chapter 4

1. As discussed in this chapter, numerous authors see extreme problems with nonnative vegetation established in riparian ecosystems of the southwestern United States. This characterization is particularly true with regard to aquatic ecosystems and native fishes (see chapter 1). In contrast, Burdick (2005) espouses the view that larger ecosystems can tolerate substantial numbers of nonnative species without severe consequences to native species.

2. S. Johnson (1987); Walker and Smith (1997); Zavaleta (2000a, 2000b); Krza (2003).

3. Shafroth and others (2005).

4. Zavaleta (2000a, 2000b).

5. Hughes (1993).

6. Busby and Schuster (1971); Carman

and Brotherson (1982); Brotherson and Winkel (1986); Di Tomaso (1998).

7. T. Robinson (1965); Crins (1989).

8. T. Robinson (1965), p. A3.

9. Kearney and Peebles (1960), p. 557.

10. Baum (1967); Benson and Darrow (1981), pp. 96–98.

11. Baum (1966) is considered to be the authoritative reference on this synonymy. Turner, Bowers, and Burgess consider *T. chinensis* and *T. ramosissima* to be synonymous, but choose to give *T. chinensis* priority (1995, 384–86). Warren and Turner (1975), J. S. Horton (1977), W. Graf (1978), Everitt (1980), Turner and Karpiscak (1980), and Hickman (1993) follow this usage. In contrast, Sanchez (1975); Benson and Darrow (1981); Brotherson and Winkel (1986); Albee, Shultz, and Goodrich (1988); L. Stevens (1989); Felger (2000); Felger, Johnson, and Wilson (2001); and Tellman (2002) use *T. ramosissima*. Note the discussion of the taxonomy of tamarisk by Brock (1994), who follows Baum (1978) and considers both *T. chinensis* and *T. ramosissima* as occurring in the region, and by Friedman and others (2005).

12. Everitt and DeLoach (1990).

13. Gaskin and Schaal (2002, 2003); DeLoach and others (2003); Lewis, DeLoach, Herr, and others (2003); Schaal, Gaskin, and Caicedo (2003).

14. Hefley (1937a).

15. J. S. Horton (1964), p. 2; T. Robinson (1965), p. 3.

16. Betancourt (1990), pp. 78–79.

17. Christensen (1962); J. S. Horton (1964); Harris (1966); T. Robinson (1965); Baum (1967); Brotherson and Winkel (1986); Di Tomaso (1998).

18. T. Robinson (1965), p. 4; Webb (1996), p. 112.

19. T. Robinson (1965), pp. 4–5.

20. Tidestrom (1925). The species is not listed in a previous flora of the region (Merriam 1893).

21. Webb, Belnap, and Weisheit (2004).

22. Christensen (1962); Welsh and others (1993).

23. Clover and Jotter (1944); Woodbury and Russell (1945).

24. Carothers and Brown (1991), pp. 120–21; Webb, Belnap, and Weisheit (2004).

25. Everitt (1980).

26. Everitt (1998).

27. T. Robinson (1965), 4.

28. R. Turner (1974).

29. W. Graf (1978), p. 1495.

30. Grinnell (1914) does not list tamarisk as an encountered species during his expedition of 1910.

31. Harris (1966).

32. T. Robinson (1965), p. 6.

33. W. Graf (1982).

34. R. Turner (1974), 7.

35. Christensen (1962). Gesink, Tomanek, and Hulett (1970) report widespread establishment on the Arkansas River in Kansas followed the 1921 flood on that river.

36. T. Robinson (1965).

37. Sanchez (1975).

38. Di Tomaso (1998). Friederici (1995) gives a figure of 1 million acres with no reference as to its source.

39. Turner and Karpiscak (1980) proposed this mechanism for Russian olive in Grand Canyon.

40. Webb (1985); Betancourt (1990).

41. W. Graf (1978). We dispute the photographic evidence of first arrival of tamarisk contained in this publication. Also see Everitt (1979).

42. Waisel (1960a); Hickman (1993).

43. Thornber (1916); Benson and Darrow (1981), p. 97.

44. Waisel (1960a); Felger (2000); Felger, Johnson, and Wilson (2001).

45. Waisel (1960a).

46. Resource managers at Lake Mead National Recreation Area have observed seedlings of Athel tamarisk in their park.

47. Gaskin and Shafroth (2005).

48. Zimmerman (1969), p. D40.

49. Welsh and others (1993).

50. Benson and Darrow (1981), p. 282.

51. Knopf and Olson (1984); Shafroth, Auble, and Scott (1995); Katz and Shafroth (2003); Friedman and others (2005).

52. Friedman and others (2005).

53. Christensen (1963).

54. Katz and Shafroth (2003).

55. Welsh and others (1993).

56. Loope and others (1988).

57. Harlan and Dennis (1976).

58. Kearney and Peebles (1960).

59. Turner and Karpiscak (1980); Olson and Knopf (1986b).

60. R. Johnson (1977).

61. Munz (1974), p. 395; Hickman (1993).

62. Shafroth, Auble, and Scott (1995).

63. Carman and Brotherson (1982); Knopf and Olson (1984).

64. Katz, Friedman, and Beatty (2001); Katz and Shafroth (2003).

65. Olson and Knopf (1986a).

66. Felger, Johnson, and Wilson (2001), p. 311.

67. Welsh and others (1993), p. 687.

68. Albee, Shultz, and Goodrich (1988).

69. Tellman (2002), p. 36.

70. Stevens and Ayers (2002).

71. Benson and Darrow (1981), p. 194.

72. Rea (1997).

73. Loope and others (1988).

74. Benson and Darrow (1981).

75. Felger (2000).

76. Tracy and DeLoach (1999).

77. Christensen (1963); Turner and Karpiscak (1980. We observed elms along the middle reaches of the Little Colorado River near Holbrook, Arizona.

78. Stevens and Ayers (2002).

79. Of the many reviews of tamarisk life history, we particularly note T. Robinson (1965); Brock (1994); and Glenn and Nagler (2005).

80. Crins (1989).

81. Stromberg (1998b).

82. Brotherson and Winkel (1986); S. D. Smith and others (1998).

83. J. S. Horton (1977).

84. J. S. Horton (1977); Busch, Ingraham, and Smith (1992); Cleverly and others (1997); Devitt, Sala, and others (1997); S. D. Smith and others (1998); J. L. Horton (2001).

85. Merkel and Hopkins (1957); see additional references in Brock (1994).

86. Warren and Turner (1975); see also Waring and Stevens (1988).

87. Warren and Turner (1975); Walker and Smith (1997); Zavaleta (2000a, 2000b).

88. Merkel and Hopkins (1957).

89. J. S. Horton (1977).

90. R. Turner (in press).

91. Lovich and de Gouvenain (1998).

92. Merkel and Hopkins (1957).

93. Gladwin and Roelle (1998).

94. Vandersande, Glenn, and Walworth (2001).

95. Vandersande, Glenn, and Walworth (2001).

96. Stromberg (1998b).

97. Waisel (1960a).

98. Wilkinson (1966); Ginzburg (1967).

99. Brock (1994).

100. Katz, Friedman, and Beatty (2001).

101. Loope and others (1988).

102. Sanchez (1975).

103. Merkel and Hopkins (1957).

104. Gary (1960).

105. L. Stevens (1989).

106. Friedman, Vincent, and Shafroth (2005).

107. Hereford (1984, 1989); Allred and Schmidt (1999); Friedman, Vincent, and Shafroth (2005).

108. Hem (1967) reported on *Tamarix ramosissima*; Berry (1970) reported on *Tamarix aphylla*; both researchers reported generally similar results.

109. Berry (1970); Busch and Smith (1995).

110. Di Tomaso (1998).

111. Hem (1967).

112. Jackson, Ball, and Rose (1990); B. Anderson (1996).

113. Gary (1963).

114. Gary (1965).

115. Merkel and Hopkins (1957); Gary (1963).

116. Gries and others (2003).

117. Busch and Smith (1995).

118. Shafroth, Stromberg, and Patten (2000).

119. Van Hylckama (1970).

120. Devitt, Piorkowski, and others (1997).

121. S. D. Smith and others (1996).

122. Root distribution diagrams in Gary (1963, p. 313) suggest that tamarisk roots may be unable to penetrate even thin clay layers.

123. Gary (1963).

124. Webb (1996); Webb, Belnap, and Weisheit (2004).

125. Irvine and West (1979).

126. Webb, Belnap, and Weisheit (2004).

127. Vandersande, Glenn, and Walworth (2001).

128. Warren and Turner (1975).

129. Stevens and Waring (1988). For a brief summary of their findings, see R. Johnson (1991), pp. 192–93.

130. Ellis, Crawford, and Molles (1998).

131. Kennedy and Hobbie (2004).

132. Kennedy, Finlay, and Hobbie (in press).

133. Shafroth and others (2005).

134. Soltz and Naiman (1978).

135. T. Robinson (1958).

136. One experimental measurement calculated an average annual water use of 234 inches (Decker, Gaylor, and Cole [1962]), which appears to be unusually high given other estimates. Another measurement suggests that water use in July was up to 0.47 inch per day with an annual water use of 68 inches per year. Water use by tamarisk near Phoenix, Arizona, reportedly is equivalent to the highest evaporation rates measured in the United States (Van Hylckama 1974).

137. Tromble (1972); S. D. Smith (1989); Glenn and others (1998); Nagler, Glenn, and Thompson (2003); Glenn and Nagler (2005). Dahm and others (2002) report tamarisk water use as density dependent, and the riparian stands with the highest water use that they measured in New Mexico were a "dense stand" of tamarisk and a cottonwood-tamarisk stand with a maximum water use that averaged about 48 inches per year.

138. Waisel (1960b); J. Anderson (1982); Gay (1985); Gay and Sammis (1977); and Di Tomaso (1998). See the chart comparing tamarisk water use with other species in Blaney (1961), p. 41.

139. J. Anderson (1982); Busch and Smith (1995); Horton (2001); Nagler, Glenn, and Thompson (2003). J. Anderson (1977) provides an extensive discussion of transpiration rates in tamarisk and their variability.

140. A. Robinson (1970); van Hylckama (1970); Davenport, Martin, and Hagan (1982); Brock (1994).

141. Fenstermaker (2003).

142. Van Hylckama (1970, 1974). In China, growth and water use of *Tamarix ramosissima* on dunes in a hyperarid environment is less related to water level than is *Populus euphratica* (Gries and others 2003).

143. Glenn and others (1998); Glenn and Nagler (2005).

144. Devitt, Sala, and others (1997).

145. S. D. Smith and others (1996); Sala, Smith, and Devitt (1996); Zavaleta (2000a, 2000b).

146. Culler and others (1982).

147. Dahm and others (2002).

148. J. Phillips and others (1998), p. 10. The species of willow is not identified; identification of black willow, as opposed to other possible species tested, was made on the basis of photographs in J. Phillips and others (1998).

149. Lesica and Miles (2001); Sexton, McKay, and Sala (2002).

150. Friedman and others (2005).

151. Sexton, McKay, and Sala (2002).

152. Friedman and others (2005).

153. Laney (1977).

154. R. Hadley (1961); W. Graf (1978), pp. 1499–501.

155. Everitt (1979); Allred and Schmidt (1999), p. 1757.

156. R. Turner (1974); Carothers and Brown (1991).

157. W. Graf (1978); Stephens and Shoemaker (1987).

158. Merritt and Cooper (2000).

159. Carothers and Brown (1991), p. 141.

160. Stevens and Ayers (2002); B. T. Brown, written communication (2003).

161. Hefley (1937a), p. 391, (1937b).

162. Hopkins and Carruth (1954).

163. Glinski and Ohmart (1984).

164. C. J. DeLoach, personal communication (2004).

165. Cohan, Anderson, and Ohmart (1978).

166. Anderson, Higgins, and Ohmart (1977).

167. Hunter, Ohmart, and Anderson (1988); Ellis (1995).

168. Zavaleta (2000a, 2000b).

169. Anderson, Higgins, and Ohmart (1977).

170. Anderson, Higgins, and Ohmart (1977); Anderson, Ohmart, and Disano (1979); Carothers and Brown (1991), pp. 10–105, 148–54.

171. B. Brown (1992).

172. Bristow (1968); Haase (1972); Anderson, Higgins, and Ohmart (1977).

173. Yard and others (2004).

174. Unitt (1987).

175. B. Brown (1988); Owen, Sogge, and Kern (2005).

176. Brown and Trosset (1989); Stevens and Ayers (2002); Owen, Sogge, and Kern (2005); B. T. Brown, written communication (2003).

177. DeLoach and others (2003).

178. Yong and Finch (1997).

179. R. Turner (2003); Turner, Webb, and Bowers (2003).

180. Phillips, Marshall, and Monson (1964), p. 212.

181. Knopf and Olson (1984).

182. Knopf and Olson (1984); Olson and Knopf (1986b).

183. Brock (1994).

184. Anderson, Higgins, and Ohmart (1977); Anderson, Ohmart, and Disano (1979); Hunter, Anderson, and Ohmart (1985); Hunter, Ohmart, and Anderson (1988).

185. Jakle and Gatz (1985); Zavaleta (2000a, 2000b).

186. R. H. Webb, personal observation (1999); Stevens and Ayers (2002); Webb, Belnap, and Weisheit (2004).

187. Lesica and Miles (2004).

188. Shafroth and others (2005).

189. B. Anderson (1996).

190. See the extensive review on control methods for tamarisk in Brock (1994) and DeLoach and others (1999).

191. Mahoney and Rood (1998); S. D. Smith and others (1998); Kearsley and Ayers (1999); Sher, Marshall, and Taylor (2002).

192. Krza (2003).

193. McDaniel and Taylor (2003).

194. Anderson, Higgins, and Ohmart (1977).

195. J. S. Horton (1977); Busch and Smith (1993); S. D. Smith and others (1998); Di Tomaso (1998).

196. Busch and Smith (1993).

197. Hughes (1993); McDaniel and Taylor (2003).

198. S. D. Smith (1989).

199. Culler and others (1970).

200. Olson and Knopf (1986a).

201. Manning, Cashore, and Szewczak (1996).

202. Tracy and DeLoach (1999).

203. DeLoach and others (1996, 2003); Milbrath, DeLoach, and Knutson (2003).

204. DeLoach (1997); Gould and DeLoach (2002); DeLoach and others (2003); Lewis, DeLoach, Knutson, and others (2003); Lewis, DeLoach, Herr, and others (2003).

205. DeLoach and others (2003); Milbrath, DeLoach, and Knutson (2003).

206. DeLoach and others (2003).

207. DeLoach and others (1999).

208. Besides tamarisk, biological control has high potential for giant reed; see Tracy and DeLoach (1999); Gould and DeLoach (2002).

209. Friederici (1995).

210. Sher, Marshall, and Taylor (2002).

211. L. Stevens (1989); Webb (1996); Webb, Belnap, and Weisheit (2004).

212. L. Stevens (1989), p. 100.

213. Kennedy and others (2002).

214. Lesica and Miles (2001).

215. Stromberg (1998a); also see Levine and Stromberg (2001).

216. Hughes (1993).

217. T. Robinson (1965).

Chapter 5

1. Naef and Wood (1975).

2. Taft (1938), p. 6.

3. Rolle credits von Egglofstein as the daguerreotypist (1991, pp. 153, 158), and Taft incorrectly credits S. N. Carvalho in that role (1938, pp. 262–66).

4. Taft (1938), p. 266; Crosby (1965).

5. Huseman (1995).

6. Taft (1938), pp. 118–19.

7. Daniels (1968); Rowe (1997).

8. Palmquist and Kailbourn (2000), pp. 201–3.

9. Palmquist and Kailbourn (2000), pp. 201.

10. Two men with the name William A. Bell worked as pioneer photographers in the region as described in this section, one in southern Arizona (1867) and one in northern Arizona (1872).

11. Bell (1869).

12. Naef and Wood (1975), pp. 52–55; Stephens and Shoemaker (1987).

13. Simmons and Simmons (1977), p. 35.

14. Wheeler (1983); Naef and Wood (1975), pp. 125–66. For an artistic interpretation of Timothy O'Sullivan's work, see Kelsey (2000).

15. Fowler (1976, 1989).

16. For examples of lithographs made from photographs by Beaman, Fennemore, C. Powell, and Hillers, see Dutton (1882) and J. Powell (1961, 1972). Dellenbaugh (1908) published many of the actual photographs.

17. Waitley (1999).

18. As an example of trading among pioneer photographers, one of Jackson's San Juan River photographs became part of the Hillers's collection (number 691). Someone (probably Frederick Dellenbaugh) surprisingly labeled this view as being on the Green River in Desolation Canyon (chapter 6), even though the Powell Expedition did not go to the San Juan River.

19. Wittick (1973).

20. Daniels (1968).

21. Naef and Wood (1975), 79–124.

22. Smith and Crampton (1987); Webb (1996).

23. Hastings and Turner (1965); Turner, Webb, and Bowers (2003).

24. Webb, Smith, and McCord (1991); Wheeler ([1875] 1983).

25. Hattersly-Smith (1966). The photographer was Sebastian Finsterwalder, who mapped changes of glaciers in the Alps in 1888–1889.

26. Webb, Belnap, and Weisheit (2004).

27. Shantz and Turner (1958).

28. W. Phillips (1963); Hastings and Turner (1965); Rogers, Malde, and Turner (1984).

29. Turner, Ochung', and Turner (1998); Rohde (1997).

30. Stephens and Shoemaker (1987). Note that the dark tent appears in camera stations 605, 850, 858, 876, 886, and 894.

31. Webb (1996).

32. Taft (1938).

33. D. Smith (1965), p. 101. Smith and Crampton give a list of photographic supplies for the second Stanton Expedition in the winter of 1889–1890 (1987, p. 97).

34. Fowler (1989) provides lists of the Hillers's views, many of which are mislabeled. See Webb, Belnap, and Weisheit (2004).

35. Webb, Belnap, and Weisheit (2004).

36. Daubenmire (1968), p. 102.

37. Use of trade names is for descriptive purposes only and does not imply endorsement by the U.S. Geological Survey.

38. Rogers, Malde, and Turner (1984).

39. Readers interested in the repeat-photography techniques used for point monitoring should see Hall (2002a, 2002b).

40. Gruell (1980, 2001); Rogers (1982); Klett and others (1984); Sallach (1986); Humphrey (1987); K. Johnson (1987); Stephens and Shoemaker (1987); R. Turner (1990); McGinnies, Shantz, and McGinnies (1991); Veblen and Lorenz (1991); Webb, Smith, and McCord (1991); Gordon, Parrott, and Smith (1992); Wright and Bunting (1994); Melis and others (1996); Webb (1996); Meagher and Houston (1998); Klement, Heitschmidt, and Kay (2001); Noel and Fielder (2001); Skovlin and others (2001); Webb, Boyer, and Berry (2001); Klett and others (2004); Webb, Belnap, and Weisheit (2004); Webb and Leake (2006). A rather superficial treatment of vegetation change in the western United States, based on repeat photography of Timothy O'Sullivan views, is given in Salamun (1990). An interesting repeat-photography project involving recovery of rivers that were severely disturbed by mining activities in the Sierra Nevada of California is given in J. Turner (1983).

41. Rogers, Malde, and Turner (1984).

42. These figures change rapidly, and this calculation was made on August 1, 2004.

43. Turner, Ochung', and Turner (1998) followed Shantz and Turner (1958).

44. Turner and Karpiscak (1980); Stephens and Shoemaker (1987); Webb, Smith, and McCord (1991); Baars and Buchanan (1994); Melis and others (1996);

Webb (1996); Webb, Melis, and others (1999).

45. Webb and others (2003).

46. Webb, Melis, and others (1999).

47. Bahre (1991).

48. Hereford (1984); Knapp, Warren, and Hutchinson (1990). R. Hereford (personal communication, 2005) asserts that tamarisk was not present in the 1930s along the Little Colorado River. This evidence was obtained from aerial photography, which could not detect the small plants that Colton (1937) documented (see chapter 10).

49. Turner, Webb, and Bowers (2003), p. 56.

50. R. Turner (1974) provides an excellent example of the conjunctive use of historical maps, repeat photography, and aerial photography to assess change quantitatively in riparian vegetation along the upper Gila River.

51. In particular, note the aerial photography discussed in Hereford (1984) for the Little Colorado River (chapter 10); the aerial photography presented in Klawon (2000) for the Big Sandy and Santa Maria Rivers (chapter 15); the discussion in Shafroth and others (1998) and in Stromberg and others (2004) of the work for the Bill Williams River (chapter 15); and the aerial photography presented in Klawon (2001) for the upper Gila River (chapter 17).

52. Kepner and others (2000); Kepner, Edmonds, and Watts (2002).

53. Nagler, Glenn, and others (2005); Nagler, Scott, and others (2005).

Chapter 6

1. Mutschler (1979).

2. Holden and Stalnaker (1975).

3. Holden and Stalnaker (1975).

4. J. Powell (1961).

5. Clover and Jotter (1944), p. 606.

6. Knipmeyer (2002), p. 19.

7. Bishop (1947); Sumner (1947); W. Powell (1949).

8. Webb, Belnap, and Weisheit (2004), 102.

9. Stephens and Shoemaker (1987).

10. W. Graf (1978); Allred and Schmidt (1999).

11. Webb, Belnap, and Weisheit (2004).

12. Orchard and Schmidt (1998).

13. Orchard and Schmidt (1998).

14. Stockton and Jacoby (1976), 31.

15. Orchard and Schmidt (1998).

16. K. Thompson (1984b).

17. Hereford and Webb (1992); Hereford, Webb, and Graham (2002).

18. W. Graf (1978).

19. Allred and Schmidt (1999).

20. Allred and Schmidt (1999).

21. W. Graf (1978); Allred and Schmidt (1999).

22. Orchard and Schmidt (1998).

23. Orchard and Schmidt (1998).

24. Allred and Schmidt (1999).

25. Webb, Belnap, and Weisheit (2004).

26. Webb, Belnap, and Weisheit (2004).

27. W. Graf (1978).

28. Allred and Schmidt (1999).

Chapter 7

1. Webb, Belnap, and Weisheit (2004).

2. Valdez, Ryel, and Williams (1986).

3. Clover and Jotter (1944).

4. Knipmeyer (2002), p. 19.

5. Sumner (1947).

6. Webb, Belnap, and Weisheit (2004).

7. Webb, Belnap, and Weisheit (2004) give a more complete interpretation of changes to the Colorado River as revealed in repeat photography. In this chapter, we summarize their results for changes in riparian vegetation.

8. K. Thompson (1984a).

9. Van Steeter and Pitlick (1998).

10. Stewart, Cayan, and Dettinger (2004).

11. Salzer and others (1996).

12. Stockton and Jacoby (1976).

13. This value is the so-called virgin or natural flow estimated for the Colorado River and accounting for the dams' influence on the Gunnison and Dolores Rivers. The data are from the Bureau of Reclamation, written communications (1990, 2005).

14. Stockton and Jacoby (1976), p. 20.

15. There is no gaging station in Cataract Canyon. Flow records for the canyon are calculated as the sum of daily flow of the Green River at the town Green River and the Colorado River near Cisco. Lag effects associated with travel time from these gaging stations to Cataract Canyon are not considered in this rudimentary calculation.

16. K. Thompson (1984a).

17. Van Steeter and Pitlick (1998).

18. Snyder and Miller (1992).

19. Kriegshauser and Somers (2004).

20. W. Graf (1978).

21. Van Steeter and Pitlick (1998).

22. Hindley and others (2000), pp. 56–57, 58–59.

23. From http://nature.org/wherewework/northamerica/states/utah/preserves/art5828.html, accessed November 4, 2005.

24. Webb, Belnap, and Weisheit (2004).

25. Albee, Shultz, and Goodrich (1988), p. 552.

26. Webb, Belnap, and Weisheit (2004).

27. Albee, Shultz, and Goodrich (1988), p. 601.

28. Salzer and others (1996).

29. Webb, Belnap, and Weisheit (2004).

Chapter 8

1. From http://www.blm.gov/utah/mon ticello/sanjuan.pdf, accessed November 4, 2005.
2. Waitley (1999), pp. 159–60.
3. D. Ross (1998).
4. Webb, Boyer, and others (2001) present preliminary information on repeat photography of the San Juan River.
5. Force and Howell (1997); Aton and McPherson (2000).
6. Oviatt (1985).
7. Stockton and Jacoby (1976), 33.
8. H. Gregory (1945); Miller (1959); McPherson (1995).
9. Perkins, Nielson, and Jones (1968), p. 60.
10. H. Gregory (1938), p. 33.
11. McPherson (1995), pp. 103–4.
12. McPherson (1995), p. 233
13. D. Ross (1998).
14. Freeman (1909).
15. Brandenburg (1911).
16. La Rue (1925).
17. Orchard (2001).
18. Miser (1924), pp. 56, 71.
19. Miser (1924), p. 67; Jensen (1966), cited in Oviatt (1985).
20. Bryan and La Rue (1927).
21. McPherson (1995), p. 233.
22. Oviatt (1985).
23. Aton and McPherson (2000), 65–83; Hindley and others (2000), 2–9.
24. Webb (1985); Hereford and Webb (1992); Hereford, Webb, and Graham (2002).
25. Roeske, Cooley, and Aldridge (1978).
26. Stakes 3768 and 3769, not shown.
27. Force and Howell (1997).
28. Aton and McPherson (2000), p. 138.
29. Aton and McPherson (2000), p. 110.
30. Hindley and others (2000), pp. 16–17, 20–21, 22–23.
31. Rink (2003).
32. H. Gregory (1917).
33. Cooley (1979).
34. Baars (1973), p. 61.
35. Pattie (2001).

Chapter 9

1. The drainage area for the Escalante River was calculated from its headwaters to the mouth of Coyote Gulch, which reaches the river at about the high lake level for Lake Powell (Webb 1985).
2. Gregory and Moore (1931).
3. Webb, O'Connor, and Baker (1988).
4. Woolsey (1964).
5. A. Thompson (1939).
6. S. Jones (1949).
7. Webb (1985).
8. Webb and Baker (1987).
9. Webb (1985); Webb and Hasbargen (1998).

10. Webb (1985; Webb and Baker (1987).
11. Beverage and Culbertson (1964); Graf, Webb, and Hereford (1991).
12. Graf, Webb, and Hereford (1991).
13. Webb, Melis, and Valdez (2002).
14. Vélez de Escalante (1995).
15. Brooks (1944), Haskell entry for October 20, 1859.
16. Crampton and Miller (1961).
17. A. Robinson (1970).
18. Vélez de Escalante (1995); Reilly (1999).
19. Reilly (1999).
20. Pitts (1987).
21. Horan (1966); Wheeler ([1875] 1983).
22. Turner and Karpiscak (1980).
23. These and other repeat photographs of Lee's Ferry appear in Hereford (2004).
24. Reilly (1999).

Chapter 10

1. Beverage and Culbertson (1964).
2. Hadley (1961).
3. Spaniards' observations of the condition of the Little Colorado River are summarized in Colton (1937).
4. Quoted in Lockett (1939).
5. Quoted in Nicholson (1974), p. 57.
6. Hereford (1984).
7. Hereford (1984); R. Hereford, written communication (2004).
8. Colton (1937); Hereford (1984), p. 657.
9. Colton (1937).
10. Colton (1937).
11. Colton (1937); Lockett (1939).
12. Hereford (1984); Geological Society of America data repository, item 8415, table B, Boulder, Colo.
13. Tellman, Yarde, and Wallace (1997), p. 112.
14. Durrenberger and Ingram (1978).
15. Granger (1960), p. 14.
16. Tellman, Yarde, and Wallace (1997).
17. Harrell and Eckel (1939).
18. J. Powell (1972), p. 15. A lithograph on page 3 of Powell (1972) appears to bear out the description of a stream lined with a thin band of woody riparian vegetation.
19. Hereford (1984, 2004).

Chapter 11

1. Webb (1996).
2. O'Connor and others (1994).
3. Webb, Belnap, and Weisheit (2004).
4. Reilly (1999). The estimate for the discharge of the 1884 flood is approximate: "This determination, though subject to some uncertainties, was accurate enough for the purpose it was intended to serve as the probable error was no greater than is inherent in flood estimates" (La Rue 1925,

p. 14). O'Connor and colleagues (1994) agreed with the approximate discharge of 300,000 ft3/s.
5. Garrett and Gellenbeck (1991), p. 133.
6. R. Martin (1989) documents the history of Glen Canyon Dam, including the political controversy surrounding the Colorado River Storage Project.
7. Reviews of the release patterns from Glen Canyon Dam appear in numerous publications, including Turner and Karpiscak (1980); Howard and Dolan (1981); Dawdy (1991); and Webb, Wegner, and others (1999).
8. Reviews of changes in riparian vegetation in Grand Canyon appear in P. Martin (1971); Turner and Karpiscak (1980); L. Stevens (1989); R. Johnson (1991); and Webb (1996).
9. Huseman (1995), pp. 120–22.
10. The controversy over who was the first photographer of Grand Canyon stems from the large amount of recognition given to J. K. Hillers in comparison to other photographers (Simmons and Simmons 1977). Simmons and Simmons credit Timothy O'Sullivan of the Wheeler Survey as the first photographer (1871), followed by E. O. Beaman (early 1872), James Fennemore and Jack Hillers of the Powell Expeditions (April 1872), and William Bell of the Wheeler Survey (October 1872). Also see Turner and Karpiscak (1980), pp. 10–12.
11. The Wheeler Expedition made hundreds of views of Grand Canyon between 1871 and 1873, most of which were from the rim (Wheeler 1983).
12. There is some confusion concerning the number of photographs that remain from the Powell Expedition. Fowler (1989, pp. 158–61) lists 259 views of the canyons between Green River, Wyoming, and Lava Falls Rapid in the 1870s catalog of the views. However, the U.S. National Archives holdings show 223 views. A total of 46 views of Marble and Grand Canyons are discussed in the 1870s; 33 remain available in archives.
13. Webb (1996).
14. These totals of matched photographs do not include nearly 150 historical photographs of Kanab Creek and Havasu Canyon, which are discussed in chapter 12.
15. As discussed in Webb, Belnap, and Weisheit (2004), the Kolb brothers matched Beaman and Hillers photographs originally taken in 1871 and 1872 during their 1909 trip from Green River, Wyoming, to Needles, California (J. Powell 1961; Kolb 1989).
16. Webb (1996).
17. Stephens and Shoemaker (1987).
18. Turner and Karpiscak (1980); Carothers and Brown (1991).
19. Cook (1987).
20. Clover and Jotter (1944). Note

discussion of other plant collections in B. Phillips and others (1987).

21. R. Johnson (1991); Webb (1996); Ralston (2005).

22. Historic photographs of Grand Canyon show essentially no perennial vegetation below the old high-water stage (Turner and Karpiscak 1980; Stephens and Shoemaker 1987; Webb 1996). Cottonwood trees were established below the old high-water stage at miles 194 and 222. A Goodding (black) willow tree present at Granite Park (mile 209) in 1923 remains but is barely alive at the time of this writing.

23. Clover and Jotter (1944), p. 601. In Clover and Jotter's specific plant lists (pp. 608–10 and 632), tamarisk is not listed at most sites in Grand Canyon. Tamarisk was noted only at Lee's Ferry, Vasey's Paradise (mile 31.8), Saddle Canyon (mile 47.0), Vulcan's Anvil (mile 178.0), Separation Rapid (mile 239.6), and the delta of Lake Mead.

24. Dodge (1936); Patraw (1936).

25. Webb, Melis, and Valdez (2002).

26. Turner and Karpiscak (1980), 14–15.

27. P. Martin's (1971) notes are sufficient to compare with Clover and Jotter's (1944) plant distributions.

28. P. Martin (1971); also see R. Johnson (1977) and Ralston (2005).

29. Stephens and Shoemaker (1987); Turner and Karpiscak (1980).

30. Turner and Karpiscak (1980), 14–15.

31. Webb (1996).

32. Brown, Carothers, and Johnson (1987).

33. U.S. Department of the Interior (1989); Carothers and Brown (1991).

34. Webb (1996).

35. Webb (1996).

36. Salzer and others (1996).

37. Turner and Karpiscak (1980) provide maps showing the distribution in the late 1970s for twenty-four plant species along the Colorado River in Grand Canyon. These maps can be used to document long-term changes in the range of certain species as well as failed incursions of nonnative species. For example, the authors' warning of increases in elm along the river near Glen Canyon Dam never materialized as a significant change in the river corridor.

38. R. Johnson (1991), 179.

39. Kearsley and Ayers (2001).

40. L. Stevens (1989).

41. Webb (1996), p. 113.

42. B. Brown (1988); Carothers and Brown (1991); Webb (1996); Webb, Melis, and Valdez (2002).

43. Kearsley and Ayers (1999, 2001).

44. Brown, Carothers, and Johnson (1987); B. Brown (1988, 1992); Stevens and others (1996).

45. B. Brown (1992).

46. Carothers and Brown (1991).

47. Hoffmeister (1971).

48. L. Stevens (1989).

49. Stevens and others (2001).

50. These plants are documented in E. C. La Rue photograph numbers 643 (Granite Park) and 660 (river mile 222), taken in September 1923 (R. H. Webb, unpublished data, 2003).

51. Stevens and Ayers (2002).

52. Webb and others (2000).

53. Stockton (1975), p. 69.

Chapter 12

1. H. Gregory (1950).

2. Webb, Smith, and McCord (1991). McKee (1946) discusses scientific expeditions involving photography in Kanab Canyon.

3. S. S. Smith (1990); Webb, Smith, and McCord (1991).

4. S. S. Smith (1990).

5. Webb, Smith, and McCord (1991).

6. J. Powell (1972), 2.

7. H. Gregory states that the start of arroyo downcutting was the July 1883 flood, but incision began earlier (1917, p. 131); see also Webb, Smith, and McCord (1991).

8. Webb, Smith, and McCord (1991).

9. Webb, Smith, and McCord (1991).

10. Webb, Smith, and McCord (1991).

11. Hughes (1993).

12. Webb, Smith, and McCord (1991).

13. Granger (1960), p. 146.

14. Melis and others (1996).

15. Wittick (1973).

16. Crosby (1965), pp. 72–73.

17. Cushing (1965). This source is a compilation of two articles published in the *Atlantic Monthly* in 1882.

18. Melis and others (1996).

19. Melis and others (1996).

20. Melis and others (1996).

21. Melis and others (1996).

22. Melis and others (1996).

23. Melis and others (1996), pp. 77–86.

Chapter 13

1. H. Gregory (1950).

2. Larson (1961), p. 23.

3. Observations of John D. Lee in 1852, quoted in Woodbury (1944).

4. J. S. Smith (1977), pp. 61–63.

5. Steen-Adams (2002).

6. J. Belnap (U.S. Geological Survey, written communication, 2004) provided this information from family history; her great-great-grandfather helped found Springdale.

7. Yeager (1944).

8. Merriam (1893), pp. 309–30, 335.

9. Yeager (1944).

10. Larson (1961), p. 30.

11. W. W. Flanigan (born 1877), written comments to the superintendent of Zion National Park (no date), Springdale, Utah, National Park Service, Resources Management Files, no numbering system.

12. Yeager (1944).

13. Fowler (1989).

14. Alder and Brooks (1996), p. 213.

15. H. Gregory (1950).

16. Hereford, Jacoby, and McCord (1996).

17. R. Warren (1999).

18. Larson (1961), pp. 26–29; Webb (1985); Engstrom (1996).

19. Hereford, Jacoby, and McCord (1996) place the initiation of downcutting at 1880 for the Virgin River in the downstream from Zion National Park.

20. Larson (1961), p. 94.

21. Larson (1961), p. 367.

22. Woolley (1946).

23. Butler and Mundorf (1970).

24. Kittredge, Finch, and Mitchell (1926), p. 3.

25. Hereford, Jacoby, and McCord (1996).

26. D. Sharrow, written communication (2005).

27. Steen-Adams (2002).

28. Hereford, Jacoby, and McCord (1996).

29. Butler and Mundorf (1970).

30. Meteorological information on this storm is in Butler and Mundorf (1970).

31. The flooding in southern California, southern Nevada, and central Arizona is documented in Rostvedt and others (1971).

32. According to S. Haile (exotic plants specialist, Zion National Park, personal communication, 2004), tamarisk eradication efforts were spotty on the East Fork until 2001, and "full eradication" would be achieved by 2004 within the Zion National Park reach.

33. S. Haile, personal communication (2004).

34. Steen-Adams (2002).

35. Hamilton (1992); D. Sharrow, written communication (2003).

36. Grater (1945); Schuster and Wieczorek (1995).

37. Hamilton (1978).

38. Jibson and Harp (1996).

39. Hughes (1993).

40. Hughes (1993), p. 152.

Chapter 14

1. La Rue (1916, 1925).

2. J. Stevens (1988); R. Martin (1989); Potter and Drake (1989).

3. C. Howard (1947); Iorns, Hembree, and Oakland (1965).

4. R. Martin (1989).

5. Ferrari (1988).

6. Webb, Belnap, and Weisheit (2004).

7. Inskip (1995); Nichols (1999).

8. Charles Goodman photograph, 1894, courtesy of the University of Utah Marriott Library, Salt Lake City, Beck 31 and 39.

9. Woodbury, Durrant, and Flowers (1959).

10. Woodbury, Durrant, and Flowers (1959), p. 28.

11. Woodbury and Russell (1945).

12. Clover and Jotter (1944); Woodbury and Russell (1945).

13. Inskip (1995); Nichols (1999).

14. L. Stevens (1989); Petroski (1993).

15. From http://www.usbr.gov/dataweb/dams/nv10122.htm, accessed November 4, 2005.

16. W. Smith and others (1960).

17. Anonymous (1931).

18. Merriam (1893), pp. 334–35.

19. Merriam (1893) uses the now discontinued names *Salix longifolia* to signify coyote willow, *Salix nigra* to describe black willow (*Salix gooddingii*), and *Pluchea sericea* to describe arrowweed (*Tessaria sericea*).

20. Turner and Karpiscak (1980).

21. Huseman (1995).

22. Horan (1966); Wheeler (1983).

Chapter 15

1. Klawon (2000), p. 9.

2. Hafen (1997), p. 123.

3. Shafroth, Stromberg, and Patten (2002).

4. Although Burro Creek is not considered in this book, repeat photographs from the U.S. 93 bridge appear to show increased riparian vegetation, notably Frémont cottonwood, but probably also tamarisk, from about the early 1960s to 1998 (Klawon 2000, pp. 43–44).

5. Klawon (2000).

6. U.S. Army Corps of Engineers, undated and unpublished "Reservoir Regulation Manual for Alamo Reservoir."

7. Klawon (2000), p. 12.

8. Klawon (2000), p. 25.

9. Klawon (2000), p. 24. Repeat photographs (p. 41) show what appears to be a change from a channel lined with young cottonwoods (unknown date between 1948 and 1975) to one lined with mostly tamarisk (1998).

10. House and Pearthree (1995).

11. Bowie and Kam (1968). For unexplained reasons, these authors referred to the watercourse as Cottonwood Wash.

12. F. A. Branson and R. S. Aro, "Vegetation," in Bowie and Kam (1968), pp. 7–11.

13. Granger (1960), p. 222.

14. Klawon (2000).

15. Shafroth, Stromberg, and Patten (2002).

16. Klawon (2000), p. 32. Two sets of aerial photography included in this report show increased floodplain vegetation (apparently mesquite) from 1953 through 1992 (pp. 30–31).

17. Shafroth, Stromberg, and Patten (2002).

18. From http://www.gf.state.az.us/w_c/eagle_closures.html, accessed September 14, 2004.

19. Granger (1960), p. 341.

20. House, Wood, and Pearthree (1999) delineate nine "relatively distinct" reaches from Alamo Dam to Lake Havasu, reflecting the geomorphic diversity of the Bill Williams River.

21. From http://southwest.fws.gov/refuges/arizona/billwill.html, accessed November 4, 2005.

22. K. Blair, U.S. Fish and Wildlife Service, personal communication (2004).

23. Tellman, Yarde, and Wallace (1997), pp. 123–24.

24. Cleland (1950).

25. Huseman (1995).

26. From http://www.spl.usace.army.mil/resreg/htdocs/almo.html, accessed November 4, 2005.

27. House, Wood, and Pearthree (1999), pp. 10–15.

28. Obtained from the U.S. Army Corps of Engineers, Los Angeles District.

29. Rivers West (1990), p. 23; House, Wood, and Pearthree (1999), pp. 12–13.

30. Rivers West (1990); K. B. Blair, personal communication (2003).

31. Shafroth and others (1998).

32. Stromberg and others (2004).

33. Granger (1960), p. 383.

34. S. Turner (1962).

35. Rivers West (1990), pp. 9, 19.

36. U.S. Fish and Wildlife Service (1994), pp. 74–75; K. B. Blair, personal communication (2003).

37. Shafroth, Stromberg, and Patten (2002).

38. Shafroth and others (1998); Shafroth, Stromberg, and Patten (2000, 2002).

39. Williams and Wolman (1984).

40. Shafroth, Stromberg, and Patten (2002) provide quantitative analyses of aerial photography of this reach. House, Wood, and Pearthree note that the 1953 aerial photographs were ideal for mapping geomorphology of the Bill Williams River because of "the lack of post-dam riparian vegetation growth" (1999, p. 18). They show a 1953 view that includes the former gaging station, and little riparian vegetation is present in this reach at that time.

41. S. Turner (1962), pp. 15–18.

42. Wolcott, Skibitzke, and Halpenny (1956), p. 311.

43. K. B. Blair, personal communication (2003).

44. P. Shafroth, written communication (2004).

45. Memo from the Lake Havasu NWR manager to the regional director, Region 2, Albuquerque, New Mexico, June 29, 1990; K. B. Blair, written communication (2003).

46. Shafroth, Stromberg, and Patten (2002); Stromberg and others (2004).

47. K. B. Blair, personal communication (2003).

48. K. B. Blair, personal communication (2003).

Chapter 16

1. Jordan and Maynard (1970).

2. Cooke and Reeves (1976).

3. Cooke and Reeves (1976), pp. 30–31.

4. Schwennesen (1919), p. 6.

5. Bryan (1925).

6. Knechtel (1938), pp. 189–90.

7. Tellman, Yarde, and Wallace (1997), p. 87.

8. From http://www.tucson.ars.ag.gov/icrw/Proceedings/Brandau.pdf, accessed November 4, 2005.

9. Schwennesen (1919), p. 2.

10. Turner, Webb, and Bowers (2003).

11. Lacey, Ogden, and Foster (1975), p. 35.

12. Granger (1960), p. 57.

13. Waters (1985).

14. Cooke and Reeves (1976), pp. 39–40.

15. Coates and Cushman (1955), p. 9.

16. Ricketts (1996), pp. 92–93.

17. Mearns (1907), pp. 95–96.

18. Mearns (1907), p. 359.

19. Cooke and Reeves (1976), p. 39.

20. Coates and Cushman (1955).

21. From http://www.fws.gov/refuges/profiles/index.cfm?id=22524, accessed November 4, 2005.

22. Tashjian (n.d.).

23. Tashjian (n.d.), p. 17.

24. Photograph X6 taken by an unknown photographer, dated April 26, 1918, of Whitewater Draw near Douglas, courtesy of the U.S. Geological Survey.

25. Meinzer and Kelton (1913), p. 89.

26. Turner, Webb, and Bowers (2003).

27. Thrapp (1967); H. Walker (1971).

28. Schwennesen (1918). The current owners use the name Diamond A, which was the original name of this ranch.

Chapter 17

1. Pattie (2001), p. 73. Pattie's 1831 account has been severely criticized because of inconsistencies in the dates and his claims of credit for certain activities that likely were really done by others. However, neither of these criticisms discredits his observations on natural history.

2. Pattie (2001), p. 75.

3. Pattie (2001), p. 95.

4. Pattie (2001), p. 109.

5. Pattie (2001), p. 170.

6. Cleland (1950).

7. Mearns (1907), p. 356.

8. Emory (1848). Observations of riparian vegetation along the upper Gila River are on pages 81, 98, and 101.

9. Corle (1951), p. 128.

10. Burkham (1972).

11. McClintock (1985), p. 234.

12. McClintock (1985), p. 242.

13. Dobyns (1981), p. 98.

14. Burkham (1970) gives a detailed history of floods before 1970 for the upper Gila River.

15. Burkham (1970), p. 24.

16. Huckleberry (1996a, 1996b). Another set of historical floods and those recorded at gaging stations are listed in Klawon (2001), p. 6.

17. Burkham (1972); R. Turner (1974).

18. R. Turner (1974), p. 11.

19. Klawon (2001), p. 11. A series of aerial photographs taken in 1935, 1953, 1967, 1978, 1992, and 1997 document channel narrowing accompanied by increases in woody riparian vegetation (probably mixtures of native and nonnative species) from 1935 through 1992, followed by renewed channel widening during the series of three large floods in the winter of 1993.

20. Roeske, Garrett, and Eychaner (1989).

21. Webb and Betancourt (1992); Turner, Webb, and Bowers (2003). Burkham (1970) recognized the effect of climate on streamflow for the Gila River and noted the upturn in precipitation and streamflow beginning in 1962.

22. Corle (1951), p. 365.

23. Rogge and others (1995), p. 11.

24. Klawon (2001), pp. 48–50.

25. Tellman, Yarde, and Wallace (1997), p. 104.

26. Minckley and Clark (1984).

27. Gatewood and others (1950).

28. Culler and others (1970).

29. Gatewood and others (1950).

30. Weist (1971).

31. Burkham (1972).

32. R. Turner (1974).

33. Burkham (1972).

34. As part of the Gila River Phreatophyte Project, vegetation maps were compiled documenting historical changes in riparian vegetation; see R. Turner (1974). Using digital orthophotograph quarter quadrangles taken in 1994 and analyzed in 2001, T. Klearman and M. B. Murov (written communication, 2001) updated these maps and compared the changes in this reach, which had an area of about 2,730 acres in 1914 and 2,300 acres in the other years.

35. R. Turner (1974), pp. 8–9.

36. Minckley and Clark (1984).

37. Turner, Webb, and Bowers (2003).

38. M. P. Collier and B. T. Brown, personal communication (1993).

Chapter 18

1. Granger (1960), p. 165.

2. Pattie (2001), p. 77.

3. Granger (1960), p. 165; Durrenberger and Ingram (1978).

4. Anonymous (1972).

5. Roeske, Garrett, and Eychaner (1989).

6. Stockton (1975).

7. Granger (1960), p. 115.

8. Pattie (2001), p. 82.

9. Granger (1960), p. 116.

Chapter 19

1. From http://www.lastgreatplaces.org/SanPedro/Index.htm, accessed November 4, 2005.

2. Skagen and others (1998).

3. Hanson (2001), 19.

4. From http://www.co.pima.az.us/cmo/sdcp/sdcp2/fsheets/swf.html, accessed November 12, 2005.

5. Stromberg, Tiller, and Richter (1996); Pool and Coes (1999).

6. Changes in the San Pedro River are reviewed in Turner, Webb, and Bowers (2003).

7. Glennon (2002), pp. 51–69.

8. Pattie (2001), p. 86.

9. Pattie (2001), p. 170.

10. Bahre and Bradbury (1978); Hanson (2001), p. 18.

11. Ricketts (1996).

12. Turner, Webb, and Bowers (2003).

13. Davis (1986); Pima County (2000).

14. Hendrickson and Minckley (1984).

15. Bryan (1928); Hastings (1959); Bahre and Bradbury (1978).

16. McClintock (1985), p. 233; Hastings (1959).

17. Bryan (1928); Hendrickson and Minckley (1984); Fonseca (1998); Turner, Webb, and Bowers (2003).

18. Davis (1986).

19. Mearns (1907), p. 350.

20. Mearns (1907), Plate X, opposite p. 92.

21. Davis (1986).

22. Waters (1985, 1988, 1992); Waters and Haynes (2001).

23. Waters and Haynes (2001).

24. Bryan (1925).

25. Cooke and Reeves (1976), pp. 41–47.

26. Turner, Webb, and Bowers (2003), pp. 29–35.

27. Hereford (1993); Huckleberry (1996a).

28. J. Bartlett (1854), quoted in Huckleberry (1996a).

29. McClintock (1985).

30. Dubois and Smith (1980), p. 56.

31. Cooke and Reeves (1976), p. 44.

32. Hereford (1993).

33. Hereford (1993).

34. Durrenberger and Ingram (1978).

35. Pool and Coes (1999), p. 38.

36. D. R. Pool, U.S. Geological Survey, written communication (2005).

37. Turner, Webb, and Bowers (2003).

38. Turner, Webb, and Bowers (2003).

39. Roeske, Garrett, and Eychaner (1989), p. 16.

40. Pool and Coes (1999).

41. Pool and Coes (1999), p. 25.

42. Hereford (1993).

43. Pool and Coes (1999), p. 26.

44. Bahre and Bradbury (1978); Humphrey (1987); Bahre (1991); Hereford (1993); Turner, Webb, and Bowers (2003).

45. Bahre (1991).

46. Kepner and others (2000).

47. Kepner, Edmonds, and Watts (2002).

48. Kepner and others (2000).

49. Hereford (1993); Turner, Webb, and Bowers (2003).

50. Bahre and Bradbury (1978), p. 152.

51. Turner, Webb, and Bowers (2003), pp. 154–55.

52. Hereford (1993), p. 19.

53. Lacey, Ogden, and Foster (1975), p, 16.

54. J. Stromberg, written communication (2005).

55. Stromberg (1998a).

56. Stromberg (1998a).

57. Stromberg (1998a).

58. Bahre (1991), pp. 68–71.

59. Dobyns (1981); Turner, Webb, and Bowers (2003).

60. Hanson (2001), 86.

61. Turner, Webb, and Bowers (2003), pp. 132–33.

62. Tellman, Yarde, and Wallace (1997), p. 133.

63. Hanson (2001), p. 123; Dollar (2002).

64. Zimmerman (1969).

65. Daniel Baker, e-mail correspondence with Raymond Turner (July 30, 2004).

66. Zimmerman (1969), p. 27.

67. Zimmerman (1984).

68. Fonseca (1998); Johnny LaVin, personal communication (2003).

69. Huckleberry (1996a); Wood (1997), p. 21.

70. Huckleberry (1996a);, Wood (1997).

71. Fonseca (1998).

72. Wood (1997), p. 13.

73. Hanson (2001), p. 147.

Chapter 20

1. Zimmerman (1969, 1984). Redfield and Buemann Canyons also would have been excellent subjects, but no old photographs are available.

2. Zimmerman (1969), pp. 21–23.

3. Zimmerman (1984).

4. Zimmerman (1984), pp. 19, 21, 23.

5. Zimmerman (1984).

6. Dan Baker, personal communication (2003).

7. Roeske, Garrett, and Eychaner (1989), p. 16.

8. Zimmerman (1984).

9. Hadley, Warshall, and Bufkin (1991).

10. Tellman, Yarde, and Wallace (1997), p. 83.

11. Hadley, Warshall, and Bufkin (1991), p. 57.

12. Bell (1869).

13. Bourke (1891), p. 4.

14. Hadley, Warshall, and Bufkin (1991).

15. Hadley, Warshall, and Bufkin (1991), p. 17.

16. Hadley, Warshall, and Bufkin (1991), pp. 248–50.

17. From http://nature.org/wherewe work/northamerica/states/arizona/ preserves/art1946.html, accessed November 4, 2005.

18. Mark Haberstich, The Nature Conservancy, personal communication (2003).

19. Tellman, Yarde, and Wallace (1997), p. 82.

20. L. Roberts (1987).

21. Bahre (1991), p. 80–81.

Chapter 21

1. Sheridan (1995), p. 56.

2. Betancourt (1990).

3. Thornber (1909); Cooke and Reeves (1976); Betancourt (1990); Wood, House, and Pearthree (1999); Mauz (2002).

4. Hendrickson and Minckley (1984).

5. Cooke and Reeves (1976); Sheridan (1986); Betancourt (1990); Wood, House, and Pearthree (1999); and Logan (2002).

6. Logan (2002), p. 65.

7. Glennon (2002), p. 38.

8. Betancourt (1990).

9. Nicholson (1974), p. 166.

10. Sykes (n.d.).

11. Logan (2002), pp. 56–57, 63.

12. Jackson (1973).

13. Halpenny and Halpenny (1988); Sykes (n.d.).

14. Pope, Rigas, and Smith (1998), p. 415.

15. R. R. Johnson and others, Johnson and Haight Environmental Consultants, written communication (2005).

16. Logan (2002), pp. 43–44.

17. Pima County (2000), p. 13.

18. Nicholson (1974), p. 171.

19. Bryan (1928), p. 475.

20. Dobyns (1981).

21. Nicholson (1974), p. 167.

22. Sykes (n.d.); Logan (2002), p. 105.

23. Sykes (n.d.).

24. Mearns (1907), pp. 105–6.

25. Betancourt and Turner (1985).

26. Waters (1988).

27. Hastings (1959); Cooke and Reeves (1976); Betancourt and Turner (1985); Betancourt (1990).

28. Betancourt (1990).

29. Betancourt (1990); Kupel (2003).

30. Betancourt (1990).

31. Webb and Betancourt (1992).

32. Hays (1984); Parker (1995).

33. Aldridge and Eychaner (1984).

34. Roeske, Garrett, and Eychaner (1989); Webb and Betancourt (1992).

35. Webb and Betancourt (1992).

36. Betancourt (1990).

37. Kupel (2003), p. 70.

38. Kupel (2003), p. 197.

39. Kupel (2003), p. 92.

40. Glennon (2002), pp. 45–50.

41. Tellman, Yarde, and Wallace (1997), p. 21.

42. Lowe (1985).

43. Minckley (1973); Pima County (2000).

44. Kupel (2003), p. 192.

45. Betancourt (1990); Betancourt and Turner (1990).

46. Bahre and Bradbury (1978), p. 161; Humphrey (1987), pp. 241, 255.

47. Logan (2002), p. 63.

48. Turner, Webb, and Bowers (2003), pp. 82–85.

49. Logan (2002), p. 209.

50. Logan (2002), pp. 182–183; Turner, Webb, and Bowers (2003), pp. 120–23.

51. Betancourt (1990).

52. Betancourt (1990); Betancourt and Turner (1990); R. Turner (2003).

53. R. R. Johnson and others, Johnson and Haight Environmental Consultants, written communication (2005). C. Olson (1940) wrote a description of this bosque, for which he used the term *forest*.

54. Arnold (1940), p. 5.

55. Glennon (2002), p. 48.

56. Photographs by J. R. Hastings, F IX 29–32, courtesy of the Desert Laboratory Collection of Repeat Photography.

57. Parker (1995); Wood, House, and Pearthree (1999), p. 53.

58. Betancourt (1990).

59. Thornber (1909); Mauz (2002).

60. Wood, House, and Pearthree (1999).

61. Schumann (1974).

Chapter 22

1. Turner, Webb, and Bowers (2003), pp. 58–59.

2. Halpenny and Halpenny (1988).

3. From http://nature.org/wherewe work/northamerica/states/arizona/ preserves/art1972.html, accessed November 4, 2005.

4. Roeske, Garrett, and Eychaner (1989).

5. Granger (1960), p. 322.

6. Hendrickson and Minckley (1984).

7. From http://nature.org/wherewe work/northamerica/states/arizona/ preserves/art1972.html, accessed November 4, 2005.

8. Bahre (1991), pp. 76–77.

9. Hendrickson and Minckley (1984).

10. From http://www.co.pima.az.us/ pksrec/parkpgs/cienega/cienega.html, accessed November 4, 2005.

11. R. M. Turner (2003), see http://www .pima.gov/cmo/sdcp/reports/WDweb.pdf, accessed November 4, 2005.

12. Zimmerman (1969), p. 43.

13. Zimmerman (1984).

14. J. Fonseca, Pima County Department of Transportation and Flood Control District, written communication (2005).

15. Pearthree and Baker (1987).

16. J. Fonseca, Pima County Department of Transportation and Flood Control District, written communication (2005); this communication (distinct from the one cited in note 14) is an unpublished manuscript by the Pima County Manager's Office entitled "Regional Overview of Land Acquisition in the Cienega Creek Watershed," dated August 14, 1990.

17. Willis (1939).

18. Pearthree and Baker (1987).

19. Stromberg and others (1992).

20. Aldridge (1970); Pearthree and Baker (1987).

21. The Sabino Creek watershed was heavily impacted by the Aspen Fire in June and July 2003.

22. Turner, Webb, and Bowers (2003), pp. 188–91.

23. Willis (1939).

24. Pima County (2000), p. 15.

25. Pima County (2000), p. 5.

26. Willis (1939).

27. Condes de la Torre (1970) claims from little historical evidence that Rillito Creek was perennial from Camp Lowell to its confluence with the Santa Cruz.

28. Pearthree and Baker (1987), p. 20.

29. W. Graf (1984), using the dates of aerial photography, gives 1937 as the year arroyo downcutting ceased.

30. Sykes (n.d.); Willis (1939).

31. Lowe (1985).

32. Pearthree and Baker (1987).

33. Pima County (2001), p. 19.

34. Schwalen and Shaw (1957, 1961). Pima County also documents groundwater fluctuations related to dry and wet years (2001, p. 12).

35. Willis (1939).

36. Pearthree and Baker (1987), p. 31.

37. Hoffmann, Ripich, and Ellett (2002), Plate 3.

38. The Cañada del Oro watershed was heavily impacted by the Aspen Fire in June and July 2003. However, the annual flood series shown in figure 22.17 is through 2002.

39. Schwalen and Shaw (1961).

40. Turner, Webb, and Bowers (2003), pp. 192–93.

Chapter 23

1. From http://www.azgfd.gov/outdoor_recreation/wildlife_area_upper.shtml, accessed November 4, 2005.
2. From http://www.gf.state.az.us/w_c/eagle_closures.html, accessed September 24, 2004.
3. From http://www.nps.gov/rivers/wsr-verde.html, accessed November 4, 2005.
4. Aldridge (1963).
5. Thomsen and Schumann (1968).
6. From http://www.gf.state.az.us/w_c/eagle_closures.html, accessed September 24, 2004.
7. Roeske, Cooley, and Aldridge (1978).
8. Thomsen and Schumann (1968).
9. Whittlesey (1997b), p. 34.
10. Granger (1960), p. 361; Byrkit (1978).
11. Pattie (2001).
12. Mearns (1907), pp. 353–59.
13. Mearns (1907), pp. 353–59.
14. Shaw (2001a, 2001b).
15. Lopez and Springer (2003), p. 5.
16. Whittlesey (1997b), p. 35.
17. McClintock (1985), p. 234.
18. Davis and Turner (1986).
19. Nicholson (1974), p. 88.
20. Allen (1937), p. 3.
21. Allen (1937); Lopez and Springer (2003), p. 7.
22. Lopez and Springer (2003), p. 16.
23. Whittlesey (1997b).
24. Whittlesey (1997b).
25. Blasch and others (2006).
26. Aldridge and Eychaner (1984); Aldridge and Hales (1984).
27. Van West and Altschul (1997), 351–53.
28. Whittlesey (1997b), p. 32.
29. Ely and Baker (1985); House, Pearthree, and Klawon (2002).
30. Pearthree (1996).
31. Tellman, Yarde, and Wallace (1997), p. 48.
32. Rogge and others (1995), p. 11.
33. From http://202.114.65.36/12/test/water/americ-dam/HorseshoeDam.htm, accessed November 4, 2005.
34. Slingluff (1993).
35. Lopez, Anderson, and Springer (2003).

Chapter 24

1. Granger (1960), p. 115.
2. From http://reference.allrefer.com/gazetteer/S/S02294-salt-river-valley.html, accessed November 4, 2005.
3. Masse (1976); Fuller (1987).
4. Hodge (1877), cited in Rea (1997).
5. Pattie (2001), 119.
6. Mearns (1907), p. 356.
7. McClintock (1985), pp. 211–14.
8. Minckley (1973), p. 121.

9. Minckley (1973), p. 121.
10. Fuller (1987).
11. Nials, Gregory, and Graybill (1989).
12. Fuller (1987).
13. This figure is obtained by adding the long-term averages for the Salt River at Roosevelt, Arizona (09498500), Tonto Creek above Gun Creek, near Roosevelt, Arizona (09499000), and the Verde River below Tangle Creek, above Horseshoe Dam, Arizona (09508500), and assuming that losses downstream from these stations are negligible.
14. Durrenberger and Ingram (1978).
15. Aldridge and Eychaner (1984). The reported discharge is for Salt River at Alma School Road near Mesa, Arizona (09512060); see figure 24.11.
16. Corle (1951), p. 269.
17. Durrenberger and Ingram (1978).
18. Rogge and others (1995), p. 11.
19. From http://www.srpnet.com/water/dams/roosevelt.aspx, accessed November 4, 2005.
20. Kupel (2003), p. 79.
21. From http://www.srpnet.com/, accessed November 4, 2005.
22. Kupel (2003), p. 92. Wolcott (1952) also discusses groundwater use in the first half of the twentieth century.
23. Robinson and Peterson (1962); Stulik and Twenter (1964); Schumann (1974); Kupel (2003), p. 211.
24. Reeter and Remick (1986); Kupel (2003).
25. Granger (1960), p. 119.
26. D. Brown and others (1976).
27. W. Graf (1983).
28. Ruff (1971), pp. 8–14.
29. Ruff (1971), p. 9.
30. Kupel (2003), p. 187.
31. Ruff (1971).

Chapter 25

1. Hansen, Shwarz, and Riedel (1977); Hansen and Shwarz (1981).
2. Roeske, Cooley, and Aldridge (1978).
3. Rogge and others (1995), p. 11.
4. Roeske, Cooley, and Aldridge (1978).
5. Granger (1960), p. 183.
6. Stromberg, Richter, and others (1993).
7. Tellman, Yarde, and Wallace (1997), p. 81.
8. Jones and Glinski (1985).
9. Durrenberger and Ingram (1978).
10. Roeske, Cooley, and Aldridge (1978).
11. Rogge and others (1995), p. 11.
12. Dill (1987).
13. Corle (1951); Dill (1987).
14. Granger (1960), pp. 196–97.
15. Stromberg, Fry, and Patten (1997).

Chapter 26

1. McNamee (1994).
2. Historical notes on Spanish observations on the Pima and their habitat come from Rea (1983, 1997).
3. Kino quoted in Rea (1983), p. 17. Whittlesey also summarizes Spanish observations of the lower Gila River (1997b, pp. 39–40).
4. Pattie (2001), p. 120.
5. Rea (1983).
6. Emory (1848). Descriptions of the lower Gila River are given on pages 103, 113, 116, 119, and 120.
7. Ricketts (1996).
8. Rea (1997).
9. Rea (1997), p. 34.
10. Haase (1972).
11. Nicholson (1974), p. 116.
12. C. Ross (1923), p. 67.
13. Couts, quoted in D. Martin (1954), p. 129.
14. Mearns (1907), p. 124.
15. C. Ross (1923).
16. McNamee (1994), pp. 125–26.
17. Burkham (1972).
18. Rea (1997), p. 38.
19. Lee (1904), p. 10, (1905), p. 135.
20. C. Ross (1923), pp. 15, 36, 65.
21. Haase (1973).
22. Office of Arid Lands Studies (1970).
23. Rea (1983), pp. 34–35.
24. Minckley (1973), pp. 120–21.
25. Rea (1983).
26. Mearns (1907), p. 359.
27. Office of Arid Lands Studies (1970).
28. Huckleberry (1995).
29. Huckleberry (1996a, 1996b).
30. McNamee (1994); Blake and Steinhart (1994).
31. Emory (1848), p. 125.
32. Ricketts (1996), pp. 107–8.
33. C. Ross (1923), p. 66; D. Martin (1954), pp. 133–35.
34. D. Martin (1954), p. 200.
35. Huckleberry (1995).
36. It is unclear from the information whether two large floods occurred in 1868 and 1869 or only one occurred within one of these years.
37. Durrenberger and Ingram (1978).
38. Huckleberry (1994).
39. Huckleberry (1995), p. 167.
40. Huckleberry (1994). This report gives misleading data on the relative sizes of the 1983 and 1993 floods; the 1993 flood was 19 percent larger than the 1993 flood at the Laveen gaging station. Attenuation of flood peaks in the regulated Gila River makes it essential to state precisely where discharges are estimated.
41. Dobyns (1981), p. 72.
42. Tellman, Yarde, and Wallace (1997), p. 101.

43. From http://www.spl.usace.army
.mil/resreg/htdocs/ptrk.html, accessed
November 4, 2005.

44. From http://www.spl.usace.army
.mil/resreg/htdocs/ptrk.html, accessed
November 4, 2005.

45. W. Graf (1981).

46. Haase (1973).

47. Rogge and others (1995), p. 11.

48. Haase (1972).

49. Office of Arid Lands Studies (1970).

50. Haase (1972).

Chapter 27

1. MacDougal (1906).

2. Fradkin (1984).

3. Hendricks (1990); Bergman (2003).

4. Mueller and Marsh (2002); Minckley
and others (2003).

5. Minckley (1973), p. 120.

6. Minckley (1973).

7. Lingenfelter (1978).

8. Much historical information on
change in riparian vegetation along the
lower Colorado River is given in Bureau of
Reclamation (1999).

9. Alvarez de Williams (1975), pp. 24,
93–96, 129.

10. Kunkel (1970).

11. J. S. Smith (1977), pp. 72–73.

12. Pattie (2001).

13. Pattie (2001), pp. 175–93.

14. Pattie (2001), p. 180.

15. Crosby (1965).

16. Huseman (1995).

17. Ives quoted in Kunkel (1970), p. 8.

18. Ives quoted in Kunkel (1970), p. 11.

19. Alvarez de Williams (1975) gives
many examples, including views on pages
12 and 93.

20. Oden (1973).

21. Lingenfelter (1978).

22. D. Martin (1954), p. 155.

23. Lingenfelter (1978).

24. D. Martin (1954), p. 190.

25. Sykes (1944), p. 247.

26. Nicholson (1974), pp. 229, 235.

27. T. Van Dyke (1895).

28. MacDougal (1906).

29. Mueller and Marsh (2002).

30. Emory (1848), pp. 125, 131.

31. Nicholson (1974), p. 230.

32. D. Brown (1985), pp. 39–50.

33. Ohmart, Deason, and Freeland (1975).

34. Sykes (1926).

35. Mearns (1907), pp. 125–29.

36. Mearns quoted in Kniffen (1931),
p. 33.

37. Mearns quoted in Kniffen (1932),
pp. 174–75.

38. The other observer quoted in
Kniffen (1932), p. 174.

39. Hendricks (1990); Bergman (2003).

40. Kniffen (1931), pp. 62, 64.

41. Grinnell (1914).

42. Castetter and Bell (1951), p. 22.

43. Beidelman (2000).

44. Leopold (1966), pp. 150–58.

45. Murphy (1917).

46. Grinnell (1914); J. Brown (1923).

47. MacDougal (1917).

48. Rosenberg and others (1991), p. 21.

49. W. Graf (1978).

50. Castetter and Bell (1951), p. 23.

51. Silsbee (1900).

52. For the 400,000-acre to 500,000-acre
values of alluvial bottomland, Mearns cites
an unknown report by the hydrographic
branch of the U.S. Geological Survey from
January 1902 (1907, p. 125). Rosenberg and
others modified this amount to 395,000
acres of riparian vegetation without explanation (1991, p. 21). Cohn (2001) cites
400,000 acres and attributes the source to
R. Ohmart of Arizona State University.

53. Holmes (1904), pp. 29–30.

54. Not discussed by Holmes (1904),
who concentrated on "timber."

55. Bureau of Reclamation (1963).

56. Anderson and Ohmart (1984), p. vi.

57. Owen-Joyce and Raymond (1996),
p. 9.

58. Younker and Andersen (1986), p. 18.

59. Cohn (2001) mistakenly compares
400,000 acres of total riparian habitat,
estimated around the turn of the twentieth
century, with 6,000 acres of cottonwood-
willow habitat measured in 1997, implying
a much larger reduction in total riparian
habitat than has actually occurred.

60. Sykes (1937), p. 65.

61. Sykes (1970), p. 173.

62. Leigh (1941).

63. E. Glenn, University of Arizona,
written communication (2003).

64. Durrenberger and Ingram (1978).

65. Durrenberger and Ingram (1978).

66. Sykes (1926, 1937, 1944, 1970).

67. Ohmart, Deason, and Freeland
(1975).

68. Van de Kamp (1973).

69. Waters (1983).

70. Fox (1936).

71. See Duke (1974) for a detailed history of the 1905 breach, the filling of the
Salton Sea, and the subsequent construction of levees to divert flow back into the
channel of the Colorado River.

72. MacDougal (1906); Sykes (1926);
Hoyt and Langbein (1955).

73. Cecil-Stephens (1891).

74. Duke (1974); J. Stevens (1988).

75. Williams and Wolman (1984); Shepherd (1961).

76. D. Martin (1954), p. 191.

77. From http://www.usbr.gov/dataweb/
dams/az10315.htm, accessed November 4,
2005.

78. Grinnell (1914).

79. From http://www.usbr.gov/dataweb/
dams/ca10159.htm, accessed November 4,
2005.

80. La Rue (1916, 1925).

81. J. Stevens (1988), pp. 16–17.

82. J. Stevens (1988).

83. For a comprehensive history of the
construction of Boulder/Hoover Dam and
its effect on the United States, see J. Stevens
(1988) and Petroski (1993).

84. From http://www.usbr.gov/dataweb/
dams/az10312.htm, accessed November 4,
2005.

85. From http://www.usbr.gov/dataweb/
dams/az10309.htm, accessed November 4,
2005.

86. Owen-Joyce and Raymond (1996);
Cohen and Henges-Jeck (2001); Cohen,
Henges-Jeck, and Castillo-Moreno (2001);
Glennon and Culp (2002).

87. See Rinne and Minckley (1991), p. 4,
for another repeat view of Black Canyon.

88. By as much as 25 feet; Williams and
Wolman (1984), pp. 21–22.

89. Metzger and Loeltz (1973); Guay
(2001).

90. Several U.S. Geological Survey
photographs looking upstream from the
bridges that cross the river near Needles,
California, were found, but they could not
be matched owing to restrictions against
pedestrians using these bridges.

91. Nagler, Glenn, and others (2005).

92. Guay (2001).

93. As noted in figure 27.8, the gaging
record of the Colorado River near Topock,
Arizona (09424000) is supplemented after
1982 with the record of the Colorado River
at Davis Dam, Arizona (09423000). Direct
combination of these two gaging records
ignores the possible tributary additions in
the intervening reach, including Sacramento Wash from the Arizona side and Piute
Wash from the California side.

94. Metzger, Loeltz, and Irelna (1973).

95. U.S. Department of the Interior (1964).

96. Oden (1973).

97. Figure 27.4 represents the combination of the original gaging station
called the Colorado River at Yuma,
Arizona (09521000; 1903–1963) and the
station at the Colorado River below Yuma
Main Canal Wasteway at Yuma, Arizona
(09521100; 1964–2002). In order to make
the two records compatible, the latter
record must be corrected for inflows from
two canals with gaging stations, namely
the Yuma Main Canal Wasteway at Yuma,
Arizona (09525000) and the Reservation
Main Drain No. 4 near Yuma, Arizona
(09530000). Because the latter two stations
have only daily values, the instantaneous
peak discharges for station 09521100 were
decreased by the combined daily inflows,
resulting in an unknown error.

98. Sykes (1926); Cohen and Henges-Jeck (2001).

99. Cory (1915); Nelson (2006).

100. Sykes (1926).

101. Tiegs and Pohl (2005).

102. Hinojosa-Huerta and others (2002).

103. Rosenberg and others (1991).

104. Rosenberg and others (1991), p. 20.

105. Luecke and others (1999).

106. Zamora-Arroyo and others (2001).

107. Luecke and others (1999).

108. Glennon and Culp (2002); Cohn (2004).

Chapter 28

1. Lines and Bilhorn (1996).

2. Lines and Bilhorn (1996).

3. Lovich and Meyer (2002).

4. Snyder (1918).

5. G. C. Lines, written communication, (2004).

6. Reviews of the climate of this drainage basin are given in Enzel (1992) and Lines (1996, 1999).

7. Enzel and others (1989); Enzel (1992).

8. Lines (1996).

9. From http://www.spl.usace.army.mil/resreg/htdocs/mojv.html, accessed November 4, 2005.

10. Lines (1996).

11. Stamos and others (2001).

12. Lines (1996); Stamos, Nishikawa, and Martin (2001).

13. D. Van Dyke (1997).

14. Stamos, Nishikawa, and Martin (2001).

15. Cox, Hillhouse, and Owen (2003); Jefferson (2003).

16. Izbicki, Radyk, and Michel (2000).

17. Izbicki, Martin, and Michel (1995).

18. Lines (1996); Stamos and others (2001); Stamos, Nishikawa, and Martin (2001).

19. Stamos, Nishikawa, and Martin (2001).

20. Stamos and others (2001).

21. Lines (1999).

22. Lines (1996).

23. Stamos, Nishikawa, and Martin (2001).

24. Lines and Bilhorn (1996), map.

25. Palmquist and Kailbourn (2000).

26. Webb, Boyer, and Berry (2001).

27. D. Thompson (1929).

28. Merriam (1893), p. 335.

29. Webb, Boyer, and Berry (2001).

30. Scott, Lines, and Auble (2000).

31. Stamos and others (2001).

32. Lines and Bilhorn (1996).

33. Stamos and others (2001).

34. Stamos and others (2001).

35. G. C. Lines, written communication (2004).

36. Egan (1997).

37. Lines (1996).

38. Durrenberger and Ingram (1978); Engstrom (1996).

39. D. Thompson (1929).

40. Egan, Chavez, and West (1993); Egan (1997).

Chapter 29

1. Webb (1996); Turner, Webb, and Bowers (2003); Griffiths, Webb, and Melis (2004); Webb, Belnap, and Weisheit (2004).

2. Webb, Melis, and others (1999). Although the photographs were of the river corridor through Grand Canyon, the subject area was the debris fan of Lava Falls Rapid with about 15 feet of total relief.

3. Turner, Webb, and Bowers (2003).

4. Bowers, Webb, and Rondeau (1995).

5. As noted in chapter 3, we combined *Baccharis salicifolia* (= *B. glutinosa*) and *B. emoryi* in our interpretations.

6. Webb, Belnap, and Weisheit (2004).

7. Friedman and others (2005).

8. D. Brown and others (1976).

Chapter 30

1. Bahre (1991), 176.

2. Whittlesey (1997b); see also Whittlesey (1997a).

3. Paleoflood records for the Southwest are summarized in Ely and others (1993). Also see House and others (2002).

4. Fonseca (1998); Turner, Webb, and Bowers (2003).

5. Turner, Webb, and Bowers (2003).

6. Melis and others (1996).

7. Webb, Belnap, and Weisheit (2004).

8. Stromberg (1998a).

9. See review in Fenstermaker (2003).

10. W. Graf (1978); Allred and Schmidt (1999).

11. Friedman and others (2005).

12. Ohmart (1996); Clary and Cruse (2004).

13. Turner, Webb, and Bowers (2003).

14. Although it generally is known that the amount of grazing in riparian areas has been reduced, reach-specific data on the timing and amounts of reductions are difficult if not impossible to obtain for the entire region.

15. Kepner and others (2000).

16. Dobyns (1981).

17. Turner, Webb, and Bowers (2003).

18. Examples of fire effects include Busch (1995) and the Bill Williams River (chapter 15).

19. Busch (1995).

20. Bahre (1991).

21. Reilly (1999).

22. Cooke and Reeves (1976); W. Graf (1983); Webb (1985).

23. Webb and Hasbargen (1998).

24. Scott, Lines, and Auble (2000).

25. Luecke and others (1999); Zamora-Arroyo and others (2001); Hinojosa-Huerta and others (2002).

26. W. Johnson (1998).

27. L. Stevens (1989).

28. Stromberg, Tiller, and Richter (1996); Merritt and Cooper (2000).

Glossary of Plant Names

Alkali sacaton	*Sporobolus airoides*
Alligator juniper	*Juniperus deppeana*
Apache plume	*Fallugia paradoxa*
Arizona alder	*Alnus oblongifolia*
Arizona ash (Velvet ash)	*Fraxinus velutina*
Arizona sycamore	*Platanus racemosa* v. *wrightii*
Arizona walnut	*Juglans microcarpa* v. *major*
Arrowweed (Cachanilla)	*Tessaria sericea*
Athel tamarisk	*Tamarix aphylla*
Big sagebrush (Great Basin sagebrush)	*Artemisia tridentata*
Black willow (Goodding willow)	*Salix gooddingii*
Blue palo verde	*Cercidium floridum*
Bonpland willow	*Salix bonplandiana*
Boxelder	*Acer negundo*
Bulrush	*Scirpus* sp.
Burrobrush	*Hymenoclea monogyra*
California palm	*Washingtonia filifera*
California scale broom	*Lepidospartum squamatum*
Camelthorn	*Alhagi camelorum*
Carrizo grass	*Phragmites australis*
Catclaw	*Acacia greggii*
Cattail	*Typha domingensis*
Cattle saltbush, cattle spinach	*Atriplex polycarpa*
Chittamwood	*Sideroxylon lanuginosum* spp. *rigidum*
Colorado River hemp (Wild flax)	*Sesbania herbacea*
Coyote willow (Sandbar willow)	*Salix exigua*
Creosotebush	*Larrea tridentata*
Date palm	*Phoenix dactylifera*
Desert broom	*Baccharis sarothroides*
Desert hackberry	*Celtis pallida*
Desert olive	*Forestiera pubescens*
Desert willow	*Chilopsis linearis*
Elm	*Ulmus minor*
Fan palm	*Washingtonia filifera*
Foothill palo verde (Littleleaf palo verde)	*Cercidium microphyllum*
Fourwing saltbush	*Atriplex canescens*
Frémont cottonwood	*Populus fremontii*
Gambel oak	*Quercus gambelii*
Giant reed	*Arundo donax*
Greasewood	*Sarcobatus vermiculatus*
Green ash	*Fraxinus pennsylvanica*
Greythorn	*Ziziphus obtusifolia*
Honey mesquite	*Prosopis glandulosa* v. *glandulosa*
Horsetail reed	*Equisetum* sp.
Long-leaf brickellbush	*Brickellia longifolia*
Mesquite mistletoe	*Phoradendron californicum*
Mexican palo verde	*Parkinsonia aculeata*
Narrowleaf cottonwood	*Populus angustifolia*
Netleaf hackberry	*Celtis reticulata*
Ocotillo	*Fouquieria splendens*
One-seed juniper	*Juniperus monosperma*

Palmer saltgrass (Wild rice)	*Distichlis palmeri*
Peachleaf willow	*Salix amygdaloides*
Pinyon pine	*Pinus edulis*
Plains cottonwood	*Populus deltoides* v. *occidentalis*
Ponderosa pine	*Pinus ponderosa*
Quailbush	*Atriplex lentiformis*
Red willow	*Salix laevigata*
Redbud	*Cercis occidentalis*
Rubber rabbitbrush	*Chrysothamnus nauseosus*
Russian olive	*Elaeagnus angustifolia*
Saguaro	*Carnegia gigantea*
Saltbush	*Atriplex lentiformis, A. polycarpa*
Screwbean mesquite	*Prosopis pubescens*
Seepwillow	*Baccharis salicifolia* = *B. emoryi*
Shrub live oak	*Quercus turbinella*
Siberian elm	*Ulmus pumila*
Silverleaf willow (Yewleaf willow)	*Salix exilifolia* = *S. taxifolia*
Singleleaf ash	*Fraxinus anomala*
Skunkbush (Squawberry)	*Rhus aromatica* = *R. trilobata*
Soaptree yucca	*Yucca elata*
Sotol (Desert spoon)	*Dasylirion wheeleri*
Squaw baccharis	*Baccharis sergiloides*
Squawbush	*Ziziphus spathulata*
Sweet bebbia	*Bebbia juncea*
Sycamore	*Platanus racemosa*
Tamarisk	*Tamarix ramosissima* = *T. chinensis*
Tree-of-heaven	*Ailanthus altissima*
Tree tobacco	*Nicotiana glauca*
Tumbleweed	*Salsola iberica*
Utah juniper	*Juniperus osteosperma*
Velvet mesquite	*Prosopis velutina*
Water birch	*Betula occidentalis*
Western honey mesquite	*Prosopis glandulosa* v. *torreyana*
Western soapberry	*Sapindus saponaria*
Wild grape (canyon grape)	*Vitis arizonica*
Wright sacaton	*Sporobolus airoides wrightii*

References

Albee, B. J., Shultz, L. M., and Goodrich, S., 1988, Atlas of the Vascular Plants of Utah: Salt Lake City, Utah Museum of Natural History, Occasional Publication no. 7, 670 pp.

Albertson, F. W., and Weaver, J. E., 1945, Injury and death or recovery of trees in prairie climate: Ecological Monographs, v. 15, pp. 393–433.

Alder, D. D., and Brooks, K. F., 1996, A History of Washington County: From Isolation to Destination: Salt Lake City, Utah State Historical Society, 418 pp.

Aldridge, B. N., 1963, Floods of August 1963 in Prescott, Arizona: U.S. Geological Survey Open-File Report, 12 pp.

———, 1970, Floods of November 1965 to January 1966 in the Gila River Basin, Arizona and New Mexico, and Adjacent Basins in Arizona: U.S. Geological Survey Water-Supply Paper no. 1850-C, 176 pp.

Aldridge, B. N., and Eychaner, J. H., 1984, Floods of October 1977 in Southern Arizona and March 1978 in Central Arizona: U.S. Geological Survey Water-Supply Paper no. 2223, 143 pp.

Aldridge, B. N., and Hales, T. A., 1984, Floods of November 1978 to March 1979 in Arizona and West-Central New Mexico: U.S. Geological Survey Water-Supply Paper no. 2241, 149 pp.

Allen, A. M. 1937, The sequence of human occupancy in the middle Rio Verde valley, Arizona: Ph.D. diss., Clark University, Worcester, Mass., 224 pp.

Allred, T. M., and Schmidt, J. C., 1999, Channel narrowing by vertical accretion along the Green River near Green River, Utah: Geological Society of America Bulletin, v. 111, pp. 1757–772.

Alvarez de Williams, A., 1975, Travelers among the Cucapa: Los Angeles, Dawson's Book Shop, 161 pp.

Anderson, B. W., 1996, Salt cedar, revegetation, and riparian ecosystems in the Southwest, in Lovich, J., Randall, J., and Kelly, M. (editors), Proceedings of the California Exotic Pest Plant Council, Symposium '95: Sacramento, California Exotic Pest Plant Council, pp. 32–41.

Anderson, B. W., Higgins, A., and Ohmart, R. D., 1977, Avian use of saltcedar communities in the lower Colorado River valley, in Johnson R. R., and Jones, D. A. (technical coordinators), Importance, Preservation, and Management of Riparian Habitat: A Symposium: U.S. Department of Agriculture, Forest Service General Technical Report no. RM-43, pp. 128–36.

Anderson, B. W., and Ohmart, R. D., 1984, Vegetation community type maps, lower Colorado River, Report on Contract no. 2-07-30-X0224 to the Bureau of Reclamation, Boulder City, Nevada: Tempe, Arizona State University, Center for Environmental Studies, 59 pp., maps.

Anderson, B. W., Ohmart, R. D., and Disano, J., 1979, Revegetating the riparian flood plain for wildlife, in Johnson, R. R., and McCormick, J. F. (editors), Strategies for Protection and Management of Floodplain Wetlands and Other Riparian Ecosystems: U.S. Department of Agriculture Forest Service Technical Report no. WO-12, pp. 318–31.

Anderson, B. W., Ohmart, R. D., and Rice, J., 1983, Avian and vegetation community structure and their seasonal relationships in the lower Colorado River valley: The Condor, v. 85, pp. 392–405.

Anderson, J. E., 1977, Transpiration and photosynthesis in saltcedar: Hydrology and Water Resources in Arizona and the Southwest, v. 7, pp. 125–31.

———, 1982, Factors controlling transpiration and photosynthesis in *Tamarix chinensis* Lour: Ecology, v. 63, pp. 48–56.

Andrade, E. R., and Sellers, W. D., 1988, El Niño and its effect on precipitation in Arizona: Journal of Climatology, v. 8, pp. 403–10.

Anning, D. W., and Duet, N. R., 1994, Summary of Ground-Water Conditions in Arizona, 1987–1990: U.S. Geological Survey Open-File Report no. 94-476, 2 sheets.

Anonymous, 1931, An historical record: Grand Canyon, Arizona: Grand Canyon Nature Notes, v. 5, pp. 123–24.

Anonymous, 1972, Flood: Graham-Greenlee, Arizona, October 1972: Safford, Ariz., Courier, 25 pp.

Apple, L. L., 1985, Riparian habitat restoration and beavers, in Johnson, R. R., Ziebell, C. D., Patton, D. R., Ffolliott, P. F., and Hamre, R. H. (editors), Riparian Ecosystems and Their Management: Reconciling Conflicting Uses: Tucson, Ariz., First North American Riparian Conference, U.S. Department of Agriculture, Forest Service General Technical Report no. RM-120, pp. 489–90.

Arizona Department of Game and Fish, 1992, Internal memo, February 14.

———, 1993, Arizona Riparian Inventory and Mapping Project: Phoenix, Arizona Department of Game and Fish, 41 pp. Available at: http://www.land.state.az.us/alris/htmls/data3.html.

Arizona State Parks, 1989, Arizona rivers, streams, and wetlands study, in Arizona State Comprehensive Outdoor Recreation Plan: Phoenix, Arizona State Parks.

Armstrong, D. M., 1982, Mammals of the Canyon Country: A Handbook of Mammals of Canyonlands National Park and Vicinity: Moab, Utah, Canyonlands Natural History Association, 263 pp.

Arnold, L., 1940, An ecological study of the vertebrate animals of the mesquite forest: M.A. thesis, University of Arizona, Tucson, 79 pp.

Aton, J. M., and McPherson, R. S., 2000, River Flowing from the Sunrise: An Environmental History of the Lower San Juan: Logan, Utah State University Press, 216 pp.

Auble, G. T., and Scott, M. L., 1998, Fluvial disturbance patches and cottonwood recruitment along the upper Missouri River, Montana: Wetlands, v. 18, pp. 546–70.

Baars, D. L., 1973, A River Runners Guide to the Canyons of the San Juan River: Durango, Colo., Four Corners Geological Society, 94 pp.

Baars, D. L., and Buchanan, R. C., 1994, The Canyon Revisited: A Rephotography of the Grand Canyon, 1923/1991: Salt Lake City, University of Utah Press, 167 pp.

Babcock, H. M., 1968, The phreatophyte problem in Arizona: Proceedings of the 12th Annual Arizona Watershed Symposium, pp. 34–36.

Bahre, C. J., 1991, A Legacy of Change: Historic Human Impact on Vegetation of the Arizona Borderlands: Tucson, University of Arizona Press, 213 pp.

Bahre, C. J., and Bradbury, D. E., 1978, Vegetation change along the Arizona-Sonora boundary: Annals of the Association of American Geographers, v. 68, pp. 145–65.

Bailey, J. K., Schweitzer, J. A., Rehill, B. J., Lindroth, R. L., Martinsen, G. D., and Whitham, T. G., 2004, Beavers as molecular geneticists: A genetic basis to the foraging of an ecosystem engineer: Ecology, v. 85, pp. 603–8.

Barnes, H. H., Jr., 1967, Roughness Characteristics of Natural Channels: U.S. Geological Survey Water-Supply Paper no. 1849, 213 pp.

Baum, B. R., 1966, Monographic Revision of the Genus *Tamarix*: Jerusalem, Hebrew University, Department of Botany, U.S. Geological Survey Final Research Report for Project no. A10-FS-9, 193 pp.

———, 1967, Introduced and naturalized tamarisks in the United States and Canada (Tamaricaceae): Baileya, v. 15, pp. 19–25.

———, 1978, The Genus *Tamarix*: Jerusalem, Israel Academy of Sciences and Humanities, 93 pp.

Beidelman, R. G., 2000, Rowboat botanizing with Willis Jepson on the Colorado River, 1912: Fremontia, v. 28, pp. 2–4.

Bell, W. A., 1869, New Tracks in North America (2 vols.): London, Chapman and Hall, 236 and 322 pp.

Belsky, A. J., Matzke, A., and Uselman, S., 1999, Survey of livestock influences on stream and riparian ecosystems in the western United States: Journal of Soil and Water Conservation, v. 54, pp. 419–31.

Benson, L., and Darrow, R. A., 1981, Trees and Shrubs of the Southwestern Deserts: Tucson, University of Arizona Press, 416 pp.

Bergman, C., 2003, Red Delta: Fighting for Life at the End of the Colorado River: Golden, Colo., Fulcrum, 312 pp.

Berry, W. L., 1970, Characteristics of salts secreted by *Tamarix aphylla*: American Journal of Botany, v. 57, pp. 1226–230.

Bessey, C. E., 1904, Botanical notes: The number and weight of cottonwood seeds: Science, v. 20, pp. 118–99.

Betancourt, J. L., 1990, Tucson's Santa Cruz River and the arroyo legacy: Ph.D. diss., University of Arizona, Tucson, 232 pp.

Betancourt, J. L., and Turner, R. M., 1985, Historic arroyo-cutting and subsequent channel changes at the Congress Street crossing, Santa Cruz River, Tucson, Arizona, *in* Whitehead, E. E., Hutchinson, C. F., Timmermann, B. N., and Varady,

R. G. (editors), Arid Lands, Today and Tomorrow: Boulder, Colo., Westview Press, pp. 1353–371.

———, 1990, Tucson's Santa Cruz River and the arroyo legacy: Tucson, U.S. Geological Survey, unpublished manuscript, 239 pp.

Beverage, J. P., and Culbertson, J. K., 1964, Hyperconcentrations of suspended sediment: Journal of the Hydraulics Division, American Society of Civil Engineers, v. 90, pp. 117–26.

Bishop, F. M., 1947, Captain Francis Marian Bishop's journal, edited by C. Kelley: Utah Historical Quarterly, v. 15, pp. 159–237.

Blake, T. A., and Steinhart, P., 1994, Two Eagles/Dos Aguilas: The Natural World of the United States–Mexico Borderlands: Berkeley, University of California Press, 202 pp.

Blaney, H. F., 1961, Consumptive use and water waste by phreatophytes: Journal of the Irrigation and Drainage Division, Proceedings of the American Society of Civil Engineers, v. 87, pp. 37–46.

Blasch, K. W., Hoffmann, J. P., Graser, L. F., Bryson, J. R., Flint, A. L., and DeWitt, E., 2006, Hydrogeology of the Upper and Middle Verde River Watersheds of Central Arizona: U.S. Geological Survey Scientific Investigations Report no. 2005-5198, 101 pp., 3 plates.

Blinn, D. W., and Cole, G. A., 1991, Algal and invertebrate biota in the Colorado River: Comparison of pre- and post-dam conditions, *in* Colorado River Ecology and Dam Management: Washington, D.C., National Academy Press, pp. 102–23.

Bornette, G., and Amoros, C., 1996, Disturbance regimes and vegetation dynamics: Role of floods in riverine wetlands: Journal of Vegetation Science, v. 7, pp. 615–22.

Bourke, J. G., 1891, On the Border with Crook: New York, Charles Scribner's Sons, 491 pp.

Bowers, J. E., Webb, R. H., and Rondeau, R. J., 1995, Longevity, recruitment, and mortality of desert plants in Grand Canyon, Arizona, U.S.A.: Journal of Vegetation Science, v. 6, pp. 551–64.

Bowie, J. E., and Kam, W., 1968, Use of Water by Riparian Vegetation, Cottonwood Wash, Arizona: U.S. Geological Survey Water-Supply Paper no. 1858, 62 pp.

Braatne, J. H., Rood, S. B., and Heilman, P. E., 1996, Life history, ecology, and conservation of riparian cottonwoods in North America, *in* Stettler, R. R., Bradshaw, H. D., Jr., Heilman, P. E., and Hinckley, T. M. (editors), Biology of *Populus* and Its Implications for Management and Conservation: Ottawa,

National Research Council of Canada, NRC Research Press, pp. 57–85.

Brady, W., Patton, D. R., and Paxson, J., 1985, The development of Southwestern riparian gallery forests, *in* Johnson, R. R., Ziebell, C. C., Patton, D. R., Ffolliott, P. F., and Hamre, R. H. (editors), Riparian Ecosystems and Their Management: Reconciling Conflicting Uses: Tucson, Ariz., First North American Riparian Conference, U.S. Department of Agriculture, Forest Service General Technical Report no. RM-120, pp. 39–43.

Brandenburg, F. H., 1911, Floods in southwestern Colorado and northwestern New Mexico: Monthly Weather Review, v. 39, pp. 1570–572.

Bredehoeft, J. D., Papadopulus, S. S., and Cooper, H. H., Jr., 1982, The Water Budget Myth in Scientific Basis of Water-Resource Management: Washington, D.C., National Academy Press, pp. 51–57.

Bristow, B., 1968, Statement by Arizona Game and Fish Dept. on phreatophyte clearing projects: Proceedings of the 12th Annual Arizona Watershed Symposium, pp. 41–43.

Brock, J. H., 1994, *Tamarix* spp. (salt cedar), an invasive exotic woody plant in arid and semi-arid riparian habitats of western USA, *in* de Waal, L. C., Child, L. E., Wade, P. M., and Brock, J. H. (editors), Ecology and Management of Invasive Riverside Plants: New York, John Wiley and Sons, pp. 27–44.

Brooks, J., 1944, Journal of Thales H. Haskell: Utah Historical Quarterly, v. 12, pp. 69–98.

Brookshire, E. N. J., Kauffman, J. B., Lytjen, D., and Otting, N., 2002, Cumulative effects of wild ungulate and livestock herbivory on riparian willows: Oecologia, v. 132, 559–66.

Brothers, T. S., 1981, Historical vegetation change in the Owens River riparian woodland: Proceedings of the California Riparian Systems Conference, Davis, California, September 17–19, pp. 75–84.

Brotherson, J. D., and Winkel, V., 1986, Habitat relationships of saltcedar (*Tamarix ramosissima*) in central Utah: Great Basin Naturalist, v. 46, pp. 535–41.

Brown, B. T., 1988, Breeding ecology of a willow flycatcher population in Grand Canyon, Arizona: Western Birds, v. 19, pp. 25–33.

———, 1992, Nesting chronology, density, and habitat use of black-chinned hummingbirds along the Colorado River, Arizona: Journal of Field Ornithology, v. 63, pp. 393–400.

Brown, B. T., Carothers, S. W., and Johnson, R. R., 1987, Grand Canyon Birds: Tucson, University of Arizona Press, 302 pp.

Brown, B. T., and Trosset, M. W., 1989, Nesting-habitat relationships of riparian birds along the Colorado River in Grand Canyon, Arizona: The Southwestern Naturalist, v. 34, pp. 260–70.

Brown, D. E., 1985, Arizona Wetlands and Waterfowl: Tucson: University of Arizona Press, 169 pp.

——— (editor), 1994, Biotic Communities, Southwestern United States and Northwestern Mexico: Salt Lake City, University of Utah Press, 342 pp.

———, 1995, We need to avoid riparian hysteria: High Country News, October 2, 1995. Available at: http://www.hcn.org/servlets/hcn.Article?article_id=1369, accessed November 4, 2005.

Brown, D. E., Carmony, N. B., Lowe, C. H., and Turner, R. M., 1976, A second locality for native California fan palms (*Washingtonia filifera*) in Arizona: Arizona Academy of Science, v. 11, pp. 37–41.

Brown, D. E., Carmony, N. B., and Turner, R. M., 1977, Inventory of riparian habitats, *in* Johnson, R. R., and Jones, D. A. (technical coordinators), Importance, Preservation, and Management of Riparian Habitat: A Symposium: U.S. Department of Agriculture, Forest Service General Technical Report no. RM-43, pp. 10–13.

———, 1981, Drainage Map of Arizona Showing Perennial Streams and Some Important Wetlands (3d ed.): Phoenix, Arizona Game and Fish Department, 1 sheet, scale 1:1,000,000.

Brown, D. E., and Lowe, C. H., 1974, A digitized computer-compatible classification for natural and potential vegetation in the Southwest with particular reference to Arizona: Journal of the Arizona Academy of Science, v. 9, pp. 1–11.

———, 1980, Biotic Communities of the Southwest: U.S. Department of Agriculture General Technical Report no. RM-78, 1 map, scale 1:1,000,000.

Brown, J. S., 1923, The Salton Sea Region, California: U.S. Geological Survey Water-Supply Paper no. 497, 292 pp.

Bryan, K., 1925, Date of channel trenching (arroyo cutting) in the arid Southwest: Science, v. 62, pp. 338–44.

———, 1928, Change in plant associations by change in ground water level: Ecology, v. 9, pp. 474–78.

Bryan, K., and La Rue, E. C., 1927, Persistence of features in an arid landscape: The Navajo Twins, Utah: Geographical Review, v. 17, pp. 251–57.

Bull, W. B., and Scott, K. M., 1974, Impact of mining gravel from urban stream beds in the southwestern United States: Geology, v. 2, pp. 171–74.

Burdick, A., 2005, The truth about invasive species: Discover, v. 26, pp. 35–42.

Bureau of Reclamation, 1963, Lower Colorado River Water Salvage Phreatophyte Control, Arizona-California-Nevada: Boulder City, Nev., Bureau of Reclamation, Reconnaissance Report, 57 pp.

———, 1999, Long-term Restoration Program for the Historical Southwestern Willow Flycatcher (*Empidonax traillii extimus*) Habitat along the Lower Colorado River: Boulder City, Nev., Bureau of Reclamation, Lower Colorado Region, 70 pp.

Burkham, D. E., 1970, Precipitation, Streamflow, and Major Floods at Selected Sites in the Gila River Drainage Basin above Coolidge Dam, Arizona: U.S. Geological Survey Professional Paper no. 655-B, 33 pp.

———, 1972, Channel Changes of the Gila River in Safford Valley, Arizona, 1846–1970: U.S. Geological Survey Professional Paper no. 655-G, 24 pp.

———, 1976a, Effects of Changes in an Alluvial Channel on the Timing, Magnitude, and Transformation of Flood Waves, Southeastern Arizona: U.S. Geological Survey Professional Paper no. 655-K, 25 pp.

———, 1976b, Hydraulic Effects of Changes in Bottom-Land Vegetation on Three Major Floods, Gila River in Southeastern Arizona: U.S. Geological Survey Professional Paper no. 655-J, 14 pp.

Busby, F. E., Jr., and Schuster, J. L., 1971, Woody phreatophyte infestation of the middle Brazos River floodplain: Journal of Range Management, v. 24, pp. 285–87.

Busch, D. E., 1995, Effects of fire on southwestern riparian plant community structure: The Southwestern Naturalist, v. 40, pp. 259–67.

Busch, D. E., Ingraham, N. L., and Smith, S. D., 1992, Water uptake in woody riparian phreatophytes of the southwestern United States: A stable isotope study: Ecological Applications, v. 2, pp. 450–59.

Busch, D. E., and S. D. Smith, 1993, Effects of fire on water and salinity relations of riparian woody taxa: Oecologia, v. 94, pp. 186–94.

———, 1995, Mechanisms associated with decline of woody species in riparian ecosystems of the southwestern U.S.: Ecological Monographs, v. 65, pp. 347–70.

Butler, E., and Mundorf, J. C., 1970, Floods of December 1966 in Southwestern Utah: U.S. Geological Survey Water-Supply Paper no. 1870-A, 40 pp.

Byrkit, J. W., 1978, A log of the Verde, the "taming" of an Arizona river: Journal of Arizona History, v. 19, pp. 31–54.

Caldwell, M. M., Dawson, T. E., and Richards, J. H., 1998, Hydraulic lift: Consequences of water efflux from the roots of plants: Oecologia, v. 113, pp. 151–61.

Campbell, C. J., and W. Green, 1968, Perpetual succession of stream-channel vegetation in a semiarid region: Journal of the Arizona Academy of Science, v. 5, pp. 86–98.

Carman, J. G., and Brotherson, J. D., 1982, Comparisons of sites infested and not infested with saltcedar (*Tamarix pentandra*) and Russian olive (*Elaeagnus angustifolia*): Weed Science, v. 30, pp. 360–64.

Carothers, S. W., and Brown, B. T., 1991, The Colorado River through Grand Canyon: Natural History and Human Change: Tucson, University of Arizona Press, 235 pp.

Carothers, S. W., Johnson, R. R., and Aitchison, S. W., 1974, Population structure and social organization of Southwestern riparian birds: American Zoology, v. 14, pp. 97–108.

Castetter, E. F., and Bell, W. H., 1951, Yuman Indian Agriculture: Primitive Subsistence on the Lower Colorado and Gila Rivers: Albuquerque, University of New Mexico Press, 274 pp.

Cayan, D. R., and Webb, R. H., 1992, El Niño/Southern Oscillation and streamflow in the western United States, *in* Diaz, H. F., and Markgraf, V. (editors), El Niño, Historical and Paleoclimatic Aspects of the Southern Oscillation: Cambridge, Cambridge University Press, pp. 29–68.

Cecil-Stephens, B. A., 1891, The Colorado Desert and its recent flooding: Journal of the American Geographical Society of New York, v. 23, pp. 367–77.

Chadde, S. W., and Kay, C. E., 1991, Tall-willow communities on Yellowstone's northern range: A test of the "natural-regulation" paradigm, *in* Keiter, R. B., and Boyce, M. S. (editors), The Greater Yellowstone Ecosystem: New Haven, Conn., Yale University Press, pp. 231–62.

Childs, C., 2000, The Secret Knowledge of Water: Discovering the Essence of the American Desert: Seattle, Sasquatch Books, 288 pp.

Christensen, E. M., 1962, The rate of naturalization of tamarisk in Utah: American Midland Naturalist, v. 68, pp. 51–57.

———, 1963, Naturalization of Russian olive (*Elaeagnus angustifolia* L.) in Utah: American Midland Naturalist, v. 70, pp. 133–37.

Clark, S., 1987, Potential for use of cottonwoods in dendrogeomorphology and paleohydrology, M.S. thesis, University of Arizona, Tucson, 52 pp.

Clary, W. P., and Kruse, W. H., 2004, Livestock grazing in riparian areas: Environmental impacts, management practices, and management implications, *in* Baker, M. B., Jr., Ffolliott, P. F., DeBano, L. F., and Neary, D. G. (editors), Riparian Areas of the Southwestern United States: Hydrology, Ecology, and Management: New York, Lewis, pp. 237–58.

Cleland, R. G., 1950, This Reckless Breed of Men: The Trappers and Fur Traders of the Southwest: New York, A. A. Knopf, 361 pp.

Cleverly, J. R., Smith, S. D., Sala, A., and Devitt, D. A., 1997, Invasive capacity of *Tamarix ramosissima* in a Mojave Desert floodplain: The role of drought: Oecologia, v. 111, pp. 12–18.

Clover, E. U., and Jotter, L., 1944, Floristic studies in the canyon of the Colorado and tributaries: American Midland Naturalist, v. 32, pp. 591–642.

Coates, D. R., and Cushman, R. L., 1955, Geology and Ground-Water Resources of the Douglas Basin, Arizona: U.S. Geological Survey Water-Supply Paper no. 1354, 56 pp.

Cohan, D. R., Anderson, B. W., and Ohmart, R. D., 1978, Avian population responses to salt cedar along the lower Colorado River, *in* Johnson, R. R., and McCormick, J. F. (editors), Strategies for Protection and Management of Floodplain Wetlands and Other Riparian Ecosystems: U.S. Department of Agriculture, Forest Service General Technical Report no. WO-12, pp. 371–82.

Cohen, M. J., and Henges-Jeck, C., 2001, Missing Water: The Uses and Flows of Water in the Colorado River Delta Region: Oakland, Calif., Pacific Institute for Studies in Development, Environment, and Security, 44 pp.

Cohen, M. J., Henges-Jeck, C., and Castillo-Moreno, G., 2001, A preliminary water balance for the Colorado River delta, 1992–1998: Journal of Arid Environments, v. 49, pp. 35–48.

Cohn, J. P., 2001, Resurrecting the dammed: A look at Colorado River restoration: BioScience, v. 51, pp. 998–1003.

————, 2004, Colorado River delta: BioScience, v. 54, pp. 386–91.

Collier, M., Webb, R. H., and Schmidt, J. C., 1996, Dams and rivers: A primer on the downstream effects of dams: U.S. Geological Survey Circular no. 1126, 94 pp.

Colton, H. S., 1937, Some notes on the original condition of the Little Colorado River: A side light on the problems of erosion: Museum of Northern Arizona, Museum Notes, v. 10, pp. 17–20.

Condes de la Torre, A., 1970, Streamflow in the Upper Santa Cruz River Basin, Santa Cruz and Pima Counties, Arizona: U.S. Geological Survey Water-Supply Paper no. 1939-A, 26 pp.

Cook, W., 1987, The *Wen*, the *Botany*, and the *Mexican Hat*: The Adventures of the First Women through Grand Canyon on the Nevills Expedition: Orangevale, Calif., Calisto Press, 151 pp.

Cooke, R. U., and Reeves, R. W., 1976, Arroyos and Environmental Change in the American Southwest: London, Oxford University Press, 213 pp.

Cooley, M. E., 1979, Depths of channels in the area of the San Juan Basin Regional Uranium Study, New Mexico, Colorado, Arizona, and Utah: U.S. Geological Survey Open-File Report no. 79-1526, 38 pp.

Cooper, D. J., Merritt, D. M., Andersen, D. C., and Chimner, R. A., 1999, Factors controlling the establishment of Fremont cottonwood seedlings on the upper Green River, USA: Regulated Rivers: Research and Management, v. 15, 419–40.

Cooper, J. R., Gilliam, J. W., Daniels, R. B., and Robarge, W. P., 1987, Riparian areas as filters for agricultural sediment: Soil Science Society of America, v. 51, pp. 416–20.

Corle, E., 1951, The Gila: River of the Southwest: Lincoln, University of Nebraska Press, 402 pp.

Cory, H. T., 1915, The Imperial Valley and the Salton Sink, with introductory monograph by W. P. Blake: San Francisco, John J. Newbegin, 1,581 pp.

Cowardin, L. M., Carter, V., Golet, F. C., and LaRoe, E. T., 1979, Classification of Wetlands and Deepwater Habitats of the United States: Washington, D.C., U.S. Department of the Interior, Fish and Wildlife Service, Office of Biological Services Report no. FWS/OBS-79/31, 131 pp.

Cox, B. F., Hillhouse, J. W., and Owen, L. A., 2003, Pliocene and Pleistocene evolution of the Mojave River, and associated tectonic development of the Transverse Ranges and Mojave Desert, based on borehole stratigraphy studies and mapping of landforms and sediments near Victorville, California, *in* Enzel, Y., Wells, S. G., and Lancaster, N. (editors), Paleoenvironments and Paleohydrology of the Mojave and Southern Great Basin Deserts: Boulder, Colo., Geological Society of America Special Paper, v. 368, pp. 1–42.

Crampton, C. G., and Miller, D. E., 1961, Journal of two campaigns by the Utah Territorial Militia against the Navajo Indians, 1869: Utah Historical Quarterly, v. 29, pp. 149–76.

Crins, W. J., 1989, The Tamaricaceae in the southeastern United States: Journal of the Arnold Arboretum, v. 70, 403–25.

Crosby, A. L. (editor), 1965, Steamboat up the Colorado: From the Journal of Lieutenant Joseph Christmas Ives, United States Topographical Engineers, 1857–1858: Boston, Little, Brown, 112 pp.

Culler, R. C., and others. 1970. Objectives, Methods, and Environment: Gila River Phreatophyte Project, Graham County, Arizona: U.S. Geological Survey Professional Paper no. 655-A, 25 pp.

Culler, R. C., Hanson, R. L., Myrick, R. M., Turner, R. M., and Kipple, F. P., 1982, Evapotranspiration before and after Clearing Phreatophytes, Gila River Flood Plain, Graham County, Arizona: U.S. Geological Survey Professional Paper no. 655-P, 52 pp.

Cushing, F. H., 1965, The Nation of the Willows: Flagstaff, Ariz., Northland Press, 75 pp.

Dahm, C. N., Cleverly, J. R., Coonrod, J. E. A., Thibault, J. R., McDonnell, D. E., and Gilroy, D. J., 2002, Evapotranspiration at the land/water interface in a semi-arid drainage basin: Freshwater Biology, v. 47, pp. 831–43.

Daniels, D., 1968, Photography's wet-plate interlude in Arizona Territory, 1864–1880: Journal of Arizona History, v. 9, pp. 171–94.

Daubenmire, R., 1968, Plant Communities: A Textbook of Plant Synecology: New York, Harper and Row, 300 pp.

Davenport, D. C., Martin, P. E., and Hagan, R. M., 1982, Evapotranspiration from riparian vegetation: Water relations and irrecoverable losses for saltcedar: Journal of Soil and Water Conservation, v. 37, pp. 233–36.

Davis, G. P., Jr., 1986, Man and Wildlife in Arizona: The American Exploration Period, 1824–1865 (edited by N. B. Carmony and D. E. Brown): Phoenix, Arizona Game and Fish Department, 231 pp.

Davis, O. K., and Turner, R. M., 1986, Palynological evidence for the historic expansion of juniper and desert shrubs in Arizona, U.S.A.: Review of Palaeobotany and Palynology, v. 49, pp. 177–93.

Dawdy, D. R., 1991, Hydrology of Glen Canyon and the Grand Canyon, *in* Colorado River Ecology and Dam Management: Washington, D.C., National Academy Press, pp. 40–53.

Dawson, T. E., and Ehleringer, J. R., 1991, Streamside trees that do not use stream water: Nature, v. 350, pp. 335–37.

DeBano, L. F., Brejda, J. J., and Brock, J. H., 1984, Enhancement of riparian vegetation following shrub control in Arizona chaparral: Journal of Soil and Water Conservation, v. 39, pp. 317–20.

DeBano, L. F., DeBano, S. J., Wooster, D. E., and Baker, M. B., Jr., 2004, Linkages between riparian corridors and surrounding watersheds, *in* Baker, M. B., Jr., Ffolliott, P. F., DeBano, L. F., and Neary, D. G. (editors), Riparian Areas of the Southwestern United States: Hydrology, Ecology, and Management: New York, Lewis, pp. 77–97.

DeBolt, A. M., and McCune, B., 1995, Ecology of *Celtis reticulata* in Idaho: Great Basin Naturalist, v. 55, pp. 2337–348.

Decker, J. P., Gaylor, W. G., and Cole, F. D., 1962, Measuring transpiration of undisturbed tamarisk shrubs: Plant Physiology, v. 37, pp. 393–97.

Dellenbaugh, F. S., 1908, A Canyon Voyage: New Haven, Conn., Yale University Press, 277 pp.

DeLoach, C. J., 1997, Biological control of weeds in the United States and Canada, *in* Luken, J. O., and Thieret, J. W. (editors), Assessment and Management of Plant Invasions: New York, Springer Verlag, pp. 172–94.

DeLoach, C. J., Carruthers, R. I., Lovich, J. E., Dudley, T. L., and Smith, S. D., 1999, Ecological interactions in the biological control of saltcedar (*Tamarix* spp.) in the United States: Toward a new understanding, *in* Spencer, N. R., (editor), Proceedings of the X International Symposium on Biological Control of Weeds, July 4–14, 1999: Bozeman, Montana State University, pp. 3–57.

DeLoach, C. J., Gerling, D., Fornasari, L., Sobhian, R., Myartseva, S., Mityaev, I. D., Lu, Q. G., Tracy, J. L, Wang, R., Wang, J. F., Kirk, A., Pemberton, R. W., Chikatunov, V., Jashenko, R. V., Johnson, J. E., Zheng, H., Jiang, S. OL., Liu, M. T., Liu, A. P., and Cisneroz, J., 1996, Biological control programme against saltcedar (*Tamarix* spp.) in the United States of America: Progress and problems, *in* Moran, V. C., and Hoffmann, J. H. (editors), Proceedings of the IX International Symposium on Biological Control of Weeds: Stellenbosch, South Africa, University of Cape Town, pp. 253–60.

DeLoach, C. J., Lewis, P. A., Herr, J. C., Carruthers, R. I., Tracy, J. L., and Johnson, J., 2003, Host specificity of the leaf beetle, *Diorhabda elongata deserticola* (Coleoptera: Chrysomelidae) from Asia, a biological control agent for saltcedars (*Tamarix*: Tamaricaceae) in the western United States: Biological Control, v. 27, pp. 117–47.

Despain, D., Houston, D., Meagher, M., and Schullery, P., 1986, Wildlife in Transition: Man and Nature on Yellowstone's Northern Range: Boulder, Colo., Roberts Rhinehart, 142 pp.

Devitt, D. A., Piorkowski, J. M., Smith, S. D., Cleverly, J. R., and Sala, A., 1997, Plant water relations of *Tamarix ramo-* *sissima* in response to the imposition and alleviation of soil moisture stress: Journal of Arid Environments, v. 36, pp. 527–40.

Devitt, D. A., Sala, A., Mace, K. A., and Smith, S. D., 1997. The effect of applied water on the water use of saltcedar in a desert riparian environment: Journal of Hydrology, v. 192, pp. 233–46.

Dick-Peddie, W. A., and Hubbard, J. P., 1977, Classification of riparian vegetation, *in* Johnson, R. R., and Jones, D. A. (technical coordinators), Importance, Preservation, and Management of Riparian Habitat: A Symposium: U.S. Department of Agriculture, Forest Service General Technical Report no. RM-43, pp. 85–90.

Dill, D. B., Jr., 1987, Terror on the Hassayampa: The Walnut Grove Dam disaster of 1890: Journal of Arizona History, v. 28, pp. 283–306.

Di Tomaso, J. M., 1998, Impact, biology, and ecology of saltcedar (*Tamarix* spp.) in the southwestern United States: Weed Technology, v. 12, pp. 326–36.

Dobyns, H. F., 1981, From Fire to Flood: Historic Human Destruction of Sonoran Desert Riverine Oases: Socorro, N.Mex., Ballena Press, 222 pp.

Dodge, N. N., 1936, Trees of Grand Canyon National Park: Grand Canyon, Ariz., Grand Canyon Natural History Association, Bulletin no. 3, 69 pp.

Dollar, T., 2002, Leave it to beavers: Wildlife Conservation, v. 105, pp. 28–35.

Drexler, J. Z., Snyder, R. L., Spano, D., and Paw U, K. T., 2004, A review of models and micrometeorological methods used to estimate wetland evapotranspiration: Hydrological Processes, v. 18, pp. 2071–101.

DuBois, S. M., and Smith, S. W., 1980, The 1887 Earthquake in San Bernardino Valley, Sonora: Historic Accounts and Intensity Patterns in Arizona: Arizona Geological Survey Special Paper no. 3, 112 pp.

Duke, A., 1974, When the Colorado River Quit the Ocean: Yuma, Ariz., Southwest Printers, 122 pp.

Durrenberger, R. W., and Ingram, R. S., 1978, Major Storms and Floods in Arizona, 1862–1977: Tempe, Ariz., Office of the State Climatologist, Climatological Publications, Precipitation Series no. 4, 44 pp.

Dutton, C. E., 1882, Tertiary History of the Grand Canyon District: U.S. Geological Survey Monograph no. 2, 264 pp.

Egan, T. B., 1997, Afton Canyon riparian restoration project, fourth year status report, *in* Proceedings of the Ninth Annual International Conference of the Society for Ecological Restoration: Fort Lauderdale, Fla., Society for Ecological Restoration, 20 pp.

Egan, T. B., Chavez, R. A., and West, B. R., 1993, Afton Canyon saltcedar removal: First year status report, *in* Young, D. D. (editor), Vegetation Management of Hot Desert Rangeland Ecosystems: Tucson, Ariz., University of Arizona, School of Renewable Natural Resources, no page numbers.

Ellis, L. M., 1995, Bird use of saltcedar and cottonwood vegetation in the Middle Rio Grande Valley of New Mexico, U.S.A.: Journal of Arid Environments, v. 30, pp. 339–49.

Ellis, L. M., Crawford, C. S., and Molles, M. C., Jr., 1998, Comparison of litter dynamics in native and exotic riparian vegetation along the middle Rio Grande of central New Mexico, U.S.A.: Journal of Arid Environments, v. 38, pp. 283–96.

Ely, L. L., and Baker, V. R., 1985, Reconstructing paleoflood hydrology with slackwater deposits, Verde River, Arizona: Physical Geography, v. 6, pp. 103–26.

Ely, L. L., Enzel, Y., Baker, V. R., and Cayan, D. R., 1993, A 5000-year record of extreme floods and climate change in the southwestern United States: Science, v. 262, pp. 410–12.

Emory, W. H., 1848, Notes of a Military Reconnoissance, from Fort Leavenworth, in Missouri, to San Diego, in California, Including Parts of the Arkansas, Del Norte, and Gila Rivers: New York, H. Long and Brother, 230 pp.

Engstrom, W. N., 1996, The California storm of January 1862: Quaternary Research, v. 46, pp. 141–48.

Enzel, Y., 1992, Flood frequency of the Mojave River and the formation of late Holocene playa lakes, southern California, USA: The Holocene, v. 2, pp. 11–18.

Enzel, Y., Cayan, D. R., Anderson, R. Y., and Wells, S. G., 1989, Atmospheric circulation during Holocene lake stands in the Mojave Desert: Evidence of regional climate change: Nature, v. 341, pp. 44–46.

Eschner, T. R., Hadley, R. F., and Crowley, K. D., 1983, Hydrologic and Morphologic Changes in Channels of the Platte River Basin in Colorado, Wyoming, and Nebraska: A Historical Perspective: U.S. Geological Survey Professional Paper no. 1277-A, 35 pp.

Everitt, B. L., 1968, Use of the cottonwood in an investigation of the recent history of a flood plain: American Journal of Science, v. 266, pp. 417–39.

———, 1979, Fluvial adjustments to the spread of tamarisk in the Colorado Plateau region: Discussion: Geological Society of America Bulletin, Part I, v. 90, p. 1183.

———, 1980, Ecology of saltcedar—A plea for research: Environmental Geology, v. 3, pp. 77–84.

———, 1995, Hydrologic factors in regeneration of Fremont cottonwood along the Fremont River, Utah, *in* Costa, J. E., Miller, A. J., Potter, K. W., and Wilcock, P. R. (editors), Natural and Anthropogenic Influences in Fluvial Geomorphology: Washington, D.C., American Geophysical Union, Geophysical Monograph no. 89, pp. 197–208.

———, 1998, Chronology of the spread of tamarisk in the central Rio Grande: Wetlands, v. 18, pp. 658–68.

Everitt, J. H., and DeLoach, C. J., 1990, Remote sensing of Chinese tamarisk (*Tamarix chinensis*) and associated vegetation: Weed Science, v. 38, pp. 273–78.

Felger, R. S., 2000, Flora of the Gran Desierto and Río Colorado of Northwestern Mexico: Tucson, University of Arizona Press, 673 pp.

Felger, R. S., Johnson, M. B., and Wilson, M. F., 2001, The Trees of Sonora, Mexico: New York, Oxford University Press, 391 pp.

Feller, J. M., 1998, Recent developments in the law affecting livestock grazing on Western riparian areas: Wetlands, v. 18, pp. 646–57.

Fenner, P., Brady, W. W., and Patton, D. R., 1984, Observations on seeds and seedlings of Fremont cottonwood: Desert Plants, v. 6, pp. 55–58.

———, 1985, Effects of regulated water flows on regeneration of Fremont cottonwood: Journal of Range Management, v. 38, pp. 135–38.

Fenstermaker, L. F., 2003, Estimation of evapotranspiration at different scales using traditional and remote sensing techniques: Ph.D. diss., University of Nevada, Las Vegas, 99 pp.

Ferrari, R. L., 1988, Lake Powell Survey: Denver, Colorado, Bureau of Reclamation Report no. REC-ERC-88-6, 67 pp.

Fleischner, T. L., 1994, Ecological costs of livestock grazing in western North America: Conservation Biology, v. 8, pp. 629–44.

Fonseca, J., 1998, Vegetation changes at Bingham Cienega, the San Pedro River valley, Pima County, Arizona, since 1879: Journal of the Arizona-Nevada Academy of Science, v. 31, pp. 103–16.

Force, E., and Howell, E., 1997, Holocene Depositional History and Anasazi Occupation in McElmo Canyon, Southwestern Colorado: Tucson, University of Arizona, Arizona State Museum Archaeological Series no. 188, 42 pp.

Fowler, D. D. (editor), 1976, "Photographed All the Best Scenery": Salt Lake City, University of Utah Press, 225 pp.

———, 1989, "Myself in the Waters": The Western Photographs of John K. Hillers: Washington, D.C., Smithsonian Institution Press, 166 pp.

Fox, C. K., 1936, The Colorado delta: A discussion of the Spanish explorations and maps, the Colorado silt load, and its seismic effect on the Southwest: Los Angeles, unpublished manuscript, no page numbers.

Fradkin, P. L., 1984, A River No More: The Colorado River and the West: Tucson, University of Arizona Press, reprint of the 1981 edition, 360 pp.

Freeman, W. B., 1909, Flood on the San Juan River September, 1909: Monthly Weather Review, v. 37, pp. 648–49.

Friederici, P., 1995, The alien saltcedar: American Forests, v. 101, pp. 45–47.

Friedman, J. M., Auble, G. T., Shafroth, P. B., Scott, M. L., Merigliano, M. F., Freehling, M. D., and Griffin, E. R., 2005, Dominance of non-native riparian trees in western USA: Biological Invasions, v. 7, pp. 747–51.

Friedman, J. M., and Lee, V. J., 2002, Extreme floods, channel change, and riparian forests along ephemeral streams: Ecological Monographs, v. 72, pp. 409–25.

Friedman, J. M., Osterkamp, W. R., and Lewis, W. M., Jr., 1996a, Channel narrowing and vegetation development following a Great Plains flood: Ecology, v. 77, pp. 2167–181.

———, 1996b, The role of vegetation and bed-level fluctuations in the process of channel narrowing: Geomorphology, v. 14, pp. 341–51.

Friedman, J. M., Osterkamp, W. R., Scott, M. L., and Auble, G. T., 1998, Downstream effects of dams on channel geometry and bottomland vegetation: Regional patterns in the Great Plains: Wetlands, v. 18, pp. 619–33.

Friedman, J. M., Vincent, K. R., and Shafroth, P. B., 2005, Dating floodplain sediments using tree-ring response to burial: Earth Surface Processes and Landforms, v. 30, pp. 1077–91.

Fuller, J. E., 1987, Paleoflood hydrology of the alluvial Salt River, Tempe, Arizona: M.S. thesis, University of Arizona, Tucson, 70 pp.

Garrett, J. M., and Gellenbeck, D. J., 1991, Basin Characteristics and Streamflow Statistics in Arizona as of 1989: U.S. Geological Survey Water-Resources Investigations Report no. 91-4041, 612 pp.

Gary, H. L., 1960, Utilization of Five-Stamen Tamarisk by Cattle: U.S. Department of Agriculture, Forest Service Research Note no. 51, 4 pp.

———, 1963, Root distribution of five-stamen tamarisk, seepwillow, and arrowweed: Forest Science, v. 9, pp. 311–14.

———, 1965, Some site relations in three flood-plain communities in central Arizona: Journal of the Arizona Academy of Science, v. 3, pp. 209–12.

Gaskin, J. F., and Schaal, B. A., 2002, Hybrid *Tamarix* widespread in U.S. invasion and undetected in native Asian range: Proceedings of the National Academy of Science, v. 99, pp. 11256–1259.

———, 2003, Molecular phylogenetic investigation of U.S. invasive *Tamarix*: Systematic Botany, v. 28, pp. 86–95.

Gaskin, J. F., and Shafroth, P. B., 2005, Hybridization of *Tamarix ramosissima* and *T. chinensis* (saltcedars) with *T. aphylla* (Athel) (Tamaricaceae) in the southwestern USA determined from DNA sequence data: Madroño, v. 52, pp. 1–10.

Gatewood, J. S., Robinson, T. W., Colby, B. R., Hem, J. D., and Halpenny, L. C., 1950, Use of Water by Bottom-Land Vegetation in Lower Safford Valley, Arizona: U.S. Geological Survey Water-Supply Paper no. 1103, 210 pp.

Gay, L. W., 1985, Evapotranspiration from saltcedar along the lower Colorado River, *in* Johnson, R. R., Ziebell, C. D., Patton, D. R., Ffolliott, P. F., and Hamre, R. H. (editors), Riparian Ecosystems and Their Management: Reconciling Conflicting Uses: Tucson, Ariz., First North American Riparian Conference, U.S. Department of Agriculture, Forest Service General Technical Report no. RM-120, 171–74.

Gay, L. W., and Sammis, T. W., 1977, Estimating phreatophyte transpiration: Hydrology and Water Resources in Arizona and the Southwest, v. 7, pp. 133–39.

Gesink, R. W., Tomanek, G. W., and Hulett, G. K., 1970, A descriptive survey of woody phreatophytes along the Arkansas River in Kansas: Transactions of the Kansas Academy of Science, v. 73, pp. 55–69.

Ginzburg, C., 1967, Organization of the adventitious root apex in *Tamarix aphylla*: American Journal of Botany, v. 54, pp. 4–8.

Gladwin, D. N., and Roelle, J. E., 1998, Survival of plains cottonwood (*Populus deltoides* subsp. *monilifera*) and saltcedar (*Tamarix ramosissima*) seedlings in response to flooding: Wetlands, v. 18, pp. 669–74.

Glenn, E. P., and Nagler, P. L., 2005, Comparative ecophysiology of *Tamarix ramosissima* and native trees in western U.S. riparian zones: Journal of Arid Environments, v. 61, pp. 419–46.

Glenn, E., Tanner, R., Mendez, S., Kehret, T., Moore, D., Garcia, J., and Valdes, C., 1998, Growth rates, salt tolerance, and water use characteristics of native and invasive riparian plants from the delta of the Colorado River, Mexico: Journal of Arid Environments, v. 40, pp. 281–94.

Glennon, R., 2002, Water Follies: Groundwater Pumping and the Fate of America's Fresh Waters: Washington, D.C., Island Press, 314 pp.

Glennon, R. J., and Culp, P. W., 2002, The last green lagoon: How and why the Bush administration should save the Colorado River delta: Ecology Law Quarterly, v. 28, pp. 903–92.

Glinski, R. L., 1977, Regeneration and distribution of sycamore and cottonwood trees along Sonoita Creek, Santa Cruz County, Arizona, *in* Johnson, R. R., and Jones, D. A. (technical coordinators), Importance, Preservation, and Management of Riparian Habitat: A Symposium: U.S. Department of Agriculture, Forest Service General Technical Report no. RM-43, pp. 116–23.

Glinski, R. L., and Brown, D. E., 1982, Mesquite (*Prosopis juliflora*) response to severe freezing in southwestern Arizona: Journal of the Arizona-Nevada Academy of Sciences, v. 17, pp. 15–18.

Glinski, R. L., and Ohmart, R. D., 1984, Factors of reproduction and population densities in the Apache cicada (*Diceroprocta apache*): The Southwestern Naturalist, v. 29, pp. 73–79.

Gordon, B. R., Parrott, G. P., and Smith, J. B., 1992, Vegetation changes in northern Arizona—the Alexander Gardner photos: Rangelands, v. 14, pp. 308–20.

Gould, F. W., 1951, Grasses of Southwestern United States: Tucson, University of Arizona Press, 343 pp.

Gould, J. R., and DeLoach, C. J., 2002, Biological control of invasive exotic plant species, *in* Tellman, B. (editor), Invasive Exotic Species in the Sonoran Region: Tucson, University of Arizona Press, pp. 284–306.

Graf, J. B., Webb, R. H., and Hereford, R., 1991, Relation of sediment load and flood-plain formation to climatic variability, Paria River drainage basin, Utah and Arizona: Geological Society of America Bulletin, v. 103, pp. 1405–415.

Graf, W. L., 1978, Fluvial adjustments to the spread of tamarisk in the Colorado Plateau region: Geological Society of America Bulletin, v. 89, pp. 1491–501.

———, 1979, Mining and channel responses: Annals of the Association of American Geographers, v. 69, pp. 262–75.

———, 1981, Channel instability in a braided, sand bed river: Water Resources Research, v. 17, pp. 1087–94.

———, 1982, Tamarisk and river-channel management: Environmental Management, v. 6, pp. 283–96.

———, 1983, Flood-related channel change in an arid-region river: Earth Surface Processes and Landforms, v. 8, pp. 125–39.

———, 1984, A probabilistic approach to the spatial assessment of river channel instability: Water Resources Research, v. 20, pp. 953–62.

Granger, B. H., 1960, Arizona Place Names: Tucson, University of Arizona Press, 519 pp.

Grater, R. K., 1945, Landslide in Zion Canyon, Zion National Park, Utah: Journal of Geology, v. 53, pp. 116–24.

Green, C. R., and Sellers, W. D., 1964, Arizona Climate: Tucson, University of Arizona Press, 503 pp.

Gregory, H. E., 1917, Geology of the Navajo Country: U.S. Geological Survey Professional Paper no. 93, 161 pp.

———, 1938, The San Juan Country: A Geographic and Geologic Reconnaissance of Southeastern Utah: U.S. Geological Survey Professional Paper no. 188, 123 pp.

———, 1945, Population of southern Utah: Economic Geography, v. 21, pp. 29–57.

———, 1950, Geology and Geography of the Zion Park Region, Utah and Arizona: U.S. Geological Survey Professional Paper no. 220, 200 pp.

Gregory, H. E., and Moore, R. C., 1931, The Kaiparowits Region: A Geographic and Geologic Reconnaissance of Parts of Utah and Arizona: U.S. Geological Survey Professional Paper no. 164, 161 pp.

Gregory, S. V., Swanson, F. J., McKee, W. A., and Cummins, K. W., 1991, An ecosystem perspective of riparian zones: Focus on links between land and water: Bioscience, v. 41, pp. 540–51.

Gries, D., Zeng, F., Foetzki, A., Arndt, S. K., Bruelheide, H., Thomas, F. M., Zhang, X., and Runge, M., 2003, Growth and water relations of *Tamarix ramosissima* and *Populus ephratica* on Taklamakan desert dunes in relation to depth to a permanent water table: Plant, Cell, and Environment, v. 26, pp. 725–36.

Griffin, E. R., and Smith, J. D., 2004, Floodplain stabilization by woody riparian vegetation during an extreme flood, *in* Bennett, S. J., and Simon, A. (editors), Riparian Vegetation and Fluvial Geomorphology: Washington, D.C., American Geophysical Union, Water Science and Application no. 8, pp. 221–36.

Griffiths, P. G., Webb, R. H., and Melis, T. S., 2004, Initiation and frequency of debris flows in Grand Canyon, Arizona: Journal of Geophysical Research, Surface Processes, v. 109, F04002, doi:10.1029/2003JF000077, 14 pp.

Grimm, N. B., 1993, Implications of climate change for stream communities, *in* Kareiva, P., Kingsolver, J., and Huey, R. (editors), Biotic Interactions and Global Change: Sunderland, Mass., Sinaver Associates, pp. 293–314.

Grinnell, J., 1914, An account of the mammals and birds of the lower Colorado Valley: University of California Publications in Zoology, v. 12, pp. 51–294.

Gruell, G. E., 1980, Fire's Influence on Wildlife Habitat on the Bridger-Teton National Forest, Wyoming, v. 1, Photographic Record and Analysis: Ogden, Utah: U.S. Department of Agriculture, Forest Service General Technical Report INT-235, 207 pp.

———, 2001, Fire in Sierra Nevada Forests: A Photographic Interpretation of Ecological Change Since 1849: Missoula, Mont., Mountain Press, 238 pp.

Guay, B. E., 2001, Preliminary hydrologic investigation of Topock Marsh, Arizona, 1995–98: Ph.D. diss., University of Arizona, Tucson, 334 pp.

Haase, E. F., 1972, Survey of floodplain vegetation along the lower Gila River in southwestern Arizona: Journal of the Arizona Academy of Science, v. 7, 75–81.

———, 1973, Draft Environmental Study: Gila River from the Confluence of the Salt River Downstream to Gillespie Dam: Tucson, University of Arizona, Office of Arid Lands Studies, report submitted to the U.S. Army Corps of Engineers, Los Angeles, 90 pp.

Haden, A., 1997, Benthic ecology of the Colorado River system through the Colorado Plateau region: M.S. thesis, Northern Arizona University, Flagstaff, 141 pp.

Hadley, D., Warshall, P., and Bufkin, D., 1991, Environmental change in Aravaipa 1870–1970: An ethnoecological survey: Phoenix, Ariz., Bureau of Land Management Cultural Resource Series no. 7, 368 pp., appendices.

Hadley, R. F., 1961, Influence of Riparian Vegetation on Channel Shape, Northeastern Arizona: U.S. Geological Survey Professional Paper no. 424C, pp. 30–31.

Hadley, R. F., Karlinger, M. R., Burns, A. W., and Eschner, T. R., 1987, Water development and associated hydrologic changes in the Platte River, Nebraska, U.S.A.: Regulated Rivers: Research and Management, v. 1, pp. 331–41.

Hafen, L. R. (editor), 1997, Fur Trappers and Traders of the Far Southwest: Twenty Biographical Sketches: Logan, Utah State University Press, reprint of the 1968 edition, 305 pp.

Hall, F. C., 2002a, Photo Point Monitoring Handbook: Part A—Field Procedures: U.S. Department of Agriculture, Forest Service General Technical Report no. PNW-GTR-526, 48 pp.

———, 2002b, Photo Point Monitoring Handbook: Part B—Concepts and Analysis: U.S. Department of Agriculture, Forest Service General Technical Report no. PNW-GTR-526, 134 pp.

Halpenny, L. C., and Halpenny, P. C., 1988, Review of the Hydrogeology of the Santa Cruz Basin in the Vicinity of the Santa Cruz–Pima County Line: Tucson, Ariz., Water Development Corporation report, variable page numbers.

Hamilton, W. L., 1978, Geological Map of Zion National Park, Utah: Springdale, Utah, Zion Natural History Association, 1 sheet, scale 1:31,680.

———, 1992, The Sculpturing of Zion: A Guide to the Geology of Zion National Park: Springdale, Utah, Zion Natural History Association, 132 pp.

Hanks, T. C., and Webb, R. H., 2006, Longitudinal profiles of rivers subject to transient aggradation and incision of debris fill: The Colorado River: Journal of Geophysical Research, Earth Surface, v. III, F02020, doi: 10.1029/2004JF000257.

Hansen, E. M., and Shwarz, F. K., 1981, Meteorology of Important Rainstorms in the Colorado River and Great Basin Drainages: U.S. Department of Commerce, National Oceanic and Atmospheric Administration, Hydrometeorological Report no. 50, 167 pp.

Hansen, E. M., Shwarz, F. K., and Riedel, J. T., 1977, Probable Maximum Precipitation Estimates in the Colorado River and Great Basin Drainages: U.S. Department of Commerce, National Oceanic and Atmospheric Administration, Hydrometeorological Report no. 49, 161 pp.

Hanson, R. B., 2001, The San Pedro River: A Discovery Guide: Tucson, University of Arizona Press, 205 pp.

Hardy, P. C., Griffin, D. J., Kuenzi, A. J., and Morrison, M. L., 2004, Occurrence and habitat use of passage neotropical migrants in the Sonoran Desert: Western North American Naturalist, v. 64, pp. 59–71.

Harlan, A., and Dennis, A. E., 1976, A preliminary plant geography of Canyon de Chelly National Monument: Journal of the Arizona Academy of Sciences, v. 11, pp. 69–78.

Harrell, M. A., and Eckel, E., 1939, Ground-Water Resources of the Holbrook Region, Arizona: U.S. Geological Survey Water-Supply Paper no. 836-B, 105 pp.

Harris, D. R., 1966, Recent plant invasions in the arid and semi-arid Southwest of the United States: Annals of the Association of American Geographers, v. 65, pp. 408–22.

Hastings, J. R., 1959, Vegetation change and arroyo cutting in southeastern Arizona: Journal of the Arizona Academy of Science, v. 1, pp. 60–67.

Hastings, J. R., and Turner, R. M., 1965, The Changing Mile: An Ecological Study of Vegetation Change with Time in the Lower Mile of an Arid and Semiarid Region: Tucson, University of Arizona Press, 317 pp.

Hattersley-Smith, G., 1966, The symposium on glacier mapping: Canadian Journal of Earth Sciences, v. 3, pp. 737–43.

Hays, M. E., 1984, Analysis of historic channel change as a method for evaluating flood hazard in the semi-arid Southwest: M.S. thesis, University of Arizona, Tucson, 41 pp.

Hefley, H. M., 1937a, Ecological studies on the Canadian River floodplain in Cleveland County, Oklahoma: Ecological Monographs, v. 7, pp. 345–402.

———, 1937b, The relations of some native insects to introduced food plants: Journal of Animal Ecology, v. 6, pp. 138–44.

Hem, J. D., 1967, Composition of Saline Residues on Leaves and Stems of Saltcedar (*Tamarix pentandra* Pallas.): U.S. Geological Survey Professional Paper no. 491-C, 9 pp.

Hendricks, W. O., 1990, The forest of the American Nile: Perspectives, Sherman Library and Gardens, no. 24, pp. 1–5.

Hendrickson, D. A., and Minckley, D. A., 1984, Ciénegas—Vanishing climax communities of the American Southwest: Desert Plants, v. 6, pp. 131–75.

Hereford, R., 1984, Climate and ephemeral-stream processes: Twentieth-century geomorphology and alluvial stratigraphy of the Little Colorado River, Arizona: Geological Society of America Bulletin, v. 95, pp. 654–68.

———, 1989, Modern alluvial history of the Paria River drainage basin, southern Utah: Quaternary Research, v. 25, pp. 293–311.

———, 1993, Entrenchment and Widening of the Upper San Pedro River, Arizona: U.S. Geological Society Special Paper no. 282, 46 pp.

———, 2002, Valley-fill alluviation during the Little Ice Age (ca. A.D. 1400–1880), Paria River basin and southern Colorado Plateau, United States: Geological Society of America Bulletin, v. 114, pp. 1550–563.

———, 2004, Map showing quaternary geology and geomorphology of the Lonely Dell Reach of the Paria River, Lees Ferry, Arizona, with accompanying pamphlet, *in* Webb, R. H., and Hereford, R., Comparative Landscape Photographs of the Lonely Dell Area and the Mouth of the Paria River: Geologic Investigations Series Map I-2771, scale 1:5,000, http://pubs.usgs.gov/imap/i2771/, accessed November 4, 2005.

Hereford, R., Jacoby, G. C., and McCord, V. A. S., 1996, Late Holocene Alluvial Geomorphology of the Virgin River in the Zion National Park Area, Southwest Utah: Boulder, Colo., Geological Society of America Special Paper no. 310, 41 pp.

Hereford, R., and Webb, R. H., 1992, Historic variation in warm-season rainfall on the Colorado Plateau, U.S.A.: Climatic Change, v. 22, pp. 239–56.

Hereford, R., Webb, R. H., and Graham, S., 2002, Precipitation History of the Colorado Plateau Region, 1900–2000: U.S. Geological Survey Fact Sheet 119-02, 4 pp.

Hereford, R., Webb, R. H., and Longpré, C. I., 2004, Precipitation History of the Mojave Desert Region, 1893–2001: U.S. Geological Survey Fact Sheet no. 117-03, 4 pp.

———, in press, Precipitation history and ecosystem response to multidecadal precipitation variability in the Mojave Desert region, 1893–2001: Journal of Arid Environments, v. 52.

Hickman, J. C., 1993, The Jepson Manual: Higher Plants of California: Berkeley, University of California Press, 1,400 pp.

Hill, M., Tillemans, B., Martin, D. W., and Platts, W., 2002, Recovery of riparian ecosystems in the upper Owens River watershed, *in* American Water Resources Association, Summer Specialty Conference, Ground Water/Surface Water Interactions, July 1–3, 2002, pp. 161–66.

Hindley, E. C., Bowns, J. E., Scherick, E. R., Curtis, P., and Forrest, J., 2000, A Photographic History of Vegetation and Stream Channel Changes in San Juan County, Utah: Logan, Utah State University Extension Service, unnumbered monograph, 121 pp.

Hinojosa-Huerta, O., Nagler, P. L., Carrillo-Guerrero, Y., Zamora-Hernández, E., García-Hernández, J., Zamora-Arroyo, F., Gillon, K., and Glenn, E. P., 2002, Andrade Mesa wetlands of the All-American Canal: Natural Resources Journal, v. 42, pp. 899–914.

Hoffmann, J. P., Ripich, M. A., and Ellett, K. M., 2002, Characteristics of Shallow Deposits beneath Rillito Creek, Pima County, Arizona: U.S. Geological Survey Water-Resources Investigations Report no. 01-4257, 51 pp.

Hoffmeister, D. F., 1971, Mammals of Grand Canyon: Chicago, University of Illinois Press, 183 pp.

Holden, P. B., and Stalnaker, C. B., 1975, Distribution and abundance of mainstream fishes of the middle and upper Colorado River basins, 1967–1973:

Transactions of the American Fisheries Society, v. 104, pp. 217–31.

Holmes, J. G., 1904, Report upon soil investigations of lands of Colorado River Land Company, Sociedad Anonima, in the Territory of Lower California, Mexico: Corona del Mar, Calif., Sherman Library, unpublished manuscript, 38 pp., table.

Hopkins, L., and Carruth, L. A., 1954, Insects associated with salt cedar in southern Arizona: Journal of Economic Entomology, v. 47, pp. 1126–129.

Horan, J. D., 1966, Timothy O'Sullivan: America's Forgotten Photographer: Garden City, N.Y., Doubleday, 334 pp.

Horton, J. L., 2001, Leaf gas exchange characteristics differ among Sonoran Desert riparian tree species: Tree Physiology, v. 21, pp. 233–41.

Horton, J. S., 1964, Notes on the Introduction of Deciduous Tamarisk: U.S. Department of Agriculture, Forest Service Research Note no. RM-16, 14 pp.

———, 1977, The development and perpetuation of the permanent tamarisk type in the phreatophyte zone of the Southwest, in Johnson, R. R., and Jones, D. A. (technical coordinators), Importance, Preservation, and Management of Riparian Habitat, A Symposium: U.S. Department of Agriculture, Forest Service General Technical Report RM-43, pp. 124–27.

House, P. K., and Pearthree, P. A., 1995, A geomorphologic and hydraulic evaluation of an extraordinary flood discharge estimate: Bronco Creek, Arizona: Water Resources Research, v. 31, pp. 3059–73.

House, P. K., Pearthree, P. A., and Klawon, J. E., 2002. Historical flood and paleoflood chronology of the lower Verde River, Arizona: Stratigraphic evidence and related uncertainties, in House, P. K., Webb, R. H., Baker, V. R., and Levish, D. R. (editors), Ancient Floods, Modern Hazards: Principles and Applications of Paleoflood Hydrology: Washington, D.C., American Geophysical Union, Water Science and Application Series, v. 5, pp. 267–93.

House, P. K., Webb, R. H., Baker, V. R., and Levish, D. R. (editors), 2002, Ancient Floods, Modern Hazards: Principles and Applications of Paleoflood Hydrology: Washington, D.C., American Geophysical Union, Water Science and Application Series, v. 5, 385 pp.

House, P. K., Wood, M. L., and Pearthree, P. A., 1999, Hydrologic and geomorphic characteristics of the Bill Williams River, Arizona: Arizona Geological Survey Open-File Report no. 99-446, 46 pp.

Howard, A., and Dolan, R., 1981, Geomorphology of the Colorado River in the Grand Canyon: Journal of Geology, v. 89, pp. 269–98.

Howard, C. S., 1947, Suspended Sediment in the Colorado River, 1925–41: U.S. Geological Survey Water-Supply Paper no. 998, 165 pp.

Hoyt, W. G., and Langbein, W. B., 1955, Floods: Princeton, N.J., Princeton University Press, 469 pp.

Huckleberry, G., 1994, Contrasting channel response to floods on the middle Gila River, Arizona: Geology, v. 22, pp. 1083–86.

———, 1995, Archaeological implications of late-Holocene channel changes on the middle Gila River, Arizona: Geoarchaeology, v. 10, pp. 159–82.

———, 1996a, Historical Channel Changes on the San Pedro River, Southeastern Arizona: Arizona Geological Survey Open-File Report no. 96-15, 22 pp.

———, 1996b, Historical Geomorphology of the Gila River: Arizona Geological Survey Open-File Report no. 96-14, 31 pp.

Hughes, L. E., 1993, "The devil's own"—tamarisk: Rangelands, v. 15, pp. 151–55.

Humphrey, R. R., 1987, 90 years and 535 miles: Vegetation Changes along the Mexican Border: Albuquerque, University of New Mexico Press, 448 pp.

Hunter, W. C., Anderson, B. W., and Ohmart, R. D., 1985, Summer avian community composition of *Tamarix* habitats in three southwestern desert riparian systems, in Johnson, R. R., Ziebell, C. D., Patton, D. R., Ffolliott, P. F., and Hamre, R. H. (editors), Riparian Ecosystems and Their Management: Reconciling Conflicting Uses: Tucson, Ariz., First North American Riparian Conference, U.S. Department of Agriculture, Forest Service General Technical Report no. RM-120, pp. 128–33.

Hunter, W. C., Ohmart, R. D., and Anderson, B. W., 1988, Use of exotic saltcedar (*Tamarix chinensis*) by birds in arid riparian systems: The Condor, v. 90, pp. 113–23.

Huseman, B. W., 1995, Wild River, Timeless Canyons: Baldwin Mollhausen's Watercolors of the Colorado: Tucson, University of Arizona Press, 232 pp.

Inskip, E. (editor), 1995, The Colorado River through Glen Canyon before Lake Powell: Historic Photo Journal, 1872 to 1964: Moab, Utah, Inskip Ink, 95 pp.

Iorns, W. V., Hembree, C. H., and Oakland, G. L., 1965, Water Resources of the Upper Colorado River Basin: U.S. Geological Survey Professional Paper no. 441, 370 pp.

Irvine, J. R., and West, N. E., 1979, Riparian tree species distribution and succession along the lower Escalante River, Utah: The Southwestern Naturalist, v. 24, 331–46.

Izbicki, J. A., Martin, P., and Michel, R. L., 1995, Source, movement, and age of groundwater in the upper part of the Mojave River basin, California, USA, in Applications of Tracers in Arid Zone Hydrology: Fontainebleau, France, International Association of Hydrological Sciences Publication no. 232, pp. 43–56.

Izbicki, J. A., Radyk, J., and Michel, R. L., 2000, Water movement through a thick unsaturated zone underlying an intermittent stream in the western Mojave Desert, southern California, USA: Journal of Hydrology, v. 238, pp. 194–217.

Jackson, E. 1973. Tumacacori's Yesterdays: Globe, Ariz.: Southwest Parks and Monuments Association, reprint of 1951 edition, 97 pp.

Jackson, J. J., Ball, J. T., and Rose, M. R., 1990, Assessment of the Salinity Tolerance of Eight Sonoran Desert Riparian Trees and Shrubs: Yuma, Ariz., Bureau of Reclamation report, 102 pp.

Jakle, M. D., and Gatz, T. A., 1985, Herpetofaunal use of four habitats of the middle Gila River drainage, Arizona, in Johnson, R. R., and McCormick, J. F. (editors), Strategies for Protection and Management of Floodplain Wetlands and Other Riparian Ecosystems: U.S. Department of Agriculture, Forest Service Technical Report no. WO-12, pp. 355–58.

Jarrell, W. M., and Virginia, R. A., 1990, Soil cation accumulation in a mesquite woodland: Sustained production and long-term estimates of water use and nitrogen fixation: Journal of Arid Environments, v. 18, pp. 51–58.

Jefferson, G. T., 2003, Stratigraphy and paleontology of the middle to late Pleistocene Manix Formation, and paleoenvironments of the central Mojave River, southern California, in Enzel, Y., Wells, S. G., and Lancaster, N. (editors), Paleoenvironments and Paleohydrology of the Mojave and Southern Great Basin Deserts: Boulder, Colo., Geological Society of America Special Paper no. 368, pp. 43–77.

Jibson, R. W., and Harp, E. L., 1996, The Springdale, Utah, landslide: An extraordinary event: Environmental and Engineering Geoscience, v. 2, pp. 137–50.

Johnson, K. L., 1987, Rangeland through Time: Laramie, University of Wyoming, Agricultural Experiment Station, Miscellaneous Publication no. 50, 188 pp.

Johnson, R. R., 1977, Synthesis and Management Implications of the Colorado River Research Program: Grand Canyon, Ariz., Grand Canyon National Park, Colorado River Research Series Contribution no. 47, 75 pp.

———, 1991, Historic changes in vegetation along the Colorado River in the Grand Canyon, *in* Colorado River Ecology and Dam Management: Washington, D.C., National Academy Press, pp. 178–206.

Johnson, R. R., and Haight, L. T., 1985, Avian use of xeroriparian ecosystems in the North American warm deserts, *in* Johnson, R. R., Ziebell, C. D., Patton, D. R., Ffolliott, P. F., and Hamre, R. H. (editors), Riparian Ecosystems and Their Management: Reconciling Conflicting Uses: Tucson, Ariz., First North American Riparian Conference, U.S. Department of Agriculture, Forest Service General Technical Report no. RM-120, pp. 156–60.

Johnson, R. R., and Simpson, J. M., 1988, Desertification of wet riparian ecosystems in arid regions of the North American Southwest, *in* Whitehead, E. E., Hutchinson, C. F., Timmermann, B. N., and Varady, R. G. (editors), Arid Lands: Today and Tomorrow: Boulder, Colo., Westview Press, pp. 1383–393.

Johnson, S., 1987, Can saltcedar be controlled? Fremontia, v. 15, pp. 19–20.

Johnson, W. C., 1992, Dams and riparian forests: Case study from the upper Missouri River: Rivers, v. 3, pp. 229–42.

———, 1994, Woodland expansion in the Platte River, Nebraska: Patterns and causes: Ecological Monographs, v. 64, 45–84.

———, 1998, Adjustment of riparian vegetation to river regulation in the Great Plains, USA: Wetlands, v. 18, pp. 608–18.

Johnson, W. C., Dixon, M. D., Simons, R., Jenson, S., and Larson, K., 1995, Mapping the response of riparian vegetation to possible flow reductions in the Snake River, Idaho: Geomorphology, v. 13, pp. 159–73.

Jones, K. B., and Glinski, P. C., 1985, Microhabitats of lizards in a southwestern riparian community, *in* Johnson, R. R., and McCormick, J. F. (editors), Strategies for Protection and Management of Floodplain Wetlands and Other Riparian Ecosystems: U.S. Department of Agriculture, Forest Service Technical Report no. WO-12, pp. 342–46.

Jones, S. V., 1949, Journal of Stephen Vandiver Jones, edited by H. E. Gregory: Utah State Quarterly, v. 17, pp. 19–174.

Jordan, G. L., and Maynard, M. L., 1970, The San Simon watershed: Historical Review: Progressive Agriculture in Arizona, v. 22, pp. 1–13.

Judd, B. I., Laughlin, J. M., Guenther, H. R., and Handegarde, R., 1971, The lethal decline of mesquite on the Casa Grande National Monument: Great Basin Naturalist, v. 31, pp. 153–59.

Junk, W. J., Bayley, P. B., and Sparks, R. E., 1989, The flood pulse concept in river-floodplain systems, *in* Dodge, D. P. (editor), Proceedings of the International Large River Symposium: Canadian Special Publication in Fisheries and Aquatic Sciences, v. 106, pp. 110–27.

Katz, G. L., Friedman, J. M., and Beatty, S. W., 2001, Effects of physical disturbance and granivory on establishment of native and alien riparian trees in Colorado, U.S.A.: Diversity and Distributions, v. 7, pp. 1–14.

Katz, G. L., and Shafroth, P. B., 2003, Biology, ecology, and management of *Elaeagnus angustifolia* L. (Russian olive) in western North America: Wetlands, v. 23, pp. 763–77.

Kay, C., 1990, Yellowstone's northern elk herd: A critical evaluation of the "natural-regulation" paradigm: Ph.D. diss., Utah State University, Logan, 490 pp.

Kearney, T. H., and Peebles, R. H., 1960, Arizona Flora: Berkeley, University of California Press, 1,085 pp.

Kearsley, M. J. C., and Ayers, T. J., 1999, Riparian vegetation responses: Snatching defeat from the jaws of victory and vice versa, *in* Webb, R. H., Schmidt, J. C., Marzolf, G. R., and Valdez, R. A. (editors), The Controlled Flood in Grand Canyon: Scientific Experiment and Management Demonstration: Washington, D.C., American Geophysical Union, Geophysical Monograph no. 110, pp. 309–27.

———, 2001, Review Assessment and Recommendations Regarding Terrestrial Riparian Vegetation Monitoring in the Colorado River Corridor of Grand Canyon: Flagstaff, Northern Arizona University, Department of Biological Sciences, Report on Cooperative Agreement CA-00-40-3180 to the Grand Canyon Monitoring and Research Center, 74 pp.

Kelsey, R. E., 2000, Photography in the field: Timothy O'Sullivan and the Wheeler Survey, 1871–1874: Ph.D. diss., Harvard University, 328 pp.

Kennedy, T. A., Finlay, J. C., and Hobbie, S. E., in press, Exotic saltcedar (*Tamarix ramosissima*) alters food web structure in a desert stream by changing resource availability: Ecological Applications.

Kennedy, T. A., and Hobbie, S. E., 2004, Saltcedar (*Tamarix ramosissima*) invasion alters organic matter dynamics in a desert stream: Freshwater Biology, v. 49, pp. 65–76.

Kennedy, T. A., Naeem, S., Howe, K. M., Knops, J. M. H., Tilman, D., and Reich, P., 2002, Biodiversity as a barrier to ecological invasion: Nature, v. 417, pp. 636–38.

Kepner, W. G., Edmonds, C. M., and

Watts, C. J., 2002, Remote Sensing and Geographic Information Systems for Decision Analysis in Public Resource Administration: A Case Study of 25 Years of Landscape Change in a Southwestern Watershed: Las Vegas, Nev., Environmental Protection Agency Report no. EPA/600/R-02/039, 23 pp.

Kepner, W. G., Watts, C. J., Edmonds, C. M., Maingi, J. K., and Marsh, S. E., 2000, A landscape approach for detecting and evaluating change in a semi-arid environment: Journal of Environmental Monitoring and Assessment, v. 64, pp. 179–95.

Kittredge, F. A., Finch, B. J., and Mitchell, R. R., 1926, Possible Routes between Zion National Park, Bryce National Park, and Grand Canyon National Park: U.S. Department of Agriculture, Bureau of Public Roads, Report of July 1926 to Dr. L. I. Hewes, Deputy Chief Engineer, San Francisco, 24 pp.

Klawon, J. E., 2000, Hydrology and Geomorphology of the Santa Maria and Big Sandy Rivers and Burro Creek, Western Arizona: Tucson, Arizona Geological Survey Open-file Report no. 00-02, 46 pp.

———, 2001, Upper Gila River Fluvial Geomorphology Study: Denver, Bureau of Reclamation, Technical Service Center, Catalog of Historical Changes in Arizona Series, 119 pp.

Klement, K. D., Heitschmidt, R. K., and Kay, C. E., 2001, Eighty Years of Vegetation and Landscape Changes in the Northern Great Plains: A Photographic Record: Miles City, Mont., U.S. Department of Agriculture, Agricultural Research Service, Conservation Research Report no. 45, 91 pp.

Klett, M., Bajakian, K., Fox, W. L., Marshall, M., Ueshina, T., and Wolfe, B., 2004, Third Views, Second Sights: A Rephotographic Survey of the American West: Santa Fe, Museum of New Mexico Press, 238 pp.

Klett, M., Manchester, E., Verburg, J., Bushaw, G., and Dingus, R., 1984, Second View: The Rephotographic Survey Project: Albuquerque, University of New Mexico Press, 211 pp.

Knapp, P. A., Warren, P. L., and Hutchinson, C. F., 1990, The use of large-scale aerial photography to inventory and monitor arid rangeland vegetation: Journal of Environmental Management, v. 31, pp. 29–38.

Knechtel, M. M., 1938, Geology and ground-water resources of the valley of Gila River and San Simon Creek, Graham County, Arizona: U.S. Geological Survey Water-Supply Paper no. 796-F, 222 pp.

Kniffen, F. B., 1931, Lower California studies. III. The primitive cultural landscape of the Colorado delta: University of California Publications in Geography, v. 5, no. 2, pp. 43–66.

———, 1932, Lower California studies. IV. The natural landscape of the Colorado delta: University of California Publications in Geography, v. 5, no. 4, pp. 149–244.

Knipmeyer, J. H., 2002, Butch Cassidy Was Here: Historic Inscriptions of the Colorado Plateau: Salt Lake City, University of Utah Press, 160 pp.

Knopf, F. L., and Olson, T. E., 1984, Naturalization of Russian-olive: Implications for Rocky Mountain wildlife: Wildlife Society Bulletin, v. 12, pp. 289–98.

Kolb, E. L., 1989, Through the Grand Canyon from Wyoming to Mexico: Tucson, University of Arizona Press, reprint of the 1914 edition, 344 pp.

Konieczki, A. D., and Heilman, J. A., 2004, Water-Use Trends in the Desert Southwest, 1950–2000: U.S. Geological Survey Scientific Investigations Report no. 2004-5148, 32 pp.

Kriegshauser, D., and Somers, P., 2004, Vegetation changes in a riparian community along the Dolores River downstream from McPhee Reservoir in southwestern Colorado, *in* van Riper, C., III, and Cole, K. L. (editors), The Colorado Plateau: Cultural, Biological, and Physical Research: Tucson, University of Arizona Press, pp. 129–36.

Krueper, D. J., 1993, Effects of land use practices on western riparian ecosystems, *in* Finch, D. M., and Stengel, P. W. (editors), Status and Management of Neotropical Migratory Birds: U.S. Department of Agriculture, Forest Service General Technical Report no. RM-229, pp. 321–30.

Krza, P., 2003, It's "bombs away" on New Mexico saltcedar: High Country News, v. 35, p. 7.

Kunkel, F., 1970, The Deposits of the Colorado River on the Fort Mojave Indian Reservation in California, 1850–1969: U.S. Geological Survey Open-File Report, 28 pp., appendixes.

Kupel, D. E., 2003, Fuel for Growth: Water and Arizona's Urban Environment: Tucson, University of Arizona Press, 294 pp.

Kusler, J. A., 1985, A call for action: Protection of riparian habitat in the arid and semi-arid West, *in* Johnson, R. R., Ziebell, C. D., Patton, D. R., Ffolliott, P. F., and Hamre, R. H. (editors), Riparian Ecosystems and Their Management: Reconciling Conflicting Uses: Tucson, Ariz., First North American Riparian Conference, U.S. Department of Agriculture, Forest Service General Technical Report no. RM-120, pp. 6–8.

Lacey, J. R., Ogden, P. R., and Foster, K. E., 1975, Southern Arizona Riparian Habitat: Spatial Distribution and Analysis: Tucson, University of Arizona, Office of Arid Lands Studies Bulletin no. 8, 148 pp.

Laney, R. L., 1977, Effects of Phreatophyte Removal on Water Quality in the Gila River Phreatophyte Project Area, Graham County, Arizona: U.S. Geological Survey Professional Paper no. 655-M, 23 pp.

Larson, A. K., 1961, "I Was Called to Dixie": Salt Lake City, Deseret News Press, 681 pp.

La Rue, E. C., 1916, Colorado River and Its Utilization: U.S. Geological Survey Water-Supply Paper no. 395, 231 pp.

———, 1925, Water Power and Flood Control of Colorado River below Green River, Utah: U.S. Geological Survey Water-Supply Paper no. 556, 176 pp.

Lee, W. T., 1904, The Underground Waters of Gila Valley, Arizona: U.S. Geological Survey Water-Supply and Irrigation Paper no. 104, Series O, Underground Waters 25, 71 pp.

———, 1905, Underground Waters of Salt River Valley, Arizona: U.S. Geological Survey Water-Supply and Irrigation Paper no. 136, Series O, Underground Waters 37, 196 pp.

Leffler, A. J., and Evans, A. S., 1999, Variation in carbon isotope composition among years in the riparian tree *Populus fremontii*, Oecologia, v. 119, pp. 311–19.

Leigh, R., 1941, Forgotten Waters: Adventure in the Gulf of California: Philadelphia, J. B. Lippincott, 321 pp.

Leopold, A., 1966, A Sand County Almanac: New York, Oxford University Press, 295 pp.

Lesica, P., and Miles, S., 2001, *Tamarix* growth at the northern margin of its naturalized range in Montana, USA: Wetlands, v. 21, pp. 240–46.

———, 2004, Beavers indirectly enhance the growth of Russian olive and tamarisk along eastern Montana Rivers: Western North American Naturalist, v. 64, pp. 93–100.

Levine, C. M., and Stromberg, J. C., 2001, Effects of flooding on native and exotic plant seedlings: Implications for restoring south-western riparian forests by manipulating water and sediment flows: Journal of Arid Environments, v. 49, pp. 111–31.

Lewis, P. A., DeLoach, C. J., Herr, J. C., Dudley, T. L., and Carruthers, R. I., 2003, Assessment of risk to native *Frankenia* shrubs from an Asian leaf beetle, *Diorhabda elongata deserticola* (Coleoptera: Chrysomelidae), introduced for biological control of saltcedars (*Tamarix* spp.)

in the western United States: Biological Control, v. 27, pp. 148–66.

Lewis, P. A., DeLoach, C. J., Knutson, A. E., Tracy, J. L., and Robbins, T. O., 2003, Biology of *Diorhabda elongata deserticola* (Coleoptera: Chrysomelidae), an Asian leaf beetle for biological control of saltcedars (*Tamarix* spp.) in the United States: Biological Control, v. 27, pp. 101–16.

Lines, G. C., 1996, Ground-Water and Surface-Water Relations along the Mojave River, Southern California: U.S. Geological Survey Water-Resources Investigations Report no. 95-4189, 43 pp.

———, 1999, Health of Native Riparian Vegetation and Its Relation to Hydrologic Conditions along the Mojave River, Southern California: U.S. Geological Survey Water-Resources Investigations Report no. 99-4112, 28 pp.

Lines, G. C., and Bilhorn, T. W., 1996, Riparian Vegetation and Its Water Use during 1995 along the Mojave River, Southern California: U.S. Geological Survey Water-Resources Investigations Report no. 96-4241, 10 pp., map.

Lingenfelter, R. E., 1978, Steamboats on the Colorado River: Tucson, University of Arizona Press, 195 pp.

Lite, S. J., and Stromberg, J. C., 2005, Surface water and ground-water thresholds for maintaining *Populus-Salix* forests, San Pedro River, Arizona: Biological Conservation, v. 125, pp. 153–67.

Lockett, H. C., 1939, Along the Beale Trail: A Photographic Account of Wasted Range Land: Lawrence, Kans., Haskell Institute, 56 pp.

Logan, M. F., 2002, The Lessening Stream: An Environmental History of the Santa Cruz River: Tucson, University of Arizona Press, 311 pp.

Loope, L. L., Sanchez, P. G., Tarr, P. W., Loope, L. L., and Anderson, R. L., 1988. Biological invasions of arid land reserves: Biological Conservation, v. 44, pp. 95–118.

Lopez, S. M., Anderson, D. E., and Springer, A., 2003, Upper Verde Valley riparian area historical analysis, *in* 2003 Verde River Almanac: Cottonwood, Ariz., Verde Watershed Association, pp. 118–19.

Lopez, S. M., and Springer, A. E., 2003, Assessment of Human Influence of Riparian Change in the Verde Valley, Arizona: Cottonwood, Ariz., Verde Watershed Research and Education Program, 42 pp. Available at: http://222.verde.nau.edu/Research/RiparianAssessment.

Lovich, J. E., and de Gouvenain, R. C., 1998, Saltcedar invasion in desert wetlands of the southwestern United States: Ecological and political implications, *in* Majumdar, S. J., Miller, E. W., and Brenner, F. J. (editors), Ecology of Wetlands and Associated Systems: Philadelphia, Pennsylvania Academy of Sciences, pp. 447–67.

Lovich, J., and Meyer, K., 2002, The western pond turtle (*Clemmys marmorata*) in the Mojave River, California, USA: Highly adapted survivor or tenuous relict? Journal of the Zoological Society of London, v. 256, pp. 537–45.

Lowe, C. H. (editor), 1980, The Vertebrates of Arizona, with Major Section on Arizona Habitats: Tucson, University of Arizona Press, 270 pp.

———, 1985, Amphibians and reptiles in Southwest riparian ecosystems, *in* Johnson, R. R., and McCormick, J. F. (editors), Strategies for Protection and Management of Floodplain Wetlands and Other Riparian Ecosystems: U.S. Department of Agriculture, Forest Service Technical Report no. WO-12, pp. 339–41.

Luecke, D. F., Pitt, J., Congdon, C., Glenn, E., Valdés-Casillas, C., and Briggs, M., 1999, A Delta Once More: Restoring Riparian and Wetland Habitat in the Colorado River Delta: Washington, D.C., Environmental Defense Fund, 52 pp., appendixes.

Lusby, G. C., Reid, V. H., and Knipe, O. D., 1971, Effects of Grazing on the Hydrology and Biology of the Badger Wash Basin in Western Colorado, 1953–66: U.S. Geological Survey Water-Supply Paper no. 1532-D, 90 pp.

MacDougal, D. T., 1906, The delta of the Rio Colorado: Bulletin of the American Geographical Society, v. 38, pp. 1–16.

———, 1917, A decade of the Salton Sea: Geographical Review, v. 3, pp. 457–73.

Magilligan, F. J., Nislow, K. H., and Graber, B. E., 2003, Scale-independent assessment of discharge reduction and riparian disconnectivity following flow regulation by dams: Geology, v. 31, pp. 569–72.

Mahoney, J. M., and Rood, S. B., 1998, Streamflow requirements for cottonwood seedling recruitment—an integrative model: Wetlands, v. 18, pp. 634–45.

Malanson, G. P., 1995, Riparian Landscapes: New York, Cambridge University Press, Cambridge Studies in Ecology, 296 pp.

Manning, S. J., Cashore, B. L., and Szewczak, J. M., 1996, Pocket gophers damage saltcedar (*Tamarix ramosissima*) roots: Great Basin Naturalist, v. 56, pp. 183–85.

Martin, D. D., 1954, Yuma Crossing: Albuquerque: University of New Mexico Press, 243 pp.

Martin, P. S., 1971, Trees and shrubs of the Grand Canyon, Lees Ferry to Diamond Creek: Tucson, University of Arizona, Desert Laboratory, unpublished manuscript, 16 pp.

Martin, R., 1989, A Story That Stands Like a Dam: New York, Henry Holt, 354 pp.

Masse, W. B., 1976, The Hohokam Expressway Project: A Study of Prehistoric Irrigation in the Salt River Valley, Arizona: Tucson, University of Arizona Press, 88 pp.

Mauz, K., 2002, Plants of the Santa Cruz Valley at Tucson: Desert Plants, v. 18, pp. 3–36.

McAuliffe, J. R., 1997, Rangeland water developments: Conservation solution or illusion? *in* Environmental, Economic, and Legal Issues Related to Rangeland Water Developments: Tempe, Arizona State University College of Law, proceedings of a symposium, November 13–15, pp. 310–38.

McClintock, J. H., 1985, Mormon Settlement in Arizona: Tucson, University of Arizona Press, reprint of the 1921 edition, 307 pp.

McDaniel, K. C., and Taylor, J. P., 2003, Saltcedar recovery after herbicide-burn and mechanical clearing practices: Journal of Range Management, v. 56, pp. 439–45.

McGinnies, W. J., Shantz, H. L., and McGinnies, W. G., 1991, Changes in Vegetation and Land Use in Eastern Colorado: U.S. Department of Agriculture, Agricultural Research Service Publication no. ARS-85, 165 pp.

McGrath, L. J., and van Riper, C., III, 2005, Influence of Riparian Tree Phenology on Lower Colorado River Spring-Migrating Birds: Implications of Flower Cueing: U.S. Geological Survey Open-File Report no. 2005-1140, 35 pp.

McKee, E. D., 1946, Kanab Canyon: The trail of scientists: Plateau, v. 18, pp. 33–42.

McLaughlin, S. P., 2004, Riparian flora, *in* Baker, M. B., Jr., Ffolliott, P. F., DeBano, L. F., and Neary, D. G. (editors), Riparian Areas of the Southwestern United States: Hydrology, Ecology, and Management: New York, Lewis, pp. 128–67.

McNamee, G., 1994, Gila: The Life and Death of an American River: Albuquerque, University of New Mexico Press, 215 pp.

McPherson, R. S., 1995, A History of San Juan County: In the Palm of Time: Salt Lake City, Utah State Historical Society, 419 pp.

McQueen, I. S., and Miller, R. F., 1972, Soil-Moisture and Energy Relationships Associated with Riparian Vegetation near San Carlos, Arizona: U.S. Geological Survey Professional Paper no. 655-E, 51 pp.

Meagher, M., and Houston, D. B., 1998, Yellowstone and the Biology of Time: Norman, University of Oklahoma Press, 287 pp.

Mearns, E. A., 1907, Mammals of the Mexican Boundary of the United States, a Descriptive Catalogue of the Species of Mammals Occurring in That Region; with a General Summary of the Natural History, and a List of Trees. Part I. Families Didelphiidae to Muridae: Washington, D.C., Smithsonian Institution, U.S. National Museum Bulletin no. 56, 530 pp.

Meehan, W. R., Swanson, F. J., and Sedell, J. R., 1977, Influences of riparian vegetation on aquatic ecosystems with particular reference to salmonid fishes and their food supply, *in* Johnson, R. R., and Jones, D. A. (technical coordinators), Importance, Preservation, and Management of Riparian Habitat, A Symposium: U.S. Department of Agriculture, Forest Service General Technical Report no. RM-43, pp. 137–45.

Meinzer, O. E., 1927, Plants as Indicators of Ground Water: U.S. Geological Survey Water-Supply Paper no. 577, 95 pp.

Meinzer, O. E., and Kelton, F. C., 1913, Geology and Water Resources of Sulphur Spring Valley, Arizona, with a Section on Agriculture by R. H. Forbes: U.S. Geological Survey Water-Supply Paper no. 320, 231 pp.

Melis, T. S., 1997, Geomorphology of debris flows and alluvial fans in Grand Canyon National Park and their influence on the Colorado River below Glen Canyon Dam, Arizona: Ph.D. diss., University of Arizona, Tucson, 490 pp.

Melis, T. S., Phillips, W. M., Webb, R. H., and Bills, D. J., 1996, When the Blue-Green Waters Turn Red: Historical Flooding in Havasu Creek, Arizona: U.S. Geological Survey Water-Resources Investigations Report no. 96-4059, 136 pp.

Merkel, D. L., and Hopkins, H. H., 1957, Life history of salt cedar (*Tamarix gallica* L.): Transactions of the Kansas Academy of Science, v. 60, pp. 360–69.

Merriam, C. H., 1893, Notes on the distribution of trees and shrubs in the deserts and desert ranges of southern California, southern Nevada, northwestern Arizona, and southwestern Utah: North American Fauna, v. 7, the Death Valley Expedition, Part II, pp. 285–343.

Merritt, D. M., and Cooper, D. J., 2000, Riparian vegetation and channel change in response to river regulation: A comparative study of regulated and unregulated streams in the Green River basin, USA: Regulated Rivers: Research and Management, v. 16, pp. 543–64.

Metzger, D. G., and Loeltz, O. J., 1973, Geohydrology of the Needles Area, Arizona, California, and Nevada: U.S. Geological Survey Professional Paper no. 486-J, 54 pp.

Metzger, D. G., Loeltz, O. J., and Irelna, B., 1973, Geohydrology of the Parker-Blythe-Cibola Area, Arizona and California: U.S. Geological Survey Professional Paper no. 486-G, 130 pp.

Meyboom, P., 1964, Three observations on streamflow depletion by phreatophytes: Journal of Hydrology, v. 2, pp. 248–61.

Milbrath, L. R., DeLoach, C. J., and Knutson, A. E., 2003, Initial results of biological control of saltcedar (*Tamarix* spp.) in the United States, *in* Proceedings of the Symposium Saltcedar and Water Resources in the West, July 16–17: San Angelo, Tex., n.p., pp. 135–41.

Miller, D. E., 1959, Hole-in-the-Rock: An Epic in the Colonization of the Great American West: Salt Lake City, University of Utah Press, 229 pp.

Minckley, W. L., 1973, Fishes of Arizona: Phoenix, Arizona Game and Fish Department, Sims Printing, 293 pp.

Minckley, W. L., and Brown, D. E., 1994, Wetlands, *in* Brown, D. E. (editor), Biotic Communities, Southwestern United States and Northwestern Mexico: Salt Lake City, University of Utah Press, pp. 223–87.

Minckley, W. L., and Clark, T. O., 1984, Formation and destruction of a Gila River mesquite bosque community: Desert Plants, v. 6, pp. 23–30.

Minckley, W. L., and Deacon, J. E. (editors), 1991, Battle against Extinction: Native Fish Management in the American West: Tucson, University of Arizona Press, 517 pp.

Minckley, W. L., Marsh, P. C., Deacon, J. E., Dowling, T. E., Hedrick, P. W., Matthews, W. J., and Mueller, G., 2003, A conservation plan for native fishes of the lower Colorado River: BioScience, v. 53, pp. 219–34.

Miser, H. D., 1924, The San Juan Canyon, Southeastern Utah: A Geographic and Hydrographic Reconnaissance: U.S. Geological Survey Water-Supply Paper no. 538, 80 pp.

Mitsch, W. J., and Gosselink, J. G., 1993, Wetlands (2d ed.): New York, Van Nostrand Reinhold, 722 pp.

Molles, M. C., Jr., Crawford, C. S., Ellis, L. M., Valett, H. M., and Dahm, C. N., 1998, Managed flooding for riparian ecosystem restoration: Bioscience, v. 48, pp. 749–56.

Moss, E. H., 1938, Longevity of seed and establishment of seedlings in species of *Populus*: Botanical Gazette, v. 99, pp. 529–42.

Mueller, G. A., and Marsh, P. C., 2002, Lost, a Desert River and Its Native Fishes:

A Historical Perspective of the Lower Colorado River: U.S. Geological Survey Information Technology Report no. USGS/BRD/ITR-2002-0010, 69 pp.

Müller-Schwarze, D., and Sun, L., 2003, The Beaver: Natural History of a Wetlands Engineer: Ithaca, N.Y., Comstock, 190 pp.

Munz, P. A., 1974, A Flora of Southern California: Berkeley, University of California Press, 1,086 pp.

Murphy, R. C., 1917, Natural history observations from the Mexican portion of the Colorado Desert: Proceedings of the Linnean Society of New York, nos. 28–29, pp. 43–101.

Mutschler, F. E., 1979, River Runners' Guide to the Canyons of the Green and Colorado Rivers with Emphasis on Geologic Features, v. 4, Desolation and Gray Canyons: Denver, Powell Society, 85 pp.

Naef, W. J., and Wood, J. N., 1975, Era of Exploration: The Rise of Landscape Photography in the American West, 1860–1885: Boston, Albright-Knox Art Gallery and the Metropolitan Museum of Art, 260 pp.

Nagler, P. L., Cleverly, J., Glenn, E., Lampkin, D., Huete, H., and Wan, Z., 2005, Predicting riparian evapotranspiration from MODIS vegetation indices and meteorological data: Remote Sensing of Environment, v. 94, pp. 17–30.

Nagler, P., Glenn, E. P., Hursh, K., Curtis, C., and Huete, A., 2005, Vegetation mapping for change detection on an arid-zone river: Environmental Monitoring and Assessment, v. 109, pp. 255–74.

Nagler, P. L., Glenn, E. P., and Thompson, T. L., 2003, Comparison of transpiration rates among saltcedar, cottonwood, and willow trees by sap flow and canopy temperature methods: Agricultural and Forest Meteorology, v. 116, pp. 73–89.

Nagler, P., Scott, R., Westenburg, C., Cleverly, J., Glenn, E., and Huete, A., 2005, Evapotranspiration on western U.S. rivers estimated using the Enhanced Vegetation Index from MODIS and data from eddy covariance and Bowen ratio flux towers: Remote Sensing of Environment, v. 97, pp. 337–51.

Naiman, R. J., and Décamps, H., 1997, The ecology of interfaces: Riparian zones: Annual Review of Ecology and Systematics, v. 28, pp. 621–58.

Naiman, R. J., Décamps, H., and Pollock, M., 1993, The role of riparian corridors in maintaining regional biodiversity: Ecological Applications, v. 3, pp. 209–12.

National Research Council, 1992, Restoration of Aquatic Ecosystems: Science, Technology, and Public Policy: Washington, D.C., National Academy Press, 552 pp.

Nelson, S., 2006, In search of El Burro, the tidal bore of the Río Colorado delta, *in* Felger, R. S., and Broyles, B. (editors), Dry Borders: Great Natural Reserves of the Sonoran Desert: Salt Lake City, University of Utah Press, pp. 448–58.

Nials, F. L., Gregory, D. A., and Graybill, D. A., 1989, Salt River streamflow and Hohokam irrigation systems, *in* The 1982–1984 Excavations at Las Colinas: Tucson, Arizona State Museum Archaeological Series no. 162, v. 5, pp. 59–76.

Nichols, T., 1999, Glen Canyon: Images of a Lost World: Santa Fe, Museum of New Mexico Press, 157 pp.

Nicholson, J. (editor), 1974, The Arizona of Joseph Pratt Allyn: Letters from a Pioneer Judge, Observations, and Travels, 1863–1866: Tucson, University of Arizona Press, 284 pp.

Noel, T. J., and Fielder, J., 2001, Colorado, 1870–2000, Revisited: The History behind the Images: Englewood, Colo., Westcliffe, 319 pp.

O'Connor, J. E., Ely, L. L., Wohl, E. E., Stevens, L. E., Melis, T. S., Kale, V. S., and Baker, V. R., 1994, A 4500-year record of large floods on the Colorado River in the Grand Canyon, Arizona: Journal of Geology, v. 102, pp. 1–9.

Oden, P., 1973, Picacho, "Life and Death of a Great Gold Mining Camp": El Centro, Calif., privately published, 44 pp.

Office of Arid Lands Studies, 1970, Environmental Study for the Gila River below Painted Rock Dam: Los Angeles, U.S. District Corps of Engineers, Contract no. DACW09-70-C-0079 with the University of Arizona, Tucson, 92 pp.

Ohmart, R. D., 1996, Historical and present impacts of livestock grazing on fish and wildlife resources in western riparian habitats, *in* Krauseman, P. R. (editor), Rangeland Wildlife: Denver, Society of Range Management, pp. 245–79.

Ohmart, R. D., and Anderson, B. W., 1982, North American desert riparian ecosystems, *in* Bender, G. L. (editor), Reference Handbook on the Deserts of North America: Westport, Conn., Greenwood Press, pp. 433–66.

Ohmart, R. D., Deason, W. O., and Burke, C., 1977, A riparian case history: The Colorado River, *in* Johnson, R. R., and Jones, D. A. (technical coordinators), Importance, Preservation, and Management of Riparian Habitat, A Symposium: U.S. Department of Agriculture, Forest Service General Technical Report no. RM-43, pp. 35–47.

Ohmart, R. D., Deason, W. O., and Freeland, S. J., 1975, Dynamics of marsh land formation and succession along the lower Colorado River and their importance and management problems as related to wildlife in the arid Southwest: Transactions of the 40th North American Wildlife and Natural Resources Conference, pp. 240–51.

Olson, C. E., 1940, Forests in the Arizona desert: Journal of Forestry, v. 38, pp. 956–59.

Olson, T. E., and Knopf, F. L., 1986a, Agency subsidization of a rapidly spreading exotic: Wildlife Society Bulletin, v. 14, pp. 492–93.

———, 1986b, Naturalization of Russian-olive in the western United States: Western Journal of Applied Forestry, v. 1, pp. 65–69.

Orchard, K. L., 2001, Paleoflood hydrology of the San Juan River, southeastern Utah: M.S. thesis, University of Arizona, Tucson, 138 pp.

Orchard, K. L, and Schmidt, J. C., 1998, A Geomorphic Assessment of the Availability of Potential Humpback Chub Habitat in the Green River in Desolation and Gray Canyons, Utah: Logan, Utah State University, Final Report for Contract 93-1070 to the Flaming Gorge Studies Program of the Upper Basin Fish Recovery Program, 68 pp.

Osterkamp, W. R., 1998, Processes of fluvial island formation, with examples from Plum Creek, Colorado, and Snake River, Idaho: Wetlands, v. 18, pp. 530–45.

Oviatt, C. G., 1985, Late Quaternary geomorphic changes along the San Juan River and its tributaries near Bluff, Utah: Utah Geological and Mineral Survey Special Studies, v. 64, pp. 33–47.

Owen, J. C., Sogge, M. K., and Kern, M. D., 2005, Habitat and sex differences in physiological condition of breeding Southwestern Willow Flycatchers (*Empidonax traillii extimus*): The Auk, v. 122, pp. 1261–270.

Owen-Joyce, S. J., and Raymond, L. H., 1996, An Accounting System for Water and Consumptive Use along the Colorado River, Hoover Dam to Mexico: U.S. Geological Survey Water-Supply Paper no. 2407, 94 p.

Palmquist, P. E., and Kailbourn, T. R., 2000, Pioneer Photographers of the Far West: A Biographical Dictionary, 1840–1865: Stanford, Calif., Stanford University Press, 679 pp.

Parker, J. T. C., 1995, Channel Change on the Santa Cruz River, Pima County, Arizona, 1936–86: U.S. Geological Survey Water-Supply Paper no. 2429, 58 pp.

Parker, M., Wood, F. J., Smith, B. H., and Elder, R. G., 1985, Erosional downcutting in lower order riparian ecosystems: Have historical changes been caused by removal of beaver? *in* Johnson, R. R., Ziebell, C. D., Patton, D. R., Ffolliott, P. F., and Hamre, R. H. (editors), Riparian

Ecosystems and Their Management: Reconciling Conflicting Uses: Tucson, Ariz., First North American Riparian Conference, U.S. Department of Agriculture, Forest Service General Technical Report no. RM-120, pp. 35–38.

Patraw, P. M., 1936, Check-list of Plants of Grand Canyon National Park: Grand Canyon, Ariz., Grand Canyon Natural History Association, Bulletin no. 6, 75 pp.

Patten, D. T., 1998, Riparian ecosystems of semi-arid North America: Diversity and human impacts: Wetlands, v. 18, pp. 498–512.

Pattie, J. O., 2001, The Personal Narrative of James O. Pattie: The True Wild West of New Mexico and California: Santa Barbara, Calif., Narrative Press, originally published in 1831, 315 pp.

Peacock, J. T., and McMillan, C., 1965, Ecotypic differentiation in *Prosopis* (mesquite): Ecology, v. 46, pp. 35–51.

Pearthree, M. S., and Baker, V. R., 1987, Channel Change along the Rillito Creek System of Southeastern Arizona, 1941 through 1983: Implications for Flood-Plain Management: Tucson, Arizona Bureau of Geology and Mineral Technology, Geological Survey Branch, Special Paper no. 6, 58 pp.

Pearthree, P. A., 1996, Historical Geomorphology of the Verde River: Tucson, Arizona Geological Survey Open-File Report no. 96-13, 10 pp.

Perkins, C. A., Nielson, M. G., and Jones, L. B., 1968, Saga of San Juan: Salt Lake City, Utah, Mercury, 367 pp.

Petroski, H., 1993, Hoover Dam: American Scientist, v. 81, pp. 517–21.

Phillips, A., Marshall, J., and Monson, G., 1964, The Birds of Arizona: Tucson, University of Arizona Press, 212 pp.

Phillips, B. G., Phillips, A. M., III, and Bernzott, M. A. S., 1987, Annotated Checklist of Vascular Plants of Grand Canyon National Park: Grand Canyon, Ariz., Grand Canyon Natural History Association Monograph no. 7, 79 pp.

Phillips, J. V., and Ingersoll, T. L., 1998, Verification of Roughness Coefficients for Selected Natural and Constructed Stream Channels in Arizona: U.S. Geological Survey Professional Paper no. 1584, 77 pp.

Phillips, J. V., McDoniel, D., Capesius, J. P., and Asquith, W., 1998, Method to Estimate Effects of Flow-Induced Vegetation Changes on Channel Conveyances of Streams in Central Arizona: U.S. Geological Survey Water-Resources Investigations Report no. 98-4040, 43 pp.

Phillips, W. S., 1963, Vegetational Changes in Northern Great Plains: Tucson, University of Arizona, Agricultural Experiment Station Report no. 214, 185 pp.

Pima County, 2000, Historical Occurrence of Native Fish in Pima County: Tucson, Pima County Flood Control District, Sonoran Desert Conservation Plan Report, 54 pp.

———, 2001, Groundwater level changes in the Tanque Verde Valley: Tucson, Ariz., Pima County Flood Control District, Sonoran Desert Conservation Plan Report, 21 pp., appendixes.

Pitts, T. R., 1987, William Bell: Philadelphia photographer: M.A. thesis, University of Arizona, Tucson, 114 pp.

Platts, W. S., and Nelson, R. L., 1985, Will the riparian pasture build good streams? Rangelands, v. 7, pp. 7–10.

Pool, D. R., and Coes, A. L., 1999, Hydrogeologic Investigations of the Sierra Vista Subwatershed of the Upper San Pedro Basin, Cochise County, Southeast Arizona: U.S. Geological Survey Water-Resources Investigations Report no. 99-4197, 41 pp.

Pope, G. L., Rigas, P. D., and Smith, C. F., 1998, Statistical Summaries of Streamflow Data and Characteristics of Drainage Basins for Selected Streamflow-Gaging Stations in Arizona through Water Year 1996: U.S. Geological Survey Water-Resources Investigations Report no. 98-4225, 907 pp.

Potter, L. D., and Drake, C. L., 1989, Lake Powell: Virgin Flow to Dynamo: Albuquerque, University of New Mexico Press, 311 pp.

Potts, D. L., 2002, Carbon isotope composition of tree ring holocellulose reveals contrasting responses to moisture availability in *Populus fremontii* at perennial and intermittent stream reaches: M.S. thesis, University of Arizona, Tucson, 53 pp.

Powell, J. W. 1961. Canyons of the Colorado: New York, Dover, originally published in 1895, 400 pp.

———, 1972, The Hopi Villages: The Ancient Province of Tusayan: Palmer Lake, Colo., Filter Press, 36 pp.

Powell, W. C., 1949, Journal of W. C. Powell, edited by C. Kelley: Utah Historical Quarterly, v. 17, pp. 257–478.

Quartarone, F., 1995, Historical Accounts of Upper Colorado River Basin Endangered Fish (edited by C. Young): Golden, Colo., U.S. Fish and Wildlife Service report, 60 pp.

Ralston, B. E., 2005, Riparian vegetation and associated wildlife, *in* Gloss, S. P., Lovich, J. E., and Melis, T. S. (editors), The State of the Colorado River Ecosystem in Grand Canyon: U.S. Geological Survey Circular no. 1282, pp. 103–21.

Rantz, S. E., and others, 1982, Measurement and Computation of Streamflow: U.S. Geological Survey Water-Supply Paper no. 2175, 631 pp.

Rea, A. M., 1983, Once a River: Bird Life and Habitat Changes on the Middle Gila: Tucson, University of Arizona Press, 284 pp.

———, 1997, At the Desert's Green Edge: An Ethnobotany of the Gila River Pima: Tucson, University of Arizona Press, 430 pp.

Reeter, R. W., and Remick, W. H., 1986, Maps Showing Groundwater Conditions in the West Salt River, East Salt River, Lake Pleasant, Carefree and Fountain Hills Sub-basins of the Phoenix Active Management Area, Maricopa, Pinal, and Yavapai Counties, Arizona—1983: Phoenix, Arizona Department of Water Resources Hydrologic Map Series Report no. 12, 3 sheets.

Reichenbacher, F. W., 1984, Ecology and evolution of Southwestern riparian plant communities: Desert Plants, no. 6, pp. 14–22.

Reilly, P. T., 1999, Lee's Ferry: From Mormon Crossing to National Park (edited by R. H. Webb): Logan, Utah State University Press, 542 pp.

Renard, K. G., Keppel, R. V, Hickey, J. J., and Wallace, D. E., 1964, Performance of local aquifers as influenced by stream transmission losses and riparian vegetation: Transactions of the American Society of Agricultural Engineers, v. 7, pp. 471–74.

Rice, J., Anderson, B. W., and Ohmart, R. D., 1980, Seasonal habitat selection by birds in the lower Colorado River valley: Ecology, v. 61, pp. 1402–411.

Ricketts, N. B., 1996, The Mormon Battalion, U.S. Army of the West, 1846–1848: Logan, Utah State University Press, 375 pp.

Rink, G. R., 2003, Vascular flora of Canyon de Chelley National Monument, Apache County, Arizona: M.S. thesis, Northern Arizona University, Flagstaff, 317 pp.

Rinne, J., and Minckley, W. L., 1991, Native Fishes of Arid Lands: A Dwindling Resource of the Desert Southwest: U.S. Department of Agriculture, Forest Service General Technical Report no. RM-206, Fort Collins, Colo., U.S. Department of Agriculture, 45 pp.

Ritzi, R. W., Bouwer, H., and Sorooshian, S., 1985, Water resource conservation by reducing phreatophyte transpiration, in Johnson, R. R., Ziebell, C. D., Patton, D. R., Ffolliott, P. F., and Hamre, R. H. (editors), Riparian Ecosystems and Their Management: Reconciling Conflicting Uses: Tucson, Ariz., First North American Riparian Conference, U.S. Department of Agriculture, Forest Service General Technical Report no. RM-120, pp. 191–96.

Rivers West, 1990, Water Resources Assessment: Bill Williams Unit, Havasu National Wildlife Refuge: Denver, Colo., final report to the U.S. Fish and Wildlife Service, Region 2, Albuquerque, N.Mex., 89 pp.

Roberts, L. K., 1987, Paleohydrologic reconstruction, hydraulics, and frequency-magnitude relationships of large flood events along Aravaipa Creek, Arizona: M.S. thesis, University of Arizona, Tucson, 63 pp.

Roberts, N., 1989, The Holocene: An Environmental History: New York, Basil Blackwell, 227 pp.

Robinson, A. F., 1970, History of Kane County: Salt Lake City, Utah Printing Company, 620 pp.

Robinson, G. M., and Peterson, D. E., 1962, Notes on Earth Fissures in Southern Arizona: U.S. Geological Survey Circular no. 466, 7 pp.

Robinson, T. W., 1958, Phreatophytes: U.S. Geological Survey Water-Supply Paper no. 1423, 84 pp.

———, 1965, Introduction, Spread, and Areal Extent of Saltcedar (Tamarix) in the Western States: U.S. Geological Survey Professional Paper no. 491-A, 12 pp.

Roeske, R. H., Cooley, M. E., and Aldridge, B. N., 1978, Floods of September 1970 in Arizona, Utah, Colorado, and New Mexico: U.S. Geological Survey Water-Supply Paper no. 2052, 135 pp.

Roeske, R. H., Garrett, J. M., and Eychaner, J. H., 1989, Floods of October 1983 in Southeastern Arizona: U.S. Geological Survey Water-Resources Investigations Report no. 85-4225-C, 77 pp.

Rogers, G., 1982, Then and Now: A Photographic History of Vegetation Change in the Central Great Basin: Salt Lake City, University of Utah Press, 152 pp.

Rogers, G. F., Malde, H. E., and Turner, R. M., 1984, Bibliography of Repeat Photography for Evaluating Landscape Change: Salt Lake City, University of Utah Press, 179 pp.

Rogge, A. E., McWaters, D. L., Keane, M., and Emanuel, R. P., 1995, Raising Arizona's Dams: Tucson, University of Arizona Press, 212 pp.

Rohde, R. F., 1997, Looking into the past: Interpretations of vegetation change in western Namibia based on matched photography: Dinteria, no. 25, pp. 121–49.

Rolle, A., 1991, John Charles Frémont: Character as Destiny: Norman, University of Oklahoma Press, 351 pp.

Rood, S. B., Kalischuk, A. R., and Mahoney, J. M., 1998, Initial cottonwood seedling recruitment following the flood of the century of the Oldman River, Alberta, Canada: Wetlands, v. 18, pp. 557–70.

Rosenberg, K. B., Ohmart, R. D., Hunter, W. C., and Anderson, B. W., 1991, Birds of the Lower Colorado River Valley: Tucson, University of Arizona Press, 416 pp.

Ross, C. P., 1923, The Lower Gila Region, Arizona: A Geographic, Geologic, and Hydrologic Reconnaissance, with a Guide to Desert Watering Places: U.S. Geological Survey Water-Supply Paper no. 498, 237 pp.

Ross, D., 1998, "I have struck it rich at last": Charles Goodman, traveling photographer: Utah Historical Quarterly, v. 66, pp. 65–83.

Rostvedt, J. O., and others, 1971, Summary of Floods in the United States during 1966: U.S. Geological Survey Water-Supply Paper no. 1870-D, D59–D91.

Rowe, J., 1997, Photographers in Arizona, 1850–1920: A History and Directory: Nevada City, Calif., Carl Mautz, 126 pp.

Ruff, P. F., 1971, A history of the Salt River channel in the vicinity of Tempe, Arizona, 1868–1969: Tempe, Arizona State University, unpublished report, 63 pp.

Saar, D. A., 2002, Riparian livestock exclosure research in the western United States: A critique and some recommendations: Environmental Management, v. 30, pp. 516–26.

Sala, A., Smith, S. D., and Devitt, D. A., 1996, Water use by Tamarix ramosissima and associated phreatophytes in a Mojave Desert floodplain: Ecological Applications, v. 6, pp. 888–98.

Salamun, P. J., 1990, An interpretation of vegetational changes along the fortieth parallel in Nevada and Utah, 1867–1981, in Murphy, E. M., and Knapp, J. M. (editors), Kaleidoscope of History: Photographic Collections in the Golda Meir Library: Milwaukee, University of Wisconsin, American Geographical Society Collection Special Publication no. 1, pp. 73–86.

Sallach, B. K., 1986, Vegetation changes in New Mexico documented by repeat photography: M.S. thesis, New Mexico State University, Las Cruces, 73 pp.

Salzer, M. W., 2000, Dendroclimatology of the San Francisco Peaks region of northern Arizona, USA: Ph.D. diss., University of Arizona, Tucson, 211 pp.

Salzer, M. W., McCord, V. A. S., Stevens, L. E., and Webb, R. H., 1996, The dendrochronology of netleaf hackberry (Celtis reticulata Torr.) in Grand Canyon: Assessing the impact of regulated river flow on tree growth, in Dean, J. S., Meko, D. M., and Swetnam, T. W. (editors), Tree Rings, Environment, and Humanity: Tucson, Ariz., Radiocarbon, pp. 273–81.

Sanchez, P. G., 1975, A tamarisk fact sheet: Desert Bighorn Council 1975 Transactions, pp. 12–14.

Sands, A., and Howe, G., 1977, An Overview of Riparian Forests in California: Their Ecology and Conservation: U.S. Department of Agriculture, Forest Service General Technical Report no. RM-43, pp. 98–115

Schaal, B. A., Gaskin, J. F., and Caicedo, A. L., 2003, Phylogeography, haplotype trees, and invasive plant species: Journal of Heredity, v. 94, pp. 197–204.

Schaeffer, S., and Williams, D. G., 1998, Transpiration of desert riparian forest canopies estimated from sap flux, *in* Wood, E. F., Chehbouni, A. G., Goodrich, D. C., Seo, D. J., and Zimmerman, J. R. (technical coordinators), Proceedings from the Special Symposium on Hydrology: Boston, American Meteorological Society, pp. 180–84.

Schaeffer, S. M., Williams, D. G., and Goodrich, D. C., 2000, Transpiration of cottonwood/willow forest estimated from sap flux: Agricultural and Forest Meteorology, v. 105, pp. 257–70.

Scheurer, T., and Molinari, P., 2003, Experimental floods in the River Spöl, Swiss National Park: Framework, objectives, and design: Aquatic Sciences, v. 65, pp. 183–90.

Schultz, B. W., 2001, Extent of vegetated wetlands at Owens Dry Lake, California, U.S.A., between 1977 and 1992: Journal of Arid Environments, v. 48, pp. 69–87.

Schulz, T. T., and Leininger, W. C., 1990, Differences in riparian vegetation structure between grazed areas and exclosures: Journal of Range Management, v. 43, pp. 295–99.

Schumann, H. H., 1974, Land Subsidence and Earth Fissures in Alluvial Deposits in the Phoenix Area, Arizona: U.S. Geological Survey Miscellaneous Investigations Map no. I-845-H, 1 sheet, scale 1:250,000.

Schumm, S. A., and Lichty, R. W., 1963, Channel Widening and Floodplain Construction along the Cimarron River in Southwestern Kansas: U.S. Geological Survey Professional Paper no. 352-D, 88 pp.

Schuster, R. L., and Wieczorek, G. F., 1995, Reconnaissance Study of the April 12, 1995, Landslide in Zion Canyon, Zion National Park, Utah: Reston, Va., U.S. Geological Survey Administrative Report, 13 pp., illustrations.

Schwalen, H. C., and Shaw, R. J., 1957, Ground Water Supplies of Santa Cruz Valley of Southern Arizona between Rillito Station and the International Boundary: Tucson, University of Arizona, Agricultural Experiment Station Bulletin no. 288, 119 pp.

———, 1961, Progress Report on Study of Water in the Santa Cruz Valley, Arizona: Tucson, University of Arizona, Agricultural Experiment Station Report, 20 pp.

Schwennesen, A. T., 1918, Ground Water in the Animas, Playas, Hachita, and San Luis Basins, New Mexico: U.S. Geological Survey Water-Supply Paper no. 422, 152 pp.

———, 1919, Ground water in San Simon Valley, Arizona and New Mexico, *in* Grover, N. C. (editor), Contributions to the Hydrology of the United States, 1917: U.S. Geological Survey Water-Supply Paper no. 425, pp. 1–35.

Scott, M. L., Auble, G. T., and Friedman, J. M., 1997, Flood dependency of cottonwood establishment along the Missouri River, Montana, USA: Ecological Applications, v. 7, pp. 677–90.

Scott, M. L., Friedman, J. M., and Auble, G. T., 1996, Fluvial process and the establishment of bottomland trees: Geomorphology, v. 14, pp. 327–39.

Scott, M. L., Lines, G. C., and Auble, G. T., 2000, Channel incision and patterns of cottonwood stress and mortality along the Mojave River, California, Journal of Arid Environments, v. 44, pp. 399–414.

Scott, M. L., Shafroth, P. B., and Auble, G. T., 1999, Responses of riparian cottonwoods to alluvial water table declines: Environmental Management, v. 23, pp. 347–58.

Scott, R., Edwards, E., Shuttleworth, W., Huxman, T., Watts, C., and Goodrich, D., 2004, Interannual and seasonal variation in fluxes of water and carbon dioxide from a riparian woodland ecosystem: Agricultural and Forest Meteorology, v. 122, pp. 65–84.

Scott, R., Shuttleworth, W., Goodrich, D., and Maddock, T., 2000, The water use of two dominant vegetation communities in a semiarid riparian ecosystem: Agricultural and Forest Meteorology, v. 105, pp. 241–56.

Sellers, W. D., and Hill, R. H. (editors), 1974, Arizona Climate, 1931–1972: Tucson, University of Arizona Press, 616 pp.

Sellers, W. D., Hill, R. H., and Sanderson-Rae, M., 1985, Arizona Climate: Tucson, University of Arizona Press, 143 pp.

Sexton, J. P., McKay, J. K., and Sala, A., 2002, Plasticity and genetic diversity may allow saltcedar to invade cold climates in North America: Ecological Applications, v. 12, pp. 1652–660.

Shafroth, P. B., Auble, G. T., and Scott, M. L., 1995, Germination and establishment of the native plains cottonwood (*Populus deltoides* Marshall subsp. *monilifera*) and the exotic Russian-olive (*Elaeagnus angustifolia* L.): Conservation Biology, v. 9, pp. 1169–175.

Shafroth, P. B., Auble, G. T., Stromberg, J. C., and Patten, D. T., 1998, Establishment of woody riparian vegetation in relation to annual patterns of streamflow, Bill Williams River, Arizona: Wetlands, v. 18, pp. 577–90.

Shafroth, P. B., Cleverly, J. R., Dudley, T. L., Taylor, J. P., van Riper, C., III, Weeks, E. P., and Stuart, J. N., 2005, Control of *Tamarix* in the western United States: Implications for water salvage, wildlife use, and riparian restoration: Environmental Management, v. 35, pp. 231–46.

Shafroth, P. B., Stromberg, J. C., and Patten, D. T., 2000, Woody riparian vegetation response to different alluvial water table regimes: Western North American Naturalist, v. 60, pp. 66–76.

———, 2002, Riparian vegetation response to altered disturbance and stress regimes: Ecological Applications, v. 12, pp. 107–23.

Shannon, J. P., and Benenati, E. P., 2002, Essentials of Aquatic Ecology in the Colorado River: Flagstaff, Northern Arizona University, Creative Communications Report no. G46703/500/02-02, 73 pp.

Shantz, H. L., and Turner, B. L., 1958, Photographic Documentation of Vegetational Changes in Africa over a Third of a Century: Tucson, University of Arizona, College of Agriculture Report no. 169, 158 pp.

Shaw, H. G., 2001a, New country: Presettlement naturalists on the upper Verde watershed: Arizona Wildlife Views, v. 44, no. 1, pp. 18–23.

———, 2001b, Territorial travels: Postsettlement 19th century naturalists on the upper Verde watershed: Arizona Wildlife Views, v. 44, no. 4, pp. 20–24.

Shepherd, J. R., 1961, Total sediment transport in the lower Colorado River: American Society of Civil Engineers Convention, Phoenix, April 10–14, 13 pp.

Sher, A. A., Marshall, D. L., and Taylor, J. P., 2002, Establishment patterns of native *Populus* and *Salix* in the presence of invasive nonnative *Tamarix*: Ecological Applications, v. 12, pp. 760–72.

Sheridan, T. E., 1986, Los Tucsonenses: The Mexican Community in Tucson, 1854–1941: Tucson, University of Arizona Press, 327 pp.

———, 1995, Arizona: A History: Tucson, University of Arizona Press, 434 pp.

Siegel, R. S., and Brock, J. H., 1990, Germination requirements of key Southwestern woody riparian species: Desert Plants, v. 10, pp. 3–8.

Silsbee, T. H., 1900, Map of that part of the Colorado Desert in California, United States and lower California, Mexico, known as the New River Country: San Diego, Calif., unpublished blue-line map, scale approximately 1:178,000.

Simmons, G. C., and Simmons, V. M., 1977, First photographers of the Grand Canyon: American West, v. 14, pp. 34–38, 61–63.

Skagen, S. K., Melcher, C. P., Howe, W. H., and Knopf, F. L., 1998, Comparative use of riparian corridors and oases by migrating birds in southeast Arizona: Conservation Biology, v. 12, pp. 896–909.

Skartvedt, P. H., 2000, Woody riparian vegetation patterns in the upper Mimbres Watershed, southwestern New Mexico: The Southwestern Naturalist, v. 45, pp. 6–14.

Skovlin, J. M., Strickler, G. S., Peterson, J. L., and Sampson, A. W., 2001, Interpreting Landscape Change in High Mountains of Northeastern Oregon from Long-Term Repeat Photography: U.S. Department of Agriculture, Forest Service, Pacific Northwest Research Station, General Technical Report no. PNW-GTR-505, 79 pp.

Slingluff, J., 1993, Verde River Recreation Guide (2d ed.): Phoenix, Golden West, 173 pp.

Smith, D. L. (editor), 1965, Down the Colorado: Norman, University of Oklahoma Press, 237 pp.

Smith, D. L., and Crampton, C. G. (editors), 1987, The Colorado River Survey: Salt Lake City, Utah, Howe Brother Books, 305 pp.

Smith, J. D., 2004, The role of riparian shrubs in preventing floodplain unraveling along the Clark Fork of the Columbia River in the Deer Lodge Valley, Montana, *in* Bennett, S. J., and Simon, A. (editors), Riparian Vegetation and Fluvial Geomorphology: Washington, D.C., American Geophysical Union, Water Science and Application no. 8, pp. 71–85.

Smith, J. S., 1977, The Southwest Expedition of Jedediah S. Smith (edited by G. R. Brooks): Lincoln, University of Nebraska Press, 259 pp.

Smith, S. D., 1989, The ecology of saltcedar (*Tamarix chinensis*) in Death Valley National Monument and Lake Mead National Recreation Area: An Assessment of Techniques and Monitoring for Saltcedar Control in the Park System: Las Vegas, National Park Service, University of Nevada, Contribution no. CPSU/UNLV 041/03, 65 pp.

Smith, S. D., Devitt, D. A., Sala, A., Cleverly, J. R., and Busch, D. E., 1998, Water relations of riparian plants from warm desert regions: Wetlands, v. 18, pp. 687–96.

Smith, S. D., Sala, A., Devitt, D. A., and Cleverly, J. R., 1996, Evapotranspiration from a saltcedar-dominated desert floodplain: A scaling approach, *in* Barrow, J.R., McArthur, E.D., Sosebee, R.E.,

Tausch, R.J. (compilers), Proceedings: Shrubland Ecosystem Dynamics in a Changing Environment: U.S. Department of Agriculture, Forest Service General Technical Report no. INT-GTR-338, pp. 199–204.

Smith, S. D., Wellington, A. B., Nachlinger, J. L., and Fox, C. A., 1991, Functional responses of riparian vegetation to streamflow diversion in the eastern Sierra Nevada: Ecological Applications, v. 1, pp. 89–97.

Smith, S. S., 1990, Large floods and rapid entrenchment, Kanab Creek, southern Utah: M.S. thesis, University of Arizona, Tucson, 82 pp.

Smith, W., 1986, The Effects of Eastern North Pacific Tropical Cyclones on the Southwestern United States: Salt Lake City, Utah, National Oceanic and Atmospheric Administration Technical Memorandum no. NWS WS-197, 229 pp.

Smith, W. O., Vetter, C. P., Cummings, G. B., and others, 1960, Comprehensive Survey of Sedimentation in Lake Mead, 1948–49: U.S. Geological Survey Professional Paper no. 295, 244 pp.

Snyder, J. O., 1918, The fishes of the Mohave River, California: Proceedings of the United States Natural History Museum, v. 54, pp. 297–99.

Snyder, K. A., and Williams, D. G., 2000, Water sources used by riparian trees varies among stream types on the San Pedro River, Arizona: Journal of Agricultural and Forest Meteorology, v. 105, pp. 227–40.

Snyder, W. D., and Miller, G. C., 1992, Changes in riparian vegetation along the Colorado River and Rio Grande, Colorado: Great Basin Naturalist, v. 52, pp. 357–63.

Soltz, D. L., and Naiman, R. J., 1978, The natural history of native fishes in the Death Valley system: Natural History Museum of Los Angeles County, Science Series, v. 30, pp. 1–76.

Sparks, R. E., Bayley, P. B., Kohler, S. L., and Osborne, L. L., 1990, Disturbance and recovery of large floodplain rivers: Environmental Management, v. 14, pp. 699–709.

Spigler, R. B., 2000, Effects of humidity and temperature on leaf and branch morphology in southwest riparian ecosystems: Undergraduate thesis, Department of Ecology and Evolutionary Biology, University of Arizona, Tucson, 32 pp.

Stahle, D. W., D'Arrigo, R. D., Krusic, P. J., Cleaveland, M. K., Cook, E. R., Allan, R. J., Cole, J. E., Dunbar, R. B., Therrell, D. A., Gay, D. A., Moore, M. D., Stokes, M. A., Burns, B. T., Lillanueva-Diaz, J., and Thompson, L. G., 1998, Experimental

dendroclimatic reconstruction of the Southern Oscillation: Bulletin of the American Meteorological Society, v. 79, pp. 2137–152.

Stamos, C. L., Martin, P., Nishikawa, T., and Cox, B. F., 2001. Simulation of Ground-Water Flow in the Mojave River Basin, California: U.S. Geological Survey Water-Resources Investigations Report no. 01-4002, 129 pp.

Stamos, C. L., Nishikawa, T., and Martin, P., 2001, Water Supply in the Mojave River Ground-Water Basin, 1931–99, and the Benefits of Artificial Recharge: U.S. Geological Survey Fact Sheet no. 122-01, 4 pp.

Steen-Adams, M., 2002, Applying environmental history to ecological restoration: A case study from Zion National Park: Ecological Restoration, v. 20, pp. 252–61.

Stephens, H. G., and Shoemaker, E. M., 1987, In the Footsteps of John Wesley Powell: Boulder, Colo., Johnson Books, 286 pp.

Stevens, J. E., 1988, Hoover Dam: Norman, University of Oklahoma Press, 326 pp.

Stevens, L. E., 1989, Mechanisms of riparian plant community organization and succession in the Grand Canyon, Arizona: Ph.D. diss., Northern Arizona University, Flagstaff, 115 pp.

Stevens, L. E., and Ayers, T., 2002, The biodiversity and distribution of exotic vascular plants and animals in the Grand Canyon region, *in* Tellman, B. (editor), Invasive Exotic Species in the Sonoran Region: Tucson, University of Arizona Press, pp. 241–65.

Stevens, L. E., Ayers, T. J., Bennett, J. B., Christensen, K., Kearsley, M. J. C., Meretsky, V. J., Phillips, A. M., III, Parnell, R. A., Spence, J., Sogge, M. K., Springer, A. E., and Wegner, D. L., 2001, Planned flooding and Colorado River riparian trade-offs downstream from Glen Canyon Dam, Arizona: Ecological Applications, v. 11, pp. 701–10.

Stevens, L. E., Brown, B. T., Simpson, J. M., and Johnson, R. R., 1977, The importance of riparian habitat to migrating birds, *in* Johnson, R. R., and Jones, D. A. (editors), Importance, Preservation, and Management of Riparian Habitat: A Symposium: U.S. Department of Agriculture, General Technical Report no. RM-43, pp. 156–64.

Stevens, L. E., Schmidt, J. C., Ayers, T. J., and Brown, B. T., 1996, Flow regulation, geomorphology, and Colorado River marsh development in the Grand Canyon, Arizona: Ecological Applications, v. 5, pp. 1025–39.

Stevens, L. E., and Waring, G. L., 1988, Effects of post-dam flooding on riparian substrate, vegetation, and invertebrate populations in the Colorado River corridor in Grand Canyon, *in* Executive Summaries of Technical Reports: Flagstaff, Ariz., Glen Canyon Environmental Studies, pp. 229–55.

Stewart, I. T., Cayan, D. R., and Dettinger, M. D., 2004, Changes in snowmelt runoff timing in western North America under a "business as usual" climate change scenario: Climatic Change, v. 62, pp. 217–32.

Stockton, C. W., 1975, Long-Term Streamflow Records Reconstructed from Tree Rings: Tucson, University of Arizona Press, Papers of the Laboratory of Tree-Ring Research no. 5, 111 pp.

Stockton, C. W., and Jacoby, G. C., Jr., 1976, Long-Term Surface-Water Supply and Streamflow Trends in the Upper Colorado River Basin Based on Tree-Ring Analyses: Lake Powell Research Project Bulletin no. 18, 70 pp.

Stromberg, J. C., 1993a, Frémont cottonwood–Goodding willow riparian forests: A review of their ecology, threats, and recovery potential: Journal of the Arizona-Nevada Academy of Science, v. 26, pp. 97–110.

———, 1993b, Riparian mesquite forests: A review of their ecology, threats, and recovery potential: Journal of the Arizona-Nevada Academy of Science, v. 27, pp. 111–24.

———, 1998a, Dynamics of Fremont cottonwood (*Populus fremontii*) and saltcedar (*Tamarix chinensis*) populations along the San Pedro River, Arizona: Journal of Arid Environments, v. 40, pp. 133–55.

———, 1998b, Functional equivalency of saltcedar (*Tamarix chinensis*) and Fremont cottonwood (*Populus fremontii*) along a free-flowing river: Wetlands, v. 18, pp. 675–86.

———, 2001, Restoration of riparian vegetation in the south-western United States: Importance of flow regimes and fluvial dynamism: Journal of Arid Environments, v. 49, pp. 17–34.

Stromberg, J. C., Briggs, M., Gourley, C., Scott, M., Shafroth, P., and Stevens, L., 2004, Human alterations of riparian ecosystems, *in* Baker, M., Jr., Ffolliott, P. F., DeBano, L. F., and Neary, D. G. (editors), Riparian Areas of the Southwestern United States: Hydrology, Ecology, and Management: New York, Lewis, pp. 99–126.

Stromberg, J. C., Fry, J., and Patten, D. T., 1997, Marsh development after large floods in an alluvial, arid-land river: Wetlands, v. 17, pp. 292–300.

Stromberg, J. C., Patten, D. T., and Richter, B. D., 1991, Flood flows and dynamics of Sonoran riparian forests: Rivers, v. 2, pp. 221–35.

Stromberg, J., Richter, B. D., Patten, D. T., and Wolden, L. G., 1993, Response of a Sonoran riparian forest to a 10-year return flood: Great Basin Naturalist, v. 53, pp. 118–30.

Stromberg, J. C., Tiller, R., and Richter, B., 1996, Effects of groundwater decline on riparian vegetation of semi-arid regions: The San Pedro, Arizona: Ecological Applications, v. 6, pp. 113–31.

Stromberg, J. C., Tress, J. A., Wilkins, S. D., and Clark, S. D., 1992, Response of velvet mesquite to groundwater decline: Journal of Arid Environments, v. 23, pp. 45–58.

Stromberg, J. C., Wilkins, S. D., and Tress, J. A., 1993, Vegetation-hydrology models: Implications for management of *Prosopis velutina* (velvet mesquite) riparian ecosystems: Ecological Applications, v. 3, pp. 307–14.

Strong, T. R., and Bock, C. E., 1990, Bird species distribution patterns in riparian habitats in southeastern Arizona: The Condor, v. 92, pp. 866–85.

Study: Native fish being allowed to die off, 2003, Arizona Daily Star, October 30, A8.

Stulik, R. S., and Twenter, F. R., 1964, Geology and Ground Water of the Luke Area, Maricopa County, Arizona: U.S. Geological Survey Water-Supply Paper no. 1779-P, 30 pp.

Sumner, J. C., 1947, J. C. Sumner's journal, edited by W. C. Darrah: Utah Historical Quarterly, v. 15, pp. 113–24.

Suttkus, R. D., Clemmer, G. H., and Jones, C., 1978, Mammals of the Riparian Region of the Colorado River in the Grand Canyon Area of Arizona: New Orleans, Tulane University Museum of Natural History, Occasional Papers no. 2, 23 pp.

Sykes, G., 1926, The delta and estuary of the Colorado River: Geographical Review, v. 16, pp. 232–55.

———, 1937, Delta, Estuary, and Lower Portion of the Channel of the Colorado River 1933 to 1935: Washington, D.C., Carnegie Institution of Washington, 70 pp.

———, 1944, A Westerly Trend: Tucson, Arizona Pioneers Historical Society, 325 pp.

———, 1970, The Colorado Delta: American Port Washington, N.Y., Kennikat Press, Geographical Society Special Publication no. 19, originally published in 1937, 193 pp.

———, n.d., The Santa Cruz: A study of a desert drainage system: Tucson, Arizona Historical Society, unpublished manuscript, no page numbers.

Szaro, R. B., and Belfit, S. C., 1987, Small Mammal Use of a Desert Riparian Island and Its Adjacent Scrub Habitat: U.S. Department of Agriculture, Forest Service Research Note no. RM-473, 5 pp.

Taft, R., 1938, Photography and the American Scene: New York, MacMillan, 546 pp.

Tashjian, P., n.d., Leslie Canyon National Wildlife Refuge in stream flow request for Leslie Creek in Leslie Canyon, U.S. Fish and Wildlife Service, Region 2, unpublished report, 66 pp.

Tellman, B., 2002, Human introduction of exotic species in the Sonoran region, *in* Tellman, B. (editor), Invasive Exotic Species in the Sonoran Region, Tucson, University of Arizona Press, pp. 23–46.

Tellman, B., Yarde, R., and Wallace, M. G., 1997, Arizona's Changing Rivers: How People Have Affected the Rivers: Tucson, University of Arizona, Water Resources Research Center Issue Paper no. 19, 198 pp.

Theis, C. V., 1940, The Source of Water Derived from Wells: Essential Factors Controlling the Response of an Aquifer to Development: Civil Engineer, v. 10, pp. 277–280.

Thompson, A. H., 1939, Diary of Almon Harris Thompson, edited by H. E. Gregory: Utah Historical Quarterly, v. 7, pp. 3–138.

Thompson, D. G., 1929, The Mohave Desert Region, California: A Geographic, Geologic, and Hydrologic Reconnaissance: U.S. Geological Survey Water-Supply Paper no. 578, 759 pp.

Thompson, K. R., 1984a, Annual Suspended-Sediment Loads in the Colorado River near Cisco, Utah, 1930–82: U.S. Geological Survey Water-Resources Investigations Report no. 85-4011, 17 pp.

———, 1984b, Annual Suspended-Sediment Loads in the Green River at Green River, Utah, 1930–82: U.S. Geological Survey Water-Resources Investigations Report no. 84-4169, 17 pp.

Thomsen, B. W., and Schumann, H. H., 1968, Water Resources of the Sycamore Creek Watershed, Maricopa County, Arizona: U.S. Geological Survey Water-Supply Paper no. 1861, 53 pp.

Thornber, J. J., 1909, Vegetation groups of the Desert Laboratory domain, *in* Spaulding, V. M. (editor), Distribution and Movements of Desert Plants: Washington, D.C., Carnegie Institution of Washington Publication no. 113, pp. 103–12.

———, 1916, Tamarisks for Southwestern Planting: Tucson, Ariz., Agricultural Experiment Station, Timely Hints for Farmers no. 121, 8 pp.

Thrapp, D. L., 1967, The Conquest of Apachería: Norman: University of Oklahoma Press, 405 pp.

Tidestrom, I., 1925, Flora of Utah and Nevada: Contributions from the U.S. National Herbarium no. 25, 665 pp.

Tiegs, S. D., and Pohl, M., 2005, Planform channel dynamics of the lower Colorado River: 1976–2000: Geomorphology, v. 69, pp. 14–27.

Tracy, J. L., and DeLoach, C. J., 1999, Suitability of classical biological control for giant reed (*Arundo donax*) in the United States, *in* Bell, C. E. (editor), *Arundo* and Saltcedar: The Deadly Duo: Ontario, Calif., Proceedings of the Arundo and Saltcedar Workshop, June 17, 1998, pp. 73–100.

Trimble, S. W., 1997, Stream channel erosion and change resulting from riparian forests: Geology, v. 25, pp. 467–69.

Tromble, J. M., 1972, Use of water by a riparian mesquite community, *in* Watersheds in Transition: Denver, Colorado State University, American Water Resources Association Special Publication, pp. 267–70.

Turner, J. H., 1983, Charles L. Weed Historic Photographs of Middle Fork American River Mining Activities: Sacramento, Calif., Bureau of Reclamation report, 47 pp.

Turner, R. M., 1963, Growth in four species of Sonoran Desert trees: Ecology, v. 44, pp. 760–65.

———, 1974, Quantitative and Historical Evidence of Vegetation Changes along the Upper Gila River, Arizona: U.S. Geological Survey Professional Paper no. 655-H, 20 pp.

———, 1990, Long-term vegetation change at a fully protected Sonoran Desert site: Ecology, v. 71, pp. 464–77.

———, 2003, Pima County's Withdrawal from Its Past: Tucson, Ariz., Pima County, Sonoran Desert Conservation Plan, 35 pp.

———, in press, Confessions of a repeat photographer, *in* Felger, R. S., and Broyles, B. (editors), Dry Borders: Great Natural Areas of the Gran Desierto and Upper Gulf of California: Salt Lake City, University of Utah Press.

Turner, R. M., Bowers, J. E., and Burgess, T. L., 1995, Sonoran Desert Plants: Tucson, University of Arizona Press, 501 pp.

Turner, R. M., and Karpiscak, M. M., 1980, Recent Vegetation Changes along the Colorado River between Glen Canyon Dam and Lake Mead, Arizona: U.S. Geological Survey Professional Paper no. 1132, 125 pp.

Turner, R. M., Ochung', H. A., and Turner, J. B., 1998, Kenya's Changing Landscape: Tucson, University of Arizona Press, 177 pp.

Turner, R. M., Webb, R. H., and Bowers, J. E., 2003, The Changing Mile Revisited: Tucson, University of Arizona Press, 334 pp.

Turner, S. F., 1962, Water resources of the Planet Ranch on Bill Williams River,

Mohave and Yuma Counties, Arizona: Scottsdale, Arizona Ranch and Metals Company report, 47 pp., maps, and appendixes.

Unitt, P., 1987, *Empidonax traillii extimus:* An endangered subspecies: Western Birds, v. 18, pp. 137–62.

Unland, H. E., Arain, A. M., Harlow, C., Houser, P. R., Garatuza-Payan, J., Scott, P., Sen, O. L., and Shuttleworth, W. J., 1998, Evaporation from a riparian system in a semi-arid environment: Hydrological Processes, v. 12, pp. 527–42.

U.S. Department of the Interior, 1964, The Lower Colorado River Land Use Plan: Washington, D.C., U.S. Department of the Interior, 187 pp.

———, 1989, Glen Canyon Environmental Studies, Final Report: Salt Lake City, Utah, Bureau of Reclamation, Upper Colorado Region (NTIS Report No. PB-88-183348/AS), 84 pp., appendixes.

U.S. Fish and Wildlife Service, 1994, Lower Colorado River National Wildlife Refuges, Comprehensive Management Plan, 1994–2014, Final Environmental Assessment: Albuquerque, U.S. Fish and Wildlife Service, 115 pp.

———, 1998, A System for Mapping Riparian Areas in the Western United States: Lakewood, Colo., U.S. Fish and Wildlife Service, National Wetlands Inventory Program, 15 pp.

Valdez, R. A., Ryel, R. J., and Williams, R., 1986, Endangered Fishes of Cataract Canyon: The Importance of the Colorado River above Lake Powell to the Colorado Squawfish, Humpback Chub, and Bonytail: Logan, Utah, Ecosystems Research Institute, Report under contract no. 5-CS-40-02820, 15 pp.

Van de Kamp, P. C., 1973, Holocene continental sedimentation in the Salton Basin, California: A reconnaissance: Geological Society of America Bulletin, v. 84, pp. 827–48.

Vandersande, M. W., Glenn, E. P., and Walworth, J. L., 2001, Tolerance of five riparian plants from the lower Colorado River to salinity, drought, and inundation: Journal of Arid Environments, v. 49, pp. 147–59.

Van Dyke, D., 1997, Daggett: Life in a Mojave Frontier Town (edited by P. Wild): Baltimore, John Hopkins University Press, 183 pp.

Van Dyke, T. S., 1895, Down the Colorado River: Land of Sunshine, v. 2, pp. 60–61.

Van Hylckama, T. E. A., 1970, Water use by salt cedar: Water Resources Research, v. 6, pp. 728–35.

———, 1974, Water Use by Saltcedar as Measured by the Water-Budget Method: U.S. Geological Survey Professional Paper no. 491-E, 30 pp.

Van Steeter, M. M., and Pitlick, J., 1998, Geomorphology and endangered fish habitats of the upper Colorado River. I. Historic changes in streamflow sediment load, and channel morphology: Water Resources Research, v. 34, pp. 287–302.

Van West, C. R., and Altschul, J. H., 1997, Environmental variability and agricultural economics along the lower Verde River, A.D. 750–1450, *in* Whittlesey, S. M., Ciolek-Torrello, R., and Altschul, J. H. (editors), Vanishing River: Landscapes and Lives of the Lower Verde Valley, Lower Verde Archaeological Project, Overview, Synthesis, and Conclusions: Tucson, Ariz., SRI Press, pp. 337–92.

Veblen, T. T., and Lorenz, D. C., 1991, The Colorado Front Range: A Century of Ecological Change: Salt Lake City, University of Utah Press, 186 pp.

Vélez de Escalante, S., 1995, The Domínguez-Escalante Journal (edited by T. J. Warner, translated by F. A. Chavez): Salt Lake City, University of Utah Press, 153 pp.

Waisel, Y., 1960a, Ecological studies on *Tamarix aphylla* (L.) Karst. I. Distribution and reproduction: ΦYTON, v. 15, pp. 7–17.

———, 1960b, Ecological studies on *Tamarix aphylla* (L.) Karst. II. The water economy: ΦYTON, v. 15, pp. 19–28.

Waitley, D., 1999, William Henry Jackson: Framing the Frontier: Missoula, Mont., Mountain Press, 217 pp.

Walker, H. P., (editor), 1971, The reluctant corporal: The autobiography of William Bladen Jett, Part 1: Journal of Arizona History, v. 12, pp. 1–50.

Walker, L. R., and Smith, S. D., 1997, Impacts of invasive plants on community and ecosystem properties, *in* Luken, J. O., and Thieret, J. W. (editors), Assessment and Management of Plant Invasions: New York, Springer-Verlag, pp. 69–86.

Waring, G. L., and Stevens, L. E., 1988, The effects of recent flooding on riparian plant establishment in Grand Canyon, *in* U.S. Department of the Interior, Glen Canyon Environmental Studies, Executive Summaries of Technical Reports: Salt Lake City, Utah, Bureau of Reclamation, Upper Colorado River District, pp. 257–70.

Warren, D. K., and Turner, R. M., 1975, Saltcedar (*Tamarix chinensis*) seed production, seedling establishment, and response to inundation: Arizona Academy of Science, v. 10, pp. 135–44.

Warren, P. L., and Schwalbe, C. R., 1985, Herpetofauna in riparian habitats along the Colorado River in Grand Canyon, *in* Johnson, R. R., and McCormick, J. F. (editors), Strategies for Protection and

Management of Floodplain Wetlands and Other Riparian Ecosystems: U.S. Department of Agriculture, Forest Service Technical Report no. WO-12, pp. 347–54.

Warren, R. J., 1999, Sixty-eight years of changes at Zion National Park: Springdale, Utah, Zion National Park, unpublished manuscript, no page numbers.

Waters, M. R., 1983, Late Holocene lacustrine chronology and archaeology of ancient Lake Cahuilla, California: Quaternary Research, v. 19, pp. 373–87.

———, 1985, Late Quaternary alluvial stratigraphy of Whitewater Draw, Arizona: Implications for regional correlation of fluvial deposits in the American Southwest: Geology, v. 13, pp. 705–8.

———, 1988, Holocene alluvial geology and geoarchaeology of the San Xavier reach of the Santa Cruz River, Arizona: Geological Society of America Bulletin, v. 100, pp. 479–91.

———, 1992, Principles of Geoarchaeology: A North American Perspective: Tucson, University of Arizona Press, 398 pp.

Waters, M. R., and Haynes, C. V., 2001, Late quaternary arroyo formation and climate change in the American Southwest: Geology, v. 29, pp. 399–402.

Webb, R. H., 1985, Late Holocene flooding on the Escalante River, south-central Utah: Ph.D. diss., University of Arizona, Tucson, 204 pp.

———, 1996, Grand Canyon: A Century of Change: Tucson, University of Arizona Press, 290 pp.

Webb, R. H., and Baker, V. R., 1987, Changes in hydrologic conditions related to large floods on the Escalante River, south-central Utah, in Singh, V. (editor), Regional Flood-Frequency Analysis: Dordrecht, Netherlands, D. Reidel, pp. 306–20.

Webb, R. H., Belnap, J., and Weisheit, J. S., 2004, Cataract Canyon: A Human and Environmental History of the Colorado River in Canyonlands: Salt Lake City, University of Utah Press, 268 pp.

Webb, R. H., and Betancourt, J. L., 1992, Climatic Variability and Flood Frequency of the Santa Cruz River, Pima County, Arizona: U.S. Geological Survey Water-Supply Paper no. 2379, 40 pp.

Webb, R. H., Blainey, J. B., and Hyndman, D. W., 2002, Paleoflood hydrology of the Paria River, southern Utah and northern Arizona, USA, in House, P. K., Webb, R. H., Baker, V. R., and Levish, D. R. (editors), Ancient Floods, Modern Hazards: Principles and Applications of Paleoflood Hydrology: American Geophysical Union, Water Science and Application Series no. 5, pp. 295–310.

Webb, R. H., Boyer, D. E., and Berry, K. H., 2001, Changes in Riparian Vegetation in the Southwestern United States: Historical Changes along the Mojave River, California: U.S. Geological Survey Open-File Report no. OF 01-245, 1 sheet.

Webb, R. H., Boyer, D. E., Orchard, K. L., and Baker, V. R., 2001, Changes in Riparian Vegetation in the Southwestern United States: Floods and Riparian Vegetation on the San Juan River, Southeastern Utah: U.S. Geological Survey Open-File Report no. OF 01-314, 1 sheet.

Webb, R. H., Griffiths, P. G., Melis, T. S., and Hartley, D. R., 2000, Sediment Delivery by Ungaged Tributaries of the Colorado River in Grand Canyon: U.S. Geological Survey Water Resources Investigations Report no. 00-4055, 67 pp.

Webb, R. H., and Hasbargen, J., 1998, Floods, groundwater levels, and arroyo formation on the Escalante River, south-central Utah, in Learning from the Land: Proceedings of the Grand Staircase–Escalante National Monument Science Symposium, November 1–5, 1995: Salt Lake City, Bureau of Land Management Report no. BLM/UT/GI-98/006 + 1220, pp. 335–57.

Webb, R. H., and Leake, S. A., 2006, Ground-water surface-water interactions and long-term change in riverine riparian vegetation in the southwestern United States: Journal of Hydrology, v. 320, pp. 302–23.

Webb, R. H., Melis, T. S., Griffiths, P. G., Elliott, J. G., Cerling, T. E., Poreda, R. J., Wise, T. W., and Pizzuto, J. E., 1999, Lava Falls Rapid in Grand Canyon: Effects of Late Holocene Debris Flows on the Colorado River: U.S. Geological Survey Professional Paper no. 1591, 90 pp.

Webb, R. H., Melis, T. S., and Valdez, R. A., 2002, Observations of Environmental Change in Grand Canyon: U.S. Geological Survey Water Resources Investigations Report no. 02-4080, 33 pp.

Webb, R. H., Murov, M. B., Esque, T. C., Boyer, D. E., DeFalco, L. A., Haines, D. F., Oldershaw, D., Scoles, S. J., Thomas, K. A., Blainey, J. B., and Medica, P. A., 2003, Perennial Vegetation Data from Permanent Plots on the Nevada Test Site, Nye County, Nevada: U.S. Geological Survey Open-File Report no. 03-336, 251 pp.

Webb, R. H., O'Connor, J. E., and Baker, V. R., 1988, Paleohydrologic reconstruction of flood frequency on the Escalante River, south-central Utah, in Baker, V. R., Kochel, R. C., and Patton, P. C. (editors), Flood Geomorphology: New York, John Wiley and Sons, pp. 403–13.

Webb, R. H., Smith, S. S., and McCord, V.

A. S., 1991, Historic Channel Change of Kanab Creek, Southern Utah and Northern Arizona: Grand Canyon, Ariz., Grand Canyon Natural History Association Monograph no. 9, 91 pp.

Webb, R. H., Wegner, D. L., Andrews, E. D., Valdez, R. A., and Patten, D. T., 1999, Downstream effects of Glen Canyon Dam on the Colorado River in Grand Canyon: A review, in Webb, R. H., Schmidt, J. C., Marzolf, G. R., and Valdez, R. A. (editors), The Controlled Flood in Grand Canyon: Scientific Experiment and Management Demonstration: Washington, D.C., American Geophysical Union, Geophysical Monograph no. 110, pp. 1–21.

Weist, W. G., Jr., 1971, Geology and Ground-Water System in the Gila River Phreatophyte Project Area, Graham County, Arizona: U.S. Geological Survey Professional Paper no. 655-D, 22 pp.

Welsh, S. L., Atwood, N. D., Goodrich, S., and Higgins, L. C. (editors), 1993, A Utah Flora (2d ed.): Provo, Utah, Brigham Young University, 986 pp.

Wheeler, Lt. G. M., 1983, Wheeler's Photographic Survey of the American West, 1871–1873, with 50 landscape photographs by Timothy O'Sullivan and William Bell: New York, Dover, originally published in 1875, folio.

Whitmore, R. C., 1975, Habitat ordination of passerine birds of the Virgin River Valley, southwestern Utah: Wilson Bulletin, v. 87, pp. 65–74.

Whittlesey, S. M., 1997a, An overview of research history and archaeology of central Arizona, in Whittlesey, S. M., Ciolek-Torrello, R., and Altschul, J. H. (editors), Vanishing River: Landscapes and Lives of the Lower Verde Valley: The Lower Verde Archaeological Project, Overview, Synthesis, and Conclusions: Tucson, Ariz., SRI Press, pp. 59–141.

———, 1997b, The vanished river: Historical-period impacts to desert landscapes and archaeological implications, in Whittlesey, S. M., Ciolek-Torrello, R., and Altschul, J. H. (editors), Vanishing River: Landscapes and Lives of the Lower Verde Valley: The Lower Verde Archaeological Project, Overview, Synthesis, and Conclusions: Tucson, Ariz., SRI Press, pp. 29–57.

Wilkinson, R. E., 1966, Adventitious shoots on saltcedar roots: Botanical Gazette, v. 127, pp. 103–4.

Williams, D. G., Brunel, J. P., Schaeffer, S. M., and Snyder, K. A., 1998, Biotic controls over the functioning of desert riparian ecosystems, in Wood, E. F., Chehbouni, A. G., Goodrich, D. C., Seo, D. J., and Zimmerman, J. R. (technical coordinators), Proceedings from the

Special Symposium on Hydrology: Boston, American Meteorological Society, pp. 43–48.

Williams, G. P., 1978, The Case of the Shrinking Channels—The North Platte and Platte Rivers in Nebraska: U.S. Geological Survey Circular no. 781, 48 pp.

Williams, G. P., and Wolman, M. G., 1984, Downstream Effects of Dams on Alluvial Rivers: U.S. Geological Survey Professional Paper no. 1286, 83 pp.

Willis, E. L., 1939, Plant associations of the Rillito floodplain in Pima County, Arizona: M.S. thesis, University of Arizona, Tucson, 48 pp.

Wilson, T. B., Webb, R. H., and Thompson, T. L., 2001, Mechanisms of Range Expansion and Removal of Mesquite (*Prosopis* spp.) in Desert Grasslands in the Southwestern United States: Tempe, Ariz., U.S. Forest Service General Technical Report no. 01-37, 23 pp.

Winter, T. C., Harvey, J. W., Franke, O. L., and Alley, W. M., 2002, Ground Water and Surface Water: A Single Resource: U.S. Geological Survey Circular no. 1139, 79 pp.

Wittick, T., 1973, An 1883 expedition to the Grand Canyon: Pioneer photographer Ben Wittick views the marvels of the Colorado: The American West, v. 10, pp. 38–47.

Wolcott, H. N., 1952, Salt River Valley area, Maricopa and Pinal Counties, *in* Halpenny, L. C., and others (editors), Ground Water in the Gila River Basin and Adjacent Areas, Arizona: U.S. Geological Survey Open-File Report, pp. 137–64.

Wolcott, H. N., Skibitzke, H. E., and Halpenny, L. C., 1956, Water Resources of Bill Williams River Valley near Alamo, Arizona: U.S. Geological Survey Water-Supply Paper no. 1360-D, pp. 291–319.

Wood, M. L., 1997, Historical Channel Changes along the Lower San Pedro River, Southeastern Arizona: Tucson, Arizona Geological Survey Open-File Report no. 97-21, 44 pp.

Wood, M. L., House, P. K., and Pearthree, P. A., 1999, Historical Geomorphology and Hydrology of the Santa Cruz River: Tucson, Arizona Geological Survey Open-File Report no. 99-13, 98 pp.

Woodbury, A. M., 1944, A history of southern Utah and its national parks: Utah Historical Quarterly, v. 12, pp. 111–209.

Woodbury, A. M., Durrant, S. D., and Flowers, S., 1959, Survey of Vegetation in the Glen Canyon Reservoir Basin: Salt Lake City, University of Utah Anthropological Papers no. 36, Glen Canyon Series no. 5, 56 pp.

Woodbury, A. M., and Russell, H. N., 1945, Birds of the Navajo country: Bulletin of the University of Utah Biological Service, v. 9, pp. 1–160.

Woolley, R. R., 1946, Cloudburst Floods in Utah, 1850–1938: U.S. Geological Survey Water-Supply Paper no. 994, 128 pp.

Woolsey, N. G., 1964, The Escalante Story: A History of the Town of Escalante, and Description of the Surrounding Territory, Garfield County, Utah, 1875–1964: Springville, Utah, Art City, 463 pp.

Wright, R. G., and Bunting, S. C., 1994, The Landscapes of Craters of the Moon National Monument: An Evaluation of Environmental Changes: Moscow, University of Idaho Press, 103 pp.

Yard, H. K., van Riper, C., III, Brown, B. T., and Kearsley, M. J., 2004, Diets of insectivorous birds along the Colorado River in Grand Canyon, Arizona: Condor, v. 106, p. 116.

Yeager, D. G., 1944, The impairment of park values in Zion National Park:

Springdale, Utah, Zion National Park, unpublished report dated March 23, 14 pp.

Yong, W., and Finch, D. M., 1997, Migration of the willow flycatcher along the Middle Rio Grande: Wilson Bulletin, v. 109, pp. 253–68.

Younker, G. L., and Andersen, C. W., 1986, Mapping Methods and Vegetation Changes along the Lower Colorado River between Davis Dam and the Border with Mexico: Salt Lake City, report prepared for the Bureau of Reclamation by AAA Engineering and Drafting, 30 pp.

Zamora-Arroyo, F., Nagler, P. L., Briggs, M., Radtke, D., Rodriguez, H., Garcia, J., Valdes, C., Huete, A., and Glenn, E. P., 2001, Regeneration of native trees in response to flood releases from the United States into the delta of the Colorado River, Mexico: Journal of Arid Environments, v. 49, pp. 49–64.

Zavaleta, E., 2000a, The economic value of controlling an invasive shrub: Ambio, v. 29, pp. 462–67.

———, 2000b, Valuing ecosystem services lost to *Tamarix* invasion in the United States, *in* Mooney, H. A., and Hobbs, R. J. (editors), Invasive Species in a Changing World: Covelo, Calif., Island Press, pp. 261–301.

Zimmerman, R. C., 1969, Plant Ecology of an Arid Basin, Tres Alamos-Redington Area, Southeastern Arizona: U.S. Geological Survey Professional Paper no. 485-D, 51 pp.

———, 1984, Change and stability in valley floor vegetation, San Pedro Valley, southeastern Arizona, 1964–1984: Tucson, Ariz., unpublished report to the U.S. Geological Survey, June 1984, 62 pp.

Index

About the Authors

Robert H. Webb has worked on long-term changes in natural ecosystems of the southwestern United States since 1976. He has degrees in engineering (B.S., University of Redlands, 1978), environmental earth sciences (M.S., Stanford University, 1980), and geosciences (Ph.D, University of Arizona, 1985). Since 1985, he has been a research hydrologist with the U.S. Geological Survey in Tucson and an adjunct faculty member of the Departments of Geosciences and Hydrology and Water Resources at the University of Arizona. Webb has authored, coauthored, or edited eight books, including *Environmental Effects of Off-Road Vehicles* (with Howard Wilshire); *Grand Canyon: A Century of Change; Floods, Droughts, and Changing Climates* (with Michael Collier); *The Changing Mile Revisited* (with Raymond Turner); and *Cataract Canyon: A Human and Environmental History of the Rivers in Canyonlands* (with Jayne Belnap and John Weisheit).

Stanley A. Leake, a native Arizonan, has worked for more than thirty-five years with the U.S. Geological Survey. He currently is a research hydrologist in Tucson, Arizona. With a background in mathematics and hydrology (B.S., Arizona State University, 1974; M.S., University of Arizona, 1984), his work has concentrated mostly on understanding the subsurface hydrology of the desert Southwest. His research interests have included quantitative investigations of the interaction of groundwater and surface water, land subsidence from groundwater pumping, and development of simulation software for groundwater investigations. He recently completed the Southwest Groundwater Resources Project, a regional study of groundwater–surface water interactions in the desert Southwest. That project included much of the work for this book as well as regional studies of natural recharge to groundwater systems and studies of effects of climate on groundwater in the Southwest. His computer codes have been incorporated into the U.S. Geological Survey MODFLOW simulation program and are in use in the United States and in many other countries. Other current work includes assessment of water availability on the lower Colorado River basin and studies of likely effects of groundwater withdrawals on perennial streams.

Raymond M. Turner, a westerner by birth, has degrees in botany (B.S., University of Utah, 1948; Ph.D., Washington State University, 1954). He taught at the University of Arizona (1954–1962) before joining the U.S. Geological Survey. His interest in desert vegetation dynamics has resulted in studies of long-term permanent vegetation study plots as well as in his use of repeat photography for documenting landscape change. His interest in repeat photography was aroused during preparation, with Rod Hastings, of *The Changing Mile* (University of Arizona Press, 1965). Turner is also author of publications describing changes in riparian vegetation along the Gila and Colorado Rivers and changes in permanent vegetation study plots at the Desert Laboratory in Tucson and in MacDougal Crater, Pinacate Preserve, Sonora, Mexico. Retired since 1989, he has subsequently coauthored three books: *Sonoran Desert Plants: An Ecological Atlas, Kenya's Changing Landscape,* and *The Changing Mile Revisited*—all published by the University of Arizona Press.